Cambridge IGCSE®

Maths

STUDENT'S BOOK

Also for Cambridge IGCSE® (9–1)

Chris Pearce

CONTENTS

Key
E = Extended

How to use this book 5

Number

Chapter 1: Number 6
1.1 Square numbers and cube numbers 8
1.2 Multiples of whole numbers 12
1.3 Factors of whole numbers 13
1.4 Prime numbers 15
1.5 Prime factorisation 16
1.6 More about HCF and LCM 18
1.7 Real numbers 20

Chapter 2: Fractions and percentages 22
2.1 Equivalent fractions 24
2.2 Fractions and decimals 26
E 2.3 Recurring decimals 27
2.4 Percentages, fractions and decimals 30
2.5 Calculating a percentage 34
2.6 Increasing or decreasing quantities by a percentage 36
2.7 One quantity as a percentage of another 40
2.8 Simple interest and compound interest 43
2.9 A formula for compound interest 45
E 2.10 Reverse percentage 47

Chapter 3: The four rules 50
3.1 Order of operations 52
3.2 Choosing the correct operation 54
3.3 Finding a fraction of a quantity 55
3.4 Adding and subtracting fractions 57
3.5 Multiplying and dividing fractions 60

Chapter 4: Directed numbers 64
4.1 Introduction to directed numbers 66
4.2 Everyday use of directed numbers 67
4.3 The number line 68
4.4 Adding and subtracting directed numbers 70
4.5 Multiplying and dividing directed numbers 73

Chapter 5: Powers and roots 76
5.1 Squares and square roots 78
5.2 Cubes and cube roots 79
5.3 More powers and roots 81
E 5.4 Exponential growth and decay 82

Chapter 6: Ordering and set notation 86
6.1 Inequalities 88
6.2 Sets and Venn diagrams 90
E 6.3 More about Venn diagrams 94

Chapter 7: Ratio, proportion and rate 100
7.1 Ratio 102
E 7.2 Increases and decreases using ratios 108
7.3 Speed 110
7.4 Rates 113
7.5 Direct proportion 116
7.6 Inverse proportion 117

Chapter 8: Estimation and limits of accuracy 120
8.1 Rounding whole numbers 122
8.2 Rounding decimals 123
8.3 Rounding to significant figures 125
8.4 Upper and lower bounds 126
E 8.5 Upper and lower bounds for calculations 128

Chapter 9: Standard form 132
9.1 Standard form 134
9.2 Calculating with standard form 136

Chapter 10: Applying number and using calculators 140
10.1 Units of measurement 142
10.2 Converting between metric units 143
10.3 Time 145
10.4 Currency conversions 147
10.5 Using a calculator efficiently 149

Examination questions: Number 151

Algebra

Chapter 11: Algebraic representation and formulae 160
11.1 The language of algebra 162
11.2 Substitution into formulae 165
11.3 Rearranging formulae 167
E 11.4 More complicated formulae 169

Chapter 12: Algebraic manipulation 172
12.1 Simplifying expressions 174
12.2 Expanding brackets 178

12.3	Factorisation	182
12.4	Multiplying two brackets: 1	184
12.5	Multiplying two brackets: 2	187
E 12.6	Expanding three brackets	190
E 12.7	Quadratic factorisation	192
E 12.8	Algebraic fractions	197

Chapter 13: Solutions of equations and inequalities — 202

13.1	Solving linear equations	204
13.2	Setting up equations	210
E 13.3	Solving quadratic equations by factorisation	212
E 13.4	Solving quadratic equations by the quadratic formula	217
E 13.5	Solving quadratic equations by completing the square	219
13.6	Simultaneous equations	222
E 13.7	Linear and non-linear simultaneous equations	229
E 13.8	Solving inequalities	232

Chapter 14: Graphs in practical situations — 236

14.1	Conversion graphs	238
14.2	Travel graphs	242
E 14.3	Speed–time graphs	246
E 14.4	Curved graphs	251

Chapter 15: Straight-line graphs — 256

15.1	Drawing straight-line graphs	258
15.2	The equation $y = mx + c$	261
E 15.3	More about straight-line graphs	265
15.4	Solving equations graphically	267
15.5	Parallel lines	270
E 15.6	Points and lines	272
E 15.7	Perpendicular lines	274

Chapter 16: Graphs of functions — 278

16.1	Quadratic graphs	280
E 16.2	Turning points on a quadratic graph	285
16.3	Reciprocal graphs	286
E 16.4	More graphs	288
E 16.5	Exponential graphs	292
E 16.6	Estimating gradients	296

Chapter 17: Number sequences — 300

17.1	Patterns in number sequences	302
17.2	The nth term of a sequence	304
17.3	General rules from patterns	309
E 17.4	Further sequences	314

Chapter 18: Indices — 318

18.1	Using indices	320
18.2	Negative indices	322
18.3	Multiplying and dividing with indices	324
E 18.4	Fractional indices	327

Chapter 19: Proportion — 332

E 19.1	Direct proportion	334
E 19.2	Inverse proportion	339

Chapter 20: Linear programming — 342

E 20.1	Graphical inequalities	344
E 20.2	More than one inequality	347
E 20.3	Linear programming	349

Chapter 21: Functions — 352

E 21.1	Function notation	354
E 21.2	Inverse functions	355
E 21.3	Composite functions	357
E 21.4	More about composite functions	359

Chapter 22: Differentiation — 362

E 22.1	The gradient of a curve	364
E 22.2	More complex curves	366
E 22.3	Turning points	369

Examination questions: Algebra — 372

Geometry and trigonometry

Chapter 23: Angle properties — 382

23.1	Angle facts	384
23.2	Parallel lines	386
23.3	Angles in a triangle	390
23.4	Angles in a quadrilateral	392
23.5	Regular polygons	395
E 23.6	Irregular polygons	398
23.7	Tangents and diameters	400
E 23.8	Angles in a circle	402
E 23.9	Cyclic quadrilaterals	405
E 23.10	Alternate segment theorem	408

Chapter 24: Geometrical terms and relationships — 412

24.1	Measuring and drawing angles	414
24.2	Bearings	417
24.3	Nets	420
24.4	Congruent shapes	423
E 24.5	Congruent triangles	424
24.6	Similar shapes	427
E 24.7	Areas of similar triangles	430
E 24.8	Areas and volumes of similar shapes	433

Chapter 25: Geometrical constructions — 438
- 25.1 Constructing shapes — 440
- 25.2 Scale drawings — 442

Chapter 26: Trigonometry — 446
- 26.1 Pythagoras' theorem — 448
- 26.2 Trigonometric ratios — 452
- 26.3 Calculating angles — 454
- 26.4 Using sine, cosine and tangent functions — 455
- 26.5 Which ratio to use — 459
- E 26.6 Applications of trigonometric ratios — 462
- E 26.7 Problems in three dimensions — 466
- E 26.8 Sine and cosine of obtuse angles — 468
- E 26.9 The sine rule and the cosine rule — 470
- E 26.10 Using sine to find the area of a triangle — 477
- E 26.11 Sine, cosine and tangent of any angle — 479

Chapter 27: Mensuration — 486
- 27.1 Perimeter and area of a rectangle — 488
- 27.2 Area of a triangle — 491
- 27.3 Area of a parallelogram — 494
- 27.4 Area of a trapezium — 495
- 27.5 Circumference and area of a circle — 498
- 27.6 Surface area and volume of a cuboid — 501
- 27.7 Volume and surface area of a prism — 503
- 27.8 Volume and surface area of a cylinder — 506
- 27.9 Sectors and arcs: 1 — 508
- E 27.10 Sectors and arcs: 2 — 510
- 27.11 Volume of a pyramid — 512
- 27.12 Volume and surface area of a cone — 514
- 27.13 Volume and surface area of a sphere — 516

Chapter 28: Symmetry — 518
- 28.1 Lines of symmetry — 520
- 28.2 Rotational symmetry — 522
- 28.3 Symmetry of special two-dimensional shapes — 523
- E 28.4 Symmetry of three-dimensional shapes — 525
- E 28.5 Symmetry in circles — 526

Chapter 29: Vectors — 530
- 29.1 Introduction to vectors — 532
- E 29.2 Using vectors — 535
- E 29.3 The magnitude of a vector — 540

Chapter 30: Transformations — 542
- 30.1 Translations — 544
- 30.2 Reflections: 1 — 546
- E 30.3 Reflections: 2 — 548
- 30.4 Rotations: 1 — 550
- E 30.5 Rotations: 2 — 553
- 30.6 Enlargements: 1 — 554
- E 30.7 Enlargements: 2 — 559
- E 30.8 Combined transformations — 561

Examination questions: Geometry — 564

Statistics and probability

Chapter 31: Statistical representation — 576
- 31.1 Frequency tables — 578
- 31.2 Pictograms — 581
- 31.3 Bar charts — 583
- 31.4 Pie charts — 587
- 31.5 Scatter diagrams — 591
- 31.6 Histograms — 596
- E 31.7 Histograms with bars of unequal width — 599

Chapter 32: Statistical measures — 606
- 32.1 The mode — 608
- 32.2 The median — 610
- 32.3 The mean — 612
- 32.4 The range — 615
- 32.5 Which average to use — 618
- 32.6 Stem-and-leaf diagrams — 620
- 32.7 Using frequency tables — 624
- E 32.8 Grouped data — 628
- E 32.9 Cumulative frequency diagrams — 631
- E 32.10 Box-and-whisker plots — 638

Chapter 33: Probability — 642
- 33.1 The probability scale — 644
- 33.2 Calculating probabilities — 646
- 33.3 Probability that an event will not happen — 649
- 33.4 Probability in practice — 651
- 33.5 Using Venn diagrams — 654
- 33.6 Possibility diagrams — 657
- 33.7 Tree diagrams — 661
- E 33.8 Conditional probability — 665

Examination questions: Statistics and probability — 672

Examination questions: Mixed type — 686

Glossary — 692
Answers — 704
Index — 766

How to use this book

Welcome to the *Collins Cambridge IGCSE® Maths Student's Book* that provides in-depth coverage of the Cambridge IGCSE Mathematics syllabus 0580 for examination from 2020. This book also provides coverage for the Cambridge IGCSE (9-1) syllabus 0980. You will find a number of features in this book that will help you with your course of study.

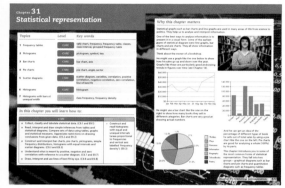

Why this chapter matters

This page is at the start of each chapter. It tells you why the mathematics in the chapter is important and how it is useful.

Chapter overviews

The overview at the start of each chapter shows what you will be studying and the key words you need to know. Syllabus references are included and if you are doing the Extended curriculum you must also cover the corresponding Core content, for example, C1.1. and E1.1.

Worked examples

Worked examples take you through questions step by step and help you understand the topic before you start the practice questions.

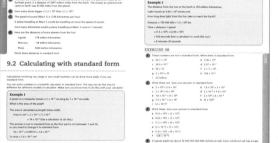

Practice questions and answers

Every chapter has extensive questions to help you practise the skills you need. You will need to be able to reason, interpret and communicate mathematically when solving problems, which are important skills to acquire.

Colour-coded levels

The colour coded panels at the side of the question pages show whether the questions are at core (blue) or extended level (yellow). The E on some topic headings shows that the content in that topic is at extended level only.

Exam preparation

Each of the four main sections in the book ends with exam questions from Cambridge International past papers. These will be useful for revision. Mark schemes, written by the author, are available in the Teacher's Pack.

Chapter 1
Number

Topics	Level	Key words
1 Square numbers and cube numbers	CORE	square, square number, square root, cube, cube number
2 Multiples of whole numbers	CORE	multiple
3 Factors of whole numbers	CORE	factor, factor pair, lowest common multiple, highest common factor
4 Prime numbers	CORE	prime number
5 Prime factorisation	CORE	product of prime factors, index (indices), prime factorisation
6 More about HCF and LCM	CORE	natural number, integer, real number, rational number, irrational number
7 Real numbers	CORE	natural number, integer, real number, rational number, irrational number, reciprocal

In this chapter you will learn how to:

CORE

- Identify and use:
    - natural numbers
    - integers (positive, negative and zero)
    - prime numbers
    - square numbers
    - cube numbers
    - common factors and common multiples
    - rational and irrational numbers (e.g. π, $\sqrt{2}$)
    - real numbers
    - reciprocals
    - Express any number as a product of its prime factors
    - Find the lowest common multiple (LCM) and highest common factor (HCF) of two numbers. (C1.1 and E1.1)
- Reason, interpret and communicate mathematically when solving problems.

Why this chapter matters

A pattern is an arrangement of repeated parts. You see patterns every day in clothes, art and home furnishings. Patterns can also occur in numbers.

There are many mathematical problems that can be solved using patterns in numbers. Some numbers have fascinating features.

Here is a pattern.

$3 + 5 = 8$ (5 miles ≈ 8 km)

$5 + 8 = 13$ (8 miles ≈ 13 km)

$8 + 13 = 21$ (13 miles ≈ 21 km)

Approximately how many kilometres are there in 21 miles?

Note: ≈ means 'approximately equal to'.

In the boxes are some more patterns. Can you work out the next line of each pattern?

Now look at these numbers and see why they are special.

$4096 = (4 + 0^9)^6$

$81 = (8 + 1)^2$

Some patterns have special names.
Can you pair up these patterns and the names?

4, 8, 12, 16, …	Prime numbers
1, 4, 9, 16, …	Multiples (of 4)
2, 3, 5, 7, …	Cube numbers
1, 8, 27, 64, …	Square numbers

You will look at these in more detail in this chapter.

Below are four sets of numbers. Think about which number links together all the other numbers in each set. (The mathematics that you cover in 1.3 'Factors of whole numbers' will help you to work this out!)

10, 5, 2, 1

18, 9, 6, 3, 2, 1

25, 5, 1

32, 16, 8, 4, 2, 1

$5^2 = 5 \times 5 = 25$
$50^2 = 50 \times 50 = 2500$
$500^2 = 500 \times 500 = 250\,000$

$10 \times 10 = 100$
$10 \times 10 \times 10 = 1000$
$10 \times 10 \times 10 \times 10 = 10\,000$

$1 \times 1 = 1$
$11 \times 11 = 121$
$111 \times 111 = 12\,321$

$1 \times 1 = 1$
$2 \times 2 = 1 + 3$
$3 \times 3 = 1 + 3 + 5$
$4 \times 4 = 1 + 3 + 5 + 7$

$1089 \times 9 = 9801$
$10\,989 \times 9 = 98\,901$
$109\,989 \times 9 = 989\,901$

1.1 Square numbers and cube numbers

What is the next number in this sequence?

1, 4, 9, 16, 25, …

Write each number as:

1 × 1, 2 × 2, 3 × 3, 4 × 4, 5 × 5, …

These factors can be represented by **square** patterns of dots:

1 × 1   2 × 2   3 × 3   4 × 4   5 × 5

From these patterns, you can see that the next pair of factors must be 6 × 6 = 36, therefore 36 is the next number in the sequence.

Because they form square patterns, the numbers 1, 4, 9, 16, 25, 36, … are called **square numbers**.

When you multiply any number by itself, the answer is called the *square of the number* or the *number squared*. This is because the answer is a square number. For example:

    the square of 5 (or 5 squared) is 5 × 5 = 25

    the square of 6 (or 6 squared) is 6 × 6 = 36

There is a short way to write the square of any number.

For example:

    5 squared (5 × 5) can be written as 5^2

    13 squared (13 × 13) can be written as 13^2

So, the sequence of square numbers, 1, 4, 9, 16, 25, 36, … , can be written as:

$1^2, 2^2, 3^2, 4^2, 5^2, 6^2, …$

The **square root** of n is the number of which the square is n. This can be written as $\sqrt{n}$. For example, the square root of 16 (4) can be written as $\sqrt{16}$.

Square numbers have exact square roots, for example:

    the square root of 9 is 3: $\sqrt{9} = 3$

    the square root of 25 is 5: $\sqrt{25} = 5$

    the square root of 100 is 10: $\sqrt{100} = 10$

EXERCISE 1A

1 The square number pattern starts:

1   4   9   16   25   …

Copy and continue the pattern above until you have written down the first 20 square numbers. You may use your calculator for this.

2 Work out the answer to each of these number sentences.

1 + 3 =

1 + 3 + 5 =

1 + 3 + 5 + 7 =

Look carefully at the pattern of the three number sentences. Then write down the next three number sentences in the pattern and work them out.

3 Find the next three numbers in each of these number patterns. (They are all based on square numbers.) You may use your calculator.

	1	4	9	16	25	36	49	64	81
a	2	5	10	17	26	37	…	…	…
b	2	8	18	32	50	72	…	…	…
c	3	6	11	18	27	38	…	…	…
d	0	3	8	15	24	35	…	…	…

Advice and Tips

Look for the connection with the square numbers on the top line.

4 a Work out the values of both expressions in each pair. You may use your calculator.

$3^2 + 4^2$ and 5^2

$5^2 + 12^2$ and 13^2

$7^2 + 24^2$ and 25^2

$9^2 + 40^2$ and 41^2

b Describe what you notice about your answers to part **a**. This will help you communicate mathematically with others.

5 a $13^2 = 169$. What is $\sqrt{169}$?

b Find $\sqrt{25}$

c Find $\sqrt{81}$

d Find $\sqrt{121}$

e Find $\sqrt{400}$

6 4 and 81 are square numbers with a sum of 85.
Find two different square numbers with a sum of 85.

The following exercise will give you some practice on multiples, factors, square numbers and prime numbers.

EXERCISE 1B

1 Write out the first three numbers that are multiples of both of the numbers shown.
   a 3 and 4   b 4 and 5   c 3 and 5   d 6 and 9   e 5 and 7

2 Here are four numbers.

   10   16   35   49

   Copy and complete the table by putting each of the numbers in the correct box.

	Square number	Factor of 70
Even number		
Multiple of 7		

3 Arrange these four number cards to make a square number.

4 An alarm flashes every 8 seconds and another alarm flashes every 12 seconds. If both alarms flash together, how many seconds will it be before they both flash together again?

5 A bell rings every 6 seconds. Another bell rings every 5 seconds. If they both ring together, how many seconds will it be before they both ring together again?

6 From this box, choose one number that fits each of these descriptions.

   a a multiple of 3 and a multiple of 4
   b a square number and an odd number
   c a factor of 24 and a factor of 18
   d a prime number and a factor of 39
   e an odd factor of 30 and a multiple of 3
   f a number with 5 factors exactly
   g a multiple of 5 and a factor of 20
   h a prime number that is one more than a square number

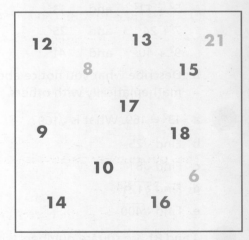

Cube numbers

What is the next number in this sequence?

1, 8, 27, …

Write each number as:

1 × 1 × 1, 2 × 2 × 2, 3 × 3 × 3, …

These factors can be represented by **cube** patterns of dots:

1 × 1 × 1   2 × 2 × 2   3 × 3 × 3

  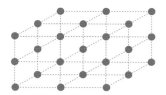

From these patterns, you can see that the next pair of factors must be 4 × 4 × 4 = 64, therefore 64 is the next number in the sequence.

Because they form cubic patterns, the numbers 1, 8, 27, 64, … are called **cube numbers**.

When you multiply any number by itself twice, the answer is called the *cube of the number* or the *number cubed*. This is because the answer is a cube number. For example: the cube of 5 (or 5 cubed) is
5 × 5 × 5 = 125.

There is a short way to write the cube of any number.

For example:

    5 cubed (5 × 5 × 5) can be written as 5^3

    10 cubed (10 × 10 × 10) can be written as 10^3

So, the sequence of cube numbers, 1, 8, 27, 64, … , can be written as:

    1^3, 2^3, 3^3, 4^3, …

You will learn more about cubes (and cube roots) in Chapter 5.

EXERCISE 1C

1 The cube number pattern starts:

    1   8   27   64 …

Copy and continue the pattern above until you have written down the first 12 cube numbers. You may use your calculator for this.

Chapter 1. Topic 2

CORE

2 Work out the answer to each of these number sentences.

$1 + 8 =$

$1 + 8 + 27 =$

$1 + 8 + 27 + 64 =$

Look carefully at the pattern of the three number sentences. What kind of numbers are these?

Now write down the next three number sentences in the pattern and work them out.

3 Find the next three numbers in each of these number patterns. (They are all based on cube numbers.) You may use your calculator.

	1	8	27	64			
a	2	9	28	65	…	…	…
b	0	7	26	63	…	…	…
c	2	16	54	128	…	…	…
d	1000	729	512	343	…	…	…

4 a Work out the values of these expressions.

$1^3 + 5^3 + 3^3$

$3^3 + 7^3 + 0^3$

$3^3 + 7^3 + 1^3$

b Describe what you notice about your answers to part **a**.

5 Work out the values of these expressions: $12^3 + 1^3$ and $9^3 + 10^3$.

Your answer is Bender's (a character in *Futurama*) serial number. It is sometimes called the Hardy–Ramanujan number after the Indian mathematician Ramanujan who noticed that this is the smallest number that can be expressed as the sum of two cubes in two different ways.

6 Work out the values of these expressions: 69^2 and 69^3 on your calculator.

What do you notice about the digits in your answers?

1.2 Multiples of whole numbers

When you multiply any whole number by another whole number, the answer is called a **multiple** of either of those numbers.

For example, 5 × 7 = 35, which means that 35 is a multiple of 5 and it is also a multiple of 7. Here are some other multiples of 5 and 7:

multiples of 5 are: 5   10   15   20   25   30   35   …

multiples of 7 are: 7   14   21   28   35   42   …

1.2 Multiples of whole numbers

Chapter 1. Topic 3
EXERCISE 1D

1 Write out the first five multiples of:

   a 3   **b** 7   **c** 9   **d** 11   **e** 16

Remember: the first multiple is the number itself.

> **Advice and Tips**
>
> There is no point testing odd numbers for multiples of even numbers such as 4 and 6.

2 Use your calculator to see which of the numbers below are:

   a multiples of 4   **b** multiples of 7   **c** multiples of 6.

   72   135   102   161   197   132   78   91   216   514

3 Find the biggest number that is smaller than 100 and that is:

   a a multiple of 2   **b** a multiple of 3   **c** a multiple of 4

   d a multiple of 5   **e** a multiple of 7   **f** a multiple of 6.

4 A party of 20 people are getting into taxis. Each taxi holds the same number of passengers. If all the taxis fill up, how many people could be in each taxi? Give two possible answers.

5 Here is a list of numbers.

   6   8   12   15   18   28

   a From the list, write down a multiple of 9.

   b From the list, write down a multiple of 7.

   c From the list, write down a multiple of both 3 and 5.

6 How many numbers between 1 and 100 are multiples of both 6 and 9? List the numbers.

1.3 Factors of whole numbers

A **factor** of a whole number is any whole number that divides into it exactly. So:

the factors of 20 are   1 2 4 5 10 20

the factors of 12 are   1 2 3 4 6 12

Factor facts

Remember these facts.

- 1 is always a factor and so is the number itself.
- When you have found one factor, there is always another factor that goes with it – unless the factor is multiplied by itself to give the number. For example, look at the number 20:

   $1 \times 20 = 20$ so 1 and 20 are both factors of 20

   $2 \times 10 = 20$ so 2 and 10 are both factors of 20

   $4 \times 5 = 20$ so 4 and 5 are both factors of 20.

   These are called **factor pairs**.

You may need to use your calculator to find the factors of large numbers.

> **Example 1**
>
> Find the factors of 36.
>
> Look for the factor pairs of 36. These are:
>
> $1 \times 36 = 36 \quad 2 \times 18 = 36 \quad 3 \times 12 = 36 \quad 4 \times 9 = 36 \quad 6 \times 6 = 36$
>
> 6 is a repeated factor so is counted only once.
>
> So, the factors of 36 are 1, 2, 3, 4, 6, 9, 12, 18, 36.

EXERCISE 1E

1 What are the factors of each of these numbers?

    a 10   **b** 28   **c** 18   **d** 17   **e** 25

    f 40   **g** 30   **h** 45   **i** 24   **j** 16

2 What is the biggest factor that is less than 100 for each of these numbers?

    a 110   **b** 201   **c** 145

    d 117   **e** 130   **f** 240

3 Find the largest common factor for each pair of numbers.

    a 2 and 4   **b** 6 and 10   **c** 9 and 12

    d 15 and 25   **e** 9 and 15   **f** 12 and 21

    g 14 and 21   **h** 25 and 30   **i** 30 and 50

    j 55 and 77

> **Advice and Tips**
>
> Look for the largest number that has both numbers in its multiplication table.

4 Find the highest odd number that is a factor of 40 and a factor of 60.

Lowest common multiple

The **lowest common multiple** (LCM) of two numbers is the *smallest* number that appears in the multiplication tables of both numbers.

For example, the LCM of 3 and 5 is 15, the LCM of 2 and 7 is 14 and the LCM of 6 and 9 is 18.

> **Example 2**
>
> Find the LCM of 18 and 24.
>
> Write out the 18 times table: 18, 36, 54, (72), 90, 108, … .
>
> Write out the 24 times table: 24, 48, (72), 96, 120, …
>
> You can see that 72 is the smallest (least) number in both (common) tables (multiples).

14   1.3 Factors of whole numbers

Chapter 1. Topic 4

Highest common factor

The **highest common factor** (HCF) of two numbers is the *biggest* number that divides exactly into both of them.

For example, the HCF of 24 and 18 is 6, the HCF of 45 and 36 is 9 and the HCF of 15 and 22 is 1.

> **Example 3**
> Find the HCF of 28 and 16.
>
> Write out the factors of 28: 1, 2, ④, 7, 14, 28
>
> Write out the factors of 16: 1, 2, ④, 8, 16
>
> You can see that 4 is the biggest (highest) number in both (common) lists (factors).

EXERCISE 1F

1 Find the LCM of each pair of numbers.

- a 24 and 56
- b 21 and 35
- c 12 and 28
- d 28 and 42
- e 12 and 32
- f 18 and 27
- g 15 and 25
- h 16 and 36

2 Find the HCF of each pair of numbers.

- a 24 and 56
- b 21 and 35
- c 12 and 28
- d 28 and 42
- e 12 and 32
- f 18 and 27
- g 15 and 25
- h 16 and 36
- i 42 and 27
- j 48 and 64
- k 25 and 35
- l 36 and 54

3 The HCF of two numbers is 6.

The LCM of the same two numbers is 72.

What are the numbers? Explain how you reached you answer.

1.4 Prime numbers

What are the factors of 2, 3, 5, 7, 11 and 13?

Notice that each of these numbers has only two factors: itself and 1. They are all examples of **prime numbers**.

Chapter 1. Topic 5

So, a prime number is a whole number that has only two factors: itself and 1.

Note: 1 is *not* a prime number, since it has only one factor – itself.

The prime numbers up to 50 are:

2, 3, 5, 7, 11, 13, 17, 19, 23, 29, 31, 37, 41, 43, 47

EXERCISE 1G

1 Write down the prime numbers between 20 and 30.

2 Write down the only prime number between 90 and 100.

3 Decide which of these numbers are **not** prime numbers.

    462   108   848   365   711

4 When three different prime numbers are multiplied together the answer is 105.

What are the three prime numbers?

5 A shopkeeper has 31 identical soap bars.

He is trying to arrange the bars on a shelf in rows, each with the same number of bars.

Is it possible?

Explain your answer.

1.5 Prime factorisation

Every whole number that is *not* prime can be written as the **product of prime factors**. For example:

$35 = 5 \times 7$

$40 = 2 \times 2 \times 2 \times 5$

$465 = 3 \times 5 \times 31$

$5929 = 7 \times 7 \times 11 \times 11$

5 and 7 are the prime factors of 35.

2 and 5 are the prime factors of 40.

You can use **indices** to write the product more easily. For example:

$40 = 2^3 \times 5$ The small 3 is an **index**.

$5929 = 7^2 \times 11^2$

$96 = 2 \times 2 \times 2 \times 2 \times 2 \times 3 = 2^5 \times 3$

Writing a number in this way is called **prime factorisation**.

You can write the numbers in a different order but you can do the factorisation in only one way. So for 96 there must be five 2s and one 3.

1.5

> **Example 4**
> Write 630 as a product of prime factors.
>
> Check the prime numbers (2, 3, 5, 7, ...) in turn to find which of them divide into 630.
>
>   2 is a factor of 630. $630 \div 2 = 315$
>
> Now do the same with 315.
>
>   2 is not a factor of 315 but 3 is. $315 \div 3 = 105$
>
>   3 is a factor of 105. $105 \div 3 = 35$
>
>   3 is not a factor of 35 but 5 is. $35 \div 5 = 7$
>
>   7 is a prime number, so stop there.
>
> It is more convenient to write the divisions in a column, like this.
>
> ```
> 2 | 6 3 0
> 3 | 3 1 5
> 3 | 1 0 5
> 5 |   3 5
>         7
> ```
>
> Now list the prime numbers, including the final 7.
>
> $630 = 2 \times 3 \times 3 \times 5 \times 7 = 2 \times 3^2 \times 5 \times 7$

EXERCISE 1H

1 Work out these numbers.

    a $2^2 \times 3^2$ **b** $3 \times 5 \times 7$ **c** 2×5^3 **d** $2^4 \times 3 \times 17$

    e $2 \times 3 \times 7 \times 17$ **f** 5×7^3 **g** $3^2 \times 11^2$ **h** $2^3 \times 13^2$

2 Write each of these numbers as a product of prime factors.

    a 90   **b** 152   **c** 64   **d** 330

    e 289   **f** 800   **g** 97   **h** 1001

3 Which of these numbers can be written as the product of exactly two prime factors?

    77   124   129   245   221   103

4 $450 = 2 \times 3^2 \times 5^2$

    a Use this fact to write 900 as a product of prime factors.

    b Write 1800 as a product of prime factors.

    c Write 1350 as a product of prime factors.

5 Match each number to the correct product of prime factors.

144	$2^2 \times 3^4$
200	$2^4 \times 3^2$
324	$2^3 \times 5^2$
500	$2^2 \times 5^3$

Chapter 1: Number

Chapter 1 . Topic 6

6   a   Find the smallest number that has four different prime factors.

    b   Find the smallest number that has five different prime factors.

7 $1224 = 2^3 \times 3^2 \times 17$

    Use this fact to write each of these numbers as a product of prime factors.

    a   612   b   306   c   408

8   Which of these numbers cannot be written as a product of prime factors?

    70   71   72   73   74   75   76   77   78   79

9 $539 = 7^2 \times 11$ $847 = 7 \times 11^2$ $539 \times 847 = 456\,533$

    Use these facts to write 456 533 as a product of prime factors.

1.6 More about HCF and LCM

You can use prime factorisation to find the LCM or the HCF of two or more numbers.

> **Example 5**
>
> $810 = 2 \times 3^4 \times 5$
>
> $252 = 2^2 \times 3^2 \times 7$
>
> a   Find the HCF of 810 and 252.
>
> b   Find the LCM of 810 and 252.
>
> ---
>
> a   To find the HCF, multiply all the prime numbers (including repeats) that occur in both lists.
>
>   There is one 2 in both numbers.
>
>   There are two 3s in both numbers.
>
>   The HCF is $2 \times 3^2 = 18$
>
> b   To find the LCM, multiply all the prime numbers (including repeats) that occur in either list.
>
>   You need to multiply two 2s, four 3s, one 5 and one 7.
>
>   The LCM is $2^2 \times 3^4 \times 5 \times 7 = 11\,340$

Why does this work?

You can check that 18 is a common factor: $810 \div 18 = 45 = 3^2 \times 5$

and: $252 \div 18 = 14 = 2 \times 7$

Since the answers to these divisions have no common factor, 18 is the *highest* common factor.

You can check that 11 340 is a common multiple.

$11\,340 \div 810 = 14 = 2 \times 7$

$11\,340 \div 252 = 45 = 3^2 \times 5$

Since the answers to these divisions have no common factor, 11 340 is the *lowest* common multiple.

> **Example 6**
>
> Find the LCM of 21, 27 and 35.
>
> ---
>
> $21 = 3 \times 7 \qquad 27 = 3^3 \qquad 35 = 5 \times 7$
>
> Looking at the prime numbers that occur in any of these, you need three 3s, one 5 and one 7.
>
> The LCM is $3^3 \times 5 \times 7 = 945$.

EXERCISE 1I

1 $\quad 72 = 2^3 \times 3^2 \qquad 162 = 2 \times 3^4$

Use these facts to find:

   a the HCF of 72 and 162   **b** the LCM of 72 and 162.

2 $\quad 105 = 3 \times 5 \times 7 \qquad 245 = 5 \times 7^2$

Use these facts to find:

   a the HCF of 105 and 245   **b** the LCM of 105 and 245.

3   **a** Write 240 as the product of prime factors.

   b Write 126 as the product of prime factors.

   c Find the HCF of 240 and 126.

   d Find the LCM of 240 and 126.

4   **a** Write 72 and 108 as the product of prime factors.

   b Find the HCF of 72 and 108.

   c Find the LCM of 72 and 108.

5   **a** Find the HCF of 64 and 144.   **b** Find the LCM of 64 and 144.

6   **a** Find the HCF of 132 and 693.   **b** Find the LCM of 132 and 693.

7 $\quad 24 = 2^3 \times 3 \qquad 60 = 2^2 \times 3 \times 5 \qquad 36 = 2^2 \times 3^2$

Use these facts to find:

   a the HCF of 24, 60 and 36   **b** the LCM of 24, 60 and 36.

8   **a** Find the HCF of 25, 35 and 45.   **b** Find the LCM of 25, 35 and 45.

9   **a** Find the HCF of 105, 135 and 375.   **b** Find the LCM of 105, 135 and 375.

10 **a** Find the HCF of 288, 432 and 648.

    b Write the LCM of 288, 432 and 648 as a product of prime factors.

Chapter 1: Number

Chapter 1 . Topic 7

11 a Find the HCF of 63 and 200.

b Find the LCM of 63 and 200.

12 In question 1 you found the HCF and LCM of 72 and 162.

a Work out the product of 72 and 162.

b Work out the product of the HCF and the LCM of 72 and 162.

c Investigate whether there is a relationship between the product of the HCF and the LCM of two numbers and the product of the numbers themselves.

d Is there a similar result for the HCF and the LCM of three numbers?

1.7 Real numbers

So far you have only been looking at **natural numbers**, which are whole, positive numbers:

0, 1, 2, 3, 4, 5....

If you also include negative whole numbers you have the **integers**:

....−3, −2, −1, 0, 1, 2, 3....

And if you include decimals too you have the **real numbers**. There are two sorts of real numbers.

Rational numbers are integers or fractions such as −17, $3\frac{5}{8}$, −8.75 or $\frac{29}{11}$.

Irrational numbers cannot be written as fractions.

Examples of irrational numbers are $\sqrt{2}$, $\sqrt{17}$ and π.

If n is a natural number that is *not* a square number, then $\sqrt{n}$ is irrational.

The reciprocal of the number n is $\frac{1}{n}$ ($n \neq 0$)

The reciprocal of the fraction $\frac{a}{b}$ is $\frac{b}{a}$

The product of a number and its reciprocal is 1

Example 7

Find the reciprocal of   a 28   b 2.8   c $\frac{1}{8}$ d $5\frac{3}{4}$

a The reciprocal is $\frac{1}{28}$

b The reciprocal is $\frac{1}{2.8} = \frac{10}{28} = \frac{5}{14}$

c The reciprocal is $\frac{8}{1} = 8$

d $5\frac{3}{4} = \frac{23}{4}$. The reciprocal is $\frac{4}{23}$

Notice that in part b you could use a calculator to get $1 \div 2.8 = 0.3571$ to 4 d.p.

This is only an approximate answer. For an exact answer you must use fractions.

EXERCISE 1J

1 State whether each of these numbers is an integer or not.

    a $\sqrt{100}$ b $85 \div 6$ c 359^2
    d -7 e $\sqrt{20}$ f 6.3×10
    g 9.27×7.23 h $\sqrt{5+11}$ i $-\sqrt{36}$

2 State whether each of these numbers is rational or irrational.

    a 12.9 b $\frac{16}{7}$ c $\sqrt{8}$
    d $\sqrt{9}$ e $\sqrt{10}$ f 2.83^2
    g 1.65×2.13 h $\pi + 2$ i $10 - \pi$

3 Find the reciprocals of   a 300 b 0.3 c $4\frac{1}{4}$ d $\sqrt{6.25}$

4 Here are some numbers and their reciprocals. Which are the odd ones out?

    2.5   1.2   7.5   0.5

5 The reciprocal of x is 0.25. The reciprocal of y is 10.

    Work out the value of xy.

6 Is it possible to find two numbers, A and B, in the following cases? If so, give an example.

    a A and B are not integers, $A + B$ is an integer.
    b A and B are not integers, $A \times B$ is an integer.
    c A and B are integers, $A \times B$ is not an integer.

7 Try to find two irrational numbers that multiply to make an integer. If this is not possible, say so.

8 Try to find two irrational numbers which add up to an integer. If this is not possible, say so.

9 Try to find two irrational numbers which add up to a rational number. If this is not possible, say so.

Check your progress

Core

- I can identify square numbers and their square roots
- I can identify cube numbers
- I can find factors, common factors and the highest common factor
- I can find multiples and the lowest common multiple
- I can identify prime numbers
- I understand the difference between rational and irrational numbers
- I can find the reciprocal of a number

Chapter 1: Number

Chapter 2
Fractions and percentages

Topics	Level	Key words
1 Equivalent fractions	CORE	numerator, denominator, cancel, lowest terms, simplest form, proper and improper fraction, mixed number, top heavy
2 Fractions and decimals	CORE	decimal, fraction
3 Recurring decimals	EXTENDED	terminating decimal, recurring decimal
4 Percentages, fractions and decimals	CORE	percentage, decimal equivalent
5 Calculating a percentage	CORE	quantity, multiplier
6 Increasing or decreasing quantities by a percentage	CORE	multiplier
7 One quantity as a percentage of another	CORE	percentage change, percentage increase, percentage decrease, percentage profit, percentage loss
8 Simple interest and compound interest	CORE	simple interest, compound interest, principal, annual rate
9 A formula for compound interest	CORE	formula
10 Reverse percentage	EXTENDED	unitary method, multiplier

In this chapter you will learn how to:

CORE	EXTENDED
• Use the language and notation of simple vulgar and decimal fractions and percentages in appropriate contexts; recognise equivalence and convert between these forms. (C1.5 and E1.5) • Calculate a given percentage of a quantity. (C1.12 and E1.12) • Express one quantity as a percentage of another. (C1.12 and E1.12) • Calculate percentage increase or decrease. (C1.12 and E1.12) • Use given data to solve problems on personal and small business finance involving earnings, simple interest and compound interest, including discount, profit and loss. (C1.16 and E1.16) • Use a formula for compound interest. (C1.16 and E1.16) • Extract data from tables and charts. (C1.16 and E1.16)	• Write recurring decimals as fractions. (E1.5) • Carry out calculations involving reverse percentages, for example, finding the cost price given the selling price and the percentage profit. (E1.12)

Why this chapter matters

We use percentages and fractions in many situations in our everyday lives.

Why use fractions and percentages?

Because:

- basic percentages and simple fractions are quite easy to understand
- they are a useful way of comparing quantities
- fractions and percentages are used a lot in everyday life.

Who uses them?

Here are some examples of what you might see:

- Shops and businesses
    - Everything at half price in the sales!
    - Special offer — 10% off!
- Banks
    - Interest rates on loans 6.25%.
    - Interest rates on savings 2.5%.
- Salespeople
    - Earn 7.5% commission on sales.
- Government
    - Half of government workers are over 55.
    - Unemployment has fallen by 1%.
- Workers
    - My pay rise is 2.3%.
    - My income tax is 20%.

Chapter 2 . Topic 1
2.1 Equivalent fractions

Equivalent fractions are two or more fractions that represent the same part of a whole.

> **Example 1**
>
> Complete the following.
>
> **a** $\dfrac{3}{4} \xrightarrow{\times 4}_{\times 4} = \dfrac{\square}{16}$ **b** $\dfrac{2}{5} = \dfrac{\square}{15}$
>
> **a** Multiplying the **numerator** by 4 gives 12. This means $\dfrac{12}{16}$ is an equivalent fraction to $\dfrac{3}{4}$.
>
> **b** To change the **denominator** from 5 to 15, you multiply by 3. Do the same thing to the numerator, which gives $2 \times 3 = 6$. So, $\dfrac{2}{5} = \dfrac{6}{15}$.

The fraction $\dfrac{3}{4}$, in Example 1a, is in its **lowest terms** or **simplest form**.
This means that the only number that is a factor of both the numerator and denominator is 1.

A fraction in which the numerator (top number) is smaller than the denominator (bottom number) is called a proper fraction. An example of a proper fraction is $\dfrac{4}{5}$.

In an improper fraction the numerator (top number) is bigger than the denominator (bottom number). An example of an improper fraction is $\dfrac{9}{5}$. It is sometimes called a **top-heavy** fraction.

A mixed number is made up of a whole number and a proper fraction. An example of a mixed number is $1\dfrac{3}{4}$.

> **Example 2**
>
> Convert $\dfrac{14}{5}$ into a mixed number.
>
> $\dfrac{14}{5}$ means $14 \div 5$.
>
> Dividing 14 by 5 gives 2 with a remainder of 4 (5 divides into 14 two times, with $\dfrac{4}{5}$ left over).
>
> This means that there are 2 whole ones and $\dfrac{4}{5}$ left over.
>
> So, $\dfrac{14}{5} = \dfrac{5}{5} + \dfrac{5}{5} + \dfrac{4}{5}$
> $\phantom{So, \dfrac{14}{5}} = 2\dfrac{4}{5}$

EXERCISE 2A

1 Copy and complete the following.

a $\dfrac{2}{5} \xrightarrow{\times 4}_{\times 4} = \dfrac{\square}{20}$ **b** $\dfrac{1}{4} \xrightarrow{\times 3}_{\times 3} = \dfrac{\square}{12}$ **c** $\dfrac{3}{8} \xrightarrow{\times 5}_{\times 5} = \dfrac{\square}{40}$

d $\dfrac{2}{3} \xrightarrow{\times \square}_{\times \square} = \dfrac{\square}{18}$ **e** $\dfrac{3}{4} \xrightarrow{\times \square}_{\times \square} = \dfrac{\square}{12}$ **f** $\dfrac{5}{8} \xrightarrow{\times \square}_{\times \square} = \dfrac{\square}{40}$

2 Copy and complete the following.

a $\dfrac{10}{15} = \dfrac{10 \div 5}{15 \div 5} = \dfrac{\Box}{\Box}$ b $\dfrac{12}{15} = \dfrac{12 \div 3}{15 \div 3} = \dfrac{\Box}{\Box}$ c $\dfrac{20}{28} = \dfrac{20 \div 4}{28 \div 4} = \dfrac{\Box}{\Box}$

d $\dfrac{12}{18} = \dfrac{12 \div \Box}{\Box \div \Box} = \dfrac{\Box}{\Box}$ e $\dfrac{15}{25} = \dfrac{15 \div 5}{\Box \div \Box} = \dfrac{\Box}{\Box}$ f $\dfrac{21}{30} = \dfrac{21 \div \Box}{\Box \div \Box} = \dfrac{\Box}{\Box}$

3 Cancel each of these fractions to its simplest form.

a $\dfrac{4}{6}$ b $\dfrac{5}{15}$ c $\dfrac{12}{18}$ d $\dfrac{6}{8}$ e $\dfrac{3}{9}$

f $\dfrac{5}{10}$ g $\dfrac{14}{16}$ h $\dfrac{28}{35}$ i $\dfrac{10}{20}$ j $\dfrac{4}{16}$

4 Put the fractions in each set in order, with the smallest first.

a $\dfrac{1}{2}, \dfrac{5}{6}, \dfrac{2}{3}$ b $\dfrac{3}{4}, \dfrac{1}{2}, \dfrac{5}{8}$ c $\dfrac{7}{10}, \dfrac{2}{5}, \dfrac{1}{2}$

d $\dfrac{2}{3}, \dfrac{3}{4}, \dfrac{7}{12}$ e $\dfrac{1}{6}, \dfrac{1}{3}, \dfrac{1}{4}$ f $\dfrac{9}{10}, \dfrac{3}{4}, \dfrac{4}{5}$

5 Here are four unit fractions.

$\dfrac{1}{2}$ $\dfrac{1}{3}$ $\dfrac{1}{4}$ $\dfrac{1}{5}$

a Which two of these fractions have a sum of $\dfrac{7}{12}$?

Show clearly how you work out your answer.

b Which fraction is the biggest? Explain your answer.

6 Change each of these improper fractions into a mixed number.

a $\dfrac{7}{3}$ b $\dfrac{8}{3}$ c $\dfrac{9}{4}$

d $\dfrac{10}{7}$ e $\dfrac{12}{5}$ f $\dfrac{7}{5}$

7 Change each of these mixed numbers into an improper fraction.

a $3\dfrac{1}{3}$ b $5\dfrac{5}{6}$ c $1\dfrac{4}{5}$

d $5\dfrac{2}{7}$ e $4\dfrac{1}{10}$ f $5\dfrac{2}{3}$

g $2\dfrac{1}{2}$ h $3\dfrac{1}{4}$ i $7\dfrac{1}{6}$

j $3\dfrac{5}{8}$ k $6\dfrac{1}{3}$ l $9\dfrac{8}{9}$

8 Check your answers to questions **1** and **2**, using the fraction buttons on your calculator.

9 Which of these improper fractions has the largest value?

$\dfrac{27}{4}$ $\dfrac{31}{5}$ $\dfrac{13}{2}$

Show your working to justify your answer.

10 Find a mixed number that is greater than $\dfrac{85}{11}$ but smaller than $\dfrac{79}{10}$.

Chapter 2 . Topic 2

2.2 Fractions and decimals

Changing a decimal into a fraction

To change a **decimal** into a **fraction**, use a place-value table.

For example, $0.32 = \frac{32}{100}$

Units	.	Tenths	Hundredths	Thousandths
0	.	3	2	

Example 3

Express 0.32 as a fraction.

$0.32 = \frac{32}{100}$

This cancels to $\frac{8}{25}$

So, $0.32 = \frac{8}{25}$

Changing a fraction into a decimal

You can change a fraction into a decimal by dividing the **numerator** by the **denominator**.

Example 4

Express $\frac{3}{8}$ as a decimal.

$\frac{3}{8}$ means $3 \div 8$. This is a division calculation.

So, $\frac{3}{8} = 3 \div 8 = 0.375$

EXERCISE 2B

1 Change each of these decimals to a fraction, cancelling where possible.

- a 0.7
- b 0.4
- c 0.5
- d 0.03
- e 0.06
- f 0.13
- g 0.25
- h 0.38
- i 0.55
- j 0.64

2 Change each of these fractions to a decimal. Where necessary, give your answer correct to three decimal places.

- a $\frac{1}{2}$
- b $\frac{3}{4}$
- c $\frac{3}{5}$
- d $\frac{9}{10}$
- e $\frac{1}{3}$
- f $\frac{5}{8}$
- g $\frac{2}{3}$
- h $\frac{7}{20}$
- i $\frac{7}{11}$
- j $\frac{4}{9}$

3 Put each set of numbers in order, with the smallest first.

a 0.6, 0.3, $\frac{1}{2}$

b $\frac{2}{5}$, 0.8, 0.3

c 0.35, $\frac{1}{4}$, 0.15

d $\frac{7}{10}$, 0.72, 0.71

e 0.8, $\frac{3}{4}$, 0.7

f 0.08, 0.1, $\frac{1}{20}$

g 0.55, $\frac{1}{2}$, 0.4

h $1\frac{1}{4}$, 1.2, 1.23

Advice and Tips

Convert the fractions to decimals first.

4 Two stores sell the same T shirts at the same price of $24.

Store A has a sale and offers $\frac{1}{3}$ off the price.

Store B has a sale and offers $\frac{1}{4}$ off the price.

Which shop has the better offer in its sale?

Give a reason for your answer.

5 During April it rained on 12 days.

a For what fraction of the month did it rain? Give your answer in its simplest form.

b Give your answer as a decimal.

6 Which is bigger, $\frac{7}{8}$ or 0.87?

Show your working.

7 Which is smaller, $\frac{2}{3}$ or 0.7?

Show your working.

2.3 Recurring decimals

You can change a fraction to a decimal by dividing the numerator by the denominator.

The decimal may work out exactly. For example:

$\frac{5}{8} = 5 \div 8 = 0.625$

$\frac{7}{20} = 7 \div 20 = 0.35$

This is called a **terminating decimal**.

Alternatively, the decimal may continue. For example:

$\frac{2}{3} = 2 \div 3 = 0.666\,66…$

$\frac{5}{11} = 5 \div 11 = 0.454\,545…$

This is called a **recurring decimal**.

There is always a repeating pattern of digits in a recurring decimal.

Here are some examples.

$\frac{5}{9} = 0.5555...$ the 5 repeats.

$\frac{7}{12} = 0.583\,333...$ the 3 repeats.

$\frac{7}{22} = 0.318\,181\,8...$ the digit pair 18 repeats.

$\frac{6}{7} = 0.857\,142\,857\,142\,85...$ the digit sequence 857 142 repeats.

You use dots over the digits to show the repeating pattern.

$\frac{5}{9} = 0.\dot{5}$

$\frac{7}{12} = 0.58\dot{3}$

$\frac{7}{22} = 0.3\dot{1}\dot{8}$ Note the dots over the 1 and the 8.

$\frac{6}{7} = 0.\dot{8}5714\dot{2}$ You just put dots over the 8 and the 2 – the first and last digits.

In the previous section, you learned how to change terminating decimals to fractions.

Changing a recurring decimal to a fraction is more difficult.

Example 5

Write 0.777... as a fraction.

Write:	$f = 0.777...$	
Multiply both sides by 10:	$10f = 7.777...$	It is easy to multiply the decimal by 10.
Now subtract:	$9f = 7$	$10f - f = 9f$
		$7.777... - 0.777... = 7$
Divide by 9:	$f = \frac{7}{9}$	

Example 6

Write $0.41\dot{6}$ as a fraction.

Write:	$f = 0.416\,66...$	
Multiply by 10:	$10f = 4.1666...$	
Subtract:	$9f = 3.75$	All the digits after 2 decimal places cancel.
Divide by 9:	$f = \frac{3.75}{9} = \frac{375}{900} = \frac{5}{12}$	Simplify the fraction.
So:	$0.41\dot{6} = \frac{5}{12}$	

2.3

In the last two examples you multiplied by 10 because there was one recurring digit. What number will you multiply by if there are two recurring digits? Or more than two?

Example 7

Write $0.0\dot{7}\dot{4}$ as a fraction.

Write: $\qquad f = 0.074074...$

There are three recurring digits.

Multiply by 1000: $\qquad 1000f = 74.074074...$

Subtract: $\qquad 999f = 74$

Divide by 999: $\qquad f = \frac{74}{999} = \frac{2}{27}$ $\qquad$ 37 is a factor of 74 and 999.

So: $\qquad 0.0\dot{7}\dot{4} = \frac{2}{27}$

EXERCISE 2C

1 Write each of these fractions as a decimal.

a $\frac{1}{3}$ $\qquad$ b $\frac{3}{4}$ $\qquad$ c $\frac{5}{6}$

d $\frac{2}{9}$ $\qquad$ e $\frac{13}{20}$ $\qquad$ f $\frac{9}{11}$

g $\frac{3}{16}$ $\qquad$ h $\frac{11}{12}$

2 a Write $\frac{7}{15}$ as a recurring decimal. $\qquad$ b Write $\frac{14}{15}$ as a recurring decimal.

3 a Write $\frac{1}{9}$ as a decimal. $\qquad$ b Write $\frac{1}{6}$ as a decimal.

c Write $\frac{1}{9} + \frac{1}{6}$ as a decimal. $\qquad$ d Write $\frac{1}{6} - \frac{1}{9}$ as a decimal.

4 Write 0.888... as a fraction.

5 Write $0.\dot{2}\dot{4}$ as a fraction.

6 Write 0.3666... as a fraction.

7 Write $0.08\dot{3}$ as a fraction.

8 Write 2.4666... as a mixed number.

9 Write $\frac{3}{13}$ as a recurring decimal.

10 a Write $\frac{1}{11}$ as a recurring decimal.

b Write $\frac{2}{11}$ as a recurring decimal.

c Use your answers to **a** and **b** to predict the values of $\frac{3}{11}, \frac{4}{11}$ and $\frac{7}{11}$ as recurring decimals. Check whether you are correct.

Chapter 2: Fractions and percentages

Chapter 2. Topic 4

EXTENDED

11 $\frac{1}{7} = 0.\dot{1}4285\dot{7}$

  a Write $\frac{2}{7}$ as a recurring decimal in a similar way.

  b Write $\frac{3}{7}$ as a recurring decimal.

  c Without doing any more calculations, write $\frac{4}{7}, \frac{5}{7}$ and $\frac{6}{7}$ as recurring decimals.

12 a Here is a list of fractions.

  $\frac{1}{5} \quad \frac{1}{6} \quad \frac{1}{7} \quad \frac{1}{8} \quad \frac{1}{9} \quad \frac{1}{10} \quad \frac{1}{11} \quad \frac{1}{12}$

  Which ones are terminating decimals?

  b N is a whole number. Try to find a rule for deciding whether $\frac{1}{N}$ is a terminating decimal or a recurring decimal.

2.4 Percentages, fractions and decimals

100% means the *whole* of something. You can express *part* of the whole as a **percentage**.

Per cent means 'out of 100'.

So, any percentage can be converted to a **fraction** with denominator 100.

For example:

$32\% = \frac{32}{100}$ which can be simplified by cancelling to $\frac{8}{25}$

Also, you can convert any percentage to a **decimal** by dividing the percentage number by 100. This means moving the digits two places to the right.

For example:

$65\% = 65 \div 100 = 0.65$

You can convert any decimal to a percentage by multiplying by 100%.

For example:

$0.43 = 0.43 \times 100\% = 43\%$

You can convert any fraction to a percentage by making the denominator into 100 and taking the numerator as the percentage.

For example:

$\frac{2}{5} = \frac{40}{100} = 40\%$

You can also convert fractions to percentages by dividing the numerator by the denominator and multiplying by 100%.

2.4

For example:

$$\frac{2}{5} = 2 \div 5 \times 100\% = 40\%$$

Knowing the percentage and **decimal equivalents** of the common fractions is extremely useful. Try to learn them.

$\frac{1}{2} = 0.5 = 50\%$ $\qquad$ $\frac{1}{4} = 0.25 = 25\%$ $\qquad$ $\frac{3}{4} = 0.75 = 75\%$ $\qquad$ $\frac{1}{8} = 0.125 = 12.5\%$

$\frac{1}{10} = 0.1 = 10\%$ $\qquad$ $\frac{1}{5} = 0.2 = 20\%$ $\qquad$ $\frac{1}{3} = 0.33... = 33\frac{1}{3}\%$ $\qquad$ $\frac{2}{3} = 0.66... = 66\frac{2}{3}\%$

These tables show how to convert from one to the other.

Convert from percentage to:	
Decimal	**Fraction**
Divide the percentage by 100, for example: $52\% = 52 \div 100$ $= 0.52$	Make the percentage into a fraction with a denominator of 100 and simplify by cancelling if possible, for example: $52\% = \frac{52}{100} = \frac{13}{25}$

Convert from decimal to:	
Percentage	**Fraction**
Multiply the decimal by 100%, for example: $0.65 = 0.65 \times 100\%$ $= 65\%$	If the decimal has 1 decimal place put it over the denominator 10. If it has 2 decimal places put it over the denominator 100, etc. Then simplify by cancelling if possible, for example: $0.65 = \frac{65}{100} = \frac{13}{20}$

Convert from fraction to:	
Percentage	**Decimal**
If the denominator is a factor of 100 multiply numerator and denominator to make the denominator 100, then the numerator is the percentage, for example: $\frac{3}{20} = \frac{15}{100} = 15\%$ or convert to a decimal and change the decimal to a percentage, for example: $\frac{7}{8} = 7 \div 8$ $= 0.875$ $= 87.5\%$	Divide the numerator by the denominator, for example: $\frac{9}{40} = 9 \div 40 = 0.225$

Chapter 2: Fractions and percentages

Example 8

Convert these numbers to decimals. **a** 78% **b** 35% **c** $\frac{3}{25}$ **d** $\frac{7}{40}$

a 78% = 78 ÷ 100
   = 0.78

b 35% = 35 ÷ 100
   = 0.35

c $\frac{3}{25}$ = 3 ÷ 25
   = 0.12

d $\frac{7}{40}$ = 7 ÷ 40
   = 0.175

Example 9

Convert these numbers to percentages. **a** 0.85 **b** 0.125 **c** $\frac{7}{20}$ **d** $\frac{3}{8}$

a 0.85 = 0.85 × 100%
   = 85%

b 0.125 = 0.125 × 100%
   = 12.5%

c $\frac{7}{20} = \frac{35}{100}$
   = 35%

d $\frac{3}{8}$ = 3 ÷ 8 × 100%
   = 0.375 × 100%
   = 37.5%

Example 10

Convert these numbers to fractions. **a** 0.45 **b** 0.4 **c** 32% **d** 15%

a 0.45 = $\frac{45}{100}$
   = $\frac{9}{20}$

b 0.4 = $\frac{4}{10}$
   = $\frac{2}{5}$

c 32% = $\frac{32}{100}$
   = $\frac{8}{25}$

d 15% = $\frac{15}{100}$
   = $\frac{3}{20}$

EXERCISE 2D

1 Write each percentage as a fraction in its simplest form.

    a 8%   **b** 50%   **c** 25%

    d 35%   **e** 90%   **f** 75%

2 Write each percentage as a decimal.

    a 27%   **b** 85%   **c** 13%

    d 6%   **e** 80%   **f** 32%

3 Write each decimal as a fraction in its simplest form.

   a 0.12   b 0.4   c 0.45

   d 0.68   e 0.25   f 0.625

4 Write each decimal as a percentage.

   a 0.29   b 0.55   c 0.03

   d 0.16   e 0.6   f 1.25

5 Write each fraction as a percentage.

   a $\frac{7}{25}$ b $\frac{3}{10}$ c $\frac{19}{20}$

   d $\frac{17}{50}$ e $\frac{11}{40}$ f $\frac{7}{8}$

6 Write each fraction as a decimal.

   a $\frac{9}{15}$ b $\frac{3}{40}$ c $\frac{19}{25}$

   d $\frac{5}{16}$ e $\frac{1}{20}$ f $\frac{1}{8}$

7 a Convert each test score into a percentage. Give each answer to the nearest whole number.

Subject	Result	Percentage
Mathematics	38 out of 60	
English	29 out of 35	
Science	27 out of 70	
History	56 out of 90	
Technology	58 out of 75	

   b If all the tests are of the same standard, which was the best result?

8 Copy and complete the table.

Percentage	Decimal	Fraction
34%		
	0.85	
		$\frac{3}{40}$
45%		
	0.3	
		$\frac{2}{3}$
84%		
	0.45	
		$\frac{3}{8}$

Chapter 2: Fractions and percentages

2.5 Calculating a percentage

To calculate a percentage of a **quantity**, you multiply the quantity by the percentage. The percentage may be expressed as either a fraction or a decimal. When finding percentages without a calculator, base the calculation on 10% (or 1%) as these are easy to calculate.

> **Example 11**
> Calculate: **a** 10% of 54 kg   **b** 15% of 54 kg.
>
> **a** 10% is $\frac{1}{10}$ so $\frac{1}{10}$ of 54 kg = 54 kg ÷ 10 = 5.4 kg
>
> **b** 15% is 10% + 5% = 5.4 kg + 2.7 kg = 8.1 kg

Using a percentage multiplier

You have already seen that percentages and decimals are equivalent so it is easier, particularly when using a calculator, to express a percentage as a decimal and use this to do the calculation.

For example, 13% is a **multiplier** of 0.13, 20% a multiplier of 0.2 (or 0.20) and so on.

> **Example 12**
> **a** Calculate 45% of 160 cm.   **b** Find 52% of $460.
>
> **a** 45% = 0.45
>
> So 45% of 160 = 0.45 × 160 = 72 cm
>
> **b** 52% = 0.52
>
> So, 0.52 × 460 = 239.2
>
> This gives $239.20.
>
> Remember always to write a money answer with 2 decimal places.

EXERCISE 2E

1 Write down the multiplier that is equivalent to each percentage.

   a 88%   **b** 30%   **c** 25%
   d 8%   **e** 115%

2 Write down the percentage that is equivalent to each multiplier.

   a 0.78   **b** 0.4   **c** 0.75
   d 0.05   **e** 1.1

3 Calculate each amount.

- a 15% of $300
- b 6% of $105
- c 23% of 560 kg
- d 45% of 2.5 kg
- e 12% of 9 hours
- f 21% of 180 cm
- g 4% of $3
- h 35% of 8.4 m
- i 95% of $8
- j 11% of 308 minutes
- k 20% of 680 kg
- l 45% of $360

4 An estate agent charges 2% commission on every house he sells. How much commission will he earn on a house that he sells for $120 500?

5 A store had 250 employees. During one week of a flu epidemic, 14% of the store's employees were absent.

- a What percentage of the employees went into work?
- b How many of the employees went into work?

6 Generally, about 20% of fans at a soccer match are women. For one match there were 42 600 fans. How many of these do you think were women?

7 At a Paris railway station, in one week 350 trains arrived. Of these trains, 5% arrived early and 13% arrived late. How many arrived on time?

8 A school estimates that 60% of the students will attend a school play. There are 1500 students in the school. The caretaker is told to put out one seat for each person expected to attend plus an extra 10% of that number in case more attend. How many seats does he need to put out?

> **Advice and Tips**
>
> It is not 70% of the number of students in the school.

9 A school had 850 students and the attendance record in one week was:

Monday 96%   Tuesday 98%   Wednesday 100%   Thursday 94%   Friday 88%

How many students were present each day?

10 Calculate each amount.

- a 12.5% of $26
- b 6.5% of 34 kg
- c 26.8% of $2100
- d 7.75% of $84
- e 16.2% of 265 m
- f 0.8% of $3000

11 Air consists of 80% nitrogen and 20% oxygen (by volume). A man's lungs have a capacity of 600 cm^3. How much of each gas will he have in his lungs when he has just taken a deep breath?

12 A factory estimates that 1.5% of all the garments it produces will have a fault in them. One week the factory produces 850 garments. How many are likely to have a fault?

13 An insurance firm sells house insurance. The annual premiums are usually 0.3% of the value of the house. What will be the annual premium for a house valued at $90 000?

14 Average prices in a shop went up by 3% last year and 3% this year. Did the actual average price of items this year rise by more, the same amount, or less than last year?

Explain how you decided.

2.6 Increasing or decreasing quantities by a percentage

Increasing by a percentage

There are two methods for increasing a quantity by a percentage.

Method 1

Work out the increase and add it on to the original amount.

> **Example 13**
>
> Increase $6 by 5%.
>
> Work out 5% of $6: (5 ÷ 100) × 6 = $0.30
>
> Add the $0.30 to the original amount: $6 + $0.30 = $6.30

Method 2

Use a **multiplier**. An increase of 6% is equivalent to the original 100% *plus* the extra 6%. This is a total of 106% and is equivalent to the multiplier 1.06.

> **Example 14**
>
> Increase $6.80 by 5%.
>
> A 5% increase is a multiplier of 1.05
>
> So $6.80 increased by 5% is $6.80 × 1.05 = $7.14

EXERCISE 2F

1 What multiplier is used to increase a quantity by:

    a 10%   **b** 3%   **c** 20%   **d** 7%   **e** 12%?

2 Increase each amount by the given percentage. (Use any method you like.)

    a $60 by 4%   **b** 12 kg by 8%   **c** 450 g by 5%
    d 545 m by 10%   **e** $34 by 12%   **f** $75 by 20%
    g 340 kg by 15%   **h** 670 cm by 23%   **i** 130 g by 95%
    j $82 by 75%   **k** 640 m by 15%   **l** $28 by 8%

3 Azwan, who was on a salary of $27 500, was given a pay rise of 7%. What is his new salary?

4 In 2005 the population of a city was 1 565 000. By 2010 it had increased by 8%. What was the population of the city in 2010?

5 A small firm made the same pay increase of 5% for all its employees.

   a Calculate the new pay of each employee listed below. Each of their salaries before the increase is given.

   Caretaker, $16 500   Supervisor, $19 500

   Driver, $17 300   Manager, $25 300

   b Explain why the actual pay increases are different for each employee.

6 A bank pays 7% interest on the money that each saver keeps in the bank for a year. Allison keeps $385 in the bank for a year. How much will she have in the bank after the year?

7 In 1980 the number of cars on the roads of a town was about 102 000. Since then it has increased by 90%. Approximately how many cars are there on the roads of the town now?

8 An advertisement for a breakfast cereal states that a special-offer packet contains 15% more cereal for the same price as a normal 500 g packet. How much breakfast cereal is in a special-offer packet?

9 A headteacher was proud to point out that, since he had arrived at the school, the number of students had increased by 35%. How many students are now in the school, if there were 680 when the headteacher started at the school?

10 At a school concert there are always about 20% more girls than boys. If at one concert there were 50 boys, how many girls were there?

11 A government adds a sales tax to the price of most goods in shops. One year it is 17.5% on all electrical equipment.

Calculate the price of the following electrical equipment when sales tax of 17.5% is added.

Equipment	Pre-sales tax price
TV set	$245
Microwave oven	$72
CD player	$115
Personal stereo	$29.50

12 A television costs $400 before sales tax at 17.5% is added.

If the rate of sales tax goes up from 17.5% to 20%, how much will the cost of the television increase?

13 Bookshop BookWorms increased its prices by 5%, then increased them by 3%. Bookshop Books Galore increased its prices by 3%, then increased them by 5%.

Which shop's prices increased by the greater percentage?

   a BookWorms
   b Books Galore
   c Both the same
   d Cannot tell

Justify your choice.

14 Shop A increased its prices by 4% and then by another 4%. Shop B increased its prices by 8%.

Which shop's prices increased by the greater percentage?

a Shop A   b Shop B   c Both the same   d Cannot tell

Justify your choice.

15 A hi-fi system was priced at $420 at the start of 2008. At the start of 2009, it was 12% more expensive. At the start of 2010, it was 15% more expensive than the price at the start of 2009. What is the price of the hi-fi at the start of 2010?

Decreasing by a percentage

There are two methods for decreasing by a percentage.

Method 1

Work out the decrease and subtract it from the original amount.

> **Example 15**
>
> Decrease $8 by 4%.
>
> Work out 4% of $8: (4 ÷ 100) × 8 = $0.32
>
> Subtract the $0.32 from the original amount: $8 − $0.32 = $7.68

Method 2

Use a **multiplier**. A 7% decrease is equivalent to 7% less than the original 100%, so it represents 100% − 7% = 93% of the original. This is a multiplier of 0.93.

> **Example 16**
>
> Decrease $8.60 by 5%.
>
> A decrease of 5% is a multiplier of 0.95.
>
> So $8.60 decreased by 5% is $8.60 × 0.95 = $8.17

EXERCISE 2G

1 What multiplier is used to decrease a quantity by:

a 8%   b 15%   c 25%   d 9%   e 12%?

2 Decrease each amount by the given percentage. (Use any method you like.)

a $10 by 6%   b 25 kg by 8%   c 236 g by 10%   d 350 m by 3%
e $5 by 2%   f 45 m by 12%   g 860 m by 15%   h 96 g by 13%
i 480 cm by 25%   j 180 minutes by 35%   k 86 kg by 5%   l $65 by 42%

3 A car valued at $6500 last year is now worth 15% less. What is its value now?

4 A new diet guarantees that you will lose 12% of your mass in the first month. What mass should the following people have after one month on the diet?

  a Gracia, who started at 60 kg
  b Pierre, who started at 75 kg
  c Greta, who started at 52 kg

5 A motor insurance firm offers no-claims discounts off the full premium, as follows.

1 year with no claims	15% discount off the full premium
2 years with no claims	25% discount off the full premium
3 years with no claims	45% discount off the full premium
4 years with no claims	60% discount off the full premium

Mr Patel and his family are all offered motor insurance from this firm.

   Mr Patel has four years' no-claims discount and the full premium would be $440.
   Mrs Patel has one year's no-claims discount and the full premium would be $350.
   Sandeep has three years' no-claims discount and the full premium would be $620.
   Priyanka has two years' no-claims discount and the full premium would be $750.

Calculate the actual amount each member of the family has to pay for the motor insurance.

6 A large factory employed 640 people. It had to streamline its workforce and lose 30% of the workers. How big is the workforce now?

7 On the last day of term, a school expects to have an absence rate of 6%. If the school population is 750 students, how many students will the school expect to see on the last day of term?

8 Most speedometers in cars have an error of about 5% from the true reading. When my speedometer says I am driving at 70 km/h:

  a what is the lowest speed I could be doing
  b what is the highest speed I could be doing?

9 Kerry wants to buy a sweatshirt ($19), a tracksuit ($26) and some running shoes ($56). If she joins the store's premium club which costs $25 to join she can get 20% off the cost of the goods.

Should she join or not? Give figures to support your answer.

10 A shop advertises garden ornaments at $50 but with 10% off in a sale. It then advertises an extra 10% off the sale price.

Show that this is not a decrease in price of 20%.

11 A computer system was priced at $1000 at the start of 2008. At the start of 2009, it was 10% cheaper. At the start of 2010, it was 15% cheaper than the price at the start of 2009. What is the price of the computer system at the start of 2010?

Chapter 2 . Topic 7

12 Show that a 10% decrease followed by a 10% increase is equivalent to a 1% decrease overall.

13 A biscuit packet normally contains 300 g of biscuit and costs $1.40.

There are two special offers.

Offer A: 20% more for the same price

Offer B: Same amount for 20% off the normal price

Which is the better offer?

a Offer A **b** Offer B **c** Both the same **d** Cannot tell

Justify your choice.

> **Advice and Tips**
>
> Choose an amount to start with.

2.7 One quantity as a percentage of another

You express one quantity as a percentage of another by setting up the first quantity as a fraction of the second. You must make sure that the *units of each are the same*. Then you convert the fraction into a percentage by multiplying by 100%.

> **Example 17**
>
> Express $6 as a percentage of $40.
>
> Set up the fraction and multiply by 100%.
>
> $\frac{6}{40} \times 100\% = 15\%$

> **Example 18**
>
> Express 75 cm as a percentage of 2.5 m.
>
> First, change 2.5 m to 250 cm to get a common unit.
>
> So, the problem now becomes: Express 75 cm as a percentage of 250 cm.
>
> Set up the fraction and multiply by 100%.
>
> $\frac{75}{250} \times 100\% = 30\%$

Percentage change

A **percentage change** may be a **percentage increase** or a **percentage decrease**.

Percentage change = $\dfrac{\text{change}}{\text{original amount}} \times 100$

Use this to calculate **percentage profit** or **percentage loss** in a financial transaction.

> **Example 19**
>
> Jake buys a car for $1500 and sells it for $1800. What is Jake's percentage profit?
>
> Jake's profit is $300, so his percentage profit is:
>
> percentage profit = $\dfrac{\text{profit}}{\text{original amount}} \times 100\% = \dfrac{300}{1500} \times 100\% = 20\%$

Using a multiplier (or decimal)

To use a multiplier, divide the increase by the original quantity and change the resulting decimal to a percentage.

> **Example 20**
>
> Express 5 as a percentage of 40.
>
> Set up the fraction or decimal: $5 \div 40 = 0.125$
>
> Convert the decimal to a percentage: $0.125 = 12.5\%$

EXERCISE 2H

1 Express each amount as a percentage. Give suitably rounded figures (see page 125) where necessary.

- **a** $5 of $20
- **b** $4 of $6.60
- **c** 241 kg of 520 kg
- **d** 3 hours of 1 day
- **e** 25 minutes of 1 hour
- **f** 12 m of 20 m
- **g** 125 g of 600 g
- **h** 12 minutes of 2 hours
- **i** 1 week of a year
- **j** 1 month of 1 year
- **k** 25 cm of 55 cm
- **l** 105 g of 1 kg

2 Liam went to school with his pocket money of $2.50. He spent 80 cents at the shop. What percentage of his pocket money had he spent?

3 In Greece, there are 3 654 000 acres of agricultural land. Olives are grown on 237 000 acres of this land. What percentage of the agricultural land is used for olives?

4 During one year, it rained in Detroit on 123 days of the year. What percentage of days were wet?

5 Find the percentage profit on each transaction. Give your answers to one decimal place.

Item	Retail price (selling price)	Wholesale price (price the shop paid)
a CD player	$89.50	$60
b TV set	$345.50	$210
c Computer	$829.50	$750

6 Before Anton started to diet, his mass was 95 kg. His mass is now 78 kg. What percentage of his original mass has he lost?

7 In 2009 a city raised $14 870 000 in local tax. In 2010 it raised $15 597 000 in tax. What was the percentage increase?

8 When Ziad's team won the soccer league in 1995, they lost only four of their 42 league games. What percentage of games did they *not* lose?

9 In one year Britain's imports were as follows.

British Commonwealth	$109 530 000
USA	$138 790 000
France	$53 620 000
Other countries	$221 140 000

a What percentage of the total imports came from each source? Give your answers to 1 decimal place.

b Add up your answers to part **a**. What do you notice? Explain your answer.

10 Imran and Nadia take the same tests. Both tests are out of the same mark.

Here are their results.

Whose result has the greater percentage increase from test A to test B? Show your working.

	Test A	Test B
Imran	12	17
Nadia	14	20

11 A supermarket advertises its cat food as shown.

A government inspector is checking the claim.

She observes that over one hour, 46 people buy cat food and 38 buy the store's own brand.

Based on these figures, is the store's claim correct?

> 8 out of 10 cat owners choose our cat food.

12 Aya buys antiques and then sells them on the internet.

Find her percentage profit or loss on each of these items.

Item	Aya bought for:	Aya sold for:
Vase	$105	$84
Radio	$72	$90
Doll	$15	$41.25
Toy train	$50	$18

2.8 Simple interest and compound interest

Erin has a loan of $500. She agrees to pay 1.6% interest each month.

This is an example of **simple interest**.

Each month she pays 1.6% of $500 = 0.016 × 500 = $8.

If she pays back the loan after six months she will pay 6 × $8 = $48 in interest.

Banks and building societies usually pay **compound interest** on savings accounts.

When compound interest is used, the interest earned each year is added to the original amount (**principal**) and the new total then earns interest at the **annual rate** in the following year. This pattern is then repeated each year while the money is in the account.

The most efficient way to calculate the total amount in the account after several years is to use a **multiplier**.

Example 21

Elizabeth invests $400 in a savings account. The account pays compound interest at 6% each year. How much will she have in the account after three years?

The amount in the account increases by 6% each year, so the multiplier is 1.06.

After 1 year she will have $400 × 1.06   = $424
After 2 years she will have $424 × 1.06   = $449.44
After 3 years she will have $449.44 × 1.06 = $476.41 (rounded)

If you calculate the differences, you can see that the amount of interest increases each year ($24, $25.44 and $26.97).

EXERCISE 2I

1. Rahul has a loan of 7000 dollars.

   He pays 2% per month simple interest.

   How much will he pay if he has the loan for three months?

2. Lee lends her friend 30 000 dollars.

   Her friend agrees to pay simple interest of 6% per year.

   How much interest will Lee earn after two years?

3. Jean has a loan of 2000 dollars.

   The rate of simple interest is 8% a year.

   He has paid 640 dollars interest.

   How many years has he had the loan?

4 Rania puts $15 000 into a savings account where it earns 4% per annum compound interest.

  a What is her investment worth after one year?
  b What is her investment worth after two years?

5 Maria invests $1200 at 6% compound interest.

  Work out the value of the investment after:

  a one year   b two years   c three years.

6 Amar invests $20 000 for two years at 8% simple interest.

  Mona invests $20 000 for two years at 8% compound interest.

  a How much does each person earn?
  b Who earned more and how much more was it?

7 Luka invests $8000 at 6% compound interest for three years.

  a How much is his investment worth after three years?
  b How much interest has he earned after three years?

8 Mikael has a loan of $40 000.

  He pays 1.6% simple interest for six months.

  How much interest does he pay altogether?

9 This table shows the amount of interest paid on a loan of $12 000:

Number of years	1	2	3	4
Interest ($)	780	1560	2340	3120

  a Is this simple interest or compound interest?
  b What is the rate of interest per year?

10 Daniel earns 15% compound interest on an investment of $12 000.

  a What will it be worth after one year?
  b What will it be worth after two years?
  c Show that it will be worth over $20 000 after four years.

11 This table shows the values of amounts invested at 3% compound interest for one, two and five years.

Investment ($)	Value after 1 year ($)	Value after 2 years ($)	Value after 5 years ($)
1000	1030	1060.90	1159.27
2500	2575		2898.18
5000	5150		5796.37

  a Work out the two numbers missing from the table.
  b Work out the interest on an amount of $5000 invested for 5 years.

2.9 A formula for compound interest

Look back at Example 21.

Elizabeth invests $400 at 6% compound interest.

After three years the value, in dollars, is:

400 × 1.06 × 1.06 × 1.06 = 476.41

You can write this calculation as 400 × 1.06^3.

Then you can use the power button on a calculator to work this out efficiently.

Alternatively, you can use a **formula** for finding the value of an investment. Here is the formula.

> Value of investment = $P\left(1 + \frac{r}{100}\right)^n$
>
> where P is the initial investment, r is the annual percentage rate and n is the number of years.

Example 22

Jasmine invests $5500 at 2.5% per annum compound interest.

a How much is the investment worth after four years?
b How much interest has she earned after four years?

a Method 1:

The multiplier for a 2.5% increase is 1.025, so the value after four years is:

$5500 \times 1.025^4 = \$6070.97$

Method 2: use the formula. $P = 5500$, $r = 2.5$ and $n = 4$

Value in dollars = $5500 \times \left(1 + \frac{2.5}{100}\right)^4$

= 5500×1.025^4

= 6070.97

You can use either of the methods in part **a** to calculate compound interest.

b Interest in dollars = 6070.97 − 5500 = $570.97

EXERCISE 2J

1 Elton invests $2000 at 4% per annum compound interest.

Work out the value of his investment after three years.

2 Carla invests $4500 at 6% compound interest.

Work out the value of her investment after four years.

3 Zak invests $4600 at 5% per annum compound interest.

Work out the value of his investment after:

a 2 years   b 4 years   c 6 years.

4 Carmen invests $2500 at 7.5% per annum compound interest.

a How much is the investment worth after 5 years?

b How much is it worth after 8 years?

5 Marco invests $25 000 at 5.4% per annum compound interest for three years.

How much interest does he receive?

6 Greta takes out a loan of $3 500. The rate of compound interest is 2% per month.

After 6 months she wants to pay off the loan and the interest.

a How much must she pay altogether?

b How much of what she pays is interest?

7 An investment is earning 10% per annum compound interest.

How long will it take for the investment to double in value?

8 A bank offers 6% per annum compound interest on deposits.

This table shows how much a deposit of $5000 will be worth at different times.

Initial amount ($)	Amount after 1 year ($)	Amount after 2 years ($)	Amount after 5 years ($)	Amount after 10 years ($)
5000	5300	5618	6691.13	

Armand works out the missing number like this.

> The interest after 5 years is $6691.13 − $5000 = $1691.13
> The interest for the next five years will be the same, $1691.13
> After ten years the investment is worth $6691.13 + $1691.13 = $8382.26

Explain why Armand is not correct.

Then work out the correct value.

9 Luisa invests $10 000 at 5% annual interest.

Work out how much is the investment worth after 10 years if this is:

a simple interest

b compound interest.

10 Credit cards usually charge monthly compound interest on any unpaid amounts.

Greta has a loan of $1000 on her credit card.

There is a 2% monthly charge.

a Work out the size of Greta's loan after 12 months if she does not pay off any money.

b Show that this is equivalent to an annual percentage of about 26.8%.

2.10 Reverse percentage

Reverse percentage questions involve working backwards from the **final amount** to find the **original amount** when you know, or can work out, the final amount as a percentage of the original amount.

Method 1: The unitary method

The **unitary method** has three steps.

Step 1: Equate the final percentage to the final value.

Step 2: Use this to calculate the value of 1%.

Step 3: Multiply by 100 to work out 100% (the original value).

Example 23

The price of a car increased by 6% to $9116. Work out the price before the increase.

106% represents $9116.

Divide by 106.

1% represents $9116 ÷ 106

Multiply by 100.

100% represents original price: $9116 ÷ 106 × 100 = $8600

So the price before the increase was $8600.

Method 2: The multiplier method

The **multiplier** method involves fewer steps.

Step 1: Write down the multiplier.

Step 2: Divide the final value by the multiplier to give the original value.

Example 24

In a sale the price of a freezer is reduced by 12%. The sale price is $220. What was the price before the sale?

A decrease of 12% gives a multiplier of 0.88.

Dividing the sale price by the multiplier gives $220 ÷ 0.88 = $250.

So the price before the sale was $250.

Chapter 2: Fractions and percentages

EXERCISE 2K

1 Find what 100% represents in these situations.

    a 40% represents 320 g   **b** 14% represents 35 m

    c 45% represents 27 cm   **d** 4% represents $123

    e 2.5% represents $5   **f** 8.5% represents $34

2 On a tiring army training session, only 28 youngsters survived the whole day. This represented 35% of the original group. How large was the original group?

3 Sales tax is added to goods and services. With sales tax at 17.5%, what is the pre-sales tax price of the following priced goods?

T-shirt	$9.87	Tights	$1.41	Shorts	$6.11
Sweater	$12.62	Trainers	$29.14	Boots	$38.07

4 Hugo spends $200 a month on food. This represents 24% of his monthly take-home pay. How much is his monthly take-home pay?

5 Zara's weekly pay is increased by 5% to $315. What was Zara's pay before the increase?

6 The number of workers in a factory fell by 5% to 228. How many workers were there originally?

7 In a sale the price of a TV is reduced to 500 dollars. This is a 7% reduction on the original price. What was the original price?

8 If 38% of plastic bottles in a production line are blue and the remaining 7750 plastic bottles are brown, how many plastic bottles are blue?

9 I received $385 back from the government, which represented the 17.5% purchase tax on a piece of equipment. How much did I pay for this equipment in the first place?

10 A company is in financial trouble. The workers are asked to take a 10% pay cut.

Tomas works out that his pay will be $1296 per month. How much is his pay now?

11 Manza buys a car and sells it for $2940. He made a profit of 20%.

What was the cost price of the car?

12 When a suit is sold in a shop the selling price is $171 and the profit is 80%.

What was the cost price?

13 Tebor buys a chair. He sells it for $63 in an auction and makes a loss of 55%.

What did he pay for the chair?

14 A woman's salary increased by 5% in one year. Her new salary was $19 845.

How much was the increase?

15 After an 8% increase, the monthly salary of a chef was $1431. What was the original monthly salary?

16 Cassie invested some money at 4% interest per annum for two years. After two years, she had $1406.08 in the bank. How much did she invest originally?

17 A teacher asked her class to work out the original price of a cooker for which, after a 12% increase, the price was 291.20 dollars.

This is Lee's answer: 12% of 291.20 = 34.94 dollars

                                Original price = 291.2 − 34.94 = 256.26 ≈ 260 dollars

When the teacher read out the answer Lee ticked his work as correct.

What errors has he made?

Check your progress

Core

- I can write simple vulgar and decimal fractions in context, and convert between these forms
- I can calculate a given percentage of a quantity
- I know how to express one quantity as a percentage of another
- I can calculate percentage increases and decreases
- I understand personal and small business finance, including earnings, simple and compound interest, discounts, profits and losses, and can use data to solve related problems
- I can use formulae for compound interest

Extended

- I can use fractions to represent recurring decimals
- I can make calculations involving reverse percentages

Chapter 3
The four rules

Topics	Level	Key words
1 Order of operations	CORE	operation, brackets, order
2 Choosing the correct operation	CORE	
3 Finding a fraction of a quantity	CORE	quantity, fraction
4 Adding and subtracting fractions	CORE	proper fraction, improper fraction, lowest terms, simplest form, denominator, mixed number, equivalent fraction
5 Multiplying and dividing fractions	CORE	numerator, reciprocal

In this chapter you will learn how to:

CORE

- Use the four rules for calculation with whole numbers, decimals, vulgar fractions and mixed numbers. (C1.8 and E1.8)
- Apply operations in the correct order, including the use of brackets. (C1.8 and E1.8)

Why this chapter matters

Most jobs will require you to use some mathematics every day. Having good number skills will help you to be more successful in your job.

The mathematics used in jobs ranges from simple calculations, such as addition, subtraction, multiplication and division, to more complex calculations involving negative numbers and approximation. You will need to select the right mathematics for the job.

Jobs using mathematics

How many jobs can you think of that require some mathematics?
Here are a few ideas.

Accountant – How much profit have they made?

Pilot – How much fuel do I need?

Engineer – What measurements do I need to take? How much of each type of material will be needed?

Cashier – What coins do I need to give as change? What is the best price to sell my goods at?

Doctor – How much medicine should I prescribe?

Delivery driver – What is the best route?

Sports commentator – How many minutes are left in the game? What is his batting average?

Baker – What quantity of flour should I order?

If you already know what job you would like to do, think of what mathematics you might need for it.

Chapter 3: The four rules

Chapter 3 . Topic 1

3.1 Order of operations

Suppose you have to work out the answer to 4 + 5 × 2. You may say the answer is 18, but the correct answer is 14.

There is an **order** of **operations** which you must follow when working out calculations like this. The × is always done *before* the +.

In 4 + 5 × 2 this gives 4 + 10 = 14.

Now suppose you have to work out the answer to (3 + 2) × (9 – 5). The correct answer is 20.

You have probably realised that the parts in the **brackets** have to be done *first*, giving 5 × 4 = 20.

So, how do you work out a problem such as 9 ÷ 3 + 4 × 2?

To answer questions like this, you *must* follow the BIDMAS (or BODMAS) rule. This tells you the **order** in which you *must* do the operations.

B	Brackets	B	Brackets
I	Indices (powers)	O	pOwers
D	Division	D	Division
M	Multiplication	M	Multiplication
A	Addition	A	Addition
S	Subtraction	S	Subtraction

For example, to work out 9 ÷ 3 + 4 × 2:

First divide:	9 ÷ 3 = 3	giving	3 + 4 × 2
Then multiply:	4 × 2 = 8	giving	3 + 8
Then add:	3 + 8 = 11		

And to work out 60 – 5 × 3² + (4 × 2):

First, work out the brackets:	(4 × 2) = 8	giving	60 – 5 × 3² + 8
Then the index (power):	3² = 9	giving	60 – 5 × 9 + 8
Then multiply:	5 × 9 = 45	giving	60 – 45 + 8
Then add:	60 + 8 = 68	giving	68 – 45
Finally, subtract:	68 – 45 = 23		

EXERCISE 3A

1 Work out each of these.

- a 2 × 3 + 5 =
- b 6 ÷ 3 + 4 =
- c 5 + 7 – 2 =
- d 4 × 6 ÷ 2 =
- e 2 × 8 – 5 =
- f 3 × 4 + 1 =
- g 3 × 4 – 1 =
- h 3 × 4 ÷ 1 =
- i 12 ÷ 2 + 6 =
- j 12 ÷ 6 + 2 =
- k 3 + 5 × 2 =
- l 12 – 3 × 3 =

2 Work these out. Remember: first work out the bracket.

  a 2 × (3 + 5) =   b 6 ÷ (2 + 1) =
  c (5 + 7) − 2 =   d 5 + (7 − 2) =
  e 3 × (4 ÷ 2) =   f 3 × (4 + 2) =
  g 2 × (8 − 5) =   h 3 × (4 + 1) =
  i 3 × (4 − 1) =   j 3 × (4 ÷ 1) =
  k 12 ÷ (2 + 2) =   l (12 ÷ 2) + 2 =

3 Copy each of these and then put brackets in, where necessary, to make the answer true.

  a 3 × 4 + 1 = 15   b 6 ÷ 2 + 1 = 4
  c 6 ÷ 2 + 1 = 2   d 4 + 4 ÷ 4 = 5
  e 4 + 4 ÷ 4 = 2   f 16 − 4 ÷ 3 = 4
  g 3 × 4 + 1 = 13   h 16 − 6 ÷ 3 = 14
  i 20 − 10 ÷ 2 = 5   j 20 − 10 ÷ 2 = 15
  k 3 × 5 + 5 = 30   l 6 × 4 + 2 = 36
  m 15 − 5 × 2 = 20   n 4 × 7 − 2 = 20
  o 12 ÷ 3 + 3 = 2   p 12 ÷ 3 + 3 = 7
  q 24 ÷ 8 − 2 = 1   r 24 ÷ 8 − 2 = 4

4 Ravi says that 5 + 6 × 7 is equal to 77.

Is he correct?

Explain your answer.

5 Three different dice give scores of 2, 3, 5. Add ÷, ×, + or − signs to make each calculation work.

  a 2 3 5 = 11   b 2 3 5 = 16
  c 2 3 5 = 17   d 5 3 2 = 4
  e 5 3 2 = 13   f 5 3 2 = 30

6 Which is smaller:

  4 + 5 × 3 or (4 + 5) × 3?

Show your working.

7 Here is a list of numbers, some signs and one pair of brackets.

  2   5   6   18   −   ×   =   (   )

Use **all** of them to make a correct calculation.

8 Here is a list of numbers, some signs and one pair of brackets.

  3   4   5   8   −   ÷   =   (   )

Use **all** of them to make a correct calculation.

3.2 Choosing the correct operation

When a problem is given in words you will need to decide the correct operation to use. Should you add, subtract, multiply or divide?

> **Example 1**
>
> A party of 613 children and 59 adults are going on a day out to a theme park.
>
> **a** How many coaches, each holding 53 people, will be needed?
>
> **b** One adult gets into the theme park free for every 15 children. How many adults will have to pay to get in?
>
> **a** Altogether there are 613 + 59 = 672 people.
>
> So the number of coaches needed is 672 ÷ 53 (number of seats on each coach) = 12.67 …
>
> 13 coaches are needed (12 will not be enough).
>
> **b** This is also a division, 613 ÷ 15 = 40.86 …
>
> 40 adults will get in free.
>
> 59 − 40 = 19 will have to pay.

EXERCISE 3B

1. There are 48 cans of soup in a crate. A shop had a delivery of 125 crates of soup.
   a How many cans of soup were in this delivery?
   b The shop is running a promotion on soup. If you buy five cans you get one free. Each can costs 39 cents. How much will it cost to get 32 cans of soup?

2. A school has 12 classes, each of which has 24 students.
   a How many students are there at the school?
   b The student–teacher ratio is 18 to 1. That means there is one teacher for every 18 students. How many teachers are there at the school?

3. A football club is organising travel for an away game. 1300 adults and 500 children want to go. Each coach holds 48 people and costs $320 to hire.
   Tickets to the match cost $18 for adults and $10 for children.
   a How many coaches will be needed?
   b The club is charging adults $26 and children $14 for travel and a ticket. How much profit does the club make out of the trip?

4. A large letter costs 39 cents to post and a small letter costs 30 cents. How many dollars will it cost to send 20 large and 90 small letters?

Chapter 3 . Topic 3

3.3

5 Kirsty collects small models of animals. Each one costs 45 cents. She saves enough to buy 23 models but when she goes to the shop she finds that the price has gone up to 55 cents. How many can she buy now?

6 Michaela wanted to save up for a bike that costs $250. She baby-sits each week for 6 hours for $2.75 an hour, and does a Saturday job that pays $27.50. She saves three-quarters of her weekly earnings. How many weeks will it take her to save enough to buy the bike?

7 The magazine *Teen Dance* comes out every month. In a newsagent's shop the magazine costs $2.45. The annual (yearly) subscription for the magazine is $21. How much cheaper is each magazine when it is bought on subscription?

8 Paula buys a sofa. She pays a deposit of 10% of the cash price and then 36 monthly payments of $12.50. In total she pays $495. How much was the cash price of the sofa?

9 There are 125 people at a wedding. They need to get to the reception.

52 people are going by coach and the rest are travelling in cars. Each car can take up to five people.

What is the least number of cars needed to take everyone to the reception?

10 Gustav's car does 8 kilometres to each litre of fuel. He does 12 600 kilometres a year of which 4600 is on company business.

Fuel costs 95 cents per litre.

Insurance and servicing costs $800 a year.

Gustav's company gives him an allowance of 40 cents for each kilometre he drives on company business.

How much does Gustav pay towards running his car each year?

3.3 Finding a fraction of a quantity

To do this, you simply multiply the **fraction** by the **quantity**, for example, $\frac{1}{2}$ of 30 is the same as $\frac{1}{2} \times 30$.

Remember: In mathematics 'of' is interpreted as ×.

For example, two lots of three is the same as 2 × 3.

> **Example 2**
> Find $\frac{3}{4}$ of $196.
>
> First, find $\frac{1}{4}$ by dividing by 4. Then find $\frac{3}{4}$ by multiplying your answer by 3.
> 196 ÷ 4 = 49   then   49 × 3 = 147
> The answer is $147.

Chapter 3: The four rules

EXERCISE 3C

1 Calculate each amount.

a $\frac{3}{5}$ of 30
b $\frac{2}{7}$ of 35
c $\frac{3}{8}$ of 48
d $\frac{7}{10}$ of 40

2 Calculate each of these quantities.

a $\frac{3}{4}$ of $2400
b $\frac{2}{5}$ of 320 grams
c $\frac{5}{8}$ of 256 kilograms
d $\frac{2}{3}$ of $174
e $\frac{5}{6}$ of 78 litres
f $\frac{3}{4}$ of 120 minutes

3 In each case, find out which is the larger number.

a $\frac{2}{5}$ of 60 or $\frac{5}{8}$ of 40
b $\frac{3}{4}$ of 280 or $\frac{7}{10}$ of 290
c $\frac{2}{3}$ of 78 or $\frac{4}{5}$ of 70
d $\frac{5}{6}$ of 72 or $\frac{11}{12}$ of 60

4 A director receives $\frac{2}{15}$ of his firm's profits. The firm made a profit of $45600 in one year. How much did the director receive?

5 A woman left $84 000 in her will.

She left $\frac{3}{8}$ of the money to charity.

How much did she leave to charity?

6 Two-thirds of a person's mass is water. Paul has a mass of 78 kg. How much of his body mass is water?

7 a Information from the first census in Singapore showed that $\frac{2}{25}$ of the population were Indian. The total population was 10 700. How many people were Indian?

b By 1990 the population of Singapore had grown to 3 002 800. Only $\frac{1}{16}$ of this population were Indian. How many Indians were living in Singapore in 1990?

8 Marc normally earns $500 a week. One week he is given a bonus of $\frac{1}{10}$ of his wage.

a Find $\frac{1}{10}$ of $500.

b How much does Marc earn altogether for this week?

9 The price of a new TV costing $360 is reduced by $\frac{1}{3}$ in a sale.

a Find $\frac{1}{3}$ of $360.

b How much does the TV cost in the sale?

10 A car is advertised at Lion Autos at $9000 including extras but with a special offer of one-fifth off this price.

The same car is advertised at Tiger Motors for $6000 but the extras add one-quarter to this price.

Which garage is the cheaper?

11 A jar of coffee normally contains 200 g and costs $2.

There are two special offers on a jar of coffee.

Offer A: $\frac{1}{4}$ extra for the same price.

Offer B: Same mass for $\frac{3}{4}$ of the original price.

Which offer is better value?

3.4 Adding and subtracting fractions

When you add two fractions with the same **denominator**, you get one of the following:

- a proper fraction that cannot be simplified, for example:
$$\frac{1}{5} + \frac{2}{5} = \frac{3}{5}$$
- a proper fraction that can be simplified to its **lowest terms** or **simplest form**, for example:
$$\frac{1}{8} + \frac{3}{8} = \frac{4}{8} = \frac{1}{2}$$
- an **improper fraction** that cannot be simplified, so it is converted to a **mixed number**, for example:
$$\frac{6}{7} + \frac{2}{7} = \frac{8}{7} = 1\frac{1}{7}$$
- an improper fraction that *can* be simplified before it is converted to a mixed number, for example:
$$\frac{5}{8} + \frac{7}{8} = \frac{12}{8} = \frac{3}{2} = 1\frac{1}{2}$$

When you subtract one fraction from another with the same denominator, you get one of the following:

- a proper fraction that cannot be simplified, for example:
$$\frac{3}{5} - \frac{1}{5} = \frac{2}{5}$$
- a proper fraction that can be simplified, for example:
$$\frac{7}{10} - \frac{1}{10} = \frac{6}{10} = \frac{3}{5}$$

Notice that $\frac{6}{10}$ and $\frac{3}{5}$ are **equivalent fractions**. They represent the same quantity.

Note: You must always simplify fractions by cancelling if possible.

Example 3

Find $\frac{1}{2} + \frac{5}{8}$

These fractions do not have the same denominator.

However $\frac{1}{2} = \frac{4}{8}$ so you can write:

$\frac{1}{2} + \frac{5}{8} = \frac{4}{8} + \frac{5}{8} = \frac{9}{8} = 1\frac{1}{8}$

EXERCISE 3D

1. Copy and complete each of these additions.

   a $\frac{3}{7} + \frac{2}{7}$ b $\frac{5}{9} + \frac{2}{9}$ c $\frac{3}{5} + \frac{1}{5}$ d $\frac{3}{7} + \frac{3}{7}$

2. Copy and complete each of these subtractions.

   a $\frac{4}{7} - \frac{1}{7}$ b $\frac{5}{9} - \frac{4}{9}$ c $\frac{7}{11} - \frac{3}{11}$ d $\frac{9}{13} - \frac{2}{13}$

3. Copy and complete each of these additions.

   a $\frac{5}{8} + \frac{1}{8}$ b $\frac{3}{10} + \frac{1}{10}$ c $\frac{2}{9} + \frac{4}{9}$ d $\frac{1}{4} + \frac{1}{4}$

4. Copy and complete each of these subtractions.

   a $\frac{7}{8} - \frac{3}{8}$ b $\frac{7}{10} - \frac{3}{10}$ c $\frac{5}{6} - \frac{1}{6}$ d $\frac{9}{10} - \frac{1}{10}$

5. Copy and complete each of these additions. Use equivalent fractions to make the denominators the same. Show your working.

   a $\frac{1}{2} + \frac{7}{10}$ b $\frac{1}{2} + \frac{5}{8}$ c $\frac{3}{4} + \frac{3}{8}$ d $\frac{3}{4} + \frac{7}{8}$

   e $\frac{1}{2} + \frac{7}{8}$ f $\frac{1}{3} + \frac{5}{6}$ g $\frac{5}{6} + \frac{2}{3}$ h $\frac{3}{4} + \frac{1}{2}$

6. Copy and complete each of these additions. Show your working.

   a $\frac{3}{8} + \frac{7}{8}$ b $\frac{3}{4} + \frac{3}{4}$ c $\frac{2}{5} + \frac{3}{5}$ d $\frac{7}{10} + \frac{9}{10}$

7. Copy and complete each of these subtractions. Use equivalent fractions to make the denominators the same. Show your working.

   a $\frac{7}{8} - \frac{1}{4}$ b $\frac{7}{10} - \frac{1}{5}$

   c $\frac{3}{4} - \frac{1}{2}$ d $\frac{5}{8} - \frac{1}{4}$

   e $\frac{1}{2} - \frac{1}{4}$ f $\frac{7}{8} - \frac{1}{2}$

   g $\frac{9}{10} - \frac{1}{2}$ h $\frac{11}{16} - \frac{3}{8}$

You can only add or subtract fractions with different denominators after you have converted them to equivalent fractions with the same denominator.

3.4

Example 4

a Find $\frac{2}{3} + \frac{1}{5}$ **b** Find $2\frac{3}{4} - 1\frac{5}{6}$

a Note that you can change both fractions to equivalent fractions with a denominator of 15. This is the lowest common multiple of 3 and 5.

This then becomes:

$$\frac{2 \times 5}{3 \times 5} + \frac{1 \times 3}{5 \times 3} = \frac{10}{15} + \frac{3}{15}$$

$$= \frac{13}{15}$$

b Split the calculation into $\left(2 + \frac{3}{4}\right) - \left(1 + \frac{5}{6}\right)$.

This then becomes:

$$= 2 - 1 + \frac{3}{4} - \frac{5}{6}$$

Note that you can change both fractions to equivalent fractions with a denominator of 12.

$$= 1 + \frac{9}{12} - \frac{10}{12}$$

$$= 1 - \frac{1}{12}$$

$$= \frac{11}{12}$$

EXERCISE 3E

1 Complete these calculations. Show your working.

- **a** $\frac{1}{3} + \frac{1}{5}$
- **b** $\frac{1}{3} + \frac{1}{4}$
- **c** $\frac{1}{5} + \frac{1}{10}$
- **d** $\frac{2}{3} + \frac{1}{4}$
- **e** $\frac{3}{4} + \frac{1}{8}$
- **f** $\frac{1}{3} + \frac{1}{6}$
- **g** $\frac{1}{2} - \frac{1}{3}$
- **h** $\frac{1}{4} - \frac{1}{5}$
- **i** $\frac{1}{5} - \frac{1}{10}$
- **j** $\frac{7}{8} - \frac{3}{4}$
- **k** $\frac{5}{6} - \frac{3}{4}$
- **l** $\frac{5}{6} - \frac{1}{2}$
- **m** $\frac{5}{12} - \frac{1}{4}$
- **n** $\frac{1}{3} + \frac{4}{9}$
- **o** $\frac{1}{4} + \frac{3}{8}$
- **p** $\frac{7}{8} - \frac{1}{2}$
- **q** $\frac{3}{5} - \frac{8}{15}$
- **r** $\frac{11}{12} + \frac{5}{8}$
- **s** $\frac{7}{16} + \frac{3}{10}$
- **t** $\frac{4}{9} - \frac{2}{21}$
- **u** $\frac{5}{6} - \frac{4}{27}$

2 Complete these calculations. Show your working.

- **a** $2\frac{1}{7} + 1\frac{3}{14}$
- **b** $6\frac{3}{10} + 1\frac{4}{5} + 2\frac{1}{2}$
- **c** $3\frac{1}{2} - 1\frac{1}{3}$
- **d** $1\frac{7}{18} + 2\frac{3}{10}$
- **e** $3\frac{2}{6} + 1\frac{9}{20}$
- **f** $1\frac{1}{8} - \frac{5}{9}$

g $1\frac{3}{16} - \frac{7}{12}$ **h** $\frac{5}{6} + \frac{7}{16} + \frac{5}{8}$ **i** $\frac{7}{10} + \frac{3}{8} + \frac{5}{6}$

j $1\frac{1}{3} + \frac{7}{10} - \frac{4}{15}$ **k** $\frac{5}{14} + 1\frac{3}{7} - \frac{5}{12}$

3 In a class of children, three-quarters are Chinese, one-fifth are Malay and the rest are Indian. What fraction of the class are Indian?

4 **a** In a class election, half of the people voted for Aminah, one-third voted for Reshma and the rest voted for Peter. What fraction of the class voted for Peter?

  b One of the following is the number of people in the class.

  25   28   30   32

  How many people are in the class?

3.5 Multiplying and dividing fractions

What is $\frac{1}{2}$ of $\frac{1}{4}$?   The diagram shows the answer is $\frac{1}{8}$.

In mathematics, you always write $\frac{1}{2}$ of $\frac{1}{4}$ as $\frac{1}{2} \times \frac{1}{4}$

So you know that $\frac{1}{2} \times \frac{1}{4} = \frac{1}{8}$

To **multiply** fractions, you multiply the **numerators** together and you multiply the **denominators** together.

> **Example 5**
> Work out $\frac{1}{4} \times \frac{2}{5}$.
>
> $\frac{1}{4} \times \frac{2}{5} = \frac{1 \times 2}{4 \times 5} = \frac{2}{20} = \frac{1}{10}$

Sometimes you can simplify by cancelling *before* you multiply.

> **Example 6**
> Find $\frac{3}{8} \times \frac{5}{9}$
>
> $\frac{3}{8} \times \frac{5}{9} = \frac{1\cancel{3}}{8} \times \frac{5}{\cancel{9}_3}$ (3 is a factor of 3 and 9.)
>
> $= \frac{5}{24}$ (5 = 1 × 5)
>
>   (24 = 8 × 3)

To multiply mixed numbers, first change them to improper fractions.

3.5

Example 7

Find $1\frac{3}{4} \times 2\frac{1}{2}$

$1\frac{3}{4} \times 2\frac{1}{2} = \frac{7}{4} \times \frac{5}{2}$

$= \frac{35}{8}$

$= 4\frac{3}{8}$

EXERCISE 3F

1 Work these out, leaving each answer in its simplest form. Show your working.

- **a** $\frac{1}{2} \times \frac{1}{3}$
- **b** $\frac{1}{4} \times \frac{2}{5}$
- **c** $\frac{3}{4} \times \frac{1}{2}$
- **d** $\frac{3}{7} \times \frac{1}{2}$
- **e** $\frac{2}{3} \times \frac{4}{5}$
- **f** $\frac{1}{3} \times \frac{3}{5}$
- **g** $\frac{1}{3} \times \frac{6}{7}$
- **h** $\frac{3}{4} \times \frac{2}{5}$
- **i** $\frac{2}{3} \times \frac{3}{4}$
- **j** $\frac{1}{2} \times \frac{4}{5}$

2 Work these out, leaving each answer in its simplest form. Show your working.

- **a** $\frac{5}{16} \times \frac{3}{10}$
- **b** $\frac{9}{10} \times \frac{5}{12}$
- **c** $\frac{14}{15} \times \frac{3}{8}$
- **d** $\frac{8}{9} \times \frac{6}{15}$
- **e** $\frac{6}{7} \times \frac{21}{30}$
- **f** $\frac{9}{14} \times \frac{35}{36}$

3 One-quarter of Lee's stamp collection was given to him by his sister. Unfortunately two-thirds of these were torn. What fraction of his collection was given to him by his sister and were not torn?

4 Bilal eats one-quarter of a cake, and then half of what is left. How much cake is left uneaten?

5 Work these out, giving each answer as a mixed number where possible. Show your working.

- **a** $1\frac{1}{4} \times \frac{1}{3}$
- **b** $1\frac{2}{3} \times 1\frac{1}{4}$
- **c** $2\frac{1}{2} \times 2\frac{1}{2}$
- **d** $1\frac{3}{4} \times 1\frac{2}{3}$
- **e** $3\frac{1}{4} \times 1\frac{1}{5}$
- **f** $1\frac{1}{4} \times 2\frac{2}{3}$
- **g** $2\frac{1}{2} \times 5$
- **h** $7\frac{1}{2} \times 4$

6 Which is larger, $\frac{3}{4}$ of $2\frac{1}{2}$ or $\frac{2}{5}$ of $6\frac{1}{2}$?

Dividing fractions

Look at the problem $3 \div \frac{3}{4}$. This is like asking, 'How many $\frac{3}{4}$s are there in 3?'

Look at the diagram.

 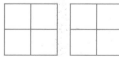 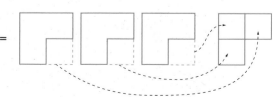

Chapter 3: The four rules

Each of the three whole shapes is divided into quarters. What is the total number of quarters divided by 3?

Can you see that you could fit the four shapes on the right-hand side of the = sign into the three shapes on the left-hand side?

i.e. $\quad 3 \div \frac{3}{4} = 4$

or $\quad 3 \div \frac{3}{4} = 3 \times \frac{4}{3} = \frac{3 \times 4}{3} = \frac{12}{3} = 4$

So, to divide by a fraction, you turn the fraction upside down (finding its **reciprocal**), and then multiply.

> **Example 8**
>
> Work out $2\frac{1}{2} \div \frac{3}{4}$
>
> $2\frac{1}{2} \div \frac{3}{4} = \frac{5}{2} \times \frac{4}{3}$ (change $2\frac{1}{2}$ to an improper fraction)
>
> $= \frac{5}{2_1} \times \frac{4^2}{3}$ (2 and 4 have 2 as a common factor)
>
> $= \frac{10}{3}$
>
> $= 3\frac{1}{3}$
>
> This means that $3\frac{1}{3} \times \frac{3}{4} = 3\frac{1}{2}$

EXERCISE 3G

1 Work these out, giving your answer as a mixed number where possible. Show your working.

a $\frac{1}{4} \div \frac{1}{3}$ b $\frac{2}{5} \div \frac{2}{7}$ c $\frac{4}{5} \div \frac{3}{4}$ d $\frac{3}{7} \div \frac{2}{5}$

e $5 \div 1\frac{1}{4}$ f $6 \div 1\frac{1}{2}$ g $7\frac{1}{2} \div 1\frac{1}{2}$ h $3 \div 1\frac{3}{4}$

i $1\frac{5}{12} \div 3\frac{3}{16}$ j $3\frac{3}{5} \div 2\frac{1}{4}$

2 A grain merchant has only thirteen and a half tonnes in stock. He has several customers who are all ordering three-quarters of a tonne. How many customers can he supply?

3 For a party, Zahar made twelve and a half litres of lemonade. His glasses could each hold $\frac{5}{16}$ of a litre. How many of the glasses could he fill from the twelve and a half litres of lemonade?

4 How many strips of ribbon, each three and a half centimetres long, can I cut from a roll of ribbon that is fifty-two and a half centimetres long?

5 Joe's stride is three-quarters of a metre long. How many strides does he take to walk the length of a bus twelve metres long?

Chapter 4 . Topic 2

3 The instructions on a bottle of de-icer say that it will stop water freezing down to −12 °C. The temperature is −4 °C.

How many more degrees does the temperature need to fall before the de-icer stops working?

4 The temperature in a room is 16 °C

The temperature in a freezer is −22 °C

How many degrees less than room temperature is the freezer?

5 Here are the temperatures at midday on January 21st in 5 cities.

City	Dubai	Helsinki	Moscow	New York	Tokyo
Temperature	24 °C	−10 °C	−8 °C	2 °C	7 °C

a Find the smallest difference between the temperatures of these cities.

b Find the largest difference between the temperatures.

4.2 Everyday use of directed numbers

There are many other situations where directed numbers are used. Here are three examples.

- When +15 m means 15 metres *above* sea level, then −15 m means 15 metres *below* sea level.
- When +2 h means 2 hours *after* midday, then −2 h means 2 hours *before* midday.
- When +$60 means a **profit** of $60, then −$60 means a **loss** of $60.

You also meet negative numbers on graphs, and you may already have plotted coordinates with negative numbers.

On bank statements and bills a negative number means you owe money. A positive number means they owe you money.

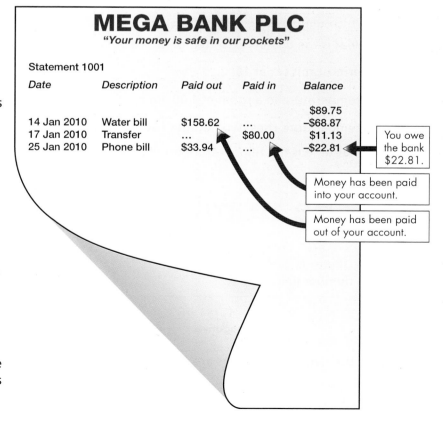

Chapter 4: Directed numbers 67

Chapter 4 . Topic 3

EXERCISE 4B

Copy and complete each statement.

1. If +$5 means a profit of five dollars, then …… means a loss of five dollars.

2. If +200 m means 200 metres above sea level, then …… means 200 metres below sea level.

3. If −100 m means one hundred metres below sea level, then +100 m means one hundred metres …… sea level.

4. If +5 h means 5 hours after midday, then …… means 5 hours before midday.

5. If +2 °C means two degrees above freezing point, then …… means two degrees below freezing point.

6. If +70 km means 70 kilometres north of the equator, then …… means 70 kilometres south of the equator.

7. If 10 minutes before midnight is represented by −10 minutes, then five minutes after midnight is represented by …… .

8. If a car moving forwards at 10 kilometres per hour is represented by +10 km/h, then a car moving backwards at 5 kilometres per hour is represented by …… .

9. In an office building, the third floor above ground level is represented by +3. So, the second floor below ground level is represented by …… .

10. The temperature on three days in Moscow was −7 °C, −5 °C and −11 °C.
    a Which temperature is the lowest?
    b What is the difference in temperature between the coldest and the warmest days?

11. A thermostat is set at 16 °C.

    The temperature in a room at 1.00 am is −2 °C.

    The temperature rises two degrees every 6 minutes.

    At what time is the temperature on the thermostat reached?

4.3 The number line

Look at the **number line**.

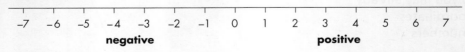

Notice that the **negative** numbers are to the left of 0 and the **positive** numbers are to the right of 0.

Numbers to the right of any number on the number line are always bigger than that number.

4.3

Numbers to the left of any number on the number line are always smaller than that number.

So, for example, you can see from a number line that:

   2 is *smaller* than 5 because 2 is to the left of 5.

You can write this as 2 < 5.

   −3 is *smaller* than 2 because −3 is to the *left* of 2.

You can write this as −3 < 2.

   7 is *bigger* than 3 because 7 is to the *right* of 3.

You can write this as 7 > 3.

   −1 is *bigger* than −4 because −1 is to the *right* of −4.

You can write this as −1 > −4.

Reminder: The **inequality** signs:

   < means 'is **less than**'

   > means 'is **greater than**' or 'is **more than**'

EXERCISE 4C

1 Copy each statement and put the correct symbol (< or >) in each space.

   a −1 …… 3   b 3 …… 2   c −4 …… −1   d −5 …… −4
   e 1 …… −6   f −3 …… 0   g −2 …… −1   h 2 …… −3
   i 5 …… −6   j 3 …… 4   k −7 …… −5   l −2 …… −4

2

Copy each statement and put the correct symbol in each space.

   a $\frac{1}{4}$ …… $\frac{3}{4}$ b $-\frac{1}{2}$ …… 0   c $-\frac{3}{4}$ …… $\frac{3}{4}$

   d $\frac{1}{4}$ …… $-\frac{1}{2}$ e -1 …… $\frac{3}{4}$ f $\frac{1}{2}$ …… 1

3 Copy these number lines and fill in the missing numbers.

   a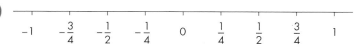

   b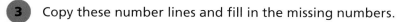

   c

   d

4 Here are some temperatures.

   2 °C   −2 °C   −4 °C   6 °C

Copy and complete the weather report, using these temperatures.

Chapter 4 . Topic 4

The hottest place today is Eastbourne with a temperature of ____, while in Barnsley a ground frost has left the temperature just below zero at ____. In Bristol it is even colder at ____. Finally, in Tenby the temperature is just above freezing at ____.

5 In each case find, if possible, an integer N with the property that

    a $N > -3$ and $N < 2$ **b** $N < -3$ and $N > 2$

    c $N > -3$ and $N > 2$ **d** $N < -3$ and $N < 2$

4.4 Adding and subtracting directed numbers

Adding and subtracting positive numbers

These two operations can be illustrated on a thermometer scale.

- **Adding** a positive number moves the marker up the thermometer scale.

  For example,

  $-2 + 6 = 4$

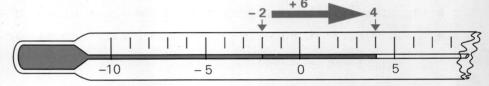

- **Subtracting** a positive number moves the marker *down* the thermometer scale.

  For example,

  $3 - 5 = -2$

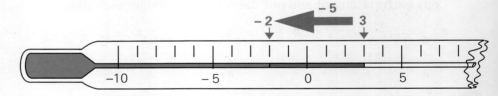

Example 1

The temperature at midnight was 2 °C but then it fell by five degrees. What was the new temperature?

Falling five degrees means the calculation is $2 - 5$, which is equal to -3. So, the new temperature is -3 °C.

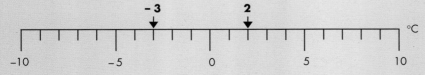

4.4

EXERCISE 4D

1 Find the answer to each of these.

   a 2° − 4° =   b 4° − 7° =   c 3° − 5° =   d 1° − 4° =
   e 6° − 8° =   f 5° − 8° =   g −2 + 5 =   h −1 + 4 =
   i −4 + 3 =   j −6 + 5 =   k −3 + 5 =   l −5 + 2 =
   m −1 − 3 =   n −2 − 4 =   o −5 − 1 =   p 3 − 4 =
   q 2 − 7 =   r 1 − 5 =   s −3 + 7 =   t 5 − 6 =
   u −2 − 3 =   v 2 − 6 =   w −8 + 3 =   x 4 − 9 =

2 At 5 am the temperature in Lisbon was −4 °C. At 11 am the temperature was 3 °C.

   a By how many degrees did the temperature rise?

   b The temperature in Madrid was two degrees lower than in Lisbon at 5 am. What was the temperature in Madrid at 5 am?

3 Here are five numbers.

   4   7   8   2   5

   a Use two of the numbers to make a calculation with an answer of −6.
   b Use three of the numbers to make a calculation with an answer of −1.
   c Use four of the numbers to make a calculation with an answer of −18.
   d Use all five of the numbers to make a calculation with an answer of −12.

4 A submarine is 600 metres below sea level.

A radar system can detect submarines down to 300 metres below sea level.

To avoid detection safely, the submarine captain keeps the submarine 50 metres below the level of detection.

How many metres can the submarine climb to be safe from detection?

Adding and subtracting negative numbers

To *subtract a negative number* ...

   ... treat the − − as +

For example: 4 − (−2) = 4 + 2 = 6

To *add a negative number* ...

   ... treat the + − as −

For example: 3 + (−5) = 3 − 5 = −2

Using your calculator

Calculations involving negative numbers can be done by using the (−) key.

Example 2

Work out −6 − −2.

Press (−) 6 − (−) 2 =

The answer should be −4.

EXERCISE 4E

1 Write down the answer to each calculation then check your answers on a calculator.

a −3 − 5 =   b −2 − 8 =   c −5 − 6 =   d 6 − 9 =
e 5 − 3 =   f 3 − 8 =   g −4 + 5 =   h −3 + 7 =
i −2 + 9 =   j −6 + −2 =   k −1 + −4 =   l −8 + −3 =
m 5 − −6 =   n 3 − −3 =   o 6 − −2 =   p 3 − −5 =
q −5 − −3 =   r −2 − −1 =   s −4 − 5 =   t 2 − 7 =

2 What is the *difference* between the temperatures in each pair?

a 4 °C and −6 °C   b −2 °C and −9 °C   c −3 °C and 6 °C

3 Find what you have to *add to* 5 to get:

a 7   b 2   c 0
d −2   e −5   f −15

4 Find what you have to *subtract from* 4 to get:

a 2   b 0   c 5
d 9   e 15   f −4

5 Find what you have to *add to* −5 to get:

a 8   b −3
c 0   d −1
e 6   f −7

6 Find what you have to *subtract from* −3 to get:

a 7   b 2
c −1   d −7
e −10   f 1

7 You have these cards.

a Which card should you choose to make the answer to this sum as large as possible? What is the answer?

$$\boxed{+6} + \boxed{} = \ldots\ldots$$

b Which card should you choose to make the answer to part **a** as small as possible? What is the answer?

c Which card should you choose to make the answer to this subtraction as large as possible? What is the answer?

$$\boxed{+6} - \boxed{} = \ldots\ldots$$

d Which card should you choose to make the answer to part **c** as small as possible? What is the answer?

8 The thermometer in a car is inaccurate by up to two degrees.

An ice alert warning comes on at 3 °C, according to the thermometer temperature.

If the actual temperature is 2 °C, will the alert come on?

Explain how you decide.

9 Two integers have a sum of 5.

One of the numbers is negative.

The other number is even.

What are the two numbers when the negative number is as large as possible?

4.5 Multiplying and dividing directed numbers

The rules for multiplying and dividing two directed numbers are very easy.
- When the signs of the two numbers are the *same*, the answer is *positive*.
- When the signs of the two numbers are *different*, the answer is *negative*.

Here are some examples.

$$2 \times 4 = 8 \qquad 12 \div -3 = -4 \qquad -2 \times -3 = 6 \qquad -12 \div -3 = 4$$

A common error is to confuse, for example, -3^2 and $(-3)^2$.

$$-3^2 = -3 \times 3 = -9$$

but:

$$(-3)^2 = -3 \times -3 = +9.$$

So, this means that if a variable is introduced, for example, $a = -5$, the calculation would be:

$$a^2 = -5 \times -5 = +25$$

Chapter 4: Directed numbers

Example 3

$a = -2$ and $b = -6$

Work out these numbers.

a a^2 b $a^2 + b^2$ c $b^2 - a^2$ d $(a - b)^2$

a a^2 $= -2 \times -2 = +4$
b $a^2 + b^2$ $= +4 + -6 \times -6 = 4 + 36 = 40$
c $b^2 - a^2$ $= 36 - 4 = 32$
d $(a - b)^2$ $= (-2 - -6)^2 = (-2 + 6)^2 = (4)^2 = 16$

EXERCISE 4F

1 Write down the answers.

a -3×5 b -2×7 c -4×6 d -2×-3 e -7×-2
f $-12 \div -6$ g $-16 \div 8$ h $24 \div -3$ i $16 \div -4$ j $-6 \div -2$
k 4×-6 l 5×-2 m 6×-3 n -2×-8 o -9×-4
p $24 \div -6$ q $12 \div -1$ r $-36 \div 9$ s $-14 \div -2$ t $100 \div 4$
u -2×-9

2 Write down the answers.

a $-3 + -6$ b -2×-8 c $2 + -5$ d 8×-4 e $-36 \div -2$
f -3×-6 g $-3 - -9$ h $48 \div -12$ i -5×-4 j $7 - -9$
k $-40 \div -5$ l $-40 + -8$ m $4 - -9$ n $5 - 18$ o $72 \div -9$
p $-7 - -7$ q $8 - -8$ r 6×-7

3 What number do you multiply by -3 to get each number?

a 6   b -90 c -45 d 81   e 21

4 What number do you divide -36 by to get each number?

a -9 b 4   c 12   d -6 e 9

5 Evaluate these.

a $-6 + (4 - 7)$ b $-3 - (-9 - -3)$ c $8 + (2 - 9)$

6 Evaluate these.

a $4 \times (-8 \div -2)$ b $-8 - (3 \times -2)$ c $-1 \times (8 - -4)$

7 What do you get if you divide -48 by each number?

a -2 b -8 c 12   d 24

8 Write down six different multiplications that give the answer -12.

9 Write down six different divisions that give the answer -4.

10 Put these calculations in order from lowest to highest.

$-5 \times 4 \qquad -20 \div 2 \qquad -16 \div -4 \qquad 3 \times -6$

11 $x = -2$, $y = -3$ and $z = -4$. Work out these numbers.

  a x^2 　　　　　　**b** $y^2 + z^2$ 　　　　　　**c** $z^2 - x^2$ 　　　　　　**d** $(x - y)^2$

12 Copy and complete this multiplication table.

×		3	
		−15	
2			−8
−6	12		

Check your progress

Core

- I can use directed numbers in practical situations
- I can put directed numbers in order by magnitude
- I can add and subtract positive and negative integers
- I can multiply and divide positive and negative integers

Chapter 5
Powers and roots

Topics	Level	Key words
1 Squares and square roots	CORE	square, square root
2 Cubes and cube roots	CORE	cube, cube root
3 More powers and roots	CORE	power
4 Exponential growth and decay	EXTENDED	exponential growth, exponential decay

In this chapter you will learn how to:

CORE	EXTENDED
Calculate: • squares • square roots • cubes and cube roots • other powers and roots of numbers. (C1.3 and E1.3)	• Use exponential growth and decay in relation to population and finance. (E1.17)

Why this chapter matters

The squares and square roots of numbers are important tools in mathematics and mathematicians have helped us to use them by inventing notation.

We often need to multiply a number by itself one or more times, for example, when finding the area of a square or the volume of a cube.

You have seen in Chapter 1 how you can write 5^2 instead of 5×5. You can also show $5 \times 5 \times 5$ by using 5^3. This short cut is called **index notation** (see Chapter 18).

The notation for the **square root** of a number (e.g. $\sqrt{25}$) is even more convenient. Without this, you would have to write 'the number which multiplies by itself to make 25'. The sign for the **cube root** of numbers is $\sqrt[3]{}$.

The root signs are especially convenient when the roots are hard to work out and difficult to express accurately. **Square numbers** such as 4, 9, 16 and 25 have whole numbers as their square roots but most numbers have fractions. If the fractions are expressed as decimals they are sometimes recurring (that is, they never end) which means that they can never be written down accurately.

The notation we now use was only introduced in the sixteenth century (CE). One of the first people to use it in print was a German mathematician called Christoph Rudolff. The notation was simple and easy to understand and was soon widely used.

Ways of working out square roots have been developed by mathematicians over the centuries. This Babylonian tablet showing how to calculate the square root of two is 2500 years ago.

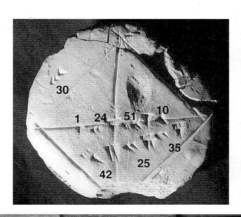

Notation can be used in calculations to represent the square roots of numbers without actually having to work them out or write them down. This makes it possible to carry out more sophisticated and accurate calculations.

Chapter 5: Powers and roots

Chapter 5 . Topic 1

5.1 Squares and square roots

You know that $6 \times 6 = 36$.

This can also be written as 6^2, which you say as 'six squared'.

Another way to describe this is to say: '6 is the **square root** of 36'.

You can write the square root of 36 as $\sqrt{36}$.

In fact 36 has *two* square roots.

$$6 \times 6 = 36$$
$$\text{and } -6 \times -6 = 36$$

So the square roots of 36 are 6 and −6 or $\sqrt{36}$ and $-\sqrt{36}$.

You can **square** decimals as well as whole numbers.

$2.5^2 = 6.25$ so $\sqrt{6.25} = 2.5$.

Most calculators have a 'square' button $\boxed{x^2}$ and a 'square root' button $\boxed{\sqrt{}}$. Check that you know how to use them.

The calculator will show the positive square root, e.g. $\sqrt{36}$.

EXERCISE 5A

1 Find the value of each expression.

a 7^2
b 10^2
c 1.2^2
d 2.5^2
e 16^2
f 20^2
g 3.1^2
h 4.5^2
i $(-3)^2$
j $(-8)^2$
k 0.5^2
l $(-0.5)^2$

2 Write down the two square roots of each number.

a 9
b 100
c 121
d 1.44
e 400
f 12.25
g 1
h 10 000

3 Find these square roots.

a $\sqrt{25}$
b $\sqrt{36}$
c $\sqrt{100}$
d $\sqrt{49}$
e $\sqrt{64}$
f $\sqrt{2.25}$
g $\sqrt{30.25}$
h $\sqrt{1.44}$
i $\sqrt{400}$
j $\sqrt{0.25}$

4 Write down the value of each of these. You will need to use your calculator for some of them. Look for the $\boxed{\sqrt{}}$ key.

a 9^2
b $\sqrt{1600}$
c 10^2
d $\sqrt{196}$
e 6^2
f $\sqrt{225}$
g 7^2
h $\sqrt{144}$
i 5^2
j $\sqrt{441}$

5 Write down each square root. You will need to use your calculator.

a $\sqrt{576}$
b $\sqrt{961}$
c $\sqrt{2025}$
d $\sqrt{1600}$
e $\sqrt{4489}$
f $\sqrt{10\,201}$
g $\sqrt{12.96}$
h $\sqrt{42.25}$
i $\sqrt{193.21}$
j $\sqrt{492.84}$

Chapter 5 . Topic 2

6 Put these in order, starting with the smallest value.

   3^2 $\sqrt{90}$ $\sqrt{50}$ 4^2

7 a Explain how you know that $\sqrt{40}$ is between 6 and 7.

   b Use your calculator to find $\sqrt{40}$. Write down all the decimal places on your display.

8 Between which two consecutive whole numbers does the square root of 20 lie?

9 Find two consecutive integers between which each of these square roots lies.

   a $\sqrt{68}$ b $\sqrt{96}$ c $\sqrt{155}$ d $\sqrt{250}$

10 Use these number cards to make the calculation below correct.

$\sqrt{\square\square\square} = \square\square$

11 A square wall in a kitchen is being tiled.

Altogether it needs 225 square tiles.
How many tiles are there in each row?

5.2 Cubes and cube roots

You can show that $6 \times 6 \times 6 = 216$.

216 is 'six cubed'. A **cube** of a number is formed when you multiply the number by itself and then by itself again. You can write this as 6^3.

   $6 \times 6 \times 6 = 6^3$ or 'six cubed'

Another way to describe this is to say: '6 is the **cube root** of 216'.

The symbol for a cube root is $\sqrt[3]{}$ so the cube root of 216 can also be written as $\sqrt[3]{216}$.

Many calculators have a button for cubes x^3 and cube roots $\sqrt[3]{}$. Check that you know how to use them.

Example 1

Find the cube roots of 64 and −64.

$4 \times 4 \times 4 = 64$ so $\sqrt[3]{64} = 4$

$-4 \times -4 \times -4 = -64$ so $\sqrt[3]{-64} = -4$

Notice that 64 and −64 have just one cube root each.

Chapter 5: Powers and roots

EXERCISE 5B

1 Work out these cubes.

a 2^3
b 3^3
c 8^3
d 10^3
e 1.1^3
f 2.5^3
g $(-3)^3$
h $(-5)^3$
i 20^3
j 4.1^3
k $(-4.1)^3$

2 Find these cube roots.

a $\sqrt[3]{8}$
b $\sqrt[3]{125}$
c $\sqrt[3]{729}$
d $\sqrt[3]{1}$
e $\sqrt[3]{27}$
f $\sqrt[3]{-27}$
g $\sqrt[3]{1000}$
h $\sqrt[3]{3.375}$
i $\sqrt[3]{91.125}$
j $\sqrt[3]{0.125}$

3 $4^3 = 64$ and $5^3 = 125$ so $\sqrt[3]{100}$ is between 4 and 5.

Find two consecutive integers between which these cube roots lie.

a $\sqrt[3]{200}$
b $\sqrt[3]{300}$
c $\sqrt[3]{500}$
d $\sqrt[3]{-500}$

4 Here are four numbers.

$2^3 \qquad 3^2 \qquad \sqrt{81} \qquad \sqrt[3]{729}$

Which is the odd one out and why?

5 Find two different positive integers, A and B, such that $A^2 = B^3$.

6 Put these numbers in order, smallest first.

$2.5^3 \qquad \sqrt{225} \qquad 4^2 \qquad \sqrt[3]{2000}$

7 Copy and complete this table.

Number	Square	Cube
	100	
5		
		64
11		
	81	

8 Write these numbers in order with the smallest first.

$\sqrt{0.8} \qquad \sqrt[3]{0.8} \qquad 0.8^2 \qquad 0.8^3$

5.2 Cubes and cube roots

5.3 More powers and roots

$2^3 = 2 \times 2 \times 2 = 8$ is a power of 2.

You can work out other powers of 2.

$2^4 = 2 \times 2 \times 2 \times 2 = 16$ This is 2 to the power 4

$2^6 = 2 \times 2 \times 2 \times 2 \times 2 \times 2 = 64$ This is 2 to the power 6

You can have powers of any number.

3 to the power $4 = 3^4 = 3 \times 3 \times 3 \times 3 = 81$

5 to the power $7 = 5^7 = 3 \times 3 \times 3 \times 3 = 78\,125$

There are also corresponding roots.

$2^4 = 16$ and so $\sqrt[4]{16} = 2$ The fourth root of 16 is 2

$3^4 = 81$ and so $\sqrt[4]{81} = 3$ The fourth root of 81 is 3

$5^7 = 78\,125$ and so $\sqrt[7]{78\,125} = 5$ The seventh root of 78 125 is 5

Your calculator probably has buttons to find powers and roots like these. Check that you know how to use them.

Example 2

Evaluate   **a** $2^6 \times \sqrt[4]{1296}$ **b** $4^5 \div \sqrt[4]{81}$

a $2^6 = 64$ and $\sqrt[4]{1296} = 6$ because $6^4 = 1296$

The answer is $64 \times 6 = 384$

b $4^5 = 1024$ and $\sqrt[4]{81} = 3$ because $3^4 = 81$

The answer is $1024 \div 3 = 341\frac{1}{3}$

EXERCISE 5C

1   Work out

     a 3^5 **b** 7^4 **c** 10^6 **d** 4^4

2   Work out

     a $2^4 - 4^2$ **b** $3^5 - 5^3$ **c** $3^6 - 6^3$

3   Evaluate

     a $3^4 \times 2^5$ **b** $3^5 - 2^4$ **c** $\dfrac{2^4 + 3^4}{4^4}$

4   Show that $1 + 2 + 2^2 + 2^3 + 2^4 = 2^5 - 1$

Chapter 5: Powers and roots

Chapter 5 . Topic 4

5 Find

a $\sqrt[4]{625}$
b $\sqrt[4]{14\,641}$
c $\sqrt[7]{2187}$
d $\sqrt[5]{3\,200\,000}$

6 Find the value of

a $\sqrt[5]{32} \times 10^4$
b $\sqrt[4]{6561} \div \sqrt[4]{4096}$

7 Write these expressions in order of size, smallest first.

$3^2 \times \sqrt[10]{1024}$ $\sqrt[3]{64} \times \sqrt[3]{125}$ $2^3 \times \sqrt[3]{8}$ $\sqrt[4]{14\,641} \times \sqrt[6]{64}$

8 Find the value of

a $4^3 \div \sqrt[4]{16}$
b $5^3 \div \sqrt[3]{1000}$

9 Evaluate

a $5^2 \div \sqrt[3]{27}$
b $\sqrt[4]{2401} \div 2^3$

10 Find the value of n if

a $4^n = 64$
b $3^n = 6561$
c $5^n = 3125$

11 Find the value of n if

a $\sqrt[n]{27} = 3$
b $\sqrt[n]{81} = 3$
c $\sqrt[n]{6561} = 3$

12 $n^3 = 125$ Find the value of

a n^6
b n^9

13 $2^x = 4^y$ where x and y are positive integers.

a Find a possible pair of values for x and y

b Can you find a different possible pair of values for x and y?

14 The number 2^{10} is called 1K (or just K in computing).

Write as a number:

a 1K
b K^2
c $\sqrt{K}$
d $\sqrt[5]{K}$
e $\sqrt[10]{K}$

5.4 Exponential growth and decay E

The number of visitors to a wildlife reserve in January is 600.

The number of visitors *doubles* every month.

In February the number of visitors is 600 × 2 = 1200

In March the number of visitors is 1200 × 2 = 2400

In April the number of visitors is 2400 × 2 = 4800

The number is multiplied by 2 from one month to the next.

The numbers for the four months can be written as a sequence.

600, 600 × 2, 600 × 2², 600 × 2³

This is an example of **exponential growth**.

> **Example 3**
>
> The cost of food increases by 10% each year.
>
> The cost of a basket of food now is $85.
>
> **a** Work out the cost after one year.
> **b** Work out the cost after two years.
> **c** Work out the cost after three years.
>
> **a** The multiplier to increase a number by 10% is 1.1   100% + 10% = 110% = 1.1
>
>   The cost after one year is $85 × 1.1 = $93.50.
>
> **b** The cost after two years is $85 × 1.1² or $93.50 × 1.1 = $102.85.
>
> **c** The cost after three years is $85 × 1.1³ or $102.85 × 1.1 = $113.14.

Raj buys a car for $56 000.

The value of his car *decreases* by 25% each year.

The multiplier for a decrease of 25% is 0.75.
100% − 25% = 75% = 0.75

The value of the car after one year is $56 000 × 0.75 = $42 000

The value of the car after two years is $42 000 × 0.75 = $31 500

The value of the car after three years is $31 500 × 0.75 = $23 625

This is an example of **exponential decay**. This is the result when the multiplier from one number to the next is less than one.

EXERCISE 5D

1 The number of bacteria on a plate is 5 million.

   a The number doubles every hour. How many bacteria are there after:

   i one hour   **ii** two hours   **iii** three hours?

   b Repeat part **a**, assuming that the number of bacteria now *triples* every hour.

2 There are 4000 people in a new town.

   The population increases by 50% each year.

   Work out what the population will be in:

   a one year   **b** two years   **c** three years   **d** four years.

Chapter 5: Powers and roots

3 There are 12 000 fish in a lake.

Because of pollution, the number of fish halves every year.

Work out the number of fish after:

a one year

b three years

c five years.

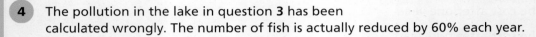

4 The pollution in the lake in question **3** has been calculated wrongly. The number of fish is actually reduced by 60% each year.

Work out the number of fish after:

a one year   b three years   c five years.

5 The population of a country is 20.0 million.

The population increases by 9% every 10 years.

Copy and complete this table. Round the numbers to one decimal place.

Number of years	0	10	20	30
Population (millions)	20.0			

6 Now suppose the population in question **5** reduces by 9% every year. Copy and complete the table in Question 5.

7 Anna has savings of $5000 in her account.

The value of her savings increases by 6% each year.

a How much will the savings be worth after four years?

b How long will it take until the savings are worth $7000?

8 Sami earns $80 000 per year.

He has been promised a pay rise of 25% each year.

a How much will he be earning after one year?

b How much will he be earning after four years?

9 The value of a flat is $150 000.

The value increases by 20% each year.

Remi says, 'The value will double in five years to $300 000 because 5 × 20% is 100%.'

Show that Remi is not correct. Then work out the correct value.

5.4 Exponential growth and decay

10 Carla saves $10 in January.

She says, 'Each month I shall save twice as much as I saved the month before.'

a How much will she save in February?

b How much will she save in December?

11 There are 4000 elephants in a reserve.

The numbers are reducing by a third each year.

a Work out the number of elephants after three years.

b Work out the number of elephants after six years.

12 In 2013 the population of Indonesia was 250 million.

The population was increasing by 1.2% per year.

Use these figures to estimate the population in 2020.

13 In Brazil in the early 1990s prices were increasing by 80% each month.

Suppose the price of a loaf of bread is $10.

a Work out the price after one month.

b Work out the price after two months.

c Work out the price after six months.

d Work out the price after a year.

Check your progress

Core

- I can calculate with squares and square roots of numbers
- I can calculate with cubes and cube roots
- I can calculate with other powers and roots

Extended

- I can use exponential growth and decay in relation to population and finance

Chapter 6
Ordering and set notation

Topics	Level	Key words
1 Inequalities	CORE	equals, greater than, less than
2 Sets and Venn diagrams	CORE	set, element, universal set, complement, union, intersection, Venn diagram
3 More about Venn diagrams	EXTENDED	subset, empty set, proper subset

In this chapter you will learn how to:

CORE	EXTENDED
• Order quantities by magnitude and demonstrate familiarity with the symbols: $=, \neq, >, <, \geq, \leq$. (C1.6 and E1.6) • Notate elements in a set, e.g.: Complement of set A A' Union of A and B $A \cup B$ Intersection of A and B $A \cap B$. (C1.2)	• Define sets, e.g. $A = \{x : x \text{ is a natural number}\}$ $B = \{(x, y): y = mx + c\}$ $C = \{x : a \leq x \leq b\}$ $D = \{a, b, c, \ldots\}$. (E1.2) • Notate elements in a set, e.g.: Number of elements in set A $n(A)$ "…is an element of…" $\in$ "…is not an element of…" $\notin$ The empty set $\emptyset$ A is a subset of B $A \subseteq B$ A is a proper subset of B $A \subset B$ A is not a subset of B $A \nsubseteq B$ A is not a proper subset of B $A \not\subset B$ (E1.2)

Why this chapter matters

Alice in Wonderland

Sets are collections of objects. Set notation gives us a way of seeing the logical connection between sets. The mathematics of sets is very useful in designing computer circuits and electronic components.

You have probably heard of Alice in Wonderland. Did you know that the author, Lewis Carroll, was actually a lecturer in mathematics at the University of Oxford, England, in the nineteenth century? His real name was Charles Dodgson.

He also wrote a mathematics book called *Symbolic Logic*. Here is a problem from it.

1. All humming birds are richly coloured.
2. No large birds can live on honey.
3. Birds that do not live on honey are dull in colour.

What conclusion follows?

Symbolic Logic was about how to write sentences in symbols so that conclusions could be seen more easily. You can use set notation and Venn diagrams to do this and you will learn about them in this chapter.

Venn diagrams were invented by the logician John Venn and may be the only mathematical invention to be celebrated in a stained glass window!

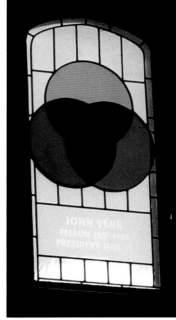

A Venn diagram in glass

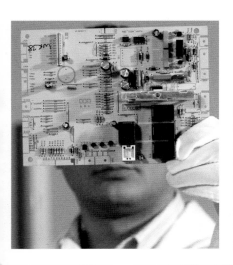

Symbolic logic has a very modern application in the design of computer circuits and the construction of electronic components. When engineers talk about NAND and NOR gates they are using ideas that were first developed in the nineteenth century for very different purposes.

Chapter 6: Ordering and set notation

6.1 Inequalities

Chapter 6 . Topic 1

You need to know the meaning of these symbols:

=	**equals**
≠	**does not equal**
>	is **greater than**
<	is **less than**
⩾	is **equal to or greater than**
⩽	is **equal to or less than**.

Advice and Tips

The arrow tips in < or > signs point towards the smaller numbers

Example 1

n is a positive integer and $n < 6$. What are the possible values of n?

Because n is positive the possible values are 1, 2, 3, 4 and 5. 6 is not included because the sign means 'less than'. If the question said $n \leqslant 6$ then 6 would be included.

Sometimes two inequalities are used together.

Example 2

Find the possible values of x if x is an integer and $-3 < x \leqslant 3$.

This means that x is a whole number between -3 and 3.

The possible values of x are $-2, -1, 0, 1, 2, 3$. Note that -3 is not in the list but 3 is included because $\leqslant 3$ means x can be 'less than or equal to' 3.

EXERCISE 6A

1 Here are three symbols: < = >.

Put the correct symbol between the numbers in each pair.

- **a** 3.5 ... 3.15
- **b** 180 cm ... 2m
- **c** 5 × 7 ... 6 × 6
- **d** 5km ... 5000m
- **e** $\frac{1}{3}$ of 27 ... $\frac{3}{4}$ of 12
- **f** 4^2 ... 8
- **g** 10 × 10 ... 8 × 12
- **h** $\sqrt{64}$ ... 3^2

2 Here are three fractions: $\frac{1}{3}, \frac{3}{5}, \frac{1}{2}$.

Use them to fill in the gaps below to list them in order, smallest first.

... < ... < ...

3 d is the number scored when a normal six-sided dice is thrown.

List the possible values of d in each case.

a $d \geq 4$
b $d < 3$
c $d > 5$
d $d \leq 5$
e $2 \leq d \leq 4$
f $3 < d < 6$
g $1 \leq d < 4$
h $5 < d \leq 6$

4 The table shows whether babies of a particular age are underweight or overweight.

Are babies of these masses underweight, overweight or normal?

a 6.3 kg
b 9.3 kg
c 7.8 kg
d 8.5 kg

Mass < 6.5 kg	Underweight
6.5 kg ≤ mass ≤ 8.5 kg	Normal
Mass > 8.5 kg	Overweight

5 e is an even number and $20 \leq e < 30$ and $e \neq 24$. List the possible values of e.

6 In a bag of balls each ball has a number n where $1 \leq n \leq 49$.

a What is the largest number on a ball?

b The number on the first ball is a multiple of 5 and $n > 40$.
What are the possible values of n?

c The number on the second ball is a multiple of 3 and $n \leq 10$.
What are the possible values of n?

d For the third ball $15 < n \leq 20$.
What are the possible values of n?

7 True or false? State which in each case.

a $3 \neq -3$

b $-3 < -5$

c $1.99 < 2$

d $2 < \sqrt{5} < 3$

e $20^2 \leq 300$

f 200 minutes $\geq$ 3 hours

Chapter 6: Ordering and set notation

Chapter 6 . Topic 2

8 List all the possible values for an integer x in the following cases.

    a $5 < x < 9$ **b** $26 \leq x \leq 28$

    c $-8 < x \leq -4$ **d** $-2 \leq x < 2$

    e $17 < x < 18$ **f** $32.5 \leq x < 33.5$

9 N is an integer. What can you say about N in the following cases?

    a $N^2 \geq 64$ **b** $N^3 \geq 64$

    c $N^3 < -64$ **d** $N^2 < -64$

6.2 Sets and Venn diagrams

The factors of 20 are 1, 2, 4, 5, 10 and 20.

The factors of 15 are 1, 3, 5 and 15.

You can show them in a **Venn diagram**.

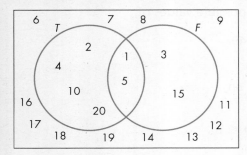

T and F are **sets**.

The items in a set are called **elements**.

You can list the elements of a set inside curly brackets, like this:

    $T = \{1, 2, 4, 5, 10, 20\}$ and $F = \{1, 3, 5, 15\}$

Or you can describe them:

    $T = \{x : x$ is a factor of $20\}$ or just $T = \{$factors of $20\}$

    $F = \{x : x$ is a factor of $15\}$ or just $F = \{$factors of $15\}$

Numbers 1 and 5 are in both sets. This is called the **intersection** of T and F.

To show this, write $T \cap F = \{1, 5\}$, which can be read as 'T intersection F'.

There are eight numbers in T or F or both. This is called the **union** of T and F.

To show this, write $T \cup F = \{1, 2, 3, 4, 5, 10, 15, 20\}$, which can be read as '$T$ union F'

In this case you are interested in the integers between 1 and 20. Nothing else can be a factor of 15 and 20.

6.2

Put all the possible elements in a rectangle called the **universal set**, ξ

$\xi = \{1, 2, 3, \ldots, 19, 20\}$

$n(A)$ is the number of elements in A.

So $n(T) = 6$, $n(F) = 4$, $n(F \cup T) = 8$ and $n(\xi) = 20$

Example 3

$\xi = \{1, 2, 3, \ldots, 16\}$ $A = \{x : x \text{ is a multiple of } 2\}$ $B = \{x : x \text{ is a multiple of } 3\}$

a Show the sets in a Venn diagram.

b List the elements of $A \cap B$.

c Find $n(A \cup B)$.

a The diagram needs two overlapping circles to show the sets.

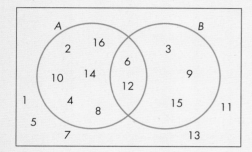

b These are the elements where the two sets overlap.

$A \cap B = \{6, 12\}$.

c These are elements of A or B or both.

$n(A \cup B) = 11$.

EXERCISE 6B

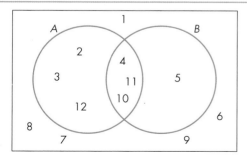

a List the elements of: **i** A **ii** B **iii** $A \cap B$

b Find: **i** $n(A)$ **ii** $n(A \cup B)$ **iii** $n(\xi)$.

c Describe the elements of ξ.

2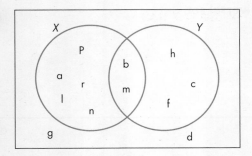

a Find i $n(X)$ ii $n(Y)$ iii $n(X \cap Y)$ iv $n(X \cup Y)$

b Write down an element in X but not in Y.

c Write down an element that is not in X and not in Y.

3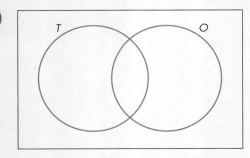

$\xi = \{1, 2, 3, \ldots, 12\}$ $T = \{\text{multiples of 3}\}$ $O = \{\text{odd numbers}\}$

Place the elements in the Venn diagram.

4 $\xi = \{\text{positive integers}\}$ $A = \{\text{multiples of 2}\}$ $B = \{\text{multiples of 3}\}$

a Is 100 in A or B?

b Find the smallest element of $A \cap B$.

c Describe the elements of $A \cap B$.

5 $\xi = \{\text{letters in the word 'Singapore'}\}$

$X = \{\text{letters in the word 'Paris'}\}$ $Y = \{\text{letters in the word 'Spain'}\}$

Show the sets in a Venn diagram.

6 This Venn diagram shows three sets: A, B and C.

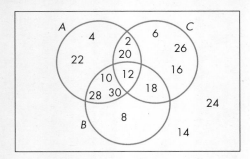

a Find the elements of   i A   ii B   iii $A \cap B$ iv $A \cap C$ v $A \cap B \cap C$

6.2 Sets and Venn diagrams

b Find   **i** $n(C)$ **ii** $n(A \cup B)$ **iii** $n(A \cup C)$ **iv** $n(A \cup B \cup C)$

c Describe the elements of ξ.

7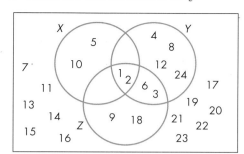

ξ = {natural number $\leq$ 24} Y = {factors of 24}

a List the elements of   **i** X **ii** Y
b Describe the elements of   **i** X **ii** Y
c Find the elements of   **i** $Y \cap Z$ **ii** $X \cap Z$ **iii** $X \cap Y \cap Z$
d Find   **i** $n(Y)$ **ii** $n(Y \cup Z)$ **iii** $n(X \cup Y \cup Z)$

8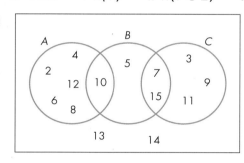

a Find the elements of   **i** $A \cap B$ **ii** $B \cap C$ **iii** $A \cap C$
b Find   **i** $n(A)$ **ii** $n(A \cup B)$ **iii** $n(B \cup C)$ **iv** $n(\xi)$ **v** $n(A \cup B \cup C)$

9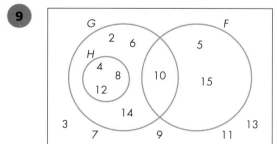

a Find the elements of   **i** H **ii** G **iii** $F \cap G$ **iv** $H \cap G$
b Find   **i** $n(F \cap G)$ **ii** $n(F \cap H)$

10 ξ = {positive integers} T = {multiples of 3}

F = {multiples of 5} S = {multiples of 7}

Find the smallest element of

a $T \cap F$ **b** $T \cap S$ **c** $F \cap S$ **d** $T \cap F \cap S$

11 ξ = {positive integers ⩽ 30}   A = {factors of 24}

B = {factors of 28}   P = {prime numbers}

Find   **a** n(A)   **b** n(P)   **c** A ∩ B   **d** P ∩ B

12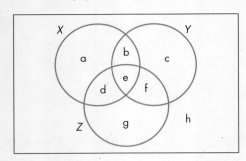

a Find the elements of

i X ∪ Y   ii (X ∪ Y) Z   iii Y ∪ Z   iv X ∪ (Y ∩ Z)

b Find n((X ∪ Z) ∩ Y)

13 R = {1, 2, 3, 4, 5, 6, 7}   S = {2, 4, 6, 8, 10}   T = {1, 4, 7, 10, 13}

Find the elements of

a R ∪ T   **b** S ∪ T   **c** (R ∪ S) ∩ T   **d** (R ∩ S) ∪ T

6.3 More about Venn diagrams

ξ = {integer x : 1 ⩽ x ⩽ 12}   A = {2, 3, 4, 5, 6}   B = {x : x is an odd number}

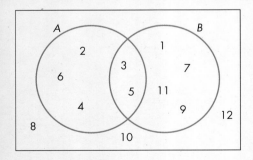

You need to know the meaning of these symbols:

∈ **is** a member of

∉ **is not** a member of.

2 ∈ A but 2 ∉ of B.   7 ∈ A ∪ B but 7 ∉ A ∩ B.

The set of elements not in A is called the **complement** of A. The symbol for this is A′

A′ = {1, 7, 8, 9, 10, 11, 12}   (A ∪ B)′ = {8, 10, 12}

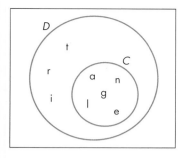

ξ {x : x is a letter of the alphabet}   C = {a, n, g, l, e}   D = {t, r, i, a, n, g, l, e}

In this diagram, all the elements of C are also in D.

You say that C is a **subset** of D. To show this, you can use the symbol ⊆, which means 'is a subset of'. Write it like this:

  C ⊆ D

{l, e, g} ⊆ C and {t, r, a, i, n} ⊆ D

The symbol ⊈ means 'is not a subset of'.

{c, a, t} ⊈ C

E = {7, 8, 9}

What are the subsets of E? You can list them.

{7}, {8} and {9} are subsets with one element.

{7, 8}, {7, 9} and {8, 9} are subsets with two elements.

E is also a subset of itself. E ⊆ E.

Sometimes we call the other ones **proper subsets** and use the symbol ⊂.

So {7}, E and {7, 9} ⊂ E but E ⊄ E

Finally there is the **empty set** or ∅, the set with no elements.

{x : x is an integer and 9 < x < 10} = ∅ because there are no integers between 9 and 10.

The empty set is a subset of any other set. In particular ∅ ⊂ E.

> **Advice and Tips**
>
> The difference between ⊂ and ⊆ is like the difference between < and ≤ for numbers.

Example 4

ξ = {integer x : 1 ≤ x ≤ 20}   A = {x : x is a factor of 20}

B = {x : x is a multiple of 4}   C = {x : x is a square number}

a Draw a Venn diagram to show the sets.

b Find i A' ∩ B   ii n(C')   iii B' ∩ B

c x ∈ A and x ∉ B and x ∈ C

What is x?

a The Venn diagram needs 3 circles to show three sets.

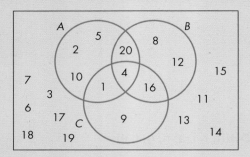

b i The elements in B and not in A are {8, 12, 16}.

  ii The number of elements not in C is 16.

  iii No element is in both B and the complement of B, and so $B' \cap B = \emptyset$.

c The element in A and C but not in B is 1.

Example 5

A, B and C are three sets.

$A \subset B$ and $B \cap C = \emptyset$

Show the relationship between the sets on a Venn diagram.

A is a proper subset of B so the circle for A must be inside the circle for B.

The intersection of B and C is the empty set so the circles for B and C must be completely separate.

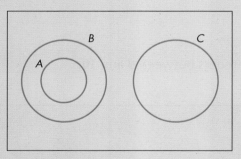

This is a possible diagram.

EXERCISE 6C

1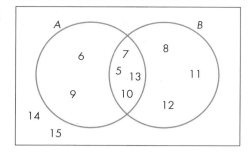

List the elements of

a A'
b B'
c $(A \cup B)'$
d $A' \cap B$
e $A \cap B'$

2 For the sets in question **1**, put the correct sign, $\in$ or $\notin$, in each of these statements.

a 6 ... B'
b 7 ... A'
c 8 ... $(A \cap B)'$
d 9 ... $A \cup B'$
e 9 ... $A' \cup B$
f 9 ... $(A \cup B)'$

3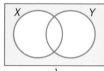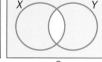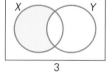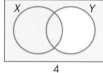

    1   2   3   4

a Match each expression with the correct diagram.

i $X \cap Y'$
ii $(X \cap Y)'$
iii $X \cup Y'$
iv $(X \cup Y)'$
v $X' \cap Y'$
vi $X' \cap Y'$

b Which pairs of expressions in part **a** are equivalent?

4 A, B and C are sets.

$A \subset C$ $B \subset C$ $A \cap B = \varnothing$

a Show the relationship between the sets in a Venn diagram.
b What is $(A \cup B) \cup C$?

5 A and B are sets with a finite number of elements.

$n(A) + n(B) = n(A \cup B)$

Explain why $A \cap B = \varnothing$

6 $\xi = \{\text{positive integers}\}$ $T = \{\text{multiples of 2}\}$

$P = \{\text{prime numbers}\}$ $F = \{\text{multiples of 4}\}$

a $x \in T \cap P$ What is x?
b Draw a Venn diagram to illustrate the sets.
c Describe the elements of T'.

7

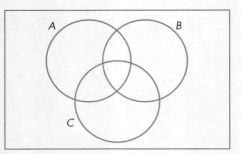

Make a copy of the diagram. Use the information below to put elements in the diagram.

a $2 \notin A$ $2 \in B$ $2 \in C$
b $3 \in A \cap C$ $3 \in B$
c $5 \notin A$ $5 \notin B$ $5 \notin C$
d $7 \notin A'$ $7 \notin (B \cup C)$
e $11 \in (A \cup B)'$ $11 \in C$

8

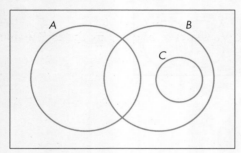

Decide which statements are true, and which are false. Correct the ones that are false.

a $C \subseteq B$ b $A \cap C = \emptyset$ c $C \cup B = C$
d $B \cap C = B$ e $A' \cap C = \emptyset$ f $C' \cup A' = \emptyset$

9 $A = \{(x, y) : x \geq 2\}$ $B = \{(x, y) : y \leq 3\}$

a Show $A \cap B$ on a coordinate grid.
b Show $A' \cap B'$ on the same grid.

10 $C = \{(x, y) : x \geq 3\}$ $D = \{(x, y) : y \geq -3\}$ $E = \{(x, y) : x + y \leq 6\}$

a Show $C \cap D \cap E$ on a coordinate grid.
b Write down the coordinate of a point on the y-axis in the region $C' \cap D \cap E$.

11 $\xi = \{\text{polygons}\}$ $S = \{\text{squares}\}$ $Q = \{\text{quadrilaterals}\}$ $T = \{\text{triangles}\}$

a Show the sets on a Venn diagram.
b $R = \{\text{regular polygons}\}$. What is $R \cap T$?

12 ξ = {real numbers} N = {natural numbers}
   S = {square numbers} P = {prime numbers}
   a Show the sets on a Venn diagram.
   b What is $S \cap P$?
   c Find the smallest element in $N \cap (P \cup S)'$.

13 $W \subset X$ $X \subset Y$ $Y \cap Z = \emptyset$

   Simplify the following expressions. a $W \cap Y$ b $W \cup Y$ c $X' \cap Z$ d $Z' \cup Y$

Check your progress

Core
- I can order quantities and use the symbols $\neq, >, <, \geq$ and $\leq$
- I can use Venn diagrams to show sets
- I can identify sets by defining or listing the elements
- I can correctly use the notation $n(A)$ for the number of elements in a set, ξ for the universal set, $A \cap B$ for the intersection and $A \cup B$ for the union

Extended
- I can use inequalities to define the elements of a set
- I can use Venn diagrams with more than two sets
- I can use the symbols $\in, \notin$, A' for a complement, $\emptyset$ for the empty set, $\subset$ and $\subseteq$

Chapter 7
Ratio, proportion and rate

Topics	Level	Key words
1 Ratio	CORE	ratio, cancel, simplest form, map scale
2 Increases and decreases using ratios	EXTENDED	increase, decrease
3 Speed	CORE	average, speed, distance, time
4 Rates	CORE	rate
5 Direct proportion	CORE	unitary method, direct proportion, single unit value
6 Inverse proportion	CORE	inverse proportion

In this chapter you will learn how to:

CORE	EXTENDED

• Demonstrate an understanding of:
  – ratio
  – direct and inverse proportion. (C1.11 and E1.11)
• Use common measures of rate. (C1.11 and E1.11)
• Use ratio and scales in practical situations. (C1.11 and E1.11)
• Calculate average speed. (C1.11 and E1.11)
	• Increase and decrease a quantity by a given ratio. (E1.11)

Why this chapter matters

We use ratio, proportion and speed in our everyday lives to help us compare two or more pieces of information.

Ratio and proportions are often used to compare sizes; speed is used to compare distances with the time taken to travel them.

A 100-m sprinter

Speed

When is a speed fast?

On 16 August 2009 Usain Bolt set a new world record for the 100-m sprint of 9.58 seconds. This is an average speed of 37.6 km/h.

The sailfish is the fastest fish and can swim at 110 km/h.

Sailfish

The cheetah is the fastest land animal and can travel at 121 km/h.

The fastest bird is the swift which can travel at 170 km/h.

Swift

Cheetah

Ratio and proportion facts

- Russia is the largest country.
- Vatican City is the smallest country.
- The area of Russia is nearly 39 million times the area of Vatican City.
- Monaco has the most people per square mile.
- Mongolia has the least people per square mile.
- The number of people per square mile in Monaco to the number of people in Mongolia is in the ratio 10 800 : 1.
- Japan has the highest life expectancy (WHO, 2016).
- Sierra Leone has the lowest life expectancy (WHO, 2016).
- On average people in Japan live over twice as long as people in Sierra Leone (WHO, 2016).

7.1 Ratio

A **ratio** is a way of comparing the sizes of two or more quantities.

A ratio can be expressed in a number of ways. For example, if Tasnim is five years old and Ziad is 20 years old, the ratio of their ages is:

$$\text{Tasnim's age : Ziad's age}$$

which is: 5 : 20

which simplifies to: 1 : 4   (dividing both sides by 5)

A ratio is usually given in one of these three ways.

Tasnim's age : Ziad's age	or	5 : 20	or	1 : 4
Tasnim's age to Ziad's age	or	5 to 20	or	1 to 4
$\dfrac{\text{Tasnim's age}}{\text{Ziad's age}}$	or	$\dfrac{5}{20}$	or	$\dfrac{1}{4}$

Common units

When working with a ratio involving different units, *always convert them to the same units*. A ratio can be simplified only when the units of each quantity are the *same*, because the ratio itself has no units. Once the units are the same, the ratio can be simplified or **cancelled** like a fraction.

For example, you must convert the ratio of 125 g to 2 kg to the ratio of 125 g to 2000 g, so that you can simplify it.

                                      125 : 2000

Divide both sides by 25:   5 : 80

Divide both sides by 5:   1 : 16   The ratio 125 : 2000 can be simplified to 1 : 16.

Example 1

Express 25 minutes : 1 hour as a ratio in its simplest form.

The units must be the same, so change 1 hour into 60 minutes.

25 minutes : 1 hour   =   25 minutes : 60 minutes

                        =   25 : 60   Cancel the units (minutes)

                        =   5 : 12   Divide both sides by 5

So, 25 minutes : 1 hour simplifies to 5 : 12

7.1

Ratios as fractions

You can express ratio in its **simplest form** as portions of a quantity by expressing the whole numbers in the ratio as fractions with the same denominator (bottom number).

> **Example 2**
>
> A garden is divided into lawn and shrubs in the ratio 3 : 2.
>
> What fraction of the garden is covered by: **a** lawn **b** shrubs?
>
> You find the denominator (bottom number) of the fraction by adding the numbers in the ratio (that is, 2 + 3 = 5).
>
> **a** The lawn covers $\frac{3}{5}$ of the garden.
>
> **b** The shrubs cover $\frac{2}{5}$ of the garden.

EXERCISE 7A

1 Express each of these ratios in its simplest form.

- **a** 6 : 18
- **b** 5 : 20
- **c** 16 : 24
- **d** 24 : 12
- **e** 20 : 50
- **f** 12 : 30
- **g** 25 : 40
- **h** 150 : 30

2 Write each of these ratios of quantities in its simplest form. (Remember that you must always express both parts in a common unit before you simplify.)

- **a** 40 minutes : 5 minutes
- **b** 3 kg : 250 g
- **c** 50 minutes to 1 hour
- **d** 1 hour to 1 day
- **e** 12 cm to 2.5 mm
- **f** 1.25 kg : 500 g
- **g** 75 cents : $2
- **h** 400 m: 2 km

3 A length of wood is cut into two pieces in the ratio 3 : 7. What fraction of the original length is the longer piece?

4 Tareq and Hassan find a bag of marbles that they share between them in the ratio of their ages. Tareq is 10 years old and Hassan is 15 years old. What fraction of the marbles did Tareq get?

5 Mona and Petra share a pizza in the ratio 2 : 3. They eat it all.

    a What fraction of the pizza did Mona eat?

    b What fraction of the pizza did Petra eat?

6 A camp site allocates space to caravans and tents in the ratio 7 : 3. What fraction of the total space is given to:

    a the caravans

    b the tents?

7 In a safari park at feeding time, the elephants, the lions and the chimpanzees are given food in the ratio 10 to 7 to 3. What fraction of the total food is given to:

    a the elephants

    b the lions

    c the chimpanzees?

8 Paula wins three-quarters of her tennis matches. She loses the rest.

What is the ratio of wins to losses?

Dividing amounts in a given ratio

To divide an amount in a given ratio, you first look at the ratio to see how many parts there are altogether.

For example, the ratio 4 : 3 has 4 parts and 3 parts giving 7 parts altogether.

    7 parts is the whole amount.

    1 part can then be found by dividing the whole amount by 7.

    3 parts and 4 parts can then be calculated from 1 part.

Example 3

Divide $28 in the ratio 4 : 3

4 + 3 = 7 parts altogether.

So 7 parts = $28.

Dividing by 7:

    1 part = $4

    4 parts = 4 × $4 = $16 and 3 parts = 3 × $4 = $12

So $28 divided in the ratio 4 : 3 is $16 : $12

Map scales

Map scales are often given as ratios in the form $1 : n$.

> **Example 4**
>
> A map of New Zealand has a scale of 1 : 900 000.
>
> The distance on the map from Auckland to Hamilton is 11.5 centimetres.
>
> What is the actual distance?
>
> 1 cm on the map = 900 000 centimetres on the ground.
>
> = 9000 metres (100 centimetres = 1 metre)
>
> = 9 kilometres (1000 metres = 1 kilometre).
>
> The distance is 11.5 × 9 kilometres
>
> = 103.5 kilometres.

EXERCISE 7B

1 Divide each according to the given ratio.

   a 400 g in the ratio 2 : 3
   b 280 kg in the ratio 2 : 5
   c 500 in the ratio 3 : 7
   d 1 km in the ratio 19 : 1
   e 5 hours in the ratio 7 : 5
   f $100 in the ratio 2 : 3 : 5
   g $240 in the ratio 3 : 5 : 12
   h 600 g in the ratio 1 : 5 : 6
   i $5 in the ratio 7 : 10 : 8
   j 200 kg in the ratio 15 : 9 : 1

2 The ratio of female to male members of a sports club is 7 : 3. The total number of members of the group is 250.

   a How many members are female?
   b What percentage of members are male?

3 A store sells small and large TV sets.

   The ratio of small : large is 2 : 3.

   The total stock is 70 sets.

   a How many small sets are in stock?
   b How many large sets are in stock?

4 When a supermarket checked a total of 357 confectionery products for sugar content, they found the ratio of products without sugar to those with sugar was 1 : 16.

   How many of those products contained no sugar?

5 Joshua, Aicha and Mariam invest $10 000 in a company.

The ratio of the amounts they invest is:

Joshua : Aicha : Mariam = 5 : 7 : 8

How much does each of them invest?

6 Rewrite each of these scales as a ratio in the form $1 : n$.

- a 1 cm to 4 km
- b 4 cm to 5 km
- c 2 cm to 5 km
- d 4 cm to 1 km
- e 5 cm to 1 km
- f 2.5 cm to 1 km
- g 8 cm to 5 km
- h 10 cm to 1 km
- i 5 cm to 3 km

7 A map has a scale of 1 cm to 10 km.

- a Rewrite the scale as a ratio in its simplest form.
- b What is the actual length of a lake that is 4.7 cm long on the map?
- c How long will a road be on the map if its actual length is 8 km?

Advice and Tips

1 km = 1000 m
= 100 000 cm

8 A map has a scale of 2 cm to 5 km.

- a Rewrite the scale as a ratio in its simplest form.
- b How long is a path that measures 0.8 cm on the map?
- c How long should a 12 km road be on the map?

9 The scale of a map is 5 cm to 1 km.

- a Rewrite the scale as a ratio in the form $1 : n$.
- b How long is a wall that is shown as 2.7 cm on the map?
- c The distance between two points is 8 km; how far will this be on the map?

10 You can simplify a ratio by changing it into the form $1 : n$. For example, 5 : 7 can be rewritten as:

$\frac{5}{5} : \frac{7}{5} = 1 : 1.4$

Rewrite each of these ratios in the form $1 : n$.

- a 5 : 8
- b 4 : 13
- c 8 : 9
- d 25 : 36
- e 5 : 27
- f 12 : 18
- g 5 hours : 1 day
- h 4 hours : 1 week
- i £4 : £5

11 Mia wants a map of the island of Sicily.

Sicily is 113 km from north to south and 134 km from west to east.

Choose the most sensible scale from this list

- a 1 : 2500
- b 1 : 25 000
- c 1 : 250 000
- d 1 : 2 500 000

Give a reason for your answer.

7.1 Ratio

7.1

Calculating with ratios when only part of the information is known

> **Example 5**
>
> Two business partners, Lubna and Adama, divided their total profit in the ratio 3 : 5. Lubna received $2100. How much did Adama get?
>
> ---
>
> Lubna's $2100 was $\frac{3}{8}$ of the total profit. (Check that you know why.)
> $\frac{1}{8}$ of the total profit = $2100 ÷ 3 = $700
> So Adama's share, which was $\frac{5}{8}$, amounted to $700 × 5 = $3500.

> **Example 6**
>
> A fruit drink is made by mixing orange squash with water in the ratio 2 : 3. How much water needs to be added to 5 litres of orange squash to make the drink?
>
> ---
>
> 2 parts is 5 litres.
>
> Dividing by 2:
>
>   1 part is 2.5 litres
>
>   3 parts = 2.5 litres × 3 = 7.5 litres
>
> So 7.5 litres of water is needed to make the drink.

EXERCISE 7C

1 Sean, aged 15, and Ricki, aged 10, shared some sweets in the same ratio as their ages. Sean had 48 sweets.

    a Simplify the ratio of their ages.

    b How many sweets did Ricki have?

    c How many sweets did they share altogether?

2 A blend of tea is made by mixing Lapsang with Assam in the ratio 3 : 5. I have a lot of Assam tea but only 600 g of Lapsang. How much Assam do I need to make the blend using all the Lapsang?

3 The ratio of male to female spectators at a hockey game is 4 : 5. 4500 men watched the match. What was the total attendance at the game?

4 A teacher always arranged the content of each of his lessons as 'teaching' and 'practising learnt skills' in the ratio 2 : 3.

    a If a lesson lasted 35 minutes, how much teaching would he do?

    b If he decided to teach for 30 minutes, how long would the lesson need to be?

Chapter 7: Ratio, proportion and rate

Chapter 7 . Topic 2

5 A 'good' children's book has pictures and text in the ratio 17 : 8. In a book I have just looked at, the pictures occupy 23 pages.

   a Approximately how many pages of text should this book have to be a 'good' children's book?

   b What percentage of a 'good' children's book will be text?

6 Three business partners, Ren, Shota and Fatima, put money into a business in the ratio 3 : 4 : 5. They shared any profits in the same ratio. Last year, Fatima received $3400 from the profits. How much did Ren and Shota receive last year?

7 **a** Iqra is making a drink from lemonade, orange and ginger ale in the ratio 40 : 9 : 1. If Iqra has only 4.5 litres of orange, how much of the other two ingredients does she need to make the drink?

   b Another drink made from lemonade, orange and ginger ale uses the ratio 10 : 2 : 1.

   Which drink has a larger proportion of ginger ale, Iqra's or this one? Show how you work out your answer.

7.2 Increases and decreases using ratios   E

Sometimes increases and decreases can be expressed in terms of ratios.

Suppose a recipe for six people requires 450 g of flour.
How much is needed for 10 people?

You need to **increase** 450 g in the ratio 10 : 6 = 5 : 3.
Think of 450 as 3 parts and you need to find 5 parts.

You need to find $\frac{5}{3}$ of 450 g.

$$450 \times \frac{5}{3} = 750 \text{ g}$$

If the recipe was to be changed to feed four people you would need to **decrease** the amount of flour in the ratio 4 : 6 = 2 : 3.

You need to find $\frac{2}{3}$ of 450 g.

$$450 \times \frac{2}{3} = 300 \text{ g}$$

EXERCISE 7D

1 Increase 200 in each ratio.

   a 3 : 1   **b** 3 : 2   **c** 10 : 1   **d** 7 : 4   **e** 6 : 5   **f** 11 : 10

2 Decrease 80 in each ratio

   a 1 : 4   **b** 3 : 4   **c** 1 : 10   **d** 7 : 10   **e** 1 : 5   **f** 4 : 5

3 A projector enlarges an image in the ratio 20 : 1.

  a What will be the size of an enlargement of a picture that is 8 cm by 6 cm?

  b What will be the size of the image if the ratio is changed to 15 : 2?

4 A photocopier enlarges a photograph in the ratio 5 : 4.

  a What is the size of the enlargement of a photograph that is 10 cm by 12 cm?

  b When the setting of the photocopier is changed, the size of the enlargement is 15 cm by 18 cm. Write the setting as a ratio, as simply as possible.

5 A photograph measures 12 cm by 20 cm.

  It is made smaller in the ratio 3 : 4.

  a What are the dimensions of the new photograph?

  b The new photograph is again made smaller in the ratio 3 : 4. What are the dimensions now?

6 $5000 is invested and after a year the value has increased in the ratio 3 : 2.

  a What is the value of the investment now?

  b What is the percentage increase over one year?

  The value of the investment continues to grow and in the second year it increases in the ratio 5 : 4.

  c What is the value after two years?

  d What is the percentage increase in the second year?

  e What is the overall percentage increase over two years?

  f Show that the overall increase could be written as a ratio as 15 : 8.

7 Prices are going up in the ratio 6 : 5.

  a Show that this is a 20% increase.

  b Write a 10% increase as a ratio.

  c Write a 10% decrease as a ratio.

8 There is a group of boys and girls waiting for school buses. 25 girls get on the first bus. The ratio of boys to girls at the stop is now 3 : 2. 15 boys get on the second bus. There are now the same number of boys and girls at the bus stop. How many students altogether were originally at the bus stop?

9 A jar contains 100 cm^3 of a mixture of oil and water in the ratio 1 : 4. Enough oil is added to make the ratio of oil to water 1 : 2. How much water must be added to make the ratio of oil to water 1 : 2?

7.3 Speed

The relationship between **speed**, **time** and **distance** can be expressed in three ways:

$$\text{speed} = \frac{\text{distance}}{\text{time}} \qquad \text{distance} = \text{speed} \times \text{time} \qquad \text{time} = \frac{\text{distance}}{\text{speed}}$$

In problems relating to speed, you usually mean **average** speed, as it would be unusual to maintain one exact speed for the whole of a journey.

This diagram will help you remember the relationships between distance (D), time (T) and speed (S).

$$D = S \times T \qquad S = \frac{D}{T} \qquad T = \frac{D}{S}$$

Units for speed include km/h (kilometres per hour, or 'the number of kilometres travelled in an hour') and m/s (metres per second).

Example 7

Paula drove a distance of 270 kilometres in 5 hours. What was her average speed?

Paula's average speed = $\frac{\text{distance she drove}}{\text{time she took}} = \frac{270}{5} = 54$ kilometres per hour (km/h)

Example 8

Renata drove from her home to Frankfurt in $3\frac{1}{2}$ hours at an average speed of 60 km/h. How far is it from Renata's home to Frankfurt?

Since:

distance = speed × time

the distance from Renata's home to Frankfurt is given by:

60 × 3.5 = 210 kilometres

Note: You need to change the time to a decimal number and use 3.5 (not 3.30).

Example 9

Maria is going to drive to Rome, a distance of 190 kilometres. She estimates that she will drive at an average speed of 50 km/h. How long will it take her?

Maria's time = $\frac{\text{distance she covers}}{\text{her average speed}} = \frac{190}{50} = 3.8$ hours

Change the 0.8 hour to minutes by multiplying by 60, to give 48 minutes.

So, the time taken for Maria's journey will be 3 hours 48 minutes.

7.3

Remember: When you calculate a time and get a decimal answer, as in Example 9, *do not mistake* the decimal part for minutes. You must either:

- leave the time as a decimal number and give the unit as hours, or
- change the decimal part to minutes by multiplying it by 60 (1 hour = 60 minutes) and give the answer in hours and minutes.

EXERCISE 7E

1 A cyclist travels a distance of 90 kilometres in 5 hours. What was her average speed?

2 How far along a road would you travel if you drove at 110 km/h for 4 hours?

3 I can drive from my home to see my aunt in about 6 hours. The distance is 315 kilometres. What is my average speed?

Advice and Tips

Remember to convert time to a decimal if you are using a calculator, for example, 8 hours 30 minutes is 8.5 hours.

4 The distance from Leeds to London is 350 kilometres. The train travels at an average speed of 150 km/h. If I catch the 9.30 am train in London, at what time should I expect to arrive in Leeds?

Advice and Tips

km/h means kilometres per hour.
m/s means metres per second.

5 How long will an athlete take to run 2000 metres at an average speed of 4 metres per second?

6 Copy and complete this table.

	Distance travelled	Time taken	Average speed
a	150 km	2 hours	
b	260 km		40 km/h
c		5 hours	35 km/h
d		3 hours	80 km/h
e	544 km	8 hours 30 minutes	
f		3 hours 15 minutes	100 km/h
g	215 km		50 km/h

7 Eliot drove a distance of 660 kilometres, in 7 hours 45 minutes.

a Change the time 7 hours 45 minutes to a decimal.

b What was the average speed of the journey? Round your answer to 1 decimal place.

8 Johan drives home from his son's house in 2 hours 15 minutes. He says that he drives at an average speed of 70 km/h.

a Change the 2 hours 15 minutes to a decimal.

b How far is it from Johan's home to his son's house?

Chapter 7: Ratio, proportion and rate

9 The distance between Paris and Le Mans is 200 km. The express train between Paris and Le Mans travels at an average speed of 160 km/h.

   a Calculate the time taken for the journey from Paris to Le Mans, giving your answer as a decimal number of hours.

   b Change your answer to part **a** to hours and minutes.

10 The distance between two cities is 420 kilometres.

   a What is the average speed of a journey from one to the other if it takes 8 hours 45 minutes?

   b If Sam covered the distance at an average speed of 63 km/h, how long would it take him?

11 A train travels at 50 km/h for 2 hours, then slows down to do the last 30 minutes of its journey at 40 km/h.

   a What is the total distance of this journey?

   b What is the average speed of the train over the whole journey?

Advice and Tips

Remember that there are 3600 seconds in an hour and 1000 metres in a kilometre. So to change from km/h to m/s multiply by 1000 and divide by 3600.

12 Suni runs and walks the 6 kilometres from home to work each day. She runs the first 4 kilometres at a speed of 16 km/h, then walks the next 2 kilometres at a steady 8 km/h.

   a How long does it take Suni to get to work?

   b What is her average speed?

13 Change these speeds to metres per second.

   a 36 km/h   **b** 12 km/h   **c** 60 km/h
   d 150 km/h   **e** 75 km/h

14 Change these speeds to kilometres per hour.

   a 25 m/s   **b** 12 m/s   **c** 4 m/s
   d 30 m/s   **e** 0.5 m/s

Advice and Tips

To change from m/s to km/h multiply by 3600 and divide by 1000.

15 A train travels at an average speed of 18 m/s.

   a Express its average speed in km/h.

   b Find the approximate time the train would take to travel 500 m.

   c The train set off at 7.30 on a 40 km journey. At approximately what time will it reach its destination?

Advice and Tips

To convert a decimal fraction of an hour to minutes, just multiply by 60.

16 A cyclist is travelling at an average speed of 24 km/h.

   a What is this speed in metres per second?

   b What distance does he travel in 2 hours 45 minutes?

   c How long does it take him to travel 2 km?

   d How far does he travel in 20 seconds?

17 How much longer does it take to travel 100 kilometres at 65 km/h than at 70 km/h?

7.4 Rates

When you travel at a constant speed, then speed = distance/time

The ratio distance : speed is constant and this ratio is the speed.

There are other examples like this.

The **density** of a substance is defined as density = mass/volume

For a particular substance, the ratio mass : volume is constant and this ratio is the density.

If the mass is in g and the volume is in cm³ then the density is in g/cm³.

Example 10

a The mass of a piece of gold is 2.4 g and the volume is 0.12 cm³.

   Work out the density of gold.

b A second piece of gold has a mass of 1.6 g.

   Work out the volume of this piece of gold.

---

a density = $\frac{\text{mass}}{\text{volume}} = \frac{2.4}{0.12} = 20$ g/cm³

b There are two ways to do this.

First method: the mass is $\frac{2}{3}$ of the first piece so the volume is $\frac{2}{3}$ of 0.12 cm³

$\frac{2}{3} \times 0.12 = 0.08$ cm³

Second method:

Use the formula density = $\frac{\text{mass}}{\text{volume}}$

$20 = \frac{1.6}{\text{volume}}$

Volume = $\frac{1.6}{20} = 0.08$ cm³

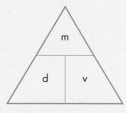

A triangle like the one for speed can help.

A box is on a table.

The weight of the box is the force due to gravity. This exerts **pressure** on the table.

The pressure is given by the formula pressure = $\frac{\text{force}}{\text{area}}$

The force is the weight of the box. The area is the base of the box in contact with the table.

If the force is in Newtons (N) and the area is in m² then the pressure is in N/m².

N/m² are usually called Pascals (Pa).

Chapter 7: Ratio, proportion and rate

Example 11

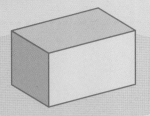

A container in the shape of a cuboid is standing on the ground.

The weight of the container is 5000 N

The area of the base of the container is 4 m²

a Find the pressure of the container on the ground.

b The end of the container has an area of 2 m²

Find the pressure if the container is turned over to stand on its end.

a The weight is the force on the ground.

pressure = $\frac{\text{force}}{\text{area}} = \frac{5000}{4}$ = 1250 Pa

b This time pressure = $\frac{\text{force}}{\text{area}} = \frac{5000}{2}$ = 2500 Pa

The area is halved; the pressure is doubled.

EXERCISE 7F

1 A plant is growing at a rate of 0.5 cm/day.

  a How much will it grow in a week?

  b How long will it take to grow 10 cm?

2 Water is flowing out of a tap at a rate of 3 litres/minute.

  a How much will flow out in 5 minutes?

  b How many seconds will it take for one litre to flow out?

3 The mass of a silver plate is 420 g and the volume is 40 cm³.

  a Calculate the density of silver in g/cm³.

  b A second plate has a mass of 525 g.

   Calculate the volume in cm³.

4 Rainforest in one area is being cleared at a rate of 50 m²/day.

  How long will it take to clear 80 000 m²?

5 The sea level in a particular country has risen by 50 cm in 1000 years.

   a Find the rate of increase in mm/year.

   Assume that this rate does not change.

   b How many millimetres will the sea level rise in the next 25 years?

   c How long will it take for the sea level to rise by 20 mm?

6 The density of copper is 9 gm/cm^3.

   a Calculate the volume of 40 g of copper.

   b Calculate the mass of 40 cm^3 of copper.

   c A piece of metal has a mass of 75.1g and a volume of 9.4cm^3.

   Can the metal be copper? Give a reason for your answer.

7 Find the pressure in the following cases:

   a A force of 750 N acts on an area of 12 m^2

   b A force of 3000 N acts on an area of 12 m^2

   c A force of 750 N acts on an area of 4 m^2

8 A car maker claims that the fuel consumption rate for a particular car is 10.6 litres/100 km.

   a How much fuel will it use to travel 300 km?

   b How much fuel will it use to travel 50 km?

   c What is the fuel consumption in litres/10 km?

   d What is the fuel consumption in litres/kilometre?

   e How far can it travel on 100 litres of fuel?

9 A car's average CO_2 (carbon dioxide) emission rate is 145 g per km.

   a How many *kilograms* of CO_2 will be emitted if the car travels 100 km?

   b A driver estimates that he will drive the car 20 000 km in a year. How much CO_2 will be emitted in that time?

   c Estimate the driver's average rate of CO_2 emission in kilograms/day.

10 A balloon is in the shape of a sphere.

   It is inflated to a radius of 12 cm.

   The pressure of the air inside the balloon is 0.4 N/cm^3

   The formula for the surface area of a sphere of radius r is $4\pi r^2$. Find the force exerted on the surface of the balloon, in Newtons.

Chapter 7: Ratio, proportion and rate

Chapter 7 . Topic 5

7.5 Direct proportion

Suppose you buy 12 items that each cost the *same*. The total amount you spend is 12 times the cost of one item.

So the total cost is in **direct proportion** to the number of items bought. The cost of a single item (the unit cost) is the constant factor that links the two quantities.

Direct proportion is not only concerned with costs. Any two related quantities can be in direct proportion to each other.

The best way to solve all problems involving direct proportion is to start by finding the **single unit value**. This method is called the **unitary method**, because it refers to a single unit value. Work through Examples 10 and 11 to see how it is done.

Remember: Before solving a direct proportion problem, think about it carefully to make sure that you know how to find the required single unit value.

Example 12

If eight pens cost $2.64, what is the cost of five pens?

First, find the cost of one pen. This is $2.64 ÷ 8 = $0.33

So, the cost of five pens is $0.33 × 5 = $1.65

Example 13

Eight loaves of bread will make packed lunches for 18 people. How many packed lunches can be made from 20 loaves?

First, find how many lunches one loaf will make.

One loaf will make 18 ÷ 8 = 2.25 lunches.

So, 20 loaves will make 2.25 × 20 = 45 lunches.

EXERCISE 7G

1. If 30 matches have a total mass of 45 g, what would be the total mass of 40 matches?

2. Five bars of chocolate together cost $2.90. Find the total cost of nine bars.

3. Eight men can chop down 18 trees in a day. How many trees can 20 men chop down in a day?

4. Find the cost of 48 eggs when 15 eggs can be bought for $2.10.

5. Seventy maths textbooks cost a total of $875.

    a. How much will 25 of the same maths textbooks cost?

    b. How many of these maths textbooks can you buy for $100?

Advice and Tips

Remember to work out the value of one unit each time. Always check that answers are sensible.

Chapter 7 . Topic 6

6 A lorry uses 80 litres of fuel on a trip of 280 kilometres.
  a How much fuel would the same lorry use on a trip of 196 kilometres?
  b How far would the lorry travel on a full tank of 100 litres of fuel?

7 During the winter, I find that 200 kg of coal keeps my fire burning for 12 weeks.
  a If I want a fire all through the winter (18 weeks), how much coal will I need to buy?
  b Last year I bought 150 kg of coal. For how many weeks did I have a fire?

8 It takes a photocopier 16 seconds to produce 12 copies. How long will it take to produce 30 copies?

9 A recipe for 12 biscuits uses:

  200 g margarine   400 g sugar

  500 g flour   300 g ground rice

  a What quantities are needed for:
   i 6 biscuits   ii 9 biscuits   iii 15 biscuits?
  b What is the maximum number of biscuits I could make if I had just 1 kg of each ingredient?

10 Pieter the baker sells bread rolls in packs of 6 for $2.30.

Paulo the baker sells bread rolls in packs of 10 for $3.50.

I have $10 to spend on bread rolls.

If I want to buy as many bread rolls as possible from one shop, which shop should I use? Show your working.

7.6 Inverse proportion

As your speed *increases*, the time you take to travel a fixed distance *decreases*. This is an example of **inverse proportion**.

Example 14

At an average speed of 6 km/h a walk takes 5 hours.

How long would it take at an average speed of 8 km/h?

The total distance is 6 × 5 = 30 kilometres.

At 8 km/h the time taken to travel 30 kilometres is:
$$\frac{30}{8} = 3\frac{3}{4}$$
= 3 hours 45 minutes.

The time taken for the journey falls as the speed rises so it is in inverse proportion to the speed.

Chapter 7: Ratio, proportion and rate

Example 15

When a teacher shares out some sweets among 12 students, each student gets 10 sweets.

How many will each get if there are 20 students?

The total number of sweets is 10 × 12 = 120.

If 20 students share 120 sweets, each student gets 120 ÷ 20 = 6 sweets.

Because there are more students, each one gets fewer sweets.

The number of sweets each student gets is in inverse proportion to the number of students.

EXERCISE 7H

1. A car travelling at 30 km/h takes 40 minutes to complete a journey.

   How long would it take if it travelled at 60 km/h?

2. Leo has some books that are 2 cm wide and he can fit 40 on a shelf.

   How many books 5 cm wide could fit on the same shelf?

3. Chairs are available at three prices: $30, $40 and $60.

   Lejla can afford to buy 24 at $60 each.

   a How many can she afford to buy at $40 each?

   b How many can she afford to buy at $30 each?

4. Some friends are sharing the cost of a taxi.

   If there are three friends it will cost them $18 each.

   How much will it cost if there are five friends?

5. Eight people can pick all the grapes in a vineyard in ten days.

   a How long will it take five people?

   b How many people are needed to pick them all in four days?

6. Copy and complete this table, showing how long it will take a cyclist to complete a journey at different average speeds.

Average speed (km/h)	10	20	30	40
Time in hours and minutes		2 hours 30 minutes		

7. The number of people that can be safely carried in a lift depends on their average mass. A lift will safely carry 8 people with an average mass of 75 kg.

   a How many people with an average mass of 100 kg can safely travel in the lift?

   b How many children with an average mass of 40 kg can safely travel in the lift?

7.6 Inverse proportion

8 Two people travel 150 kilometres in a car.

Paying for the petrol costs each person $12.

Work out the cost if five people travel 250 kilometres.

Check your progress

Core
- I can understand and use ratios
- I can solve numerical problems involving direct proportion
- I can solve numerical problems involving indirect proportion
- I can calculate average speed
- I can use common measures of rates including density and pressure

Chapter 8
Estimation and limits of accuracy

Topics	Level	Key words
1 Rounding whole numbers	CORE	approximation, rounded up, rounded down
2 Rounding decimals	CORE	round, digit, decimal place
3 Rounding to significant figures	CORE	significant figure
4 Upper and lower bounds	CORE	upper bound, lower bound, limits of accuracy
5 Upper and lower bounds for calculations	EXTENDED	upper bound, lower bound

In this chapter you will learn how to:

CORE	EXTENDED
• Make estimates of numbers, quantities and lengths. (C1.9 and E1.9) • Give approximations to specified numbers of significant figures and decimal places and round off answers to reasonable accuracy in the context of a given problem. (C1.9 and E1.9) • Give appropriate upper and lower bounds for data given to a specified accuracy (e.g. measured lengths). (C1.10 and E1.10)	• Obtain appropriate upper and lower bounds to solutions of simple problems (e.g. the calculation of the perimeter or the area of a rectangle) given data to a specified accuracy. (E1.10)

Why this chapter matters

How accurate are we?

In real life it is not always sensible to use exact values. Sometimes it would be impossible to have exact measurements. People often round values without realising it. Rounding is done so that values are sensible.

Is it exactly 23 km to Utrecht and exactly 54 km to Amsterdam?

Does this box contain exactly 750 g of rice when full?

Was her time exactly 13.4 seconds?

Does the school have exactly 1500 students?

Imagine if people tried to use exact values all the time. Would life seem strange?

Chapter 8: Estimation and limits of accuracy

8.1 Rounding whole numbers

You use rounded information, or **approximations** all the time. Look at the examples on the right. Each actual figure is either above or below the approximation shown here.

How do you round numbers up or down?

If you want to round a number to the nearest multiple of ten, you round it up if it ends in 5, 6, 7, 8 or 9, and round it down if it ends in 1, 2, 3 or 4. For example:

- 25, 26, 27, 28 and 29 are **rounded up** to 30.
- 24, 23, 22 and 21 are **rounded down** to 20.

So a box with approximately 30 mints could contain any number from 25 to 34.

You can round numbers to the nearest multiple of 10, 100, 1000 and so on. The number of runners (23 000) in the report on the marathon is rounded to the nearest 1000:

- the smallest number of people actually running would be 22 500 (22 500–22 999 are rounded up to 23 000)
- the largest number of people running would be 23 499 (23 500 would be rounded up to 24 000).

So, there could actually be from 22 500 to 23 499 people in the marathon.

EXERCISE 8A

1 Round each of these numbers to the nearest 10.

a 24 b 57 c 78 d 54 e 96
f 21 g 88 h 66 i 14 j 26

2 Round each of these numbers to the nearest 100.

a 240 b 570 c 780 d 504 e 967
f 112 g 645 h 358 i 998 j 1050

3 Round each of these numbers to the nearest 1000.

a 2400 b 5700 c 7806 d 5040 e 9670
f 1120 g 6450 h 3499 i 9098 j 1500

4

Welcome to Elsecar	Welcome to Hoyland	Welcome to Jump
Population 800 (to the nearest 100)	Population 1200 (to the nearest 100)	Population 600 (to the nearest 100)

Chapter 8 . Topic 2

Which of these sentences could be true and which must be false?

a There are 789 people living in Elsecar.
b There are 1278 people living in Hoyland.
c There are 550 people living in Jump.
d There are 843 people living in Elsecar.
e There are 1205 people living in Hoyland.
f There are 650 people living in Jump.

5 These are the average attendance figures in four football leagues in 2008–9:

England Premier League	35 600
Germany Bundesliga	42 565
Italy Serie A	25 303
Spain La Liga	29 124

a Which were the highest and lowest?

b Round each number to the nearest thousand.

c The figure for Ligue 1 in France was 25 000 to the nearest thousand. What were the largest and smallest actual values for Ligue 1?

6 Matthew and Vincenza are playing a game with whole numbers.

a What is the smallest number Matthew could be thinking of?

I am thinking of a number. Rounded to the nearest 10 it is 380.

I am thinking of a different number that is smaller than Matthew's. Rounded to the nearest 100 it is 400.

b Matthew's number is the smallest possible. How many possible values are there for Vincenza's number?

7 The number of adults attending a comedy show is 80 to the nearest 10. The number of children attending is 50 to the nearest 10.

Katya says that 130 adults and children attended the comedy show. Give an example to show that she may **not** be correct.

8.2 Rounding decimals

Decimal places

When a number is written in decimal form, the **digits** to the right of the decimal point are called **decimal places**. For example:

79.4 is written 'with one decimal place'

6.83 is written 'with two decimal places'

0.526 is written 'with three decimal places'.

Chapter 8: Estimation and limits of accuracy

To **round** a decimal number to a particular number of decimal places, take these steps.

- Count along the decimal places from the decimal point and look at the first digit to be removed.
- When the value of this digit is less than five, just remove the unwanted places.
- When the value of this digit is five or more, add 1 onto the digit in the last decimal place then remove the unwanted places.

Here are some examples.

5.852 rounds to 5.85 to two decimal places

7.156 rounds to 7.16 to two decimal places

0.274 rounds to 0.3 to one decimal place

15.3518 rounds to 15.4 to one decimal place

EXERCISE 8B

1 Round each number to one decimal place.

a 4.83   b 3.79   c 2.16   d 8.25
e 3.673   f 46.935   g 23.883   h 9.549
i 11.08   j 33.509

Advice and Tips

Just look at the value of the digit in the second decimal place.

2 Round each number to two decimal places.

a 5.783   b 2.358   c 0.977   d 33.085   e 6.007
f 23.5652   g 91.7895   h 7.995   i 2.3076   j 23.9158

3 Round each of these to the number of decimal places (dp) indicated.

a 4.568 (1 dp)   b 0.0832 (2 dp)   c 45.715 93 (3 dp)   d 94.8531 (2 dp)
e 602.099 (1 dp)   f 671.7629 (2 dp)   g 7.1124 (1 dp)   h 6.903 54 (3 dp)
i 13.7809 (2 dp)   j 0.075 11 (1 dp)

4 Round each of these to the nearest whole number.

a 7.82   b 3.19   c 7.55   d 6.172   e 3.961
f 7.388   g 1.514   h 46.78   i 23.19   j 96.45

5 Anna puts the following items in her shopping basket: bread $3.20, meat $8.95, cheese $6.16 and butter $3.90.

By rounding each price to the nearest dollar, work out an estimate of the total cost of the items.

6 Which of these are correct roundings of the number 3.456?

3   3.0   3.4   3.40   3.45   3.46   3.47   3.5   3.50

7 When a number is rounded to three decimal places the answer is 4.728.

Which of these could be the number?

4.71   4.7275   4.7282   4.73

8.2 Rounding decimals

8.3 Rounding to significant figures

You will often use **significant figures** (sf) when you want to *approximate* a number with a lot of digits in it. You frequently use this technique with calculator answers.

A calculator gives √800 as 28.284 271 25…

Rounded to four sf this is 28.28

three sf it is 28.3

two sf it is 28

one sf it is 30

The steps taken to round a number to a given number of significant figures are very similar to those used for rounding to a given number of decimal places.

- Count the digits from the left. If you are rounding to 2 sf, count two digits, for 3 sf count three digits, and so on. When the original number is less than 1, start counting from the first non-zero digit.
- Look at the next digit to the right. When the value of this next digit is less than 5, leave the digit you counted to the same. However, if the value of this next digit is equal to or greater than 5, add 1 to the digit you counted to.
- Ignore all the other digits, but put in enough zeros to keep the number the right size (value).

For example, look at this table, which shows some numbers rounded to one, two and three significant figures, respectively.

Number	Rounded to 1 sf	Rounded to 2 sf	Rounded to 3 sf
45 281	50 000	45 000	45 300
568.54	600	570	569
7.3782	7	7.4	7.38
8054	8000	8100	8050
99.8721	100	100	99.9
0.7002	0.7	0.70	0.700

EXERCISE 8C

1 Round each number to 1 significant figure.

- a 46 313
- b 57 123
- c 30 569
- d 94 558
- e 85 299
- f 0.5388
- g 0.2823
- h 0.005 84
- i 0.047 85
- j 0.000 876
- k 9.9
- l 89.5
- m 90.78
- n 199
- o 999.99

2 Round each number to 2 significant figures.

- a 56 147
- b 26 813
- c 79 611
- d 30 578
- e 14 009
- f 1.689
- g 4.0854
- h 2.658
- i 8.0089
- j 41.564
- k 0.8006
- l 0.458
- m 0.0658
- n 0.9996
- o 0.009 82

Chapter 8 . Topic 4

3 Round each of these to the number of significant figures (sf) indicated.
- **a** 57 402 (1 sf)
- **b** 5288 (2 sf)
- **c** 89.67 (3 sf)
- **d** 105.6 (2 sf)
- **e** 8.69 (1 sf)
- **f** 1.087 (2 sf)
- **g** 0.261 (1 sf)
- **h** 0.732 (1 sf)
- **i** 0.42 (1 sf)
- **j** 0.758 (1 sf)
- **k** 0.185 (1 sf)
- **l** 0.682 (1 sf)

4 What are the lowest and the highest numbers of sweets that may be in these jars?

a

70 sweets (to 1 sf)

b

100 sweets (to 1 sf)

c

1000 sweets (to 1 sf)

5 What are the least and the greatest numbers of people that live in these towns?

Satora   population 800 (to 1 significant figure)

Nimral   population 1200 (to 2 significant figures)

Korput   population 165 000 (to 3 significant figures)

6 There are 500 fish in a pond, correct to 1 sf. What is the least possible number of fish that could be taken from the pond so that there are 400 fish in the pond, correct to 1 sf?

7 Rani says that the population of Bikran is 132 000 to the nearest thousand. Vashti says that the population of Bikran is 130 000. Explain why Vashti could also be correct.

8.4 Upper and lower bounds

Any recorded measurements have usually been rounded.

The true value will be somewhere between the **lower bound** and the **upper bound**.

The lower and upper bounds are sometimes known as the **limits of accuracy**.

A journey of D kilometres that is measured as 26 kilometres to the nearest kilometre could be anything between 25.5 and 26.5 kilometres:

   25.5 would round up to 26.

   26.5 would round up to 27 but anything less would round down to 26.

This means that 26.5 is the upper bound and 25.5 is the lower bound of D.

You can write $25.5 \leq D < 26.5$ which means that D is greater than or equal to 25.5 but less than 26.5.

Example 1

A stick of wood measures 32 cm, to the nearest centimetre.

What are the lower and upper bounds of the actual length of the stick?

The lower bound is 31.5 cm as this is the lowest value that rounds to 32 cm to the nearest centimetre.

The upper bound is 32.5 cm as anything lower rounds to 32 cm to the nearest centimetre. 32.5 cm would round to 33 cm.

You can write:

$31.5 \leqslant$ length of stick in cm < 32.5

Note the use of $<$ for the upper bound.

Example 2

A time of 53.7 seconds is accurate to 1 decimal place.

What are the upper and lower bounds for the time?

The lower bound is 53.65 seconds and the upper bound is 53.75 seconds.

So $53.65 \leqslant$ time in seconds < 53.75

EXERCISE 8D

1 Write down the upper and lower bounds of each measurement.

  a A length measured as 7 cm to the nearest cm.
  b A mass measured as 120 g to the nearest 10 g.
  c A length measured as 3400 km to the nearest 100 km.
  d A speed measured as 50 km/h to the nearest km/h.
  e An amount given as $6 to the nearest dollar.
  f A length given as 16.8 cm to the nearest tenth of a centimetre.
  g A mass measured as 16 kg to the nearest kg.
  h A football crowd of 14 500 to the nearest 100.
  i A distance given as 55 km to the nearest km.
  j A distance given as 55 km to the nearest 5 km.

2 Write down the upper and lower bounds for each of these values, which are rounded to the given degree of accuracy. Use inequalities to show your answer.
For example, part **a** should be $5.5 \leqslant$ length in cm < 6.5.

  a 6 cm (1 significant figure)   b 17 kg (2 significant figures)
  c 32 min (2 significant figures)   d 238 km (3 significant figures)
  e 7.3 m (1 decimal place)   f 25.8 kg (1 decimal place)

Chapter 8: Estimation and limits of accuracy

g 3.4 h (1 decimal place)
h 87 g (2 significant figures)
i 4.23 mm (2 decimal places)
j 2.19 kg (2 decimal places)
k 12.67 minutes (2 decimal places)
l 25 m (2 significant figures)
m 40 cm (1 significant figure)
n 600 g (2 significant figures)
o 30 minutes (1 significant figure)
p 1000 m (2 significant figures)
q 4.0 m (1 decimal place)
r 7.04 kg (2 decimal places)
s 12.0 s (1 decimal place)
t 7.00 m (2 decimal places)

3 Write down the lower and upper bounds of each of these values, which are rounded to the accuracy stated.

a 8 m (1 significant figure)
b 26 kg (2 significant figures)
c 25 minutes (2 significant figures)
d 85 g (2 significant figures)
e 2.40 m (2 decimal places)
f 0.2 kg (1 decimal place)
g 0.06 s (2 decimal places)
h 300 g (1 significant figure)
i 0.7 m (1 decimal place)
j 366 d (3 significant figures)
k 170 weeks (2 significant figures)
l 210 g (2 significant figures)

4 A chain is 30 m long, to the nearest metre.

A chain is needed to fasten a boat to a harbour wall. The distance to the wall is also 30 m, to the nearest metre.

Which statement is definitely true? Explain your decision.

   A: The chain will be long enough.
   B: The chain will not be long enough.
   C: It is impossible to tell whether or not the chain is long enough.

5 A bag contains 2.5 kg of soil, to the nearest 100 g.
What is the least amount of soil in the bag?
Give your answer in kilograms and grams.

6 Chang has 40 identical marbles. Each marble has a mass of 65 g (to the nearest gram).

a What is the greatest possible mass of one marble?
b What is the least possible mass of one marble?
c What is the greatest possible mass of all the marbles?
d What is the least possible mass of all the marbles?

8.5 Upper and lower bounds for calculations

When rounded values are used for a calculation, you can find upper and lower bounds for the result of the calculation.

Example 3

The dimensions of this rectangle are given to the nearest centimetre.

Find the lower bound for the perimeter and the upper bound for the area.

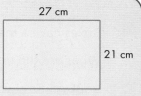

The upper and lower bounds for the sides are:

$26.5 \leq$ length in cm < 27.5

$20.5 \leq$ width in cm < 21.5

Perimeter = 2 × (length + width)

The lower bound will be found from the lower bounds of the length and width.

Lower bound of perimeter = 2 × (26.5 + 20.5) = 94 cm.

Area = length × width

The upper bound will be found from the upper bounds of the length and width.

Upper bound of area = 27.5 × 21.5 = 591.25 cm^2

Example 4

A car travels 125 km (to the nearest km) and uses 16.1 litres of fuel (correct to one decimal place).

Find the upper and lower bounds of the fuel consumption in km/litre.

$124.5 \leq$ distance in kilometres < 125.5

$16.05 \leq$ fuel in litres < 16.15

The fuel consumption is distance ÷ fuel used.

To find the upper bound of this, use:

upper bound of the distance ÷ *lower* bound of the fuel used

upper bound is 125.5 ÷ 16.05 = 7.8193......

To find the lower bound of fuel consumption, use:

lower bound of distance ÷ *upper* bound of fuel used:

$$= 124.5 \div 16.15 = 7.7089......$$

So $7.709 \leq$ fuel consumption in km/litre < 7.819 when the answers are rounded to three decimal places.

When solving a problem, write down the upper and lower bounds for the values given and then decide which to use to find the solution.

EXERCISE 8E

1 Boxes have a mass of 7 kg, to the nearest kilogram.

What are the upper and lower bounds for the total mass of 10 of these boxes?

2 Books each have a mass of 1200 g, to the nearest 100 g.

   a What is the greatest possible mass of 10 books?
   Give your answer in kilograms.

   b A trolley can safely hold up to 25 kg of books.
   How many books can safely be put on the trolley?

3 Jasmine says, 'My mass is 45 kilos.' Yolander says, 'My mass is 53 kilos.' Both are measured to the nearest kilogram.

What is the greatest possible difference between their masses?

Show how you worked out your answer.

4 For each of these rectangles, find the upper and lower bounds for the perimeter.
The measurements are shown to the level of accuracy indicated in brackets.

   a 5 cm × 9 cm (nearest cm)

   b 4.5 cm × 8.4 cm (1 decimal place)

   c 7.8 cm × 18 cm (2 significant figures)

5 Calculate the upper and lower bounds for the areas of each rectangle in question **4**.

6 A cinema screen is measured as 6 m by 15 m, to the nearest metre. Calculate the upper and lower bounds for the area of the screen.

7 The measurements, to the nearest centimetre, of a box are given as 10 cm × 7 cm × 4 cm. Calculate the upper and lower bounds for the volume of the box.

8 Mr Sparks is an electrician. He has a 50-m roll of cable, correct to the nearest metre.

He uses 10 m on each job, to the nearest metre.

If he does four jobs, what is the maximum amount of cable he could have left?

9 Jon and Matt are exactly 7 kilometres apart. They are walking towards each other.

Jon is walking at 4 km/h and Matt is walking at 2 km/h.

Both speeds are given to the nearest kilometre per hour.

Without doing any time calculations, decide whether it is possible for them to meet in 1 hour. Justify your answer.

10 The area of a rectangular field is given as 350 m², to the nearest 10 m². One length is given as 16 m, to the nearest metre. Find the upper and lower bounds for the other length of the field.

11 A stopwatch records the time for the winner of a 100-metre race as 14.7 seconds, measured to the nearest one-tenth of a second.

   a What are the upper and lower bounds of the winner's time?

   b The length of the 100-metre track is correct to the nearest 1 m. What are the upper and lower bounds of the lengths of the track?

   c What is the greatest possible average speed of the winner, with a time of 14.7 seconds in the 100-metre race?

12 A model car travels 40 m, measured to one significant figure, at a speed of 2 m/s, measured to one significant figure. Find the upper and lower bounds of the time taken.

13 The population of Japan is 127 000 000 to the nearest million.

The area of Japan is 378 000 km² to the nearest 1000 km².

The population density for any country is the total population divided by the area.

Find the upper and lower bounds for the population density of Japan in people/km². Round your answers to two decimal places.

14 An engineer testing a car's CO_2 (carbon dioxide) emissions measures 26 kg of CO_2 when it is driven 150 km.

The mass is given to the nearest kilogram.

The distance is given to the nearest kilometre.

Find the upper and lower bounds for the CO_2 emissions in grams/kilometre. Round your answers to one decimal place.

Check your progress

Core

- I can make estimates of numbers, quantities and lengths
- I can give estimates to a given number of decimal places
- I can give estimates to a given number of significant figures
- I can round off answers to a reasonable level of accuracy in the context of a problem
- I can give upper and lower bounds for data given to a specified accuracy

Extended

- I can give upper and lower bounds for the results of calculations with data given to a specified accuracy

Chapter 9
Standard form

Topics	Level	Key words
1 Standard form	CORE	standard form, index
2 Calculating with standard form	CORE	

In this chapter you will learn how to:

CORE
• Use the standard form $A \times 10^n$ when n is a positive or negative integer and $1 \leqslant A < 10$. (C1.7 and E1.7)

Why this chapter matters

Very large and very small numbers can often be difficult to read. Scientists use standard form as a shorthand way of representing numbers.

The planets

Mercury is the closest planet to the Sun (and is very hot). It orbits 60 million km (6×10^7 km) away from the Sun.

Venus rotates the opposite way to the other planets and has a diameter of 12 100 km (1.21×10^4 km).

Earth takes 365 days to orbit the Sun and 24 hours to complete a rotation.

Mars has the largest volcano in the solar system. It is almost 600 km across and rises 24 km above the surface. This is five times bigger than the biggest volcano on Earth.

Jupiter is made of gas. It has no solid land so visiting it is not recommended! It has a huge storm which rages across its surface. This is about 8 km high, 40 000 km long and 14 000 km wide. It looks like a red spot and is called 'the Great Red Spot'.

Saturn is the largest planet in the solar system. It is about 120 000 km across (1.2×10^5 km) and 1400 million km from the Sun (1.4×10^9 km).

Uranus takes 84 days to orbit the Sun.

Neptune is similar to Jupiter in that it is a gas planet and has violent storms. Winds can blow at up to 2000 km per hour, so a cloud can circle Neptune in about 16 hours.

Pluto is the furthest planet from the Sun. Some astronomers dispute whether it can be classed as a planet. The average surface temperature on Pluto is about −230 °C.

The mass of an electron is about 0.000 000 000 000 000 000 000 000 000 91 kg. This is written 9.1×10^{-31} kg.

The mass of the Earth is about 5 970 000 000 000 000 000 000 000 kg. This is written 5.97×10^{24} kg.

9.1 Standard form

Powers of ten:

$100 = 10 \times 10 = 10^2$

$1000 = 10 \times 10 \times 10 = 10^3$

Extending this idea:

$10\,000 = 10 \times 10 \times 10 \times 10 = 10^4$

$100\,000 = 10^5$

$1\,000\,000 = 10^6$

and so on.

The power of 10 is called the **index**.

Standard form is a way of writing very large and very small numbers using powers of 10. In this form, a number is given a value between 1 and 10 multiplied by a power of 10. That is,

$A \times 10^n$ where $1 \leqslant A < 10$, and n is a whole number.

Follow through these examples to see how numbers are written in standard form.

$52 = 5.2 \times 10 = \mathbf{5.2 \times 10^1}$

$73 = 7.3 \times 10 = \mathbf{7.3 \times 10^1}$

$625 = 6.25 \times 100 = \mathbf{6.25 \times 10^2}$ The numbers in bold are in standard form.

$389 = 3.89 \times 100 = \mathbf{3.89 \times 10^2}$

$3147 = 3.147 \times 1000 = \mathbf{3.147 \times 10^3}$

When writing a number in this way, you must always follow two rules.

- The first part must be a number between 1 and 10 (1 is allowed but 10 isn't).
- The second part must be a whole-number (negative or positive) power of 10. Note that you would *not normally* write the power 1.

Standard form on a calculator

A number such as 123 000 000 000 is obviously difficult to key into a calculator. Instead, you enter it in standard form (assuming you are using a scientific calculator):

$123\,000\,000\,000 = 1.23 \times 10^{11}$

The key strokes to enter this into your calculator could be something like this:

1 **.** **2** **3** **×10^x** **1** **1**

Your calculator display will display the number either as an ordinary number, if there is enough space, or in standard form. Make sure you know how to use standard form on your calculator.

Some calculators may use a different symbol for **×10^x**, for example, **EXP**

9.1

Standard form of numbers less than 1

You will need to use a negative index for numbers between 0 and 1:

$0.1 = 10^{-1}$

$0.01 = 10^{-2}$

$0.001 = 10^{-3}$

$0.0001 = 10^{-4}$

and so on.

For example:

$0.000\,729 = 7.29 \times 0.0001$

$= 7.29 \times 10^{-4}$ in standard form

These numbers are written in standard form. Make sure that you understand how they are formed.

a $0.4 = 4 \times 10^{-1}$
b $0.05 = 5 \times 10^{-2}$
c $0.007 = 7 \times 10^{-3}$
d $0.123 = 1.23 \times 10^{-1}$
e $0.007\,65 = 7.65 \times 10^{-3}$
f $0.9804 = 9.804 \times 10^{-1}$
g $0.0098 = 9.8 \times 10^{-3}$
h $0.000\,0078 = 7.8 \times 10^{-6}$

On a calculator you would enter 1.23×10^{-6}, for example, as:

[1] [•] [2] [3] [×10ˣ] [(−)] [6]

Try entering some of the numbers **a** to **h** (above) into your calculator for practice.

EXERCISE 9A

1 These numbers are in standard form. Write them out in full.

a 2.5×10^2
b 3.45×10
c 4.67×10^{-3}
d 3.46×10
e 2.0789×10^{-2}
f 5.678×10^3
g 2.46×10^2
h 7.6×10^3
i 8.97×10^5
j 8.65×10^{-3}
k 6×10^7
l 5.67×10^{-4}

2 Write these numbers in standard form.

a 250
b 0.345
c 46 700
d 3 400 000 000
e 20 780 000 000
f 0.000 567 8
g 2460
h 0.076
i 0.000 76
j 0.999
k 234.56
l 98.7654
m 0.0006
n 0.005 67
o 56.0045

In questions **3** to **5**, write the numbers given in each statement in standard form.

3 One year, 27 797 runners completed the New York marathon.

4 The largest number of dominoes ever toppled by one person is 321 197, although a team set up and toppled 4 491 863.

Chapter 9. Topic 2

5 The asteroid *Phaethon* comes within 12 980 000 miles of the Sun. The asteroid *Pholus*, at its furthest point, is a distance of 2997 million miles from the Earth. The closest an asteroid ever came to Earth was 93 000 miles from the planet.

6 How many times bigger is 3.2×10^6 than 3.2×10^4?

7 The speed of sound (Mach 1) is 1236 kilometres per hour.

A plane travelling at Mach 2 would be travelling at twice the speed of sound.

How many kilometres would a plane travelling at Mach 3 cover in 1 minute?

8 Here are the distances of some planets from the Sun:

Jupiter	778 million kilometres
Mercury	58 million kilometres
Pluto	5920 million kilometres

Write these distances in standard form.

9.2 Calculating with standard form

Calculations involving very large or very small numbers can be done more easily if you use standard form.

You can enter numbers in a scientific calculator in standard form. The way you do this may be different for different models of calculator. Make sure you know how to do this with your calculator.

Example 1

A pixel on a computer screen is 2×10^{-2} cm long by 7×10^{-3} cm wide.

What is the area of the pixel?

The area is calculated as length times width.

Area in cm² = $2 \times 10^{-2} \times 7 \times 10^{-3}$

= 14×10^{-5} (Use a calculator to do this.)

The answer is not in standard form as the first part is not between 1 and 10, so you need to change it to standard form.

$14 \times 10^{-5} = 0.000\,14 = 1.4 \times 10^{-4}$

So area = 1.4×10^{-4} cm²

If you use a calculator you can enter the numbers directly without any rearranging. Your calculator may give you the answer in standard form.

9.2

Example 2

The distance from the Sun to the Earth is 150 million kilometres.

Light travels at 3.00×10^5 km/second.

How long does light from the Sun take to reach the Earth?

Distance = 150 000 000 = 1.5×10^8 km

Time = distance ÷ speed

= $(1.5 \times 10^8) \div (3.00 \times 10^5)$

= 500 seconds (Use a calculator to work this out.)

= 8 minutes 20 seconds

EXERCISE 9B

1 These numbers are not in standard form. Write them in standard form.
- a 56.7×10^2
- b 0.06×10^4
- c 34.6×10^{-2}
- d 0.07×10^{-2}
- e 56×10
- f $2 \times 3 \times 10^5$
- g $2 \times 10^2 \times 35$
- h 160×10^{-2}
- i 23 million

2 Work these out. Give your answers in standard form.
- a $2 \times 10^4 \times 5.4 \times 10^3$
- b $1.6 \times 10^2 \times 3 \times 10^4$
- c $2 \times 10^4 \times 6 \times 10^4$
- d $2 \times 10^{-4} \times 5.4 \times 10^3$
- e $1.6 \times 10^{-2} \times 4 \times 10^4$
- f $2 \times 10^4 \times 6 \times 10^{-4}$
- g $7.2 \times 10^{-3} \times 4 \times 10^2$
- h $(5 \times 10^3)^2$
- i $(2 \times 10^{-2})^3$

3 Work these. Give your answers in standard form.
- a $(5.4 \times 10^4) \div (2 \times 10^3)$
- b $(4.8 \times 10^2) \div (3 \times 10^4)$
- c $(1.2 \times 10^4) \div (6 \times 10^4)$
- d $(2 \times 10^{-4}) \div (5 \times 10^3)$
- e $(1.8 \times 10^4) \div (9 \times 10^{-2})$
- f $\sqrt{36 \times 10^{-4}}$

4 A typical adult has about 20 000 000 000 000 red blood cells. Each red blood cell has a mass of about 0.000 000 000 1 g. Write both of these numbers in standard form and work out the total mass of red blood cells in a typical adult.

Chapter 9: Standard form

5 A man puts one grain of rice on the first square of a chess board, two on the second square, four on the third, eight on the fourth and so on.

   a How many grains of rice will he put on the 64th square of the board?

   b How many grains of rice will there be altogether?

   Give your answers in standard form.

> **Advice and Tips**
>
> Compare powers of 2 with the running totals.
>
> By the fourth square you have 15 grains altogether, and $2^4 = 16$.

6 The surface area of the Earth is approximately 3.2×10^8 square kilometres. The area of the Earth's surface that is covered by water is approximately 2.2×10^8 square kilometres.

   a Calculate the area of the Earth's surface *not* covered by water. Give your answer in standard form.

   b What percentage of the Earth's surface is not covered by water?

7 The Moon is a sphere with a radius of 1.74×10^3 kilometres. The formula for working out the surface area of a sphere is:

   surface area = $4\pi r^2$

   Calculate the surface area of the Moon.

8 Evaluate $\dfrac{E}{M}$ when $E = 1.5 \times 10^3$ and $M = 3 \times 10^{-2}$, giving your answer in standard form.

9 Work out the value of $\dfrac{3.2 \times 10^7}{1.4 \times 10^2}$ giving your answer in standard form, correct to 2 significant figures.

10 In one year, British Airways carried 33 million passengers. Of these, 70% passed through Heathrow Airport. On average, each passenger carried 19.7 kg of luggage. Calculate the total mass of the luggage carried by these passengers.

11 In 2013 the world population was approximately 7.14×10^9. In 2014 the world population was approximately 7.24×10^9.

   a By how much did the population rise? Give your answer as an ordinary number.

   b What was the percentage increase?

12 Here are four numbers written in standard form.

   1.6×10^4 4.8×10^6 3.2×10^2 6.4×10^3

   a Work out the smallest answer when two of these numbers are multiplied together.

   b Work out the largest answer when two of these numbers are added together.

   Give your answers in standard form.

13 The mass of Saturn is 5.686×10^{26} tonnes. The mass of the Earth is 6.04×10^{21} tonnes. How many times heavier is Saturn than the Earth? Give your answer in standard form to a suitable degree of accuracy.

14 A number is greater than 100 million and less than 1000 million.

   Write down a possible value of the number, in standard form.

Here are some population figures for some countries.

Country	Population
Tunisia	1.10×10^7
Denmark	5.60×10^6
Senegal	1.39×10^7
Jamaica	2.71×10^6
Mexico	1.19×10^8
India	1.29×10^9

a Which country has the largest population?

b Which two countries have the largest difference in population?

c Find the total population of Senegal, Denmark and Jamaica. Give your answer in standard form to two significant figures.

d Complete this sentence:
The population of Mexico is approximately … times larger than the population of Denmark.

e Complete this sentence:
The population of India is approximately … times larger than the population of Jamaica.

 This table shows the populations and the areas of five different countries.

Country	Population	Area
Russian Federation	1.44×10^8	1.71×10^7
Sri Lanka	2.07×10^7	6.56×10^4
Thailand	6.64×10^7	5.13×10^5
Togo	6.98×10^6	5.68×10^5
Iran	7.74×10^7	1.65×10^6

a Which country has the smallest population?

b Which country has the smallest area?

The population density is the population divided by the area.

c Which country has the largest population density?

d Which country has the smallest population density?

e What fraction of the area of the Russian Federation is the area of Sri Lanka?
Give your answer in the form $\frac{1}{N}$.

Check your progress

Core

- I can understand and write numbers in standard form
- I can calculate using numbers written in standard form

Chapter 10
Applying number and using calculators

Topics	Level	Key words
1 Units of measurement	CORE	metric system, length, mass, volume, capacity
2 Converting between metric units	CORE	centimetre, millimetre, kilometre, gram, kilogram, tonne, litre, millilitre, centilitre
3 Time	CORE	24-hour clock, 12-hour clock, timetable
4 Currency conversions	CORE	exchange rate
5 Using a calculator efficiently	CORE	

In this chapter you will learn how to:

CORE

- Use current units of mass, length, area, volume and capacity in practical situations and express quantities in terms of larger or smaller units. (C5.1 and E5.1)
- Calculate times in terms of the 24-hour and 12-hour clock. (C1.14 and E1.14)
- Read clocks, dials and timetables. (C1.14 and E1.14)
- Calculate using money and convert from one currency to another. (C1.15 and E1.15)
- Use a calculator efficiently. (C1.13 and E1.13)
- Apply appropriate checks of accuracy. (C1.13 and E1.13)

Why this chapter matters

Technology is increasingly important in our lives. It helps us do many things more efficiently than we could without it.

Hundreds of years ago people in different countries had different systems of measurement. They were often based on the human body, for example, the length of people's hands, arms or feet, but they all varied and all had different names.

Now the world has an official standard system of measurement – the metric system. This is especially important for scientists so they can work together worldwide. It is also helpful for everyone who needs to compare lengths, masses, volumes and so on between different countries.

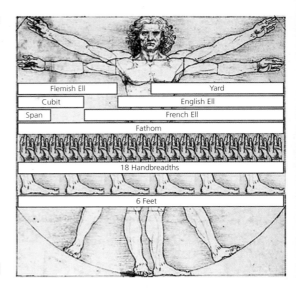

We also have more help now in calculating complicated measurements such as volume. Calculating aids have been used for thousands of years. In about 2000 BCE the abacus was being used in Egypt and China.

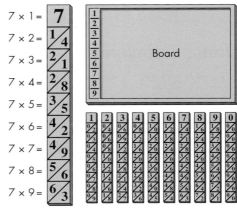

Abacuses are still widely used in China today and they were used everywhere for almost 3500 years, until John Napier devised a calculating aid called **Napier's bones**.

Napier's bones

These led to the invention of the **slide rule** by William Oughtred in 1622. This was in use until the mid-1960s. Engineers working on the first ever moon landings used slide rules to do some of their calculations.

The first **electronic computers** were produced in the mid-20th century. When the transistor was invented, the power increased and the cost and size decreased until the point where the average scientific calculator that students use in schools has more computing power than the first craft that went into space.

Chapter 10: Applying number and using calculators

Chapter 10 . Topic 1

10.1 Units of measurement

The **metric system** of measurement is now used in most countries of the world, except for the United States. Here is a list of the most common metric units.

Unit	How to estimate it
Length	
1 metre	A long stride for an average person
1 kilometre	Two and a half times round a school track
1 centimetre	The distance across a fingernail
Mass	
1 gram	A small coin weighs a few grams
1 kilogram	A bag of sugar
1 tonne	A saloon car
Volume/capacity	
1 litre	A full carton of orange juice
1 centilitre	A small glass is about 10 centilitres
1 millilitre	A full teaspoon is about 5 millilitres

Volume and capacity

The term 'capacity' is normally used to refer to the volume of a liquid or a gas.

For example, when referring to the volume of petrol that a car's fuel tank will hold, people may say its **capacity** is 60 litres.

EXERCISE 10A

1 Decide the metric unit you would be most likely to use to measure each of these.

   a The height of your classroom
   b The distance from Athens to Vienna
   c The thickness of your little finger
   d The mass of this book
   e The amount of water in a fish tank
   f The mass of an aircraft
   g A spoonful of medicine
   h The length of a football pitch
   i The mass of your head teacher
   j The thickness of a piece of wire

2 Estimate the approximate metric length, mass or capacity of each of these.

   a This book (both length and mass)
   b The length of your school hall
   c The capacity of a bucket
   d The diameter of a coin, and its mass
   e The mass of a cat
   f The amount of water in one raindrop
   g The dimensions of the room you are in
   h Your own height and mass

Chapter 10 . Topic 2

10.2

3 Milo was asked to put up some decorative bunting from the top of each lamp post in his street. He had three sets of ladders he could use: a 2 metre, a 3.5 metre and a 5 metre ladder.

He looked at the lamp posts and estimated that they were about three times his height. He is slightly below average height for an adult male.

Which of the ladders should he use? Give a reason for your choice.

10.2 Converting between metric units

You should already know the relationships between these metric units.

Length
10 **millimetres** = 1 **centimetre**
1000 **millimetres** = 100 **centimetres**
= 1 **metre**
1000 **metres** = 1 **kilometre**

Mass
1000 **grams** = 1 **kilogram**
1000 **kilograms** = 1 **tonne**

Capacity
10 **millilitres** = 1 **centilitre**
1000 **millilitres** = 100 **centilitres**
= 1 **litre**

Volume
1000 litres = 1 metre^3
1 millilitre = 1 centimetre^3

Note the equivalence between the units of capacity and volume:

1 litre = 1000 cm^3 which means   1 ml = 1 cm^3

You need to be able to convert from one metric unit to another.

Since the metric system is based on powers of 10, it should be easy for you to multiply or divide to change units. Work through the following examples.

Example 1
Change:

a 732 cm to metres   **b** 840 mm to metres.

To change small units to larger units, always divide.

a 732 ÷ 100 = 7.32 so 732 cm = 7.32 m
b 840 ÷ 1000 = 0.84 so 840 mm = 0.84 m

Chapter 10: Applying number and using calculators   **143**

Example 2

Change:

a 1.2 m to centimetres **b** 0.62 cm to millimetres.

To change large units to smaller units, always multiply.

a 1.2 × 100 = 120 so 1.2 m = 120 cm

b 0.62 × 10 = 6.2 so 0.62 cm = 6.2 mm

EXERCISE 10B

1. Fill in the gaps, using the information in this section.
   - **a** 125 cm = … m
   - **b** 82 mm = … cm
   - **c** 550 mm = … m
   - **d** 4200 g = … kg
   - **e** 5750 kg = … t
   - **f** 85 ml = … cl
   - **g** 755 g = … kg
   - **h** 800 ml = … l
   - **i** 200 cl = … l
   - **j** 1035 l = … m³
   - **k** 530 l = … m³
   - **l** 34 km = … m

2. Fill in the gaps, using the information in this section.
   - **a** 3.4 m = … mm
   - **b** 13.5 cm = … mm
   - **c** 0.67 m = … cm
   - **d** 0.64 km = … m
   - **e** 2.4 l = … ml
   - **f** 5.9 l = … cl
   - **g** 3.75 t = … kg
   - **h** 0.94 cm³ = … l
   - **i** 21.6 l = … cl
   - **j** 15.2 kg = … g
   - **k** 14 m³ = … l
   - **l** 0.19 cm³ = … ml

3. Sarif wanted to buy two lengths of wood, each 2 m long, and 1.5 cm by 2 cm. He went to the local store where the types of wood were described as:

   2000 mm × 15 mm × 20 mm

   200 mm × 15 mm × 20 mm

   200 mm × 150 mm × 2000 mm

   200 mm × 15 mm × 2000 mm

   Should he choose any of these? If so, which one?

4. How many square metres are there in a square kilometre?

   Advice and Tips

   The answer is not 1000.

5. The circumference of Earth is approximately 4×10^7 metres.

   If you travel at 100 kilometres per hour, how many hours does it take to go round planet Earth?

6. In 2017, the population of Earth was estimated to be 7.5 billion people.

   One billion is 1000 million.

   Write the population in standard form.

10.3 Time

Times can be given using the **12-hour** or **24-hour clock**.

The 12-hour clock starts at midnight and runs to 12:00 at midday. After 12:59 it goes back to 1:00 and runs through to 12 again. So 7:45 could be 7:45 in the morning (am) or 7:45 in the evening (pm).

In everyday life you will usually use the 12-hour clock and add 'am' or 'pm' to indicate whether you mean before or after midday.

The 24-hour clock indicates the number of hours and minutes after midnight and has four digits. The first two digits are hours and the last two digits are minutes. So 1:45 pm in the 12-hour clock is 1345 in the 24-hour clock, meaning 13 hours and 45 minutes after midnight. 7:30am is 0730 but 7:30 pm is 1930 and so on.

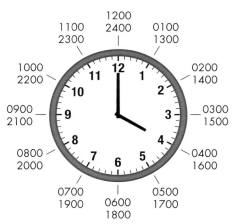

Fifteen minutes after midnight is 0015 in the 24-hour clock and 12:15 am in the 12-hour clock.

Timetables usually use the 24-hour clock to avoid confusion.

    7: 35 am = 0735   1: 42 pm = 1342   9:30 pm = 2130

Example 3

A train left at 1135 and arrived at 1415.

How long did the journey take?

Do not use a calculator for this sort of question. Calculating 1415 − 1135 will not give the correct answer!

Break the journey into sections.

Here is one way:

1135 →(25 minutes)→ 1200 →(2 hours)→ 1400 →(15 minutes)→ 1415

Total time = 2 hours 40 minutes.

EXERCISE 10C

1 Here is the timetable for four trains from Rome to Naples.

Rome (depart)	0900	0927	1027	1045
Naples (arrive)	1010	1130	1236	1230

a How long does each of the journeys take in hours and minutes?

b Which train was the high-speed express?

2 Here are the times of two trains from Rome to Venice.

Rome (depart)	0945	1036
Venice (arrive)	1333	1649

   a Write the four times as 12-hour clock times.
   b Find the length of each journey in hours and minutes.

3 A man arrived at his office at 0835 and left at 1520.
   a His journey home took 45 minutes. What time did he arrive home?
   b His journey to work in the morning took one hour and 20 minutes. What time did he leave home?
   c How long was he in the office?

4 Sunetra went on a coach trip to a forest park.

The coach left at 0830 and returned at 1855.

The journey took 2 hours and 20 minutes each way.
   a What time did Sunetra arrive at the forest park?
   b What time did she leave the forest park?
   c How long did she spend there?

5 a A car left at 0845 and arrived at its destination 3 hours and 25 minutes later.

   What time did it arrive?

   b On the return journey the car left at 1835 and arrived at 2125.

   How long did this return journey take?

6 Here is the timetable for a bus journey.

Lympstone	1729
Exton	1741
Topsham	1757
Digby	1809
Sowton	1823

How long did the journey take from Lympstone to:
   a Exton
   b Digby
   c Sowton?

7 The express train left at 1050 and arrived at 1324.

The slow train left at 1242 and arrived at 1629.

How much faster was the express train?

8 Pierre flew from Paris to Doha on a day when the clocks in Doha were one hour ahead of the clocks in Paris.

He left at 0740 and the flight took 5 hours and 35 minutes.

What was the local time when he arrived in Doha?

9 Boston is 5 hours behind London.

A flight from Boston to London left at 1935 and took 6 hours and 40 minutes.

What time did it arrive in London?

10.4 Currency conversions

Exchange rates are used to convert between one currency and another. They vary all the time, depending on what happens in the world's stock exchanges.

> **Example 4**
> Assume the exchange rate is 1 US dollar = 0.7775 euros.
>
> **a** How many euros is 210 US dollars?
>
> **b** How many dollars is 850 euros?
>
> **a** 210 US dollars = 210 × 0.7775 euros
>
> = 163.275 euros
>
> **b** 850 euros = 850 ÷ 0.7775 US dollars
>
> = 1093.25 US dollars, rounding the answer to 2 decimal places.

EXERCISE 10D

1 The exchange rate is 1 euro = 9.9919 Hong Kong dollars.

Change 320 euros into Hong Kong dollars, giving your answer to 2 decimal places.

2 The exchange rate is 1 Russian rouble = 0.0328 US dollars.

How many US dollars could you get for 5000 Russian roubles?

3 Copy and complete this guide for changing US dollars into euros.

$5.55	$10	$50	$100	$250	$500	$1000
			€77.55			

4 1 US dollar = 85.7 Pakistani rupees.

If a gift costs 3686 Pakistani rupees, how many US dollars is that?

5 The exchange rate is 1 British pound to 1.2128 euros.

  a A flight from London to Paris costs £185.45.
  How many euros is that?

  b A flight from Paris to London costs €209.50.
  How many British pounds is that?

6 This table shows the conversion rates on one day between three currencies.

Currency	euro	US dollar	Japanese yen
1 euro =	1	1.2863	109.7406
1 US dollar =	0.7774	1	85.315
1 Japanese yen =	0.0091	0.0117	1

  a Use the table to convert:

   i 450 US dollars to euros

   ii 225 euros to Japanese yen (Give your answer to the nearest yen.)

   iii 37 000 Japanese yen to US dollars.

  b Here are three amounts: 500 euros, 650 dollars, 54 000 Japanese yen.

  Use the values in the table to put them in order, from smallest to largest, and complete this statement.

  .... < .... < ....

7 On one day the exchange rate is 1 US dollar = 31.885 Taiwan dollars = 46.53 Indian rupees.

  a What is 75 US dollars in Taiwan dollars?

  b What is 75 US dollars in Indian rupees?

  c Which is worth more, a Taiwan dollar or an Indian rupee?

  d Complete this exchange rate: 1 Taiwan dollar = ... Indian rupees.

8 On 1 July 2005 the exchange rate was 1 US dollar = 8.2765 Chinese yuan.

On 1 July 2010 the exchange rate was 1 US dollar = 6.78099 Chinese yuan.

  a How many *fewer* Chinese yuan could you buy for $50 on 1 July 2010 compared to 1 July 2005?

  b Write the exchange rate on 1 July 2010 in the form:
  1 Chinese yuan = ... US dollars.

Chapter 10 . Topic 5

10.5 Using a calculator efficiently

The aim of this topic is to make you aware of some of the keys on your calculator and how to use them to make calculations as efficiently as possible.

Consider the calculation $\dfrac{3.7 + 9.5}{0.38 + 0.16}$

If you just key in to your calculator 3.7 + 9.5 ÷ 0.38 + 0.16 you will not get the correct answer.

One method is to calculate the numerator (that is, 3.7 + 9.5) first, then the denominator and finally divide one by the other. Using the bracket keys, you can do it all in one operation:

(3.7 + 9.5) ÷ (0.38 + 0.16) gives 24.44 (to 2 decimal places)

It is also useful to check whether the answer is reasonable.

$\dfrac{3.7 + 9.5}{0.38 + 0.16} \approx \dfrac{4 + 10}{0.5} = \dfrac{14}{0.5} = 28$ so the answer seems reasonable.

The check should be a calculation you can easily do in your head.

EXERCISE 10E

Use your calculator for this exercise. Try to key in each calculation as one continuous set, without writing down any intermediate values.

1 Work these out.

   a (10 − 2) × 180 ÷ 10

   b 180 − (360 ÷ 5)

2 Work these out.

   a $\dfrac{1}{2} \times (4.6 + 6.8) \times 2.2$

   b $\dfrac{1}{2} \times (2.3 + 9.9) \times 4.5$

3 Work out the value of each of these, when $a = 3.4$, $b = 5.6$, and $c = 8.8$.

   a $2(ab + ac + bc)$

   b $\dfrac{a + b}{c}$

   c $\dfrac{a}{b + c}$

   d $\sqrt{a} + b + c$

4 Work these out, giving your answers to 2 decimal places.

   a $\sqrt{3.2^2 + 1.6^2}$

   b $\sqrt{4.8^2 + 3.6^2}$

5 Work these out.

   a $7.8^3 + 3 \times 7.8$

   b $5.45^3 - 2 \times 5.45 - 40$

6 Do these calculations as efficiently as you can. Check that your answers are sensible.

   a $\dfrac{15.89}{3.24 + 1.86}$

   b $\dfrac{27}{18.1 + 17.95}$

   c $\dfrac{383 + 936}{1.47 + 13.11}$

   d $\dfrac{0.34^2}{0.025^2}$

   e $\sqrt{3.8 + 9.7 \times 2.8}$

   f $\sqrt{32.4^2 - 17.1^2}$

Check your progress

Core

- I can use units of mass, length, area, volume and capacity
- I can convert between units, including units of area and volume
- I can calculate times using the 24-hour or the 12-hour clock
- I can read timetables, clocks and dials
- I can calculate with money and convert from one currency to another
- I can use a calculator efficiently

Examination questions: Number

Past paper questions reproduced by permission of Cambridge Assessment International Education.
Other exam-style questions have been written by the authors.

PAPER 1

1 a Work out $\frac{5}{12}$ of 168. [1]

 b Write $\frac{3}{8}$ as a decimal. [1]

Cambridge International IGCSE Mathematics 0580 *Paper 11 Q4 Oct/Nov 2015*

2 Calculate.

 a $3.2 \times (5.7 - 1.3) + 4.8$ [1]

 b $\sqrt{2.54 - 0.85}$

Cambridge International IGCSE Mathematics 0580 *Paper 11 Q5 Oct/Nov 2015*

3 Pip and Ali share $785 in the ratio Pip : Ali = 4 : 1.

Work out Pip's share. [2]

Cambridge International IGCSE Mathematics 0580 *Paper 11 Q8 Oct/Nov 2015*

4 a Sara works for 28 hours each week.

 She earns $12.45 per hour.
 Calculate how much she earns in one week. [1]

 b Sara invests $750 for 3 years at a rate of 2.4% per year compound interest.
 Calculate the total amount she will have at the end of the 3 years. [3]

Cambridge International IGCSE Mathematics 0580 *Paper 11 Q21 Oct/Nov 2015*

5 The temperature in Berlin is −7°C and the temperature in Istanbul is −3°C.

 a Write down how many degrees colder it is in Berlin than it is in Istanbul. [1]

 b Sydney is 23 degrees warmer than Berlin.
 Write down the temperature in Sydney. [1]

Cambridge International IGCSE Mathematics 0580 *Paper 11 Q2 May/June 2015*

6 a A mass of 300 kg is increased by 8%.

 Work out the increase in mass. [1]

 b Nelson scores 27 out of 40 in a history test.
 Work out his score as a percentage. [1]

Cambridge International IGCSE Mathematics 0580 *Paper 11 Q3 May/June 2015*

7 The total mass of 38 spoons is 1824 g.

Work out the mass of 53 spoons. [2]

Cambridge International IGCSE Mathematics 0580 *Paper 11 Q4 May/June 2015*

Examination questions: Number

8 a Write 30 as a product of its prime factors. [2]

b Find the lowest common multiple (LCM) of 30 and 45. [2]

Cambridge International IGCSE Mathematics 0580 Paper 11 Q21 May/June 2015

9 Write the following in order of size, smallest first. [2]

π 3.14 $\dfrac{22}{7}$ 3.142   3

Cambridge International IGCSE Mathematics 0580 Paper 11 Q6 Oct/Nov 2014

10 a Write down a 2-digit odd number that is a factor of 182. [1]

b Find all the prime factors of 182. [2]

Cambridge International IGCSE Mathematics 0580 Paper 11 Q12 Oct/Nov 2014

11 a Write 2.8×10^2 as an ordinary number. [1]

b Work out $2.5 \times 10^8 \times 2 \times 10^{-2}$

Give your answer in standard form. [2]

Cambridge International IGCSE Mathematics 0580 Paper 11 Q13 Oct/Nov 2014

12 Dominic invests $850 at a rate of 3.5% per year compound interest.

Calculate the **total** amount he has after 3 years. [3]

Cambridge International IGCSE Mathematics 0580 Paper 11 Q17 Oct/Nov 2014

13 $p = \dfrac{4.8 \times 1.98276}{16.83}$

a In the spaces provided, write each number in this calculation correct to 1 significant figure. [1]

$$\dfrac{\ldots\ldots \times \ldots\ldots}{\ldots\ldots}$$

b Use your answer to **part (a)** to to estimate the value of p. [1]

Cambridge International IGCSE Mathematics 0580 Paper 11 Q5 May/June 2014

14 Use your calculator to work out $\sqrt{\dfrac{3}{4}} + 2^{-1}$.

Give your answer correct to 2 decimal places. [2]

Cambridge International IGCSE Mathematics 0580 Paper 11 Q10 May/June 2014

15 Carlo changed 800 euros (€) into dollars for his holiday when the exchange rate was €1 = $1.50.

His holiday was then cancelled.

He changed all his dollars back into euros and he received €750.

Find the new exchange rate. [3]

Cambridge International IGCSE Mathematics 0580 Paper 11 Q15 May/June 2014

16 In this question, do not use your calculator and show all the steps in your working.

a Show that $3\dfrac{1}{5} - 2\dfrac{5}{8} = \dfrac{23}{40}$. [2]

b Work out $\dfrac{7}{8} \div \dfrac{23}{40}$.

Give your answer as a mixed number in its simplest form. [2]

Cambridge International IGCSE Mathematics 0580 Paper 11 Q18 May/June 2014

Examination questions: Number

PAPER 3

1 a Write down a number between 20 and 30 that is
   i   a multiple of 6, [1]
   ii a square number, [1]
   iii a cube number, [1]
   iv a prime number. [1]

  b Find
   i $\sqrt[3]{4913}$, [1]
   ii 3^5, [1]
   iii 6^0, [1]
   iv 2^{-4}. [1]

  c i Write 84 as a product of its prime factors. [2]
   ii Find the highest common factor (HCF) of 84 and 126. [2]

  Cambridge International IGCSE Mathematics 0580 Paper 31 Q2 Oct/Nov 2015

2 A carton of fruit juice contains apple, orange, pineapple and tropical juices.

  a They are mixed in the ratio
   apple : orange : pineapple : tropical = 9 : 7 : 4 : 5.
   The carton contains 540 millilitres of apple juice.
   i   Show that the total amount of fruit juice in the carton is 1.5 **litres**. [3]
   ii Calculate the amount of tropical juice in the carton.
           Give your answer in millilitres. [2]
   iii 70% of the tropical juice is mango.
           Calculate the amount of mango juice in the carton. [2]

  b A shopkeeper pays $36 for 16 cartons.
   i   How much does he pay for one carton? [1]
   ii He sells $\frac{7}{8}$ of the 16 cartons for $3.40 each and the rest for $2.50 each. [2]
           Calculate the total amount he receives from selling the cartons.
   iii Calculate his percentage profit. [3]

  Cambridge International IGCSE Mathematics 0580 Paper 31 Q1 Oct/Nov 2014

3 a Write down
   i   two factors of 12, [1]
   ii the next prime number after 19, [1]
   iii the cube root of 64, [1]
   iv two million five hundred and seven in figures, [1]
   v   two multiples of 75, [1]
   vi the value of π correct to 5 significant figures. [1]

Examination questions: Number

   b Write as a percentage.
   - **i** 1.63 [1]
   - **ii** $\frac{3}{40}$ [1]

   c i Write 63 521.769 correct to 1 decimal place. [1]

   ii Write 63 521.769 correct to the nearest hundred. [1]

   d i Change 234 mm into metres. [1]

   ii Change 876 m² into square centimetres. [1]

Cambridge International IGCSE Mathematics 0580 Paper 31 Q1 May/June 2015

4 a Luka earns $475 each week.
   - **i** He works for 38 hours each week.
     How much does he earn for each hour he works? [1]
   - **ii** Luka pays $175 in rent each week.
     Write the amount he pays in rent as a fraction of his weekly earnings.
     Give your answer in its lowest terms. [2]
   - **iii** He spends $\frac{7}{20}$ of his weekly earnings on bills.
     How much money does he have left after paying rent and bills? [2]

   b Luka's weekly earnings of $475 are increased by 6%.
   Calculate his new weekly earnings. [2]

   c Luka has saved $350.
   He invests this for 2 years at a rate of 4% per year compound interest.
   How much interest does he receive after 2 years? [3]

Cambridge International IGCSE Mathematics 0580 Paper 31 Q2 Oct/Nov 2012

5 a From the integers 50 to 100, find
   - **i** a multiple of 43, [1]
   - **ii** a factor of 165, [1]
   - **iii** an odd number that is also a square number, [1]
   - **iv** a number which is a square number and also a cube number. [1]

   b i Find the square root of 5929. [1]

   ii Find the lowest common multiple of 24 and 30. [2]

   c Elena goes on a journey to the North Pole.
   She leaves home at 7 am on 15 July and arrives at the North Pole at 10 pm on 27 July.
   How long, in days and hours, did her journey take? [2]

Cambridge International IGCSE Mathematics 0580 Paper 31 Q2 May/June 2014

Examination questions: Number

6 Work out the highest common factor (HCF) of 36 and 90. [2]

Cambridge International IGCSE Mathematics 0580 *Paper 21 Q7 Oct/Nov 2015*

7 Hazel invests $1800 for 7 years at a rate of 1.5% per year compound interest.

Calculate how much interest she will receive after the 7 years.
Give your answer correct to the nearest dollar. [4]

Cambridge International IGCSE Mathematics 0580 *Paper 21 Q16 Oct/Nov 2015*

8 Write 30 as a product of its prime factors. [2]

Cambridge International IGCSE Mathematics 0580 *Paper 21 Q17 May/June 2015*

9 On a ship, the price of a gift is 24 euros (€) or $30.

What is the difference in the price on a day when the exchange rate is €1 = $1.2378?
Give your answer in dollars, correct to the nearest cent. [3]

Cambridge International IGCSE Mathematics 0580 *Paper 21 Q8 Oct/Nov 2014*

Examination questions: Number

PAPER 2

1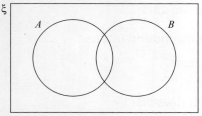

In the Venn diagram shade the region $A \cup B'$. [1]

Cambridge International IGCSE Mathematics 0580 *Paper 21 Q2 Oct/Nov 2015*

2 Write the recurring decimal $0.1\dot{5}$ as a fraction. [2]

[$0.1\dot{5}$ means 0.1555...]

Cambridge International IGCSE Mathematics 0580 *Paper 21 Q10 Oct/Nov 2015*

3 V is directly proportional to the cube of $(r + 1)$.

When $r = 1$, $V = 24$.

Work out the value of V when $r = 2$. [3]

Cambridge International IGCSE Mathematics 0580 *Paper 21 Q12 Oct/Nov 2015*

4 Rice is sold in 75 gram packs and 120 gram packs.

The masses of both packs are given correct to the nearest gram.

Calculate the lower bound for the difference in mass between the two packs. [2]

Cambridge International IGCSE Mathematics 0580 *Paper 21 Q6 May/June 2015*

5 A car travels a distance of 1280 **metres** at an average speed of 64 kilometres per hour.

Calculate the time it takes for the car to travel this distance.

Give your answer in **seconds**. [3]

Cambridge International IGCSE Mathematics 0580 *Paper 21 Q13 May/June 2015*

6 y varies directly with $\sqrt{x + 5}$.

$y = 4$ when $x = -1$.

Find y when $x = 11$. [3]

Cambridge International IGCSE Mathematics 0580 *Paper 21 Q13 Oct/Nov 2014*

7 $y = \dfrac{2}{x^2} + \dfrac{x^2}{2}$

Find the value of y when $x = 6$.

Give your answer as a mixed number in its simplest form. [2]

Cambridge International IGCSE Mathematics 0580 *Paper 21 Q2 May/June 2014*

Examination questions: Number

8 Write down all your working to show that the following statement is correct. [2]

$$\frac{1 + \frac{8}{9}}{2 + \frac{1}{2}} = \frac{34}{45}$$

Cambridge International IGCSE Mathematics 0580 *Paper 21 Q4 Oct/Nov 2012*

9 A large water bottle holds 25 litres of water correct to the nearest litre.

A drinking glass holds 0.3 litres correct to the nearest 0.1 litre.

Calculate the lower bound for the number of glasses of water which can be filled from the bottle. [3]

Cambridge International IGCSE Mathematics 0580 *Paper 21 Q10 Oct/Nov 2012*

10 The electrical resistance, R, of a length of cylindrical wire varies inversely as the square of the diameter, d, of the wire.

$R = 10$ when $d = 2$.

Find R when $d = 4$. [3]

Cambridge International IGCSE Mathematics 0580 *Paper 21 Q11 Oct/Nov 2012*

Examination questions: Number

PAPER 4

1 David sells fruit at the market.

   a In one week, David sells 120 kg of tomatoes and 80 kg of grapes.

      i Write 80 kg as a fraction of the total mass of tomatoes and grapes.

         Give your answer in its lowest terms. [1]

      ii Write down the ratio mass of tomatoes : mass of grapes.

         Give your answer in its simplest form. [1]

   b **i** One day he sells 28 kg of oranges at $1.56 per kilogram.

        He also sells 35 kg of apples.

        The total he receives from selling the oranges and the apples is $86.38.

        Calculate the price of 1 kilogram of apples. [2]

      ii The price of 1 kilogram of oranges is $1.56.

         This is 20% more than the price two weeks ago.

         Calculate the price two weeks ago. [3]

   c On another day, David received a total of $667 from all the fruit he sold.

     The cost of the fruit was $314.20.

     David worked for $10\frac{1}{2}$ hours on this day.

     Calculate David's rate of profit in dollars per hour. [2]

Cambridge International IGCSE Mathematics 0580 Paper 41 Q1 Oct/Nov 2013

2 **a** One day, Maria took 27 minutes to walk 1.8km to school.

     She left home at 07 48.

      i Write down the time Maria arrived at school. [1]

      ii Show that Maria's average walking speed was 4 km/h. [2]

   b Another day, Maria cycled the 1.8 km to school at an average speed of 15 km/h.

      i Calculate the percentage **increase** that 15 km/h is on Maria's walking speed of 4 km/h. [3]

      ii Calculate the percentage **decrease** that Maria's cycling time is on her walking time of 27 minutes. [3]

      iii After school, Maria cycled to her friend's home.

         This took 9 minutes, which was 36% of the time Maria takes to walk to her friend':

         Calculate the time Maria takes to walk to her friend's home. [2]

Cambridge International IGCSE Mathematics 0580 Paper 41 Q1 May/June 2013

3 12000 vehicles drive through a road toll on one day.

   The ratio cars : trucks : motorcycles = 13 : 8 : 3.

   a **i** Show that 6500 cars drive through the road toll on that day. [1]

      ii Calculate the number of trucks that drive through the road toll on that day. [1]

Examination questions: Number

b The toll charges in 2014 are shown in the table.

Vehicle	Charge
Cars	$2
Trucks	$5
Motorcycles	$1

Show that the total amount paid in tolls on that day is $34500. [2]

c This total amount is a decrease of 8% on the total amount paid on the same day in 2013.

Calculate the total amount paid on that day in 2013. [3]

d 2750 of the 6500 car drivers pay their toll using a credit card.

Write down, in its simplest terms, the fraction of car drivers who pay using a credit card. [2]

e To the nearest thousand, 90000 cars drive through the road toll in one week. Write down the lower bound for this number of cars. [1]

Cambridge International IGCSE Mathematics 0580 *Paper 41 Q1 May/June 2015*

4 Distances from the Sun can be measured in astronomical units, AU.

Earth is a distance of 1 AU from the Sun.

One AU is approximately 1.496×10^8 km.

The table shows distances from the Sun.

Name	Distance from the Sun in AU	Distance from the Sun in kilometres
Earth	1	1.496×10^8
Mercury	0.387	....
Jupiter	....	7.79×10^8
Pluto	....	5.91×10^9

a Complete the table. [3]

b Light travels at approximately 300 000 kilometres per second.

  i How long does it take light to travel from the Sun to Earth?

   Give your answer in seconds. [2]

  ii How long does it take light to travel from the Sun to Pluto?

   Give your answer in minutes. [2]

c One light year is the distance that light travels in one year (365 days).

   How far is one light year in kilometres? Give your answer in Standard form. [3]

d How many astronomical units (AU) are equal to one light year? [2]

Cambridge International IGCSE Mathematics 0580 *Paper 41 Q9 Oct/Nov 2012*

Chapter 11
Algebraic representation and formulae

Topics	Level	Key words
1 The language of algebra	CORE	expression, symbol, variable, formula, formulae, equation, term, solving
2 Substitution into formulae	CORE	substitute
3 Rearranging formulae	CORE	rearrange, subject
4 More complicated formulae	EXTENDED	

In this chapter you will learn how to:

CORE	EXTENDED
• Use letters to express generalised numbers and express basic arithmetic processes algebraically. (C2.1 and E2.1) • Substitute numbers for words and letters in formulae. (C2.1 and E2.1) • Rearrange simple formulae. (C2.1) • Construct simple expressions and set up simple equations. (C2.1)	• Construct and transform complicated formulae and equations. (E2.1)

Why this chapter matters

Where is mathematics used? Is it used in:

Art   Science   Sport   Language?

In fact, it is important for them all.

Art

Mathematicians think that famous works of art are often based on the 'golden ratio'. This is the ratio of one part of the art to another. It seems that human brains find the 'golden ratio' very attractive.

Science

Science needs mathematics. In 1962 a space probe went off course because someone had got a mathematical formula wrong in its programming.

Sport

Is mathematics a sport? There are national and international competitions each year that use mathematics. University students compete in the annual 'Mathematics Olympiad' and there is a world Sudoku championship each year. Lots of sporting activities require maths too, such as throwing a javelin (angles).

Language

But the best description of mathematics is that it is a language.

It is the only language which people in all countries understand. Everyone understands the numbers on this stamp even if they do not speak the language of the country.

Algebra is an important part of the language of mathematics. It comes from the Arabic *al-jabr*. It was first used in a book written in 820 CE by a Persian mathematician called al-Khwarizmi.

The use of symbols grew until the 17th century when a French mathematician called Descartes developed them into the sort of algebra we use today.

11.1 The language of algebra

Algebra is based on the idea that if something works with numbers, it will work with letters. The main difference is that when you work only with numbers, the answer is also a number. When you work with letters, you get an **expression** as the answer.

Algebra follows the same rules as arithmetic, and uses the same **symbols** (+, −, × and ÷). Below are seven important algebraic rules.

- Write '4 more than x' as $4 + x$ or $x + 4$.
- Write '6 less than p' or 'p minus 6' as $p - 6$.
- Write '4 times y' as $4 \times y$ or $y \times 4$ or $4y$. The last one of these is the neatest way to write it.
- Write 'b divided by 2' as $b \div 2$ or $\frac{b}{2}$.
- When a number and a letter or a letter and a letter appear together, there is a hidden multiplication sign between them. So, $7x$ means $7 \times x$ and ab means $a \times b$.
- Always write '$1 \times x$' as x.
- Write 't times t' as $t \times t$ or t^2.

Here are some algebraic words that you need to know.

Variable: This is what the letters used to represent numbers are called. These letters can take on any value, so they 'vary'.

Expression: This is any combination of letters and numbers. For example, $2x + 4y$ and $\frac{p-6}{5}$ are expressions.

Equation: An equation contains an equals sign and at least one variable. The important fact is that a value can be found for the variable. This is called **solving** the equation. You will learn more about equations in another chapter.

Formula: These are like equations in that they contain an equals sign, but there is more than one variable and they are rules for working out amounts such as area or the cost of taxi fares. You refer to more than one formula as formulae.

For example, $V = x^3$, $A = \frac{1}{2}bh$ and $C = 3 + 4m$ are all **formulae**.

Term: These are the separate parts of expressions, equations or formula.

For example, in $3x + 2y - 7$, there are three terms: $3x$, $+2y$ and -7.

Example 1

One side of this rectangle is three centimetres longer than the other.

Find a formula for the area (A) in square centimetres, and the perimeter (P) in centimetres.

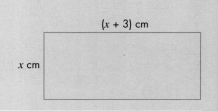

Area = width × length

$A = x(x + 3)$ (Leave out the multiplication sign.)

Perimeter = distance around the outside

$P = x + (x + 3) + x + (x + 3)$

You can simplify this to:

$P = 4x + 6$

EXERCISE 11A

1 Write down the algebraic expression for:

a 2 more than x
b 6 less than x
c k more than x
d x minus t
e x added to 3
f d added to m
g y taken away from b
h p added to t added to w
i 8 multiplied by x
j h multiplied by j
k x divided by 4
l 2 divided by x
m y divided by t
n w multiplied by t
o a multiplied by a
p g multiplied by itself.

2 Asha, Bernice and Charu are three sisters. Bernice is x years old. Asha is three years older than Bernice. Charu is four years younger than Bernice.

a How old is Asha?
b How old is Charu?

3 An approximation method of converting from degrees Celsius to degrees Fahrenheit is given by this rule:

Multiply by 2 and add 30.

Using C to stand for degrees Celsius and F to stand for degrees Fahrenheit, complete this formula.

$F = \ldots\ldots$

4 Cows have four legs. Which of these formulae connects the number of legs (L) and the number of cows (C)?

a $C = 4L$
b $L = C + 4$
c $L = 4C$
d $L + C = 4$

5 a Lakmini has three bags of marbles. Each bag contains *n* marbles. How many marbles does she have altogether?

  b Rushani gives her another three marbles. How many marbles does Lakmini have now?

  c Lakmini puts one of her new marbles in each bag. How many marbles are there now in each bag?

  d Lakmini takes two marbles out of each bag. How many marbles are there now in each bag?

6 Lee has *n* cubes.

- Anil has twice as many cubes as Lee.
- Reza has two more than Lee.
- Dale has three fewer than Lee.
- Chen has three more than Anil.

How many cubes does each person have?

Advice and Tips

Remember that you do not have to write down a multiplication sign between numbers and letters, or letters and letters.

7 a I go shopping with $10 and spend $6. How much do I have left?

  b I go shopping with $10 and spend $*x*. How much do I have left?

  c I go shopping with $*y* and spend $*x*. How much do I have left?

  d I go shopping with $3*x* and spend $*x*. How much do I have left?

8 Give the total cost of:

  a five books at $15 each

  b *x* books at $15 each

  c four books at $*A* each

  d *y* books at $*A* each.

9 A boy went shopping with *A* dollars. He spent *B* dollars. How much did he have left?

10 Five ties together cost $*A*. What is the cost of one tie?

11 a My dad is 72 years old and I am *T* years old. How old shall we each be in *x* years' time?

  b My mum is 64 years old. In two years' time she will be twice as old as I am. What age am I now?

12 I am twice as old as my son. I am *T* years old.

  a How old is my son?

  b How old will my son be in four years' time?

  c How old was I *x* years ago?

13 Write down the perimeter of each of these figures.

a
Square, $2x$

b
Equilateral triangle, $4m$

c
Regular hexagon, $3t$

14 Write down the number of marbles each student ends up with.

Student	Action	Marbles
Andrea	Start with three bags each containing n marbles and give away one marble from each bag	
Barak	Start with three bags each containing n marbles and give away one marble from one bag	
Ahmed	Start with three bags each containing n marbles and give away two marbles from each bag	
Dina	Start with three bags each containing n marbles and give away n marbles from each bag	
Emma	Start with three bags each containing n marbles and give away n marbles from one bag	
Hana	Start with three bags each containing n marbles and give away m marbles from each bag	

15 The answer to $3 \times 4m$ is $12m$.
Write down two *different* expressions for which the answer is $12m$.

11.2 Substitution into formulae

A formula expresses the value of one variable as the other variables in the formula change.

Example 2

The formula for the area of a trapezium is:
$$A = \frac{(a + b)h}{2}$$

Find the area of the trapezium when $a = 5$, $b = 9$ and $h = 3$.

$$A = \frac{(5 + 9) \times 3}{2} = \frac{14 \times 3}{2} = 21$$

Always **substitute** the numbers for the letters before trying to work out the value of the expression. You are less likely to make a mistake this way.

EXERCISE 11B

1 Find the value of $3x + 2$ when:
  a $x = 2$ **b** $x = 5$ **c** $x = 10$.

Advice and Tips

It helps to put the numbers in brackets.
$3(2) + 2 = 6 + 2 = 8$
$3(5) + 2 = 15 + 2 = 17$
etc ...

2 Find the value of $4k - 1$ when:
  a $k = 1$ **b** $k = 3$ **c** $k = 11$.

3 Find the value of $5 + 2t$ when:
  a $t = 2$ **b** $t = 5$ **c** $t = 12$.

4 Evaluate $15 - 2f$ when: **a** $f = 3$ **b** $f = 5$ **c** $f = 8$.

5 Evaluate $5m + 3$ when: **a** $m = 2$ **b** $m = 6$ **c** $m = 15$.

6 Evaluate $3d - 2$ when: **a** $d = 4$ **b** $d = 5$ **c** $d = 20$.

7 A taxi company uses this rule to calculate their fares.

  Fare = $2.50 plus $0.50 per kilometre.

  a How much is the fare for a journey of 3 km?
  b Farook pays $9.00 for a taxi ride. How far was the journey?
  c Mayra knows that her house is 5 kilometres from town. She has $5.50 left in her purse after a night out. Has she got enough for a taxi ride home?

8 Kaz knows that x, y and z have the values 2, 8 and 11, but she does not know which variable has which value.

  a What is the maximum value that the expression $2x + 6y - 3z$ could have?
  b What is the minimum value that the expression $5x - 2y + 3z$ could have?

Advice and Tips

You could just try all combinations, but if you think for a moment you will find that the $6y$ term must give the largest number. This will give you a clue to the other terms.

9 The formula for the area, A, of a rectangle with length l and width w is $A = lw$.

  The formula for the area, T, of a triangle with base b and height h is $T = \frac{1}{2}bh$.

  Find values of l, w, b and h so that $A = T$.

10 Find the value of $\frac{8 \times 4h}{5}$ when: **a** $h = 5$ **b** $h = 10$ **c** $h = 25$.

11 Find the value of $\frac{25 - 3p}{2}$ when: **a** $p = 4$ **b** $p = 8$ **c** $p = 10$.

12 Evaluate $\frac{x}{3}$ when: **a** $x = 6$ **b** $x = 24$ **c** $x = -30$.

13 Evaluate $\frac{A}{4}$ when: **a** $A = 12$ **b** $A = 10$ **c** $A = -20$.

14 Evaluate $\frac{12}{y}$ when: **a** $y = 2$ **b** $y = 4$ **c** $y = 6$.

15 Evaluate $\frac{24}{x}$ when: **a** $x = 2$ **b** $x = 3$ **c** $x = 16$.

Chapter 11 . Topic 3

16 A holiday cottage costs 150 dollars per day to rent.

A group of friends decide to rent the cottage for seven days.

a Which formula represents the cost of the rental for each person if there are n people in the group? Assume that they share the cost equally.

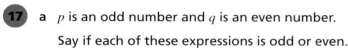
$$\frac{150}{n} \qquad \frac{150}{7n} \qquad \frac{1050}{n} \qquad \frac{150n}{n}$$

Advice and Tips

To check your choice in part **a**, make up some numbers and try them in the formula. For example, take $n = 5$.

b Eventually 10 people go on the holiday. When they get the bill, they find that there is a discount for a seven-day rental.

After the discount, they each find it cost them 12.50 dollars less than they expected.

How much does a 7-day rental cost?

17 a p is an odd number and q is an even number.

Say if each of these expressions is odd or even.

i $p + q$ **ii** $p^2 + q$ **iii** $2p + q$ **iv** $p^2 + q^2$

b x, y and z are all odd numbers.

Write an expression, using x, y and z, so that the value of the expression is always even.

Advice and Tips

There are many answers for b, and a should give you a clue.

18 A formula for the cost of delivery, in dollars, of orders from a warehouse is:

$$D = 2M - \frac{C}{5}$$

where D is the cost of the delivery, M is the distance in kilometres from the store and C is the cost of the goods to be delivered.

Advice and Tips

Note: a rebate is a refund of some of the money that someone has already paid for goods or services.

a How much does the delivery cost when $M = 30$ and $C = 200$?

b Rahim buys goods worth $300 and lives 10 kilometres from the store.

  i The formula gives the cost of delivery as a negative value. What is this value?

  ii Explain why Rahim will not get a rebate from the store.

c Maya buys goods worth $400. She calculates that her cost of delivery will be zero. What is the greatest distance that Maya could live from the store?

11.3 Rearranging formulae

The **subject** of a formula is the **variable** (letter) in the formula that stands on its own, usually on the left-hand side of the equals sign. For example, x is the subject of each of these equations.

$$x = 5t + 4 \qquad x = 4(2y - 7) \qquad x = \frac{1}{t}$$

To change the existing subject to a different variable, you have to **rearrange** the formula to get that variable on the left-hand side.

Example 3

Make m the subject of this formula. $T = m - 3$

Add 3 to both sides. $T + 3 = m$

Reverse the formula. $m = T + 3$

Example 4

From the formula $P = 4t$, express t in terms of P.

(This is another common way of asking you to make t the subject.)

Divide both sides by 4: $\dfrac{P}{4} = \dfrac{4t}{4}$

$\dfrac{P}{4} = t$

Reverse the formula: $t = \dfrac{P}{4}$

Example 5

From the formula $C = 2m^2 + 3$, make m the subject.

Subtract 3 from both sides so that the $2m^2$ is on its own. $C - 3 = 2m^2$

Divide both sides by 2: $\dfrac{C - 3}{2} = \dfrac{2m^2}{2}$

Reverse the formula: $m^2 = \dfrac{C - 3}{2}$

Take the square root on both sides: $m = \sqrt{\dfrac{C - 3}{2}}$

EXERCISE 11C

1. $T = 3k$ Make k the subject.
2. $X = y - 1$ Express y in terms of X.
3. $Q = \dfrac{p}{3}$ Express p in terms of Q.
4. $A = 4r + 9$ Make r the subject.
5. $W = 3n - 1$ Make n the subject.
6. $p = m + t$ a Make m the subject.   b Make t the subject.
7. $g = \dfrac{m}{v}$ Make m the subject.
8. $t = m^2$ Make m the subject.
9. $C = 2\pi r$ Make r the subject.

Advice and Tips

Remember about inverse operations, and the rule 'change sides, change signs'.

Chapter 11 . Topic 4

10 $A = bh$ — Make b the subject.

11 $P = 2l + 2w$ — Make l the subject.

12 $m = p^2 + 2$ — Make p the subject.

13 The formula for converting temperatures in degrees Fahrenheit (F) to temperatures in degrees Celsius (C) is $C = \dfrac{5}{9}(F - 32)$.
  a Show that when $F = -40$, C is also equal to -40.
  b Find the value of C when $F = 68$.
  c Show that the formula can be rearranged as $F = \dfrac{9C}{5} + 32$

14 $v = u + at$ — **a** Make a the subject. **b** Make t the subject.

15 $A = \dfrac{1}{4}\pi d^2$ — Make d the subject.

16 $W = 3n + t$ — **a** Make n the subject. **b** Express t in terms of n and W.

17 $x = 5y - w$ — **a** Make y the subject. **b** Express w in terms of x and y.

18 $k = 2p^2$ — Make p the subject.

19 $v = u^2 - t$ — **a** Make t the subject. **b** Make u the subject.

20 $k = m + n^2$ — **a** Make m the subject. **b** Make n the subject.

21 $T = 5r^2$ — Make r the subject.

22 $K = 5n^2 + w$ — **a** Make w the subject. **b** Make n the subject.

11.4 More complicated formulae

To find the value of a letter you need to make it the subject of the formula.

You often need to rearrange the formula to do this.

Some formulae will need a number of separate steps to rearrange them.

Example 6

$a = c + \dfrac{d}{y^2}$

Make y the subject.

Subtract c from both sides: $\quad a - c = \dfrac{d}{y^2}$

Multiply both sides by y^2: $\quad y^2(a - c) = d$

Divide both sides by $a - c$: $\quad y^2 = \dfrac{d}{a - c}$

Take the square root of y^2: $\quad y = \sqrt{\dfrac{d}{a - c}}$

EXERCISE 11D

1 $a^2 + b^2 = c^2$
  a Find the value of a when $b = 6.0$ and $c = 6.5$.
  b Make a the subject.

2 $s = ut + \frac{1}{2}at^2$ is a formula used in mechanics.
  a Find the value of s when $u = -5$, $t = 4$ and $a = 10$.
  b Make a the subject.

3 $a = \dfrac{b+2}{c}$
  a Make b the subject.
  b Make c the subject.

4 $p = \dfrac{r}{t-3}$
  Make t the subject.

5 $d = \dfrac{12}{1+\sqrt{e}}$
  Make e the subject.

6 $v^2 = u^2 + 2as$
  a Find the value of v when $u = 3$, $a = 2$ and $s = 4$.
  b Make u the subject.
  c Make s the subject.

7 $T = 2\pi\sqrt{\dfrac{L}{G}}$
  a Make L the subject.
  b Show that $G = L\left(\dfrac{2\pi}{T}\right)^2$

8 $D = \pi R^2 - \pi r^2$
  a Make R the subject.
  b Make r the subject.
  c Make π the subject.

9 $3x^2 - 4y^2 = 11$
  a Find the value of x when $y = 4$.
  b Make x the subject.
  c Make y the subject.

10 $T = 2\sqrt{\dfrac{a}{c+3}}$
  a Make a the subject.
  b Make c the subject.

11 $a^2 = b^2 + c^2 - 2bcT$

Make T the subject.

12 $uv = fu + fv$ is a formula used in optics.

   a Find the value of f when $u = 20$ and $v = 30$.

   b Make f the subject.

   c Make u the subject.

   d Make v the subject.

Check your progress

Core

- I can use letters to express arithmetic processes
- I can substitute numbers into formulae
- I can rearrange formulae
- I can use information to set up simple equations

Extended

- I can construct and transform complicated equations and formulae

Chapter 12
Algebraic manipulation

Topics	Level	Key words
1 Simplifying expressions	CORE	simplifying, coefficient, like terms
2 Expanding brackets	CORE	expand, simplification
3 Factorisation	CORE	common factors, factorise, factorisation
4 Multiplying two brackets: 1	CORE	quadratic expression, linear, quadratic expansion
5 Multiplying two brackets: 2	CORE	
6 Expanding three brackets	EXTENDED	
7 Quadratic factorisation	EXTENDED	difference of two squares, brackets
8 Algebraic fractions	EXTENDED	cancel, single fractions

In this chapter you will learn how to:

CORE	EXTENDED
• Manipulate directed numbers. (C2.2 and E2.2) • Use brackets and extract common factors. (C2.2 and E2.2) • Expand products of algebraic expressions. (C2.2 and E2.2)	• Factorise where possible expressions of the form: $ax + bx + kay + kby$, $a^2x^2 - b^2y^2$; $a^2 + 2ab + b^2$; $ax^2 + bx + c$. (E2.2) • Manipulate algebraic fractions such as: $\dfrac{x}{3} + \dfrac{x-4}{2}$, $\dfrac{2x}{3} - \dfrac{3(x-5)}{2}$, $\dfrac{3a}{4} \times \dfrac{5ab}{3}$, $\dfrac{3a}{4} \div \dfrac{9a}{10}$, $\dfrac{1}{x-2} + \dfrac{2}{x-3}$. (E2.3) • Factorise and simplify rational expressions such as: $\dfrac{x^2 - 2x}{x^2 - 5x + 6}$. (E2.3)

Why this chapter matters

How can algebra save your life? Read on...

When you drive a car you must leave a safe distance between you and the car in front. What is a safe distance? How long will it take you to stop?

The **stopping distance** has two parts:

- **thinking distance** is the distance travelled before your brain reacts and you apply the brakes
- **braking distance** is the time it takes the car to come to a complete stop once the brakes have been applied.

Stopping distances vary according to the weather conditions, the road conditions and vehicle. This table shows stopping distances for a car with good brakes on a dry, level road with good visibility.

Typical stopping distances

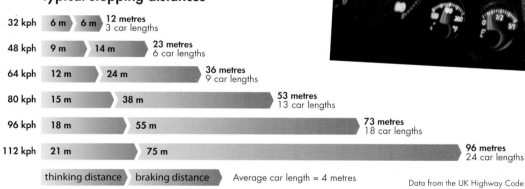

Data from the UK Highway Code

You will notice that the thinking distance increases steadily with speed. It increases by 3 metres for every 16 km/h increase in speed. If you double the speed you double the thinking distance. You can show this in a simple algebraic expression:

thinking distance in metres = $0.1875x$
(where x is the speed in kilometres per hour).

The relationship between speed and braking distance is more complicated. It can be shown as:

braking distance = $0.006x^2$

So:

total stopping distance = $0.1875x + 0.006x^2$

This is a quadratic expression and you can use it to work out stopping distance at any speed (value of x). You will learn about both simple and quadratic expressions in this chapter.

Chapter 12 . Topic 1

12.1 Simplifying expressions

Simplifying an algebraic expression means making it neater and, usually, shorter by combining its terms where possible.

Multiplying expressions

When you multiply algebraic expressions, first you multiply the numbers, then the letters.

> **Example 1**
> Simplify each expression.
>
> **a** $2 \times t$ **b** $m \times t$ **c** $2t \times 5$ **d** $3y \times 2m$
>
> The convention is to write the number first then the letters. The number in front of the letter is called the **coefficient**.
>
> **a** $2 \times t = 2t$ **b** $m \times t = mt$ **c** $2t \times 5 = 10t$ **d** $3y \times 2m = 6my$

You would not normally be penalised for writing $2ba$ instead of $2ab$, but you should never write $ab2$ as this can be confused with powers, so *always* write the number first.

> **Example 2**
> Simplify each expression.
>
> **a** $t \times t$ **b** $3t \times 2t$ **c** $3t^2 \times 4t$ **d** $2t^3 \times 4t^2$
>
> Multiply the same variables, using powers. The indices are added together.
>
> **a** $t \times t = t^2$ (Remember: $t = t^1$) **b** $3t \times 2t = 6t^2$
> **c** $3t^2 \times 4t = 12t^3$ **d** $2t^3 \times 4t^2 = 8t^5$

EXERCISE 12A

1 Simplify these expressions.

a $2 \times 3t$	**b** $5y \times 3$	**c** $2w \times 4$
d $5b \times b$	**e** $2w \times w$	**f** $4p \times 2p$
g $3t \times 2t$	**h** $5t \times 3t$	**i** $m \times 2t$
j $5t \times q$	**k** $n \times 6m$	**l** $3t \times 2q$
m $5h \times 2k$	**n** $3p \times 7r$	

Advice and Tips

Remember to multiply numbers and add indices.

12.1

2 a Which of these expressions are equivalent?

   $2m \times 6n$ $4m \times 3n$ $2m \times 6m$ $3m \times 4n$

b The expressions $2x$ and x^2 have the same value for only two values of x. What are these values?

3 A square and a rectangle have the same area.

The rectangle has sides of length $2x$ cm and $8x$ cm.

What is the length of a side of the square?

4 Simplify these expressions.

- **a** $y^2 \times y$
- **b** $3m \times m^2$
- **c** $4t^2 \times t$
- **d** $3n \times 2n^2$
- **e** $t^2 \times t^2$
- **f** $h^3 \times h^2$
- **g** $3n^2 \times 4n^3$
- **h** $3a^4 \times 2a^3$
- **i** $k^5 \times 4k^2$
- **j** $-t^2 \times -t$
- **k** $-4d^2 \times -3d$
- **l** $-3p^4 \times -5p^2$
- **m** $3mp \times p$
- **n** $3mn \times 2m$
- **o** $4mp \times 2mp$

Collecting like terms

Like terms are those that are multiples of the same variable or of the same combination of variables. For example, a, $3a$, $9a$, $\frac{1}{4}a$ and $-5a$ are all like terms.

So are $2xy$, $7xy$ and $-5xy$, and so are $6x^2$, x^2 and $-3x^2$.

Collecting like terms generally involves two steps.

- Collect like terms into groups.
- Then combine the like terms in each group.

Only like terms can be added or subtracted to simplify an expression. For example:

   $a + 3a + 9a - 5a$ simplifies to $8a$

   $2xy + 7xy - 5xy$ simplifies to $4xy$

Note that the variable does not change. You just have to combine the coefficients.

For example:

   $6x^2 + x^2 - 3x^2 = (6 + 1 - 3)x^2 = 4x^2$

However, an expression such as $4p + 8t + 5x - 9$ cannot be simplified, because $4p$, $8t$, $5x$ and 9 are *not like terms*, which *cannot* be combined.

Chapter 12: Algebraic manipulation

Example 3

Simplify this expression.

$7x^2 + 3y - 6z + 2x^2 + 3z - y + w + 9$

Write out the expression: $7x^2 + 3y - 6z + 2x^2 + 3z - y + w + 9$

Then collect like terms: $\boxed{7x^2 + 2x^2}$ $\boxed{+3y - y}$ $\boxed{-6z + 3z}$ $\boxed{+w}$ $\boxed{+9}$

Then combine them: $9x^2$ $+$ $2y$ $-$ $3z$ $+ w + 9$

So, the expression in its simplest form is:

$9x^2 + 2y - 3z + w + 9$

EXERCISE 12B

1. Jared is given $t, Jatan has $3 more than Jared, and Jasmin has $2t.
   a How much more money has Jasmin than Jared?
   b How much do the three of them have altogether?

2. Write down an expression for the perimeter of each of these shapes.

   a 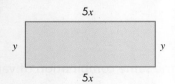   b 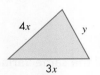   c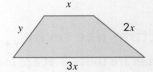

3. Write each of these expressions in a shorter form.

   a $a + a + a + a + a$ b $c + c + c + c + c + c$
   c $4e + 5e$ d $f + 2f + 3f$
   e $5j + j - 2j$ f $9q - 3q - 3q$
   g $3r - 3r$ h $2w + 4w - 7w$
   i $5x^2 + 6x^2 - 7x^2 + 2x^2$ j $8y^2 + 5y^2 - 7y^2 - y^2$
   k $2z^2 - 2z^2 + 3z^2 - 3z^2$

   Advice and Tips

   The term a has a coefficient of 1, i.e. $a = 1a$, but you do not need to write the 1.

4. Simplify each of these expressions.

   a $3x + 4x$ b $5t - 2t$
   c $-2x - 3x$ d $-k - 4k$
   e $m^2 + 2m^2 - m^2$ f $2y^2 + 3y^2 - 5y^2$

   Advice and Tips

   Remember that only **like terms** can be added or subtracted.

   If all the terms cancel out, just write 0 rather than $0x^2$, for example.

5. Simplify each of these expressions.

   a $5x + 8 + 2x - 3$ b $7 - 2x - 1 + 7x$
   c $4p + 2t + p - 2t$ d $8 + x + 4x - 2$
   e $3 + 2t + p - t + 2 + 4p$ f $5w - 2k - 2w - 3k + 5w$
   g $a + b + c + d - a - b - d$ h $9k - y - 5y - k + 10$

6 Simplify these expressions. (Be careful – two of them will not simplify.)

a $c + d + d + d + c$
b $2d + 2e + 3d$
c $f + 3g + 4h$
d $5u - 4v + u + v$
e $4m - 5n + 3m - 2n$
f $3k + 2m + 5p$
g $2v - 5w + 5w$
h $2w + 4y - 7y$
i $5x^2 + 6x^2 - 7y + 2y$
j $8y^2 + 5z - 7z - 9y^2$
k $2z^2 - 2x^2 + 3x^2 - 3z^2$

7 Find the perimeter of each of these shapes, giving it in its simplest form.

a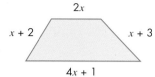

b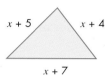

c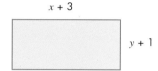

8 $3x + 5y + 2x - y = 5x + 4y$

Write down two other *different* expressions that are equal to $5x + 4y$.

9 Find the missing terms to make these equations true.

a $4x + 5y + \ldots\ldots - \ldots\ldots = 6x + 3y$
b $3a - 6b - \ldots\ldots + \ldots\ldots = 2a + b$

10 ABCDEF is an L-shape.

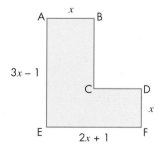

$AB = DF = x$
$AE = 3x - 1$ and $EF = 2x + 1$

Advice and Tips

Make sure your explanation uses expressions. Do not try to explain in words alone.

a Explain why the length $BC = 2x - 1$.
b Find the perimeter of the shape in terms of x.
c If $x = 2.5$ cm, what is the perimeter of the shape?

Chapter 12: Algebraic manipulation

11 A teacher asks her class to work out the perimeter of this L-shape.

Tia says: 'There is information missing so you cannot work out the perimeter.'

Maria says: 'The perimeter is $4x - 1 + 4x - 1 + 3x + 2 + 3x + 2$.'

Who is correct?

Explain your answer.

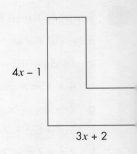

12.2 Expanding brackets

In mathematics, to '**expand**' usually means 'multiply out'. For example, expressions such as $3(y + 2)$ and $4y^2(2y + 3)$ can be expanded by multiplying them out.

Remember that there is an invisible multiplication sign between the outside number and the opening bracket. So $3(y + 2)$ is really $3 \times (y + 2)$ and $4y^2(2y + 3)$ is really $4y^2 \times (2y + 3)$.

You expand by multiplying *everything inside* the brackets by what is outside the brackets.

So in the case of the two examples above:

$3(y + 2) = 3 \times (y + 2) = 3 \times y + 3 \times 2 = 3y + 6$

$4y^2(2y + 3) = 4y^2 \times (2y + 3) = 4y^2 \times 2y + 4y^2 \times 3 = 8y^3 + 12y^2$

Look at these next examples of expansion, which show clearly how the term outside the brackets has been multiplied with the terms inside them.

$2(m + 3) = 2m + 6$ $y(y^2 - 4x) = y^3 - 4xy$

$3(2t + 5) = 6t + 15$ $3x^2(4x + 5) = 12x^3 + 15x^2$

$m(p + 7) = mp + 7m$ $-3(2 + 3x) = -6 - 9x$

$x(x - 6) = x^2 - 6x$ $-2x(3 - 4x) = -6x + 8x^2$

$4t(t^3 + 2) = 4t^4 + 8t$ $3t(2 + 5t - p) = 6t + 15t^2 - 3pt$

Note: The signs change when the quantity outside the brackets is negative. For example:

$a(b + c) = ab + ac$ $a(b - c) = ab - ac$

$-a(b + c) = -ab - ac$ $-a(b - c) = -ab + ac$

$-(a - b) = -a + b$ $-(a + b - c) = -a - b + c$

Note: A minus sign on its own in front of the brackets actually means -1, so:

$-(x + 2y - 3) = -1 \times (x + 2y - 3) = -1 \times x + -1 \times 2y + -1 \times -3 = -x - 2y + 3$

The effect of a minus sign outside the brackets is to change the sign of everything inside the brackets.

EXERCISE 12C

1 Expand these expressions.

- **a** $2(3 + m)$
- **b** $5(2 + l)$
- **c** $3(4 - y)$
- **d** $4(5 + 2k)$
- **e** $3(2 - 4f)$
- **f** $2(5 - 3w)$
- **g** $5(2k + 3m)$
- **h** $4(3d - 2n)$
- **i** $t(t + 3)$
- **j** $k(k - 3)$
- **k** $4t(t - 1)$
- **l** $2k(4 - k)$
- **m** $4g(2g + 5)$
- **n** $5h(3h - 2)$
- **o** $y(y^2 + 5)$
- **p** $h(h^3 + 7)$
- **q** $k(k^2 - 5)$
- **r** $3t(t^2 + 4)$
- **s** $3d(5d^2 - d^3)$
- **t** $3w(2w^2 + t)$
- **u** $5a(3a^2 - 2b)$
- **v** $3p(4p^3 - 5m)$
- **w** $4h^2(3h + 2g)$
- **x** $2m^2(4m + m^2)$

2 The local shop is offering $1 off the price of a large tin of biscuits. Maddox wants five tins.

a If the price of one tin is $t, which of the expressions below represents how much it will cost Maddox to buy five tins?

$5(t - 1)$ $5t - 1$ $t - 5$ $5t - 5$

b Maddox has $20 to spend. Will he have enough money for five tins?
Let $t = \$4.50$. Show working to justify your answer.

3 Dylan wrote the following.

$3(5x - 4) = 8x - 4$

Dylan has made two mistakes.

Explain the mistakes that Dylan has made.

> **Advice and Tips**
>
> It is not enough to give the right answer. You must try to explain why Dylan wrote 8 for 3 × 5 instead of 15.

4 The expansion $2(x + 3) = 2x + 6$ can be shown by this diagram.

	x	3
2	$2x$	6

a What expansion is shown in this diagram?

3	$6y$	9

b Write down an expansion that is shown on this diagram.

	$12z$	8

Simplification

Simplification means writing down an expression as simply as possible, combining any like terms. Like terms are terms that have the same letter(s) raised to the same power and can differ only in their numerical coefficients (numbers in front). For example:

m, $3m$, $4m$, $-m$ and $76m$ are all like terms in m

t^2, $4t^2$, $7t^2$, $-t^2$, $-3t^2$ and $98t^2$ are all like terms in t^2

pt, $5tp$, $-2pt$, $7pt$, $-3tp$ and $103pt$ are all like terms in pt.

Note: All the terms in tp are also like terms to all the terms in pt.

When simplifying an expression, you can only add or subtract like terms. For example:

$4m + 3m = 7m$

$4h - h = 3h$

$2m + 6 + 3m = 5m + 6$

$3ab + 2ba = 5ab$

$10g - 4 - 3g = 7g - 4$

$3y + 4y + 3 = 7y + 3$

$2t^2 + 5t^2 = 7t^2$

$7t + 8 - 2t = 5t + 8$

$5k - 2k = 3k$

Expand and simplify

When you expand the product of two brackets there are often like terms that you can collect together. You should always simplify algebraic expressions as much as possible.

Example 4

Simplify the expression $3(4 + m) + 2(5 + 2m)$.

$3(4 + m) + 2(5 + 2m) = 12 + 3m + 10 + 4m = 22 + 7m$

Example 5

Simplify the expression $3t(5t + 4) - 2t(3t - 5)$.

$3t(5t + 4) - 2t(3t - 5) = 15t^2 + 12t - 6t^2 + 10t = 9t^2 + 22t$

EXERCISE 12D

1. Simplify these expressions.

   a $4t + 3t$

   b $3d + 2d + 4d$

   c $5e - 2e$

   d $3t - t$

   e $2t^2 + 3t^2$

   f $6y^2 - 2y^2$

   g $3ab + 2ab$

   h $7a^2d - 4a^2d$

2. Find the missing terms to make these equations true.

   a $4x + 5y + \ldots - \ldots = 6(x - y)$

   b $3a - 6b - \ldots + \ldots = 2(a + b)$

3. The length of AB is 3 cm less than twice the length of AD.

   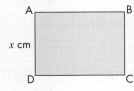

   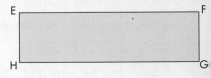

   a Write an expression for the length of AB, in centimetres.

   EF is twice as long as AB.

   EH is 2 cm shorter than AD.

   b Find an expression for the perimeter of EFGH. Give your answer in cm.

4 Expand and simplify each expression.

  a $3(4 + t) + 2(5 + t)$
  b $5(3 + 2k) + 3(2 + 3k)$
  c $4(3 + 2f) + 2(5 - 3f)$
  d $5(1 + 3g) + 3(3 - 4g)$

5 Expand and simplify each expression.

  a $4(3 + 2h) - 2(5 + 3h)$
  b $5(3g + 4) - 3(2g + 5)$
  c $5(5k + 2) - 2(4k - 3)$
  d $4(4e + 3) - 2(5e - 4)$

Advice and Tips

Be careful with minus signs. For example, $-2(5e - 4) = -10e + 8$

6 Expand and simplify each expression.

  a $m(4 + p) + p(3 + m)$
  b $k(3 + 2h) + h(4 + 3k)$
  c $4r(3 + 4p) + 3p(8 - r)$
  d $5k(3m + 4) - 2m(3 - 2k)$

7 Expand and simplify each expression.

  a $t(3t + 4) + 3t(3 + 2t)$
  b $2y(3 + 4y) + y(5y - 1)$
  c $4e(3e - 5) - 2e(e - 7)$
  d $3k(2k + p) - 2k(3p - 4k)$

8 Expand and simplify each expression.

  a $4a(2b + 3c) + 3b(3a + 2c)$
  b $3y(4w + 2t) + 2w(3y - 4t)$
  c $5m(2n - 3p) - 2n(3p - 2m)$
  d $2r(3r + r^2) - 3r^2(4 - 2r)$

9 Fill in whole-number values so that this expansion is true.

  $3(\ldots x + \ldots y) + 2(\ldots x + \ldots y) = 11x + 17y$

Advice and Tips

There is more than one answer. You don't have to give them all.

10 A rectangle with sides 5 and $3x + 2$ has a smaller rectangle with sides 3 and $2x - 1$ cut from it.

Work out the remaining area.

Advice and Tips

Write out the expression for the difference between the areas of the two rectangles and then work it out.

12.3 Factorisation

Factorisation is the opposite of expansion. When you factorise an expression you put it back into the brackets it may have come from.

In factorisation, you have to look for the common factors in *every* term of the expression.

Example 6

Factorise each expression.

a $6t + 9m$ b $6my + 4py$ c $5k^2 - 25k$ d $10a^2b - 15ab^2$

a First look at the numerical coefficients 6 and 9.

These have a common factor of 3.

Then look at the letters, t and m.

Neither of these is a common factor as they do not appear in both terms.

You can think of the expression as $3 \times 2t + 3 \times 3m$, which gives the factorisation:

$6t + 9m = 3(2t + 3m)$

Note: You can always check a factorisation by expanding the answer.

b First look at the numbers.

These have a common factor of 2. m and p do not occur in both terms but y does, and is a common factor, so the factorisation is:

$6my + 4py = 2y(3m + 2p)$

c 5 is a common factor of 5 and 25 and k is a common factor of k^2 and k.

$5k^2 - 25k = 5k(k - 5)$

d 5 is a common factor of 10 and 15, a is a common factor of a^2 and a, b is a common factor of b and b^2.

$10a^2b - 15ab^2 = 5ab(2a - 3b)$

Note: If you multiply out each answer, you will get the expressions you started with.

EXERCISE 12E

1 Factorise these expressions.

a $6m + 12t$ b $9t + 3p$
c $8m + 12k$ d $4r + 8t$
e $mn + 3m$ f $5g^2 + 3g$
g $4w - 6t$ h $3y^2 + 2y$

i $4t^2 - 3t$
j $3m^2 - 3mp$
k $6p^2 + 9pt$
l $8pt + 6mp$
m $8ab - 4bc$
n $5b^2c - 10bc$
o $8abc + 6bde$
p $4a^2 + 6a + 8$
q $6ab + 9bc + 3bd$
r $5t^2 + 4t + at$
s $6mt^2 - 3mt + 9m^2t$
t $8ab^2 + 2ab - 4a^2b$
u $10pt^2 + 15pt + 5p^2t$

2 Three friends have a meal together. They each have a main meal costing $6.75 and a dessert costing $3.25.

Carla says that the bill will be 3 × 6.75 + 3 × 3.25.

Suni says that she has an easier way to work out the bill as 3 × (6.75 + 3.25).

a Explain why Carla's and Suni's methods both give the correct answer.
b Explain why Suni's method is better.   c What is the total bill?

3 Factorise these expressions where possible. List those that do not factorise.

a $7m - 6t$
b $5m + 2mp$
c $t^2 - 7t$
d $8pt + 5ab$
e $4m^2 - 6mp$
f $a^2 + b$
g $4a^2 - 5ab$
h $3ab + 4cd$
i $5ab - 3b^2c$

4 Three students are asked to factorise the expression $12m - 8$. These are their answers.

Ahmed	Bernice	Craig
$2(6m - 4)$	$4(3m - 2)$	$4m(3 - \frac{2}{m})$

All the answers are accurately factorised, but only one is the normally accepted answer.

a Which student gave the correct answer?
b Explain why the other two students' answers are not acceptable as correct answers.

5 Explain why $5m + 6p$ cannot be factorised.

6 Alvin has correctly factorised the top and bottom of an algebraic fraction and cancelled out the terms to give a final answer of $2x$. Unfortunately some of his work has had coffee spilt on it. What was the original fraction?

$$\frac{4x}{2} = \frac{4x}{2(x-3)} = 2x$$

Chapter 12. Topic 4

7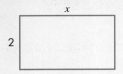

These shapes are made from rectangles x cm long and 2 cm wide.

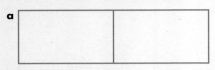

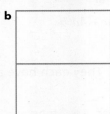

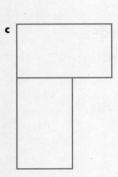

Find an expression for the perimeter of each shape in centimetres.

Factorise each expression as much as possible.

12.4 Multiplying two brackets: 1

A **quadratic expression** is one in which the highest power of the variables is 2.

For example:

y^2 $3t^2 + 5t$ $5m^2 + 3m + 8$

An expression such as $(3y + 2)(4y - 5)$ can be expanded to give a quadratic expression.

Multiplying out such pairs of brackets is usually called **quadratic expansion**.

The rule for expanding expressions such as $(t + 5)(3t - 4)$ is similar to that for expanding single brackets: multiply everything in one set of brackets by everything in the other set of brackets.

12.4

There are several methods for doing this. Examples 7 to 9 show the three main methods: expansion, FOIL and the box method.

In the expansion method, split the terms in the first set of brackets, make each of them multiply both terms in the second set of brackets, then simplify the outcome.

Example 7

Expand $(x + 3)(x + 4)$.

$$(x + 3)(x + 4) = x(x + 4) + 3(x + 4)$$
$$= x^2 + 4x + 3x + 12$$
$$= x^2 + 7x + 12$$

FOIL stands for First, Outer, Inner and Last. This is the order of multiplying the terms from each set of brackets.

Example 8

Expand $(t + 5)(t - 2)$.

First terms give: $t \times t = t^2$
Outer terms give: $t \times -2 = -2t$
Inner terms give: $5 \times t = 5t$
Last terms give: $+5 \times -2 = -10$

$$(t + 5)(t - 2) = t^2 - 2t + 5t - 10$$
$$= t^2 + 3t - 10$$

The 'box method' can be used to lay out the multiplication.

Example 9

Expand $(k - 3)(k - 2)$.

$$(k - 3)(k - 2) = k^2 - 2k - 3k + 6$$
$$= k^2 - 5k + 6$$

	k	-3
k	k^2	$-3k$
-2	$-2k$	$+6$

Warning: Be careful with the signs. This is the main place where mistakes are made in questions involving the expansion of brackets.

Chapter 12: Algebraic manipulation

EXERCISE 12F

Expand the expressions in questions **1–17**.

1 $(x + 3)(x + 2)$
2 $(t + 4)(t + 3)$
3 $(w + 1)(w + 3)$
4 $(m + 5)(m + 1)$
5 $(k + 3)(k + 5)$
6 $(a + 4)(a + 1)$
7 $(x + 4)(x - 2)$
8 $(t + 5)(t - 3)$
9 $(w + 3)(w - 1)$
10 $(f + 2)(f - 3)$
11 $(g + 1)(g - 4)$
12 $(y + 4)(y - 3)$
13 $(x - 3)(x + 4)$
14 $(p - 2)(p + 1)$
15 $(k - 4)(k + 2)$
16 $(y - 2)(y + 5)$
17 $(a - 1)(a + 3)$

Advice and Tips

A common error is to get minus signs wrong.
$-2x - 3x = -5x$ but
$-2x \times -3 = +6x$

The expansions of the expressions in questions **18–26** follow a pattern. Work out the first few and try to spot the pattern that will allow you immediately to write down the answers to the rest.

18 $(x + 3)(x - 3)$
19 $(t + 5)(t - 5)$
20 $(m + 4)(m - 4)$
21 $(t + 2)(t - 2)$
22 $(y + 8)(y - 8)$
23 $(p + 1)(p - 1)$
24 $(5 + x)(5 - x)$
25 $(7 + g)(7 - g)$
26 $(x - 6)(x + 6)$

27 This rectangle is made up of four parts with areas of x^2, $2x$, $3x$ and 6 square units.

Work out expressions for the sides of the rectangle, in terms of x.

28 This square has an area of x^2 square units.
It is split into four rectangles.

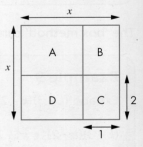

a Copy and complete the table below to show the dimensions and area of each rectangle.

Rectangle	Length	Width	Area
A	$x - 1$	$x - 2$	$(x - 1)(x - 2)$
B			
C			
D			

b Add together the areas of rectangles B, C and D.
Expand any brackets and collect terms together.

c Use the results to explain why $(x - 1)(x - 2) = x^2 - 3x + 2$.

12.4 Multiplying two brackets: 1

29 a Expand $(x - 3)(x + 3)$.

  b Use the result in a to write down the answers to these. (Do not use a calculator or do a long multiplication.)

  i 97×103 ii 197×203

12.5 Multiplying two brackets: 2

All the algebraic 'squared' terms (such as x^2) in Exercise 12F have a **coefficient** of 1 or −1. The next two examples show what to do if you have to expand brackets containing terms with coefficients that are not 1 or −1.

Example 10

Expand $(2t + 3)(3t + 1)$.

$(2t + 3)(3t + 1) = 6t^2 + 2t + 9t + 3$

$= 6t^2 + 11t + 3$

	$2t$	$+3$
$3t$	$6t^2$	$+9t$
$+1$	$+2t$	$+3$

Example 11

Expand $(4x - 1)(3x - 5)$.

$(4x - 1)(3x - 5) = 4x(3x - 5) - (3x - 5)$ [**Note:** $(3x - 5)$ is the same as $1(3x - 5)$.]

$= 12x^2 - 20x - 3x + 5$

$= 12x^2 - 23x + 5$

EXERCISE 12G

Expand the expressions in questions **1–21**.

1 $(2x + 3)(3x + 1)$

2 $(3y + 2)(4y + 3)$

3 $(3t + 1)(2t + 5)$

4 $(4t + 3)(2t - 1)$

5 $(5m + 2)(2m - 3)$

6 $(4k + 3)(3k - 5)$

7 $(3p - 2)(2p + 5)$

8 $(5w + 2)(2w + 3)$

9 $(2a - 3)(3a + 1)$

10 $(4r - 3)(2r - 1)$

11 $(3g - 2)(5g - 2)$

12 $(4d - 1)(3d + 2)$

Advice and Tips

Always give answers in the form $\pm ax^2 \pm bx \pm c$ even if the quadratic coefficient is negative.

13 $(5 + 2p)(3 + 4p)$ **14** $(2 + 3t)(1 + 2t)$ **15** $(4 + 3p)(2p + 1)$

16 $(6 + 5t)(1 - 2t)$ **17** $(4 + 3n)(3 - 2n)$ **18** $(2 + 3f)(2f - 3)$

19 $(3 - 2q)(4 + 5q)$ **20** $(1 - 3p)(3 + 2p)$ **21** $(4 - 2t)(3t + 1)$

22 Expand these expressions.

a $(x + 1)(x - 1)$

b $(2x + 1)(2x - 1)$

c $(2x + 3)(2x - 3)$

d Use the results in parts **a**, **b** and **c** to write down the expansion of $(3x + 5)(3x - 5)$.

23 a Without expanding the brackets, match each expression on the left with an expression on the right. One is done for you.

$(3x - 2)(2x + 1)$ $4x^2 - 4x + 1$

$(2x - 1)(2x - 1)$ $6x^2 - x - 2$

$(6x - 3)(x + 1)$ $6x^2 + 7x + 2$

$(4x + 1)(x - 1)$ $6x^2 + 3x - 3$

$(3x + 2)(2x + 1)$ $4x^2 - 3x - 1$

b Taking any expression on the left, explain how you can match it with an expression on the right without expanding the brackets.

EXERCISE 12H

Try to spot the pattern in each of the expressions in questions **1–15** so that you can immediately write down the expansion.

1 $(2x + 1)(2x - 1)$ **2** $(3t + 2)(3t - 2)$ **3** $(5y + 3)(5y - 3)$

4 $(4m + 3)(4m - 3)$ **5** $(2k - 3)(2k + 3)$ **6** $(4h - 1)(4h + 1)$

7 $(2 + 3x)(2 - 3x)$ **8** $(5 + 2t)(5 - 2t)$ **9** $(6 - 5y)(6 + 5y)$

10 $(a + b)(a - b)$ **11** $(3t + k)(3t - k)$ **12** $(2m - 3p)(2m + 3p)$

13 $(5k + g)(5k - g)$ **14** $(ab + cd)(ab - cd)$ **15** $(a^2 + b^2)(a^2 - b^2)$

16 Imagine a square of side a units with a square of side b units cut from one corner.

a What is the area remaining after the small square is cut away?

b The remaining area is cut into rectangles, A, B and C, and rearranged as shown.

Write down the dimensions and area of the rectangle formed by A, B and C.

c Explain why $a^2 - b^2 = (a + b)(a - b)$.

17 Explain why the areas of the shaded regions are the same.

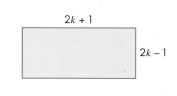

Expanding squares

Whenever you see a **linear** bracketed term squared you must write the brackets down twice and then use whichever method you prefer to expand.

Example 12
Expand $(x + 3)^2$.

$(x + 3)^2 = (x + 3)(x + 3)$
$= x(x + 3) + 3(x + 3)$
$= x^2 + 3x + 3x + 9$
$= x^2 + 6x + 9$

Example 13
Expand $(3x - 2)^2$.

$(3x - 2)^2 = (3x - 2)(3x - 2)$
$= 9x^2 - 6x - 6x + 4$
$= 9x^2 - 12x + 4$

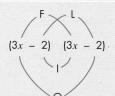

EXERCISE 12I

Expand the squares in questions **1–24** and simplify.

Advice and Tips

Remember always write down the brackets twice. Do not try to take any short cuts.

1. $(x + 5)^2$
2. $(m + 4)^2$
3. $(6 + t)^2$
4. $(3 + p)^2$
5. $(m - 3)^2$
6. $(t - 5)^2$
7. $(4 - m)^2$
8. $(7 - k)^2$
9. $(3x + 1)^2$
10. $(4t + 3)^2$
11. $(2 + 5y)^2$
12. $(3 + 2m)^2$
13. $(4t - 3)^2$
14. $(3x - 2)^2$
15. $(2 - 5t)^2$
16. $(6 - 5r)^2$
17. $(x + y)^2$
18. $(m - n)^2$
19. $(2t + y)^2$
20. $(m - 3n)^2$
21. $(x + 2)^2 - 4$
22. $(x - 5)^2 - 25$
23. $(x + 6)^2 - 36$
24. $(x - 2)^2 - 4$

25 A teacher asks her class to expand $(3x + 1)^2$.

Marcela's answer is $9x^2 + 1$.

Paulo's answer is $3x^2 + 6x + 1$.

  a Explain the mistakes that Marcela has made.

  b Explain the mistakes that Paulo has made.

  c Work out the correct answer.

26 Use the diagram to show algebraically and diagrammatically that:

$(2x - 1)^2 = 4x^2 - 4x + 1$

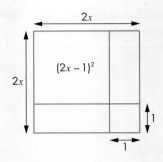

12.6 Expanding three brackets

Here is an expression with three brackets. $(x + 2)(x - 4)(2x + 3)$

You can expand three brackets in two stages.

First multiply two brackets. You can use **FOIL** or the box method.

$(x + 2)(x - 4) = x^2 + 2x - 4x - 8$

$ = x^2 - 2x - 8$

	x	$+2$
x	x^2	$+2x$
-4	$-4x$	-8

Now multiply this expression by the third bracket.

If you use the box method you will have six cells in the second box.

To use a similar method to FOIL, you multiply each term in the first bracket by each term in the second bracket and then simplify.

$(2x + 3)(x^2 - 2x - 8) = 2x^3 - 4x^2 - 16x$

$ + 3x^2 - 6x - 24$

$ = 2x^3 - x^2 - 22x - 24$

	x^2	$-2x$	-8
$2x$	$2x^3$	$-4x^2$	$-16x$
$+3$	$+3x^2$	$-6x$	-24

You can multiply the brackets in any order. For example, you could start with the second and third and then multiply the result by the first. You will get the same answer.

EXERCISE 12J

1 Expand
  a $(x + 3)(x - 1)$
  b $x(x + 3)(x - 1)$
  c $(x + 2)(x + 3)(x - 1)$

2 Expand
  a $(x - 2)(x - 5)$
  b $(x + 1)(x - 2)(x - 5)$
  c $(2x + 1)(x - 2)(x - 5)$

3 Expand
  a $(x - 1)(x + 3)(x - 5)$
  b $(x + 2)(x + 7)(3x + 4)$
  c $(x - 8)(x - 1)(x - 5)$

4 Expand
  a $(x + 2)^2$
  b $(x + 2)^3$
  c $(2x + 1)^3$

5 Expand
  a $(x + 2)(x - 2)(x + 1)$
  b $(2x - 1)(x + 2)(x - 3)$
  c $(x - 2)(x + 4)(x + 2)$

6 Expand
  a $(x + 1)(x + 2)(x + 3)$
  b $(x - 1)(x - 2)(x - 3)$

7 Expand
  a $(x + 3)^2 (x - 2)$
  b $(x - 5)^2 (x + 4)$
  c $(3x - 2)^2 (x + 10)$

8 a Show that $(x + 1)^3 - (x - 1)^3 = 2(3x^2 + 1)$
  b Find a similar expression for $(x + 2)^3 - (x - 2)^3$

9 A cube with side $x + 1$ is cut into eight pieces.

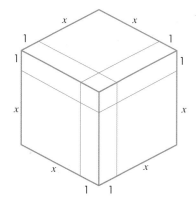

Explain how this illustrates the expansion of $(x + 1)^3$

10 $(x + 1)(x - 2)(x + a) = x^3 + bx^2 + cx - 12$

Find the values of a, b and c

11 a Expand $(x^2 + x + 1)(x - 1)$
  b Expand $(x^2 + 2x + 4)(x - 2)$
  c Hence write $x^3 - 27$ as the product of two brackets.

Chapter 12 . Topic 7

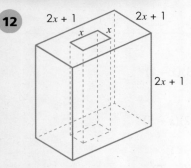

Each side of a wooden cube is $2x + 1$ cm

A hole with a square cross-section and a side of x cm is cut through the cube.

Find an expression for the amount of wood remaining.

12.7 Quadratic factorisation

Factorisation involves putting a quadratic expression back into its **brackets** (if possible). Start with the factorisation of quadratic expressions of the type:

$$x^2 + ax + b$$

where a and b are integers.

Sometimes it is easy to put a quadratic expression back into its brackets, at other times it seems hard. However, there are some simple rules that will help you to factorise.

- The expression inside each set of brackets will start with an algebraic unknown such as x, and the signs in the quadratic expression show which signs to put after the unknowns.
- When the second sign in the expression is a plus, the signs in both sets of brackets are the same as the first sign.

    $x^2 + ax + b = (x + ?)(x + ?)$ Since everything is positive.

    $x^2 - ax + b = (x - ?)(x - ?)$ Since negative × negative = positive

- When the *second* sign is a *minus*, the signs in the brackets are *different*.

    $x^2 + ax - b = (x + ?)(x - ?)$ Since positive × negative = negative

    $x^2 - ax - b = (x + ?)(x - ?)$

- Next, look at the *last* number, b, in the expression. When multiplied together, the two numbers in the brackets must give b.
- Finally, look at the coefficient *of x*, which is a. The *sum* of the two *numbers* in the brackets will give a.

12.7

Example 14

Factorise $x^2 - x - 6$.

Because of the signs you know the brackets must be $(x + ?)(x - ?)$.

Two numbers that have a product of -6 and a sum of -1 are -3 and $+2$.

So, $x^2 - x - 6 = (x + 2)(x - 3)$

Example 15

Factorise $x^2 - 9x + 20$.

Because of the signs you know the brackets must be $(x - ?)(x - ?)$.

Two numbers that have a product of $+20$ and a sum of -9 are -4 and -5.

So, $x^2 - 9x + 20 = (x - 4)(x - 5)$

EXERCISE 12K

Factorise the expressions in questions **1–40**.

1. $x^2 + 5x + 6$
2. $t^2 + 5t + 4$
3. $m^2 + 7m + 10$
4. $k^2 + 10k + 24$
5. $p^2 + 14p + 24$
6. $r^2 + 9r + 18$
7. $w^2 + 11w + 18$
8. $x^2 + 7x + 12$
9. $a^2 + 8a + 12$
10. $k^2 + 10k + 21$
11. $f^2 + 22f + 21$
12. $b^2 + 20b + 96$
13. $t^2 - 5t + 6$
14. $d^2 - 5d + 4$
15. $g^2 - 7g + 10$
16. $x^2 - 15x + 36$
17. $c^2 - 18c + 32$
18. $t^2 - 13t + 36$
19. $y^2 - 16y + 48$
20. $j^2 - 14j + 48$
21. $p^2 - 8p + 15$
22. $y^2 + 5y - 6$
23. $t^2 + 2t - 8$
24. $x^2 + 3x - 10$
25. $m^2 - 4m - 12$
26. $r^2 - 6r - 7$
27. $n^2 - 3n - 18$
28. $m^2 - 7m - 44$
29. $w^2 - 2w - 24$
30. $t^2 - t - 90$
31. $h^2 - h - 72$
32. $t^2 - 2t - 63$
33. $d^2 + 2d + 1$
34. $y^2 + 20y + 100$
35. $t^2 - 8t + 16$
36. $m^2 - 18m + 81$
37. $x^2 - 24x + 144$
38. $d^2 - d - 12$
39. $t^2 - t - 20$
40. $q^2 - q - 56$

Advice and Tips

First decide on the signs in the brackets, then look at the numbers.

41 This rectangle is made up of four parts. Two of the parts have areas of x^2 and 6 square units.

The sides of the rectangle are of the form $x + a$ and $x + b$.

There are two possible answers for a and b.

Work out both answers and copy and complete the areas in the other parts of the rectangle.

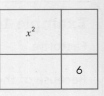

Difference of two squares

In Exercise 12H, you multiplied out, for example, $(a + b)(a - b)$ and obtained $a^2 - b^2$. This type of quadratic expression has only two terms, both of which are perfect squares separated by a minus sign. This type of expression is called the **difference of two squares**. You should have found that all the expansions in Exercise 12H involved the differences of two squares.

The exercise illustrates a system of factorisation that will *always* work for the difference of two squares such as these.

$$x^2 - 9 \qquad x^2 - 25 \qquad x^2 - 4 \qquad x^2 - 100$$

There are three conditions that must be met if the difference of two squares works.

- There must be two terms.
- They must separated by a negative sign.
- Each term must be a perfect square, such as x^2 and n^2.

When these three conditions are met, the factorisation is:

$$x^2 - n^2 = (x + n)(x - n)$$

Example 16

Factorise $x^2 - 36$.

- Recognise the difference of two squares x^2 and 6^2.
- So it factorises to $(x + 6)(x - 6)$.

Expanding the brackets shows that they do come from the original expression.

Example 17

Factorise $9x^2 - 169$.

- Recognise the difference of two squares $(3x)^2$ and 13^2.
- So it factorises to $(3x + 13)(3x - 13)$.

EXERCISE 12L

Each of the expressions in questions **1–9** is the difference of two squares. Factorise them.

1. $x^2 - 9$
2. $t^2 - 25$
3. $m^2 - 16$
4. $9 - x^2$
5. $49 - t^2$
6. $k^2 - 100$
7. $4 - y^2$
8. $x^2 - 64$
9. $t^2 - 81$

10. **a** A square has a side of x units.

    What is the area of the square?

    b A rectangle, A, 2 units wide, is cut from the square and placed at the side of the remaining rectangle, B.

    A square, C, is then cut from the bottom of rectangle A to leave a final rectangle, D.

    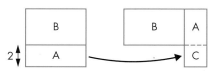
    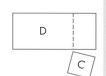

       i What is the height of rectangle B?

       ii What is the width of rectangle D?

       iii What is the area of rectangle B?

       iv What is the area of square C? plus rectangle A?

    c By working out the area of rectangle D, explain why $x^2 - 4 = (x + 2)(x - 2)$.

11. **a** Expand and simplify: $(x + 2)^2 - (x + 1)^2$.

    b Factorise: $a^2 - b^2$.

    c In your answer for part **b**, replace a with $(x + 2)$ and b with $(x + 1)$. Expand and simplify the answer.

    d What can you say about the answers to parts **a** and **c**?

    e Simplify: $(x + 1)^2 - (x - 1)^2$.

Each of the expressions in questions **12–20** is the difference of two squares. Factorise them.

12. $x^2 - y^2$
13. $x^2 - 4y^2$
14. $x^2 - 9y^2$

15 $9x^2 - 1$

16 $16x^2 - 9$

17 $25x^2 - 64$

18 $4x^2 - 9y^2$

19 $9t^2 - 4w^2$

20 $16y^2 - 25x^2$

Factorising $ax^2 + bx + c$

We can adapt the method for factorising $x^2 + ax + b$ to take into account the factors of the coefficient of x^2.

Example 18

Factorise $3x^2 + 8x + 4$.

- First, note that both signs are positive. So the signs in the brackets must be $(?x + ?)(?x + ?)$.
- As 3 has only 3 and 1 as factors, the brackets must be $(3x + ?)(x + ?)$.
- Next, note that 4 can be factorised as 4×1 and 2×2.
- Now find which pair of factors of 4 combine with 3 and 1 to give 8.

    3   4 ②
    ①   1 ②

    You can see that the combination 3×2 and 1×2 adds up to 8.

- So, the complete factorisation becomes $(3x + 2)(x + 2)$.

Example 19

Factorise $6x^2 - 7x - 10$.

- First, note that both signs are negative. So the signs in the brackets must be $(?x + ?)(?x - ?)$.
- As 6 can be factorised as 6×1 and 3×2, the brackets could be $(6x \pm ?)(x \pm ?)$ or $(3x \pm ?)(2x \pm ?)$.
- Next, note that 10 can be factorised as 5×2 and 1×10.
- Now find which pair of factors of 10 combine with the factors of 6 to give -7.

    3 ⑥   ±1   ⟨±2⟩
    2 ①   ±10 ⟨±5⟩

You can see that the combination 6×-2 and 1×5 adds up to -7.

- So, the complete factorisation becomes $(6x + 5)(x - 2)$.

Although this seems to be very complicated, it becomes quite easy with practice and experience.

Chapter 12. Topic 8

EXERCISE 12M

Factorise the expressions in questions **1–12**.

1. $2x^2 + 5x + 2$
2. $7x^2 + 8x + 1$
3. $4x^2 + 3x - 7$
4. $24t^2 + 19t + 2$
5. $15t^2 + 2t - 1$
6. $16x^2 - 8x + 1$
7. $6y^2 + 33y - 63$
8. $4y^2 + 8y - 96$
9. $8x^2 + 10x - 3$
10. $6t^2 + 13t + 5$
11. $3x^2 - 16x - 12$
12. $7x^2 - 37x + 10$

13. This rectangle is made up of four parts, with areas of $12x^2$, $3x$, $8x$ and 2 square units.

    Work out expressions for the sides of the rectangle, in terms of x.

14. Three students are asked to factorise the expression $6x^2 + 30x + 36$. These are their answers.

Adam	Bella	Cara
$(6x + 12)(x + 3)$	$(3x + 6)(2x + 6)$	$(2x + 4)(3x + 9)$

    All the answers are correctly factorised.

    a Explain why one quadratic expression can have three different factorisations.

    b Which of these is the most complete factorisation?

    $2(3x + 6)(x + 3)$

    $6(x + 2)(x + 3)$

    $3(x + 2)(2x + 6)$

    Explain your choice.

    c What is the geometrical significance of the answers to parts **a** and **b**?

12.8 Algebraic fractions E

The following four rules are used to work out the value of fractions.

Addition: $\quad \dfrac{a}{b} + \dfrac{c}{d} = \dfrac{ad + bc}{bd}$

Subtraction: $\quad \dfrac{a}{b} - \dfrac{c}{d} = \dfrac{ad - bc}{bd}$

Multiplication: $\dfrac{a}{b} \times \dfrac{c}{d} = \dfrac{ac}{bd}$

Note that a, b, c and d can be numbers, other letters or algebraic expressions. Remember:

- use brackets, if necessary
- **factorise** if you can
- **cancel** if you can.

Example 20

Simplify each expression.

a $\dfrac{1}{x} + \dfrac{x}{2y}$

b $\dfrac{2}{b} - \dfrac{a}{2b}$

a Using the addition rule: $\dfrac{1}{x} + \dfrac{x}{2y} = \dfrac{(1)(2y) + (x)(x)}{(x)(2y)} = \dfrac{2y + x^2}{2xy}$

b Using the subtraction rule: $\dfrac{2}{b} - \dfrac{a}{2b} = \dfrac{(2)(2b) - (a)(b)}{(b)(2b)} = \dfrac{4b - ab}{2b^2}$

$$= \dfrac{b(4 - a)}{2b^2} = \dfrac{4 - a}{2b}$$

Note: There are different ways of working out fraction calculations. Part **b** could have been done by making the denominator of each fraction the same.

$$\dfrac{(2)2}{(2)b} - \dfrac{a}{2b} = \dfrac{4 - a}{2b}$$

Example 21

Simplify each expression.

a $\dfrac{x}{3} \times \dfrac{x + 2}{x - 2}$

b $\dfrac{x}{3} \div \dfrac{2x}{7}$

a Using the multiplication rule: $\dfrac{x}{3} \times \dfrac{x + 2}{x - 2} = \dfrac{(x)(x + 2)}{(3)(x - 2)} = \dfrac{x^2 + 2x}{3x - 6}$

Remember that the line that separates the top from the bottom of an algebraic fraction acts as brackets as well as a division sign. Note that it is sometimes preferable to leave an algebraic fraction in a factorised form.

b Using the division rule: $\dfrac{x}{3} \div \dfrac{2x}{7} = \dfrac{(x)(7)}{(3)(2x)} = \dfrac{7}{6}$

Example 22

Write $\dfrac{3}{x - 1} - \dfrac{2}{x + 1}$ as a **single fraction** as simply as possible.

Using the subtraction rule:

$$\dfrac{3}{x - 1} - \dfrac{2}{x + 1} = \dfrac{3(x + 1) - 2(x - 1)}{(x - 1)(x + 1)}$$

$$= \dfrac{3x + 3 - 2x + 2}{(x - 1)(x + 1)}$$

$$= \dfrac{x + 5}{(x - 1)(x + 1)}$$

12.8

Example 23
Simplify this expression. $\dfrac{2x^2 + x - 3}{4x^2 - 9}$

Factorise the numerator and denominator: $\dfrac{(2x + 3)(x - 1)}{(2x + 3)(2x - 3)}$

Denominator is the difference of two squares.

Cancel any common factors: $\dfrac{\cancel{(2x + 3)}(x - 1)}{\cancel{(2x + 3)}(2x - 3)}$

If at this stage there isn't a common factor on top and bottom, you should check your factorisations.

The remaining fraction is the answer: $\dfrac{(x - 1)}{(2x - 3)}$

EXERCISE 12N

1 Simplify each of these.

a $\dfrac{x}{2} + \dfrac{x}{3}$
b $\dfrac{3x}{4} + \dfrac{x}{5}$
c $\dfrac{3x}{4} + \dfrac{2x}{5}$
d $\dfrac{x}{2} + \dfrac{y}{3}$
e $\dfrac{xy}{4} + \dfrac{2}{x}$
f $\dfrac{x + 1}{2} + \dfrac{x + 2}{3}$
g $\dfrac{2x + 1}{2} + \dfrac{3x + 1}{4}$
h $\dfrac{x}{5} + \dfrac{2x + 1}{3}$
i $\dfrac{x - 2}{2} + \dfrac{x - 3}{4}$
j $\dfrac{x - 4}{4} + \dfrac{2x - 3}{2}$

2 Simplify each of these.

a $\dfrac{x}{2} - \dfrac{x}{3}$
b $\dfrac{3x}{4} - \dfrac{x}{5}$
c $\dfrac{3x}{4} - \dfrac{2x}{5}$
d $\dfrac{x}{2} - \dfrac{y}{3}$
e $\dfrac{xy}{4} - \dfrac{2}{y}$
f $\dfrac{x + 1}{2} - \dfrac{x + 2}{3}$
g $\dfrac{2x + 1}{2} - \dfrac{3x + 1}{4}$
h $\dfrac{x}{5} - \dfrac{2x + 1}{3}$
i $\dfrac{x - 2}{2} - \dfrac{x - 3}{4}$
j $\dfrac{x - 4}{4} - \dfrac{2x - 3}{2}$

3 Simplify each of these.

a $\dfrac{x}{2} \times \dfrac{x}{3}$
b $\dfrac{2x}{7} \times \dfrac{3y}{4}$
c $\dfrac{4x}{3y} \times \dfrac{2y}{x}$
d $\dfrac{4y^2}{9x} \times \dfrac{3x^2}{2y}$
e $\dfrac{x}{2} \times \dfrac{x - 2}{5}$
f $\dfrac{x - 3}{15} \times \dfrac{5}{2x - 6}$

EXTENDED

Chapter 12: Algebraic manipulation

g $\dfrac{2x+1}{2} \times \dfrac{3x+1}{4}$ h $\dfrac{x}{5} \times \dfrac{2x+1}{3}$

i $\dfrac{x-2}{2} \times \dfrac{4}{x-3}$ j $\dfrac{x-5}{10} \times \dfrac{5}{x^2-5x}$

4 Simplify each of these. Factorise and cancel where appropriate.

a $\dfrac{3x}{4} + \dfrac{x}{4}$ b $\dfrac{3x}{4} - \dfrac{x}{4}$

c $\dfrac{3x}{4} \times \dfrac{x}{4}$ d $\dfrac{3x}{4} \div \dfrac{x}{4}$

e $\dfrac{3x+1}{2} + \dfrac{x-2}{5}$ f $\dfrac{3x+1}{2} - \dfrac{x-2}{5}$

g $\dfrac{3x+1}{2} \times \dfrac{x-2}{5}$ h $\dfrac{x^2-9}{10} \times \dfrac{5}{x-3}$

i $\dfrac{2x^2}{9} - \dfrac{2y^2}{3}$

5 Write these expressions as single fractions as simply as possible.

a $\dfrac{2}{x+1} + \dfrac{5}{x+2}$ b $\dfrac{4}{x-2} + \dfrac{7}{x+1}$

c $\dfrac{3}{4x+1} - \dfrac{4}{x+2}$ d $\dfrac{2}{2x-1} - \dfrac{6}{x+1}$

e $\dfrac{3}{2x-1} - \dfrac{4}{3x-1}$

6 For homework a teacher asks her class to simplify the expression $\dfrac{x^2-x-2}{x^2+x-6}$.

This is Tom's answer:

$$\dfrac{\cancel{x^2-x-2}^{-1}}{\cancel{x^2+x-6}_{+3}}$$

$$= \dfrac{-x-1}{x+3} = \dfrac{x+1}{x+3}$$

When she marked the homework, the teacher was in a hurry and only checked the answer, which was correct.

Tom made several mistakes. What are they?

7 An expression of the form $\dfrac{ax^2+bx-c}{dx^2-e}$ simplifies to $\dfrac{x-1}{2x-3}$.

What was the original expression?

8 Write these expressions as single fractions.

a $\dfrac{4}{x+1} + \dfrac{5}{x+2}$ b $\dfrac{18}{4x-1} - \dfrac{1}{x+1}$ c $\dfrac{2x-1}{2} - \dfrac{6}{x+1}$ d $\dfrac{3}{2x-1} - \dfrac{4}{3x-1}$

9 Simplify these expressions.

a $\dfrac{x^2+2x-3}{2x^2+7x+3}$ b $\dfrac{4x^2-1}{2x^2+5x-3}$ c $\dfrac{6x^2+x-2}{9x^2-4}$ d $\dfrac{4x^2+x-3}{4x^2-7x+3}$ e $\dfrac{4x^2-25}{8x^2-22x+5}$

Check your progress

Core
- I can simplify expressions
- I can expand brackets
- I can find common factors and factorise

Extended
- I can expand expressions with two sets of brackets
- I can expand expressions with three sets of brackets
- I can work out the value of fractions using algebraic expressions

Chapter 13
Solutions of equations and inequalities

Topics	Level	Key words
1 Solving linear equations	CORE	equation, variable, solution, brackets, solve
2 Setting up equations	CORE	
3 Solving quadratic equations by factorisation	EXTENDED	linear equations, quadratic expressions, quadratic equation, factors, difference of two squares
4 Solving quadratic equations by the quadratic formula	EXTENDED	quadratic formula, coefficients, constant, soluble
5 Solving quadratic equations by completing the square	EXTENDED	completing the square
6 Simultaneous equations	CORE	simultaneous linear equations, eliminate, substitute
7 Linear and non-linear simultaneous equations	EXTENDED	linear, non-linear
8 Solving inequalities	EXTENDED	inequality

In this chapter you will learn how to:

CORE	EXTENDED
• Derive and solve simple linear equations in one unknown. (C2.5 and E2.5) • Derive and solve simultaneous linear equations in two unknowns. (C2.5 and E2.5)	• Derive and solve simultaneous equations, involving one linear and one quadratic. (E2.5) • Derive and solve quadratic equations by factorisation, completing the square and by use of the formula. (E2.5) • Derive and solve linear inequalities. (E2.5)

Why this chapter matters

We use equations to explain some of the most important things in the world.

Three of the most important are shown on this page.

Why does the Moon keep orbiting the Earth and not fly off into space?

This is explained by Newton's law of universal gravitation, which describes the gravitational attraction between two bodies:

$$F = G \times \frac{m_1 \times m_2}{r^2}$$

where F is the force between the bodies, G is the gravitational constant, m_1 and m_2 are the masses of the two bodies and r is the distance between them.

Why don't planes fall out of the sky?

This is explained by Bernoulli's principle, which states that as the speed of a fluid increases, its pressure decreases. This is what causes the difference in air pressure between the top and bottom of an aircraft wing, as shown in the diagram on the left.

In its simplest form, the equation can be written as:

$$p + q = p_0$$

where p = static pressure, q = dynamic pressure and p_0 is the total pressure.

How can a small amount of plutonium have enough energy to wipe out a city?

This is explained by Einstein's theory of special relativity, which connects mass and energy in the equation:

$$E = mc^2$$

where E is the energy, m is the mass and c is the speed of light. As the speed of light is nearly 300 000 kilometres per second, the amount of energy in a small mass is huge. If this can be released, it can be used for good (as in nuclear power stations) or harm (as in nuclear bombs).

Chapter 13 . Topic 1
13.1 Solving linear equations

A teacher gave these instructions to her class.

What algebraic expression represents the teacher's statement?

- Think of a number.
- Double it.
- Add 3.

This is what two of her students said.

Can you work out Kim's answer and the number that Freda started with?

Kim's answer will be:

$2 \times 5 + 3 = 13$.

I chose the number 5.

Kim

Freda's answer can be set up as an **equation**.

An equation is formed when an expression is put equal to a number or another expression. You should be able to deal with equations that have only one **variable** or letter.

The **solution** to an equation is the value of the variable that makes the equation true.
For example, the equation for Freda's answer is

$2x + 3 = 10$

where x represents Freda's number.

The value of x that makes this true is $x = 3\frac{1}{2}$.

My final answer was 10.

Freda

To **solve** an equation such as $2x + 3 = 10$, do the same thing to each side of the equation until you have x on its own.

$$2x + 3 = 10$$

Subtract 3 from both sides: $\quad 2x + 3 - 3 = 10 - 3$

$$2x = 7$$

Divide both sides by 2: $\quad \dfrac{2x}{2} = \dfrac{7}{2}$

$$x = 3\frac{1}{2}$$

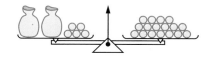

Here is another example.

Mary had two bags, each of which contained the same number of marbles. She also had five spare marbles.

She put the two bags and the five spare marbles on scales and balanced them with 17 single marbles.

How many marbles were there in each bag?

If x is the number of marbles in each bag, then the equation representing Mary's balanced scales is:

$$2x + 5 = 17$$

Take five marbles from each pan:

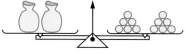

$$2x + 5 - 5 = 17 - 5$$
$$2x = 12$$

Now halve the number of marbles on each pan.

That is, divide both sides by 2:

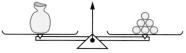

$$\frac{2x}{2} = \frac{12}{2}$$
$$x = 6$$

Checking the answer gives $2 \times 6 + 5 = 17$, which is correct.

Example 1

Solve each of these equations by 'doing the same to both sides'.

a $3x - 5 = 16$

b $\frac{x}{2} + 2 = 10$

a Add 5 to both sides

$$3x - 5 + 5 = 16 + 5$$
$$3x = 21$$

Divide both sides by 3.

$$\frac{3x}{3} = \frac{21}{3}$$
$$x = 7$$

Checking the answer gives:

$$3 \times 7 - 5 = 16$$

which is correct.

b Subtract 2 from both sides

$$\frac{x}{2} + 2 - 2 = 10 - 2$$
$$\frac{x}{2} = 8$$

Multiply both sides by 2.

$$\frac{x}{2} \times 2 = 8 \times 2$$
$$x = 16$$

Checking the answer gives:

$$16 \div 2 + 2 = 10$$

which is correct.

EXERCISE 13A

1 Solve each of these equations by 'doing the same to both sides'. Remember to check that each answer works for its original equation.

a $x + 4 = 60$ **b** $3y - 2 = 4$

c $3x - 7 = 11$ **d** $5y + 3 = 18$

e $7 + 3t = 19$ **f** $5 + 4f = 15$

g $3 + 6k = 24$ **h** $4x + 7 = 17$ **i** $5m - 3 = 17$

j $\dfrac{w}{3} - 5 = 2$ **k** $\dfrac{x}{8} + 3 = 12$ **l** $\dfrac{m}{7} - 3 = 5$

m $\dfrac{x}{5} + 3 = 3$ **n** $\dfrac{h}{7} + 2 = 1$ **o** $\dfrac{w}{3} + 10 = 4$

p $\dfrac{x}{3} - 5 = 7$ **q** $\dfrac{y}{2} - 13 = 5$ **r** $\dfrac{f}{6} - 2 = 8$

s $\dfrac{x+3}{2} = 5$ **t** $\dfrac{t-5}{2} = 3$ **u** $\dfrac{3x+10}{2} = 8$

v $\dfrac{2x+1}{3} = 5$ **w** $\dfrac{5y-2}{4} = 3$ **x** $\dfrac{6y+3}{9} = 1$

y $\dfrac{2x-3}{5} = 4$ **z** $\dfrac{5t+3}{4} = 1$

Advice and Tips

Be careful with negative numbers.

2

Teacher: Think of a number. Divide it by 3 and subtract 6.

Mandy: My answer is −1.

Andy: My starting number is 6.

a What answer did Andy get?

b What number did Mandy start with?

3 A teacher asked her class to solve the equation $2x - 1 = 7$.

Mustafa wrote:

$2x - 1 = 7$
$2x - 1 - 1 = 7 - 1$
$2x = 6$
$2x - 2 = 6 - 2$
$x = 4$

Elif wrote:

$2x - 1 = 7$
$2x - 1 + 1 = 7 + 1$
$2x = 8$
$2x \div 2 = 8 \div 2$
$x = 4$

When the teacher read out the correct answer of 4, both students ticked their work as correct.

a Which student used the correct method?

b Explain the mistakes the other student made.

13.1

Brackets

When you have an equation that contains **brackets**, you first must multiply out the brackets and then solve the resulting equation.

Example 2

Solve the equation $5(x + 3) = 25$.

First multiply out the brackets to get:

$$5x + 15 = 25$$

Subtract 15: $5x = 25 - 15 = 10$

Divide by 5: $\dfrac{5x}{5} = \dfrac{10}{5}$

$$x = 2$$

An alternative method is to divide by the number outside the brackets.

Example 3

Solve the equation $3(2x - 7) = 15$.

Divide both sides by 3: $2x - 7 = 5$

Add 7: $2x = 12$

Divide by 2: $x = 6$

Make sure you can use both methods.

EXERCISE 13B

 Solve each of these equations. Some of the answers may be decimals or negative numbers. Remember to check that each answer works for its original equation. Use your calculator if necessary.

a $2(x + 5) = 16$
b $5(x - 3) = 20$
c $3(t + 1) = 18$
d $4(2x + 5) = 44$
e $2(3y - 5) = 14$
f $5(4x + 3) = 135$
g $4(3t - 2) = 88$
h $6(2t + 5) = 42$
i $2(3x + 1) = 11$
j $4(5y - 2) = 42$
k $6(3k + 5) = 39$

Advice and Tips

Once the brackets have been expanded the equations become straightforward. Remember to multiply *everything* inside the brackets with what is outside.

Chapter 13: Solutions of equations and inequalities

l $5(2x + 3) = 27$
m $9(3x - 5) = 9$
n $2(x + 5) = 6$
o $5(x - 4) = -25$
p $3(t + 7) = 15$
q $2(3x + 11) = 10$
r $4(5t + 8) = 12$

2 Fill in values for a, b and c so that the answer to this equation is $x = 4$.

$a(bx + 3) = c$

3 My son is x years old. In five years' time, I will be twice his age and both our ages will be multiples of 10. The sum of our ages will be between 55 and 100. How old am I now?

Advice and Tips

Set up an equation and put it equal to 60, 70, 80, etc. Solve the equation and see if the answer fits the conditions.

Equations with the variable on both sides

When a letter (or variable) appears on both sides of an equation, collect all the terms containing the letter on one side. This is usually the left-hand side of the equation. When there are more of the letters on the right-hand side, it may be easier to turn the equation round. When an equation contains brackets, you must multiply them out first.

Example 4

Solve this equation. $\qquad 5x + 4 = 3x + 10$

There are more xs on the left-hand side, so leave the equation as it is.

Subtract $3x$ from both sides: $\quad 2x + 4 = 10$

Subtract 4 from both sides: $\quad 2x = 6$

Divide both sides by 2: $\quad x = 3$

Example 5

Solve this equation. $\qquad 2x + 3 = 6x - 5$

There are more xs on the right-hand side, so turn the equation round.

$$6x - 5 = 2x + 3$$

Subtract $2x$ from both sides: $\quad 4x - 5 = 3$

Add 5 to both sides: $\quad 4x = 8$

Divide both sides by 4: $\quad x = 2$

Example 6

Solve this equation.

$3(2x + 5) + x = 2(2 - x) + 2$

Multiply out both brackets: $6x + 15 + x = 4 - 2x + 2$

Simplify both sides: $7x + 15 = 6 - 2x$

There are more xs on the left-hand side, so leave the equation as it is.

Add $2x$ to both sides: $9x + 15 = 6$

Subtract 15 from both sides: $9x = -9$

Divide both sides by 9: $x = -1$

EXERCISE 13C

1 Solve each of these equations.

a $2x + 3 = x + 5$
b $5y + 4 = 3y + 6$
c $4a - 3 = 3a + 4$
d $5t + 3 = 2t + 15$
e $7p - 5 = 3p + 3$
f $6k + 5 = 2k + 1$
g $4m + 1 = m + 10$
h $8s - 1 = 6s - 5$

Advice and Tips

Remember: 'Change sides, change signs'. Show all your working. Rearrange before you simplify. If you try to do these at the same time you could get it wrong.

2 Hasan says:

 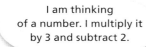

I am thinking of a number. I multiply it by 3 and subtract 2.

Miriam says:

I am thinking of a number. I multiply it by 2 and add 5.

Hasan and Miriam find that they both thought of the same number and both got the same final answer.

What number did they think of?

Advice and Tips

Set up expressions; make them equal and solve.

3 Solve each of these equations.

a $2(d + 3) = d + 12$
b $5(x - 2) = 3(x + 4)$
c $3(2y + 3) = 5(2y + 1)$
d $3(h - 6) = 2(5 - 2h)$
e $4(3b - 1) + 6 = 5(2b + 4)$
f $2(5c + 2) - 2c = 3(2c + 3) + 7$

4 Explain why the equation $3(2x + 1) = 2(3x + 5)$ cannot be solved.

Advice and Tips

Expand the brackets and collect terms on one side as usual. What happens?

5 Explain why these are an infinite number of solutions to this equation.

$2(6x + 9) = 3(4x + 6)$

13.2 Setting up equations

You can use equations to represent situations, so that you can solve real-life problems. You can solve many real-life problems by setting them up as linear equations and then solving the equation.

Example 7

The rectangle shown has a perimeter of 40 cm.

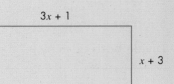

Find the value of x.

The perimeter of the rectangle is:

$3x + 1 + x + 3 + 3x + 1 + x + 3 = 40$

This simplifies to: $8x + 8 = 40$

Subtract 8 from both sides: $8x = 32$

Divide both sides by 8: $x = 4$

EXERCISE 13D

Set up an equation to represent each situation described below. Then solve the equation. Remember to check each answer.

1. Every day, from Monday to Saturday a man buys a daily paper for d cents. He buys a Sunday paper for 1.80 dollars. His weekly paper bill is 7.20 dollars.

   What is the price of his daily paper?

2. The diagram shows a rectangle.

   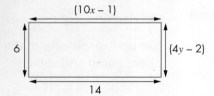

   Advice and Tips

   Use the letter x for the variable unless you are given a letter to use. Once the equation is set up solve it by the methods shown above.

   a What is the value of x?
   b What is the value of y?

3. In this rectangle, the length is 3 cm more than the width. The perimeter is 12 cm.

   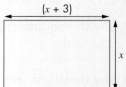

   a What is the value of x?
   b What is the area of the rectangle?

4. Masha has two bags, each of which contains the same number of sweets. She eats four sweets. She then finds that she has 30 sweets left. How many sweets were there in each bag to start with?

5 Flooring costs $12.75 per square metre.

The shop charges $35 for fitting. The final bill was $137.

How many square metres of flooring were fitted?

6 Moshin bought eight garden chairs. When he got to the till he used a $10 voucher as part payment. His final bill was $56.

   a Set this problem up as an equation, using c as the cost of one chair.

   b Solve the equation to find the cost of one chair.

7 This diagram shows the traffic flow through a one-way system in a town centre.

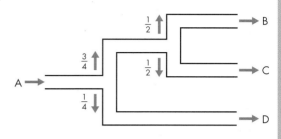

Cars enter at A and at each junction the fractions show the proportion of cars that take each route.

   a 1200 cars enter at A. How many come out of each of the exits, B, C and D?

   b If 300 cars exit at B, how many cars entered at A?

   c If 500 cars exit at D, how many exit at B?

8 A rectangular room is 3 m longer than it is wide. The perimeter is 16 m.

Floor tiles cost $9 for a pack that covers a square metre. How much will it cost to cover the floor?

> **Advice and Tips**
>
> Set up an equation to work out the length and width, then calculate the area.

9 A boy is Y years old. His father is 25 years older than he is. The sum of their ages is 31. How old is the boy?

10 Another boy is X years old. His sister is twice as old as he is. The sum of their ages is 27. How old is the boy?

11 The diagram shows a square.

Find the value of x if the perimeter is 44 cm.

$(4x - 1)$

12 Max thought of a number. He then multiplied his number by 3. He added 4 to the answer. He then doubled that answer to get a final value of 38. What number did he start with?

13 The angles of a triangle, in degrees, are $2x$, $x + 5$ and $x + 35$.

   a Write down an equation to show that the angles add up to 180 degrees.

   b Solve your equation to find the value of x.

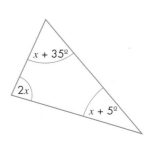

14 Five friends went for a meal in a restaurant. The bill was x.
They decided to add a $10 tip and split the bill equally between them.

Each person paid $9.50.

  a Set up this problem as an equation.

  b Solve the equation to work out the bill before the tip was added.

15 A teacher asked her class to find three angles of a triangle that were consecutive even numbers.

Tammy wrote:
$$x + x + 2 + x + 4 = 180$$
$$3x + 6 = 180$$
$$3x = 174$$
$$x = 58$$

So the angles are 58°, 60° and 62°.

The teacher then asked the class to find four angles of a quadrilateral that are consecutive even numbers.

Can this be done? Explain your answer.

> **Advice and Tips**
>
> Do the same type of working as Tammy did for a triangle. Work out the value of x. What happens?

16 Maria has a large and a small bottle of cola. The large bottle holds 50 cl more than the small bottle.

From the large bottle she fills four cups and has 18 cl left over.

From the small bottle she fills three cups and has 1 cl left over.

How much cola was there in each bottle?

> **Advice and Tips**
>
> Set up equations for both using x as the amount of cola in a cup. Put them equal but remember to add 50 to the small bottle equation to allow for the difference. Solve for x, then work out how much is in each bottle.

13.3 Solving quadratic equations by factorisation

So far, all the equations you have worked with have been **linear equations**. Now you will look at equations that involve **quadratic expressions** such as $x^2 - 2x - 3$, which contain the square of the variable.

Solving the quadratic equation $x^2 + ax + b = 0$

To solve a **quadratic equation** such as $x^2 - 2x - 3 = 0$, you first must be able to factorise it. Work through Examples 8 to 10 below to see how this is done.

13.3

Example 8

Solve the equation $x^2 + 6x + 5 = 0$.

This factorises into $(x + 5)(x + 1) = 0$.

The only way this expression can ever equal 0 is if the value of one of the expressions in the brackets is 0.

Hence either $(x + 5) = 0$ or $(x + 1) = 0$

$\Rightarrow x + 5 = 0$ or $x + 1 = 0$

$\Rightarrow x = -5$ or $x = -1$

So the solution is $x = -5$ or $x = -1$.

There are *two* possible values for x.

Example 9

Solve the equation $x^2 + 3x - 10 = 0$.

This factorises into $(x + 5)(x - 2) = 0$.

Hence either $(x + 5) = 0$ or $(x - 2) = 0$

$\Rightarrow x + 5 = 0$ or $x - 2 = 0$

$\Rightarrow x = -5$ or $x = 2$.

So the solution is $x = -5$ or $x = 2$.

Example 10

Solve the equation $x^2 - 6x + 9 = 0$.

This factorises into $(x - 3)(x - 3) = 0$.

The equation has repeated roots.

That is: $(x - 3)^2 = 0$

Hence, there is only one solution, $x = 3$.

EXERCISE 13E

Solve the equations in questions **1–12**.

1. $(x + 2)(x + 5) = 0$
2. $(t + 3)(t + 1) = 0$
3. $(a + 6)(a + 4) = 0$
4. $(x + 3)(x - 2) = 0$
5. $(x + 1)(x - 3) = 0$
6. $(t + 4)(t - 5) = 0$
7. $(x - 1)(x + 2) = 0$
8. $(x - 2)(x + 5) = 0$
9. $(a - 7)(a + 4) = 0$
10. $(x - 3)(x - 2) = 0$
11. $(x - 1)(x - 5) = 0$
12. $(a - 4)(a - 3) = 0$

First factorise, then solve the equations in questions **13–26**.

13 $x^2 + 5x + 4 = 0$

14 $x^2 + 11x + 18 = 0$

15 $x^2 - 6x + 8 = 0$

16 $x^2 - 8x + 15 = 0$

17 $x^2 - 3x - 10 = 0$

18 $x^2 - 2x - 15 = 0$

19 $t^2 + 4t - 12 = 0$

20 $t^2 + 3t - 18 = 0$

21 $x^2 - x - 2 = 0$

22 $x^2 + 4x + 4 = 0$

23 $m^2 + 10m + 25 = 0$

24 $t^2 - 8t + 16 = 0$

25 $t^2 + 8t + 12 = 0$

26 $a^2 - 14a + 49 = 0$

27 A woman is x years old. Her husband is three years younger.

The product of their ages is 550.

  a Set up a quadratic equation to represent this situation.

  b How old is the woman?

> **Advice and Tips**
>
> If one solution to a real-life problem is negative, reject it and only give the positive answer.

28 A rectangular field is 40 m longer than it is wide.
The area is 48 000 square metres.

The farmer wants to place a fence all around the field.

How long will the fence be?

> **Advice and Tips**
>
> Let the width be x, set up a quadratic equation and solve it to get x.

First rearrange the equations in questions **29–37**, then solve them.

29 $x^2 + 10x = -24$

30 $x^2 - 18x = -32$

31 $x^2 + 2x = 24$

32 $x^2 + 3x = 54$

33 $t^2 + 7t = 30$

34 $x^2 - 7x = 44$

35 $t^2 - t = 72$

36 $x^2 = 17x - 72$

37 $x^2 + 1 = 2x$

> **Advice and Tips**
>
> You cannot solve a quadratic equation by factorisation unless it is in the form
> $x^2 + ax + b = 0$

38 A teacher asks her class to solve $x^2 - 3x = 4$.

This is Mario's answer.

$x^2 - 3x - 4 = 0$

$(x - 4)(x + 1) = 0$

Hence $x - 4 = 0$ or $x + 1 = 0$

$x = 4$ or -1

13.3 Solving quadratic equations by factorisation

This is Sylvan's answer.

$x(x - 3) = 4$

Hence $x = 4$ or $x - 3 = 4 \Rightarrow x = -3 + 4 = -1$

When the teacher reads out the answer of $x = 4$ or -1, both students mark their work as correct.

Who used the correct method and what mistakes did the other student make?

Solving the general quadratic equation by factorisation

The general quadratic equation is of the form $ax^2 + bx + c = 0$ where a, b and c are positive or negative whole numbers. (It is easier to make sure that a is always positive.) Before you can solve any quadratic equation by factorisation, you must rearrange it to this form.

The method is similar to that used to solve equations of the form $x^2 + ax + b = 0$. That is, you have to find two **factors** of $ax^2 + bx + c$ with a product of 0.

Example 11

Solve these quadratic equations.   **a** $12x^2 - 28x = -15$ **b** $30x^2 - 5x - 5 = 0$

a First, rearrange the equation to the general form.

$12x^2 - 28x + 15 = 0$

This factorises into $(2x - 3)(6x - 5) = 0$.

The only way this product can equal 0 is if the value of the expression in one of the brackets is 0. Hence:

either $2x - 3 = 0$ or $6x - 5 = 0$

$\Rightarrow 2x = 3$ or $6x = 5$

$\Rightarrow x = \frac{3}{2}$ or $x = \frac{5}{6}$

So the solution is $x = 1\frac{1}{2}$ or $x = \frac{5}{6}$

Note: It is almost always the case that if a solution is a fraction which is then changed into a rounded decimal number, the original equation cannot be evaluated exactly, using that decimal number. So it is preferable to leave the solution in its fraction form. This is called the *rational form*.

b This equation is already in the general form and it will factorise to $(15x + 5)(2x - 1) = 0$ or $(3x + 1)(10x - 5) = 0$.

Look again at the equation. There is a common factor of 5 which can be taken out to give:

$5(6x^2 - x - 1 = 0)$

This is much easier to factorise to $5(3x + 1)(2x - 1) = 0$, which can be solved to give $x = -\frac{1}{3}$ or $x = \frac{1}{2}$.

Special cases

Sometimes the value of b or c may be zero. (Note that if a is zero the equation is no longer a quadratic equation but a linear equation.)

> **Example 12**
>
> Solve these quadratic equations.
>
> **a** $3x^2 - 4 = 0$
>
> **b** $4x^2 - 25 = 0$
>
> **c** $6x^2 - x = 0$
>
> **a** Rearrange to get $3x^2 = 4$.
>
>   Divide both sides by 3: $\quad x^2 = \frac{4}{3}$
>
>   Take the square root on both sides: $x = \pm\sqrt{\frac{4}{3}} = \pm\frac{2}{\sqrt{3}} = \pm\frac{2\sqrt{3}}{3}$
>
>   **Note**: A square root can be positive or negative. The symbol ± s means that the square root has a positive and a negative value, *both* of which must be used in solving for x.
>
> **b** You can use the method of part **a** or you can treat this as the **difference of two squares**. This can be factorised to $(2x - 5)(2x + 5) = 0$.
>
>   Each set of brackets can be put equal to zero.
>
> $2x - 5 = 0 \Rightarrow x = +\frac{5}{2}$
>
> $2x + 5 = 0 \Rightarrow x = -\frac{5}{2}\quad$ So the solution is $x = \pm\frac{5}{2}$.
>
> **c** There is a common factor of x, so factorise as $x(6x - 1) = 0$.
>
>   There is only one set of brackets this time but each factor can be equal to zero, so $x = 0$ or $6x - 1 = 0$.
>
>   Hence, $x = 0$ or $\frac{1}{6}$.

EXERCISE 13F

Give your answers either in rational form or as mixed numbers.

1 Solve these equations.

a $3x^2 + 8x - 3 = 0$

b $6x^2 - 5x - 4 = 0$

c $5x^2 - 9x - 2 = 0$

d $4t^2 - 4t - 35 = 0$

e $18t^2 + 9t + 1 = 0$

f $3t^2 - 14t + 8 = 0$

g $6x^2 + 15x - 9 = 0$

h $12x^2 - 16x - 35 = 0$

i $15t^2 + 4t - 35 = 0$

j $28x^2 - 85x + 63 = 0$

k $24x^2 - 19x + 2 = 0$

l $16t^2 - 1 = 0$

m $4x^2 + 9x = 0$

n $25t^2 - 49 = 0$

o $9m^2 - 24m - 9 = 0$

Advice and Tips

Look out for the special cases where b or c is zero.

Chapter 13. Topic 4

2 Rearrange these equations into the general form and then solve them.

a $x^2 - x = 42$
b $8x(x + 1) = 30$
c $(x + 1)(x - 2) = 40$
d $13x^2 = 11 - 2x$
e $(x + 1)(x - 2) = 4$
f $10x^2 - x = 2$
g $8x^2 + 6x + 3 = 2x^2 + x + 2$
h $25x^2 = 10 - 45x$
i $8x - 16 - x^2 = 0$
j $(2x + 1)(5x + 2) = (2x - 2)(x - 2)$
k $5x + 5 = 30x^2 + 15x + 5$
l $2m^2 = 50$
m $6x^2 + 30 = 5 - 3x^2 - 30x$
n $4x^2 + 4x - 49 = 4x$
o $2t^2 - t = 15$

3 Here are three equations.

  A: $(x - 1)^2 = 0$ B: $3x + 2 = 5$ C: $x^2 - 4x = 5$

a Give some mathematical fact that equations A and B have in common.
b Give a mathematical reason why equation B is different from equations A and C.

4 Pythagoras' theorem states that the sum of the squares of the two short sides of a right-angled triangle equals the square of the long side (hypotenuse).

A right-angled triangle has sides $5x - 1$, $2x + 3$ and $x + 1$ cm.

a Show that $20x^2 - 24x - 9 = 0$.
b Find the area of the triangle.

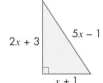

13.4 Solving quadratic equations by the quadratic formula

Many quadratic equations cannot be solved by factorisation because they do not have simple factors. Try to factorise, for example, $x^2 - 4x - 3 = 0$ or $3x^2 - 6x + 2 = 0$. You will find it is impossible.

One way to solve this type of equation is to use the **quadratic formula**. This formula can be used to solve *any* quadratic equation that can be solved (is **soluble**).

The solution of the equation $ax^2 + bx + c = 0$ is given by:

$$x = \frac{-b \pm \sqrt{b^2 - 4ac}}{2a}$$

where a and b are the **coefficients** of x^2 and x respectively and c is the **constant** term.

This is the quadratic formula, which you should memorise.

Example 13

Solve $5x^2 - 11x - 4 = 0$, giving solutions correct to 2 decimal places.

Take the quadratic formula:
$$x = \frac{-b \pm \sqrt{b^2 - 4ac}}{2a}$$
and put $a = 5$, $b = -11$ and $c = -4$, which gives:
$$x = \frac{-(-11) \pm \sqrt{(-11)^2 - 4(5)(-4)}}{2(5)}$$

Note that the values for a, b and c have been put into the formula in brackets. This is to avoid mistakes in calculation. It is a very common mistake to get the sign of b wrong or to think that -11^2 is -121. Using brackets will help you do the calculation correctly.

$$x = \frac{11 \pm \sqrt{121 + 80}}{10} = \frac{11 \pm \sqrt{201}}{10}$$

$\Rightarrow x = 2.52$ or -0.32

Note: The calculation has been done in stages. With a calculator it is possible just to work out the answer, but make sure you can use your calculator properly. Otherwise, break the calculation down. Remember the rule: 'if you try to do two things at once, you will probably get one of them wrong'.

Note: If you are asked to solve a quadratic equation to one or two decimal places, you can be sure that it can be solved *only* by the quadratic formula.

EXERCISE 13G

Use the quadratic formula to solve the equations in questions **1–15**. Give your answers to 2 decimal places.

Advice and Tips

Use brackets when substituting and do not try to work two things out at the same time.

1 $2x^2 + x - 8 = 0$
2 $3x^2 + 5x + 1 = 0$
3 $x^2 - x - 10 = 0$
4 $5x^2 + 2x - 1 = 0$
5 $7x^2 + 12x + 2 = 0$
6 $3x^2 + 11x + 9 = 0$
7 $4x^2 + 9x + 3 = 0$
8 $6x^2 + 22x + 19 = 0$
9 $x^2 + 3x - 6 = 0$
10 $3x^2 - 7x + 1 = 0$
11 $2x^2 + 11x + 4 = 0$
12 $4x^2 + 5x - 3 = 0$

Chapter 13 . Topic 5 13 . 5

13 $4x^2 - 9x + 4 = 0$

14 $7x^2 + 3x - 2 = 0$

15 $5x^2 - 10x + 1 = 0$

16 A rectangular lawn is 2 m longer than it is wide.

The area of the lawn is 21 m². The gardener wants to edge the lawn with edging strips, which are sold in lengths of $1\frac{1}{2}$ m. How many will she need to buy?

17 Shaun is solving a quadratic equation, using the formula.

He correctly substitutes values for a, b and c to get:

$x = \dfrac{3 \pm \sqrt{37}}{2}$

What is the equation Shaun is trying to solve?

18 Hasan uses the quadratic formula to solve $4x^2 - 4x + 1 = 0$.

Miriam uses factorisation to solve $4x^2 - 4x + 1 = 0$.

They both find something unusual in their solutions.

Explain what this is, and why.

13.5 Solving quadratic equations by completing the square   E

Another method for solving quadratic equations is **completing the square**. You can use this method as an alternative to the quadratic formula.

You will remember that:

$(x + a)^2 = x^2 + 2ax + a^2$

which can be rearranged to give:

$x^2 + 2ax = (x + a)^2 - a^2$

This is the basic principle behind completing the square.

There are three basic steps in rewriting $x^2 + px + q$ in the form $(x + a)^2 + b$.

Step 1: Ignore q and just look at the first two terms, $x^2 + px$.

Step 2: Rewrite $x^2 + px$ as $\left(x + \dfrac{p}{2}\right)^2 - \left(\dfrac{p}{2}\right)^2$.

Step 3: Bring q back to get $x^2 + px + q = \left(x + \dfrac{p}{2}\right)^2 - \left(\dfrac{p}{2}\right)^2 + q$.

Chapter 13: Solutions of equations and inequalities

Example 14

Rewrite each expression in the form $(x \pm a)^2 \pm b$.

a $x^2 + 6x - 7$

b $x^2 - 8x + 3$

a Ignore −7 for the moment.

Rewrite $x^2 + 6x$ as $(x + 3)^2 - 9$.

(Expand $(x + 3)^2 - 9 = x^2 + 6x + 9 - 9 = x^2 + 6x$. The 9 is subtracted to get rid of the constant term when the brackets are expanded.)

Now bring the −7 back, so $x^2 + 6x - 7 = (x + 3)^2 - 9 - 7$

Combine the constant terms to get the final answer:

$x^2 + 6x - 7 = (x + 3)^2 - 16$

b Ignore +3 for the moment.

Rewrite $x^2 - 8x$ as $(x - 4)^2 - 16$.

(Note that you still subtract $(-4)^2$, as $(-4)^2 = +16$.)

Now bring the +3 back, so $x^2 - 8x + 3 = (x - 4)^2 - 16 + 3$.

Combine the constant terms to get the final answer:

$x^2 - 8x + 3 = (x - 4)^2 - 13$

Example 15

Rewrite $x^2 + 4x - 7$ in the form $(x + a)^2 - b$. Hence solve the equation $x^2 + 4x - 7 = 0$, giving your answers to 2 decimal places.

Note that:
$x^2 + 4x = (x + 2)^2 - 4$

So:
$x^2 + 4x - 7 = (x + 2)^2 - 4 - 7 = (x + 2)^2 - 11$

When $x^2 + 4x - 7 = 0$, you can rewrite the equations completing the square as:
$(x + 2)^2 - 11 = 0$

Rearranging gives $(x + 2)^2 = 11$.

Taking the **square root** of both sides gives:

$x + 2 = \pm\sqrt{11}$

$\Rightarrow x = -2 \pm \sqrt{11}$

This answer could be left like this, but you are asked to calculate it to 2 decimal places.

$\Rightarrow x = 1.32$ or -5.32 (to 2 decimal places)

13.5

To solve $ax^2 + bx + c = 0$ when a is not 1, start by dividing through by a.

Example 16

Solve by completing the square.

$2x^2 - 6x - 7 = 0$

Divide by 2: $\quad x^2 - 3x - 3.5 = 0$

$\qquad\qquad x^2 - 3x = (x - 1.5)^2 - 2.25$

So: $\qquad\quad x^2 - 3x - 3.5 = (x - 1.5)^2 - 5.75$

When: $\qquad x^2 - 3x - 3.5 = 0$

then: $\qquad\quad (x - 1.5)^2 = 5.75$

$\qquad\qquad\quad x - 1.5 = \pm\sqrt{5.75}$

$\qquad\qquad\qquad x = 1.5 \pm \sqrt{5.75}$

$\qquad\qquad\qquad x = 3.90 \text{ or } x = -0.90$

EXERCISE 13H

1 Write an equivalent expression in the form $(x \pm a)^2 - b$.

   a $x^2 + 4x$ **b** $x^2 + 14x$ **c** $x^2 - 6x$ **d** $x^2 + 6x$

   e $x^2 - 3x$ **f** $x^2 - 9x$ **g** $x^2 + 13x$ **h** $x^2 + 10x$

   i $x^2 + 8x$ **j** $x^2 - 2x$ **k** $x^2 + 2x$

2 Write an equivalent expression in the form $(x \pm a)^2 - b$.
Question **1** will help with **a** to **h**.

   a $x^2 + 4x - 1$ **b** $x^2 + 14x - 5$ **c** $x^2 - 6x + 3$ **d** $x^2 + 6x + 7$

   e $x^2 - 3x - 1$ **f** $x^2 + 6x + 3$ **g** $x^2 - 9x + 10$ **h** $x^2 + 13x + 35$

   i $x^2 + 8x - 6$ **j** $x^2 + 2x - 1$ **k** $x^2 - 2x - 7$ **l** $x^2 + 2x - 9$

3 Solve each equation by completing the square. Leave a square root sign in your answer where appropriate. The answers to question **2** will help.

   a $x^2 + 4x - 1 = 0$ **b** $x^2 + 14x - 5 = 0$ **c** $x^2 - 6x + 3 = 0$

   d $x^2 + 6x + 7 = 0$ **e** $x^2 - 3x - 1 = 0$ **f** $x^2 + 6x + 3 = 0$

   g $x^2 - 9x + 10 = 0$ **h** $x^2 + 13x + 35 = 0$ **i** $x^2 + 8x - 6 = 0$

   j $x^2 + 2x - 1 = 0$ **k** $x^2 - 2x - 7 = 0$ **l** $x^2 + 2x - 9 = 0$

4 Solve by completing the square. Give your answers to 2 decimal places.

   a $x^2 + 2x - 5 = 0$ **b** $x^2 - 4x - 7 = 0$ **c** $x^2 + 2x - 9 = 0$

5 Solve these equations by completing the square. Leave your answer in square root form.

   a $2x^2 - 6x - 3 = 0$ **b** $4x^2 - 8x + 1 = 0$ **c** $2x^2 + 5x - 10 = 0$

   d $0.5x^2 - 7.5x + 8 = 0$

Chapter 13: Solutions of equations and inequalities

Chapter 13 . Topic 6

6 Ahmed rewrites the expression $x^2 + px + q$ by completing the square. He does this correctly and gets $(x - 7)^2 - 52$.

What are the values of p and q?

7 a Jorge writes the steps to solve $x^2 + 6x + 7 = 0$ by completing the square. He writes them on sticky notes. Unfortunately he drops the sticky notes and they get out of order. Try to put the notes in the correct order.

| Add 2 to both sides | Subtract 3 from both sides | Write $x^2 + 6x + 7 = 0$ as $(x + 3)^2 - 2 = 0$ | Take the square root of both sides |

13.6 Simultaneous equations

All the equations we have looked at so far have just one unknown.

Sometimes there is more than one unknown variable in a problem. In these cases, we will have several **simultaneous equations** to solve.

Example 17

Tariq is twice as old as Meera. Their total age is 39 years. How old are they?

Suppose Tariq is x years old and Meera is y years old.

Tariq is twice as old as Meera: $x = 2y$ (equation 1)

Their total age is 39 years: $x + y = 39$ (equation 2)

We have two unknowns and two equations to use to find them.

Substitute $2y$ for x in equation 2:

$2y + y = 39$

$\Rightarrow 3y = 39$

$\Rightarrow y = 13$

Now use equation (1) to find x: $x = 2 \times 13 = 26$

Tariq is 26 and Meera is 13.

EXERCISE 13I

1 Solve each of these pairs of simultaneous equations.

a $x + y = 15$
$y = 2x$

b $x = 3y$
$x + y = 24$

c $x + y = 60$
$y = 4x$

2 Solve each of these pairs of simultaneous equations.

   a $y = x + 12$ **b** $y = x - 10$ **c** $x + 4 = y$
        $y = 3x$ $x = 5y$ $y = 9x$

3 Solve each of these pairs of simultaneous equations.

   a $x + y = 20$ **b** $y + x = 23$ **c** $x + y = 6$
        $x - y = 6$ $y - x = 5$ $x - y = 14$

4 Solve each of these pairs of simultaneous equations.

   a $y = 2x + 3$ **b** $x + y = 20$ **c** $y = 2x + 4$
        $y = 8x$ $y = 3x - 2$ $y = 10 - x$

5 Carmen and Anish are carrying some books. There are 40 books altogether.

Carmen has 4 times as many as Anish.

How many do they each have?

6 Ari writes down two numbers. The total is 37. The difference between them is 14.

What are the numbers?

7 Luis records the temperature at midday and again at midnight.

He notices that the temperatures add up to 5 and the difference between them is 11.

What are the temperatures?

8 Carlos has x and Sarah has y.

   a Together they have $75. Write an equation to show this.
   b Sarah has twice as much as Carlos. Write an equation to show this.
   c Solve the equations to find x and y.

9 The mass of a plate is x g and the mass of a cup is y g.

   a The total mass is 300 g. Write an equation to show this.
   b The plate is 60 g heavier than the cup. Write an equation to show this.
   c Find the values of x and y.

10 Ahmed is x years old. His mother is y years old.

   a Ahmed is 26 years younger than his mother. Write an equation to show this.
   b The total of Ahmed's age and his mother's age is 50 years. Write an equation to show his.
   c Find the age of Ahmed and his mother.

11 The length of Chen's car is x m. The length of Ari's car is y m.

   a Chen's car is 0.4 m shorter than Paola's. Write an equation to show this.
   b The total length of the two cars is 8.6 m. Write an equation to show this.
   c Find the length of Paola's car.

A pair of **simultaneous linear equations** is exactly that — two linear equations for which you want the same solution, and which you therefore *solve together*. For example,

$x + y = 10$ has many solutions:

$x = 2, y = 8$

$x = 4, y = 6$

$x = 5, y = 5$ …

and $2x + y = 14$ has many solutions:

$x = 2, y = 10$

$x = 3, y = 8$

$x = 4, y = 6$ …

But only *one* solution, $x = 4$ and $y = 6$, satisfies both equations at the same time.

In the last section you looked at some simple examples. You can now look at ways of solving more complicated examples of simultaneous equations.

Elimination method

One way to solve simultaneous equations is by the *elimination method*. There are six steps in this method.

Step 1 is to balance the coefficients of one of the variables.

Step 2 is to **eliminate** this variable by adding or subtracting the equations.

Step 3 is to solve the resulting linear equation in the other variable.

Step 4 is to **substitute** the value found back into one of the previous equations.

Step 5 is to solve the resulting equation.

Step 6 is to check that the two values found satisfy the original equations.

Example 18

Solve the equations: $6x + y = 15$ and $4x + y = 11$.

Label the equations so that the method can be clearly explained.

$6x + y = 15$ (1)

$4x + y = 11$ (2)

Step 1: Since the y-term in both equations has the same coefficient there is no need to balance them.

Step 2: Subtract one equation from the other. (Equation (1) minus equation (2) will give positive values.)

    (1) − (2) $2x = 4$

Step 3: Solve this equation: $x = 2$

Step 4: Substitute $x = 2$ into one of the original equations. (Usually it is best to the one with smallest numbers involved.)

So substitute into: $4x + y = 11$

which gives: $8 + y = 11$

Step 5: Solve this equation: $y = 3$

Step 6: Test the solution in the original equations. So substitute $x = 2$ and $y = 3$ into $6x + y$, which gives $12 + 3 = 15$ and into $4x + y$, which gives $8 + 3 = 11$. These are correct, so you can confidently say the solution is $x = 2$ and $y = 3$.

Example 19

Solve these equations. $5x + y = 22$ (1)

$2x - y = 6$ (2)

Step 1: Both equations have the same y-coefficient but with *different* signs so there is no need to balance them.

Step 2: As the signs are different, *add* the two equations, to eliminate the y-terms.

(1) + (2) $7x = 28$

Step 3: Solve this equation: $x = 4$

Step 4: Substitute $x = 4$ into one of the original equations, $5x + y = 22$,

which gives: $20 + y = 22$

Step 5: Solve this equation: $y = 2$

Step 6: Test the solution by putting $x = 4$ and $y = 2$ into the original equations, $2x - y$, which gives $8 - 2 = 6$ and $5x + y$ which gives $20 + 2 = 22$. These are correct, so the solution is $x = 4$ and $y = 2$.

Substitution method

This is an alternative method. The method you use depends very much on the coefficients of the variables and the way that the equations are written in the first place. There are five steps in the substitution method.

Step 1 is to rearrange one of the equations into the form $y = \ldots$ or $x = \ldots$.

Step 2 is to substitute the right-hand side of this equation into the other equation in place of the variable on the left-hand side.

Step 3 is to expand and solve this equation.

Step 4 is to substitute the value into the $y = \ldots$ or $x = \ldots$ equation.

Step 5 is to check that the values work in both original equations.

Example 20

Solve the simultaneous equations: $y = 2x + 3$, $3x + 4y = 1$.

Because the first equation is in the form $y = \ldots$ you can use the substitution method.

Again label the equations to help with explaining the method.

$y = 2x + 3$ (1)

$3x + 4y = 1$ (2)

Step 1: As equation (1) is in the form $y = \ldots$ there is no need to rearrange an equation.

Step 2: Substitute the right-hand side of equation (1) into equation (2) for the variable y.

$$3x + 4(2x + 3) = 1$$

Step 3: Expand and solve the equation. $3x + 8x + 12 = 1$, $11x = -11$, $x = -1$

Step 4: Substitute $x = -1$ into $y = 2x + 3$: $y = -2 + 3 = 1$

Step 5: Test the values in $y = 2x + 3$, which gives $1 = -2 + 3$ and $3x + 4y = 1$, which gives $-3 + 4 = 1$. These are correct so the solution is $x = -1$ and $y = 1$.

EXERCISE 13J

1 Solve these simultaneous equations.

In question **1** parts **a** to **i** the coefficients of one of the variables are the same so there is no need to balance them. Subtract the equations when the identical terms have the same sign. Add the equations when the identical terms have opposite signs. In parts **j** to **l** use the substitution method.

a $4x + y = 17$ $2x + y = 9$	**b** $5x + 2y = 13$ $x + 2y = 9$	**c** $2x + y = 7$ $5x - y = 14$	
d $3x + 2y = 11$ $2x - 2y = 14$	**e** $3x - 4y = 17$ $x - 4y = 3$	**f** $3x + 2y = 16$ $x - 2y = 4$	
g $x + 3y = 9$ $x + y = 6$	**h** $2x + 5y = 16$ $2x + 3y = 8$	**i** $3x - y = 9$ $5x + y = 11$	
j $2x + 5y = 37$ $y = 11 - 2x$	**k** $4x - 3y = 7$ $x = 13 - 3y$	**l** $4x - y = 17$ $x = 2 + y$	

2 In this sequence, the next term is found by multiplying the previous term by a and then adding b. a and b are positive whole numbers.

   3   14   47   …   …

a Explain why $3a + b = 14$.

b Set up another equation in a and b.

c Solve the equations to solve for a and b.

d Work out the next two terms in the sequence.

13.6

Balancing coefficients in one equation only

You could solve all the examples in Exercise 13I, question **1** by adding or subtracting the equations in each pair, or by substituting without rearranging. This does not always happen. The next examples show what to do when there are no identical terms, or when you need to rearrange.

Example 21

Solve these equations.

$$3x + 2y = 18 \quad (1)$$
$$2x - y = 5 \quad (2)$$

Step 1: Multiply equation (2) by 2. There are other ways to balance the coefficients but this is the easiest and leads to less work later. With practice, you will get used to which will be the best way to balance the coefficients.

$$2 \times (2) \qquad 4x - 2y = 10 \quad (3)$$

Label this equation as number (3).

Be careful to multiply every term and not just the y-term. You could write:

$$2 \times (2x - y = 5) \Rightarrow 4x - 2y = 10 \quad (3)$$

Step 2: As the signs of the y-terms are opposite, add the equations.

$$(1) + (3) \qquad 7x = 28$$

Be careful to add the correct equations. This is why labelling them is useful.

Step 3: Solve this equation: $x = 4$

Step 4: Substitute $x = 4$ into any equation, say $2x - y = 5 \Rightarrow 8 - y = 5$

Step 5: Solve this equation: $y = 3$

Step 6: Check: (1), $3 \times 4 + 2 \times 3 = 18$ and (2), $2 \times 4 - 3 = 5$, which are correct so the solution is $x = 4$ and $y = 3$.

Example 22

Solve the simultaneous equations:

$$3x + y = 5 \quad (1)$$
$$5x - 2y = 12 \quad (2)$$

Step 1: Multiply the first equation by 2: $6x + 2y = 10 \quad (3)$

Step 2: Add (2) + (3): $11x = 22$

Step 3: Solve: $x = 2$

Step 4: Substitute back: $3 \times 2 + y = 5$

Step 5: Solve: $y = -1$

Step 6: Check: (1) $3 \times 2 - 1 = 5$ and (2) $5 \times 2 - 2 \times -1 = 10 + 2 = 12$, which are correct.

EXERCISE 13K

1 Solve parts **a** to **c** by the substitution method and the rest by first changing one of the equations in each pair to obtain identical terms for one unknown, and then adding or subtracting the equations to eliminate those terms.

a $5x + 2y = 4$ $4x - y = 11$	**b** $4x + 3y = 37$ $2x + y = 17$	**c** $x + 3y = 7$ $2x - y = 7$
d $2x + 3y = 19$ $6x + 2y = 22$	**e** $5x - 2y = 26$ $3x - y = 15$	**f** $10x - y = 3$ $3x + 2y = 17$
g $3x + 5y = 15$ $x + 3y = 7$	**h** $3x + 4y = 7$ $4x + 2y = 1$	**i** $5x - 2y = 24$ $3x + y = 21$
j $5x - 2y = 4$ $3x - 6y = 6$	**k** $2x + 3y = 13$ $4x + 7y = 31$	**l** $3x - 2y = 3$ $5x + 6y = 12$

2 a Francesca is solving the simultaneous equations $4x - 2y = 8$ and $2x - y = 4$.

She finds a solution of $x = 5$, $y = 6$ which works for both equations.

Explain why this is not a unique solution.

b Dimitri is solving the simultaneous equations $6x + 2y = 9$ and $3x + y = 7$.

Why is it impossible to find a solution that works for both equations?

Balancing coefficients in both equations

In some cases, you will need to change *both* equations to obtain identical terms. The next example shows you how to do this.

Note: The substitution method is not suitable for these types of equation as you end up with fractional terms.

Example 23

Solve these equations. $4x + 3y = 27$ (1)

$5x - 2y = 5$ (2)

You need to change both equations to obtain identical terms in either x or y. However, you can see that if you make the y-coefficients the same, you will add the equations. Addition is always safer than subtraction, so this is obviously the better choice. Do this by multiplying the first equation by 2 (the y-coefficient of the second equation) and the second equation by 3 (the y-coefficient of the first equation).

Step 1: (1) × 2 or 2 × ($4x + 3y = 27$) $\Rightarrow$ $8x + 6y = 54$ (3)

(2) × 3 or 3 × ($5x - 2y = 5$) $\Rightarrow$ $15x - 6y = 15$ (4)

Label the new equations (3) and (4).

Step 2: Eliminate one of the variables: (3) + (4): $23x = 69$

Step 3: Solve the equation: $x = 3$

Step 4: Substitute into equation (1): $12 + 3y = 27$

Step 5: Solve the equation: $y = 5$

Step 6: Check: (1), $4 \times 3 + 3 \times 5 = 12 + 15 = 27$ and (2), $5 \times 3 - 2 \times 5 = 15 - 10 = 5$, which are correct so the solution is $x = 3$ and $y = 5$.

Chapter 13. Topic 7

EXERCISE 13L

1 Solve these simultaneous equations.

a	$2x + 5y = 15$ $3x - 2y = 13$	**b**	$2x + 3y = 30$ $5x + 7y = 71$	**c**	$2x - 3y = 15$ $5x + 7y = 52$
d	$3x - 2y = 15$ $2x - 3y = 5$	**e**	$5x - 3y = 14$ $4x - 5y = 6$	**f**	$3x + 2y = 28$ $2x + 7y = 47$
g	$2x + y = 4$ $x - y = 5$	**h**	$5x + 2y = 11$ $3x + 4y = 8$	**i**	$x - 2y = 4$ $3x - y = -3$
j	$3x + 2y = 2$ $2x + 6y = 13$	**k**	$6x + 2y = 14$ $3x - 5y = 10$	**l**	$2x + 4y = 15$ $x + 5y = 21$
m	$3x - y = 5$ $x + 3y = -20$	**n**	$3x - 4y = 4.5$ $2x + 2y = 10$	**o**	$x - 5y = 15$ $3x - 7y = 17$

2 Here are four equations.

A: $5x + 2y = 1$

B: $4x + y = 9$

C: $3x - y = 5$

D: $3x + 2y = 3$

Here are four sets of (x, y) values.

(1, −2), (−1, 3), (2, 1), (3, −3)

Match each pair of (x, y) values to a pair of equations.

> **Advice and Tips**
>
> You could solve each possible set of pairs but there are six to work out. Alternatively you can substitute values into the equations to see which work.

3 Find the area of the triangle enclosed by these three equations.

$y - x = 2$ $x + y = 6$ $3x + y = 6$

4 Find the area of the triangle enclosed by these three equations.

$x - 2y = 6$ $x + 2y = 6$ $x + y = 3$

> **Advice and Tips**
>
> Find the point of intersection of each pair of equations, plot the points on a grid and use any method to work out the area of the resulting triangle.

13.7 Linear and non-linear simultaneous equations

You have already seen the method of substitution for solving **linear** simultaneous equations.

You can use a similar method when you need to solve a pair of equations, one of which is linear and the other of which is **non-linear**. But you must always substitute from the linear into the non-linear.

Example 24

Solve these simultaneous equations.

$x^2 + y^2 = 5$

$x + y = 3$

Call the equations (1) and (2):

$x^2 + y^2 = 5$ (1)

$x + y = 3$ (2)

Rearrange equation (2) to obtain:

$x + = 3 - y$

Substitute this into equation (1), which gives:

$(3 - y)^2 + y^2 = 5$

Expand and rearrange into the general form of the quadratic equation:

$9 - 6y + y^2 + y^2 = 5$

$2y^2 - 6y + 4 = 0$

Divide by 2:

$y^2 - 3y + 2 = 0$

Factorise:

$(y - 1)(y - 2) = 0$

$\Rightarrow y = 1$ or 2

Substitute for y in equation (2):

When $y = 1$, $x = 2$ and when $y = 2$, $x = 1$

Note that you should always give answers as a pair of values in x and y.

Example 25

Find the solutions of the pair of simultaneous equations: $y = x^2 + x - 2$ and $y = 2x - 4$

This example is slightly different, as both equations are given in terms of y,

$2x + 4 = x^2 + x - 2$

Rearranging into the general quadratic:

$x^2 - x - 6 = 0$

Factorising and solving gives:

$(x + 2)(x - 3) = 0$

$x = -2$ or 3

Substituting back to find y:

When $x = -2$, $y = 0$

When $x = 3$, $y = 10$

So the solutions are $(-2, 0)$ and $(3, 10)$.

EXERCISE 13M

1. Solve these pairs of linear simultaneous equations using the substitution method.

   a $2x + y = 9$
      $x - 2y = 7$

   b $3x - 2y = 10$
      $4x + y = 17$

   c $x - 2y = 10$
      $2x + 3y = 13$

2. Solve these pairs of simultaneous equations.

   a $xy = 2$
      $y = x + 1$

   b $xy = -4$
      $2y = x + 6$

3. Solve these pairs of simultaneous equations.

   a $x^2 + y^2 = 25$
      $x + y = 7$

   b $x^2 + y^2 = 9$
      $y = x + 3$

   c $x^2 + y^2 = 13$
      $5y + x = 13$

4. Solve these pairs of simultaneous equations.

   a $y = x^2 + 2x - 3$
      $y = 2x + 1$

   b $y = x^2 - 2x - 5$
      $y = x - 1$

   c $y = x^2 - 2x$
      $y = 2x - 3$

5. Solve these pairs of simultaneous equations.

   a $y = x^2 + 3x - 3$ and $y = x$

   b $x^2 + y^2 = 13$ and $x + y = 1$

   c $x^2 + y^2 = 5$ and $y = x + 1$

   d $y = x^2 - 3x + 1$ and $y = 2x - 5$

   e $y = x^2 - 3$ and $y = x + 3$

   f $y = x^2 - 3x - 2$ and $y = 2x - 6$

   g $x^2 + y^2 = 41$ and $y = x + 1$

6. Ravi's phone number has 6 digits: XY1290

   He notices that the sum of the first 2 digits is 12 and the sum of the squares of the first two digits is 90.

   a Write down two equations using X and Y.
   b What is Ravi's phone number?

7. Samara says:

   I am 4 years older than my brother. The difference between the squares of our ages is 80.

   How old is Samara?

8. Salman is thinking of two numbers.

   The sum of the squares of the numbers is 85. The square of the sum of the numbers is 121

   a If the numbers are x and y, write down two equations connecting x and y.
   b Work out the two numbers.

Chapter 13 . Topic 8

13.8 Solving inequalities

Inequalities behave similarly to equations. You use the same rules to solve linear inequalities as you use for linear equations.

There are four inequality signs:

  $<$ means 'less than'

  $>$ means 'greater than'

  $\leqslant$ means 'less than or equal to'

  $\geqslant$ means 'greater than or equal to'.

Be careful. Never replace the inequality sign with an equals sign.

Example 26

Solve $2x + 3 < 14$.

Rewrite this as: $2x < 14 - 3$

        $2x < 11$

Divide both sides by 2: $\dfrac{2x}{2} < \dfrac{11}{2}$

        $\Rightarrow x < 5.5$

This means that x can take any value below 5.5 but *not* the value 5.5.

You can use a number line to show the solution to the last example.

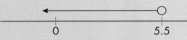

The open circle shows that 5.5 is not included in the solution.

Example 27

a Solve $\dfrac{x}{2} + 13 \geqslant 4$.   **b** Show the solution on a number line.

a Solve as you would an equation but leave the inequality sign in place of the equals sign.

 Subtract 4 from both sides: $\dfrac{x}{2} \geqslant -9$

 Multiply both sides by 2: $x \geqslant -18$

This means that x can take any value above and including -18.

b
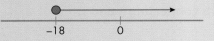

13.8

If you multiply or divide by a negative number when you are solving an inequality you must change the sign.

'less than' becomes 'more than'

'more than' becomes 'less than'

> **Example 28**
>
> **a** Solve the inequality $10 - 2x \leq 3$
>
> **b** Show the solution on a number line.
>
> **a** Subtract 10 from both sides: $-2x \leq -7$
>
> Divide both sides by -2: $\quad x \geq 3.5$
>
> Note that you must reverse inequality signs when multiplying or dividing both sides by a negative number. So the inequality has changed in the last line.
>
> You could solve example 25 in a different way:
>
> $$10 - 2x \leq 3$$
>
> Add $2x$ to both sides: $\quad 10 \leq 2x + 3$
>
> Subtract 3 from both sides: $\quad 7 \leq 2x$
>
> Divide both sides by 2: $\quad 3.5 \leq x$
>
> The sign does not change this time.
>
> $3.5 \leq x$ is equivalent to $x \geq 3.5$
>
> **b**

EXERCISE 13N

1 Solve these linear inequalities. Show each solution on a number line.

   a $x + 4 < 7$
   b $t + 5 > 3$
   c $p + 12 \geq 2$
   d $2x - 3 < 7$
   e $4y + 5 \leq 17$
   f $3t + 4 > 13$
   g $\frac{x}{2} + 4 < 7$
   h $\frac{y}{5} + 6 \leq 3$
   i $\frac{t}{3} - 2 \geq 4$
   j $3(x - 2) < 15$
   k $5(2x + 1) \leq 35$
   l $2(4t - 3) \geq 36$

2 Write down the largest integer value of x that satisfies each inequality.

   a $x - 3 \leq 5$, where x is positive
   b $x + 2 < 9$, where x is positive and even
   c $3x - 11 < 40$, where x is a square number
   d $5x - 8 \leq 15$, where x is positive and odd
   e $2x + 1 < 19$, where x is positive and prime

3 Write down the smallest integer value of x that satisfies each inequality.

  a $x - 2 \geqslant 9$, where x is positive
  b $x - 2 > 13$, where x is positive and even
  c $2x - 11 \geqslant 19$, where x is a square number

4 Ahmed went to town with $20 to buy two CDs. His bus fare was $3. The CDs were both the same price. When he reached home he still had some money in his pocket. What was the most each CD could cost?

Advice and Tips

Set up an inequality and solve it.

5 **a** Explain why you cannot make a triangle with three sticks of length 3 cm, 4 cm and 8 cm.

  b Three sides of a triangle are x, $x + 2$ and 10 cm.

  x is a whole number.

  What is the smallest value x can take?

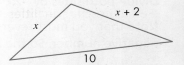

6 Five cards have inequalities and equations marked on them.

  | $x > 0$ | $x < 3$ | $x \geqslant 4$ | $x = 2$ | $x = 6$ |

The cards are shuffled, laid face down and then turned over, one at a time.
If the possible values on two consecutive cards have any numbers in common, then a point is scored.
If they do not have any numbers in common, then a point is deducted.

  a The first two cards below score –1 because $x = 6$ and $x < 3$ have no numbers in common.

  Explain why the total for this combination scores 0.

  | $x = 6$ | $x < 3$ | $x > 0$ | $x = 2$ | $x \geqslant 4$ |

  b What does this combination score?

  | $x > 0$ | $x = 6$ | $x \geqslant 4$ | $x = 2$ | $x < 3$ |

  c Arrange the cards to give a maximum score of 4.

7 Solve these linear inequalities.

  a $4x + 1 \geqslant 3x - 5$ **b** $5t - 3 \leqslant 2t + 5$ **c** $3y - 12 \leqslant y - 4$
  d $2x + 3 \geqslant x + 1$ **e** $5w - 7 \leqslant 3w + 4$ **f** $2(4x - 1) \leqslant 3(x + 4)$

8 Solve these linear inequalities.

  a $\dfrac{x + 4}{2} \leqslant 3$ **b** $\dfrac{x - 3}{5} > 7$ **c** $\dfrac{2x + 5}{3} < 6$
  d $\dfrac{4x - 3}{5} \geqslant 5$ **e** $\dfrac{2t - 2}{7} > 4$ **f** $\dfrac{5y + 3}{5} \leqslant 2$

9 In this question n is always an integer.

   a Find the largest possible value of n if $2n + 3 < 12$.
   b Find the largest possible value of n if $\frac{n}{5} < 20$.
   c Find the smallest possible value of n if $3(n - 7) \geqslant 10$.
   d Find the smallest possible value of n if $\frac{6n - 2}{7} \geqslant 9$.
   e Find the smallest possible value of n if $3n + 14 \leqslant 8n - 13$.

10 **a** If $20 - x > 4$, which of these numbers are possible values of x?

   $-10 \quad 0 \quad 10 \quad 20 \quad 30$

   b Solve the inequality $20 - x > 4$.

11 Solve these inequalities.

   a $15 - x > 6$
   b $18 - x \leqslant 7$
   c $6 \geqslant 9 - x$

12 Solve these inequalities.

   a $20 - 2x \leqslant 5$ **b** $3 - 4x \geqslant 11$
   c $25 - 3x > 7$ **d** $2(6 - x) < 9$
   e $\frac{10 - 2x}{5} \leqslant 4$ **f** $\frac{8 - 4x}{3} > 2$

Check your progress

Core
- I can derive simple linear equations in one unknown
- I can solve simple linear equations in one unknown
- I can derive simple simultaneous linear equations in two unknowns
- I can solve simple simultaneous linear equations in two unknowns

Extended
- I can derive quadratic equations
- I can solve quadratic equations by factorisation, by completing the square or by use of the formula
- I can derive and solve simultaneous equations where one equation is linear and the other is quadratic
- I can derive and solve linear inequalities
- I can interpret and represent inequalities on a number line

Chapter 14
Graphs in practical situations

Topics	Level	Key words
1 Conversion graphs	CORE	scale, estimate, conversion graph
2 Travel graphs	CORE	average speed, distance–time graph
3 Speed–time graphs	EXTENDED	speed–time, constant speed, distance travelled, rate of increase, acceleration, deceleration
4 Curved graphs	EXTENDED	

In this chapter you will learn how to:

CORE	EXTENDED

• Interpret and use graphs in practical situations including travel graphs and conversion graphs. (C2.10 and E2.10)
• Draw graphs from given data. (C2.10 and E2.10)
	• Apply the idea of rate of change to easy kinematics involving:
	  – distance–time and speed–time graphs
	  – acceleration and deceleration. (E2.10)
	• Calculate distance travelled as area under a linear speed–time graph. (E2.10)

Why this chapter matters

Line graphs are used in many media, including newspapers and the textbooks of most of the subjects that you learn in school.

Graphs show the relationship between variables. Often one of these variables is time and the graph shows how the other variable changes over time.

For example, this graph on the right shows how the exchange rate between the dollar and the pound changed over six months in 2015.

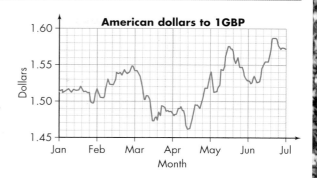

Graphs like this make it easy to see what is happening to a variable – much easier than looking at lists of data. Here you can see instantly that the value of the pound, compared to US dollars, went up and down, but increased overall, over the five months.

A graph can show several variables to make it easier to compare them. The graph below shows data about a racing car going round a circuit. It compares the driver's acceleration and deceleration (speeding up and slowing down) with his steering. The green line is speed and the pink line is steering.

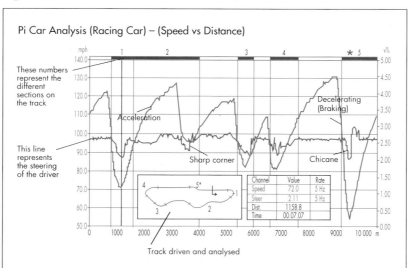

The graph gives the team engineers and trainers an instant picture of the way the driver goes round the course. It would be difficult to compare all this data in any other way.

Chapter 14: Graphs in practical situations

Chapter 14. Topic 1
14.1 Conversion graphs

Look at Examples 1 and 2, and make sure that you can understand the conversions. You need to be able to read these types of graph by finding a value on one axis and following it through to the other axis. Make sure you understand the **scales** on the axes to help you **estimate** the answers.

Example 1

This is a **conversion graph** between litres and gallons.

a How many litres are there in 5 gallons?

b How many gallons are there in 15 litres?

From the graph you can see that:

a 5 gallons are approximately equivalent to 23 litres

b 15 litres are approximately equivalent to $3\frac{1}{4}$ gallons.

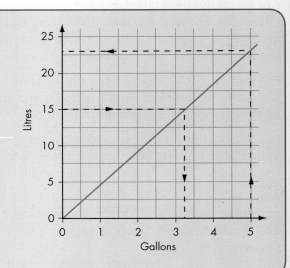

Example 2

This is a graph of the charges made for units of electricity used in the home.

a How much will a customer who uses 500 units of electricity be charged?

b How many units of electricity will a customer who is charged $20 have used?

From the graph you can see that:

a a customer who uses 500 units of electricity will be charged $45

b a customer who is charged $20 will have used about 150 units.

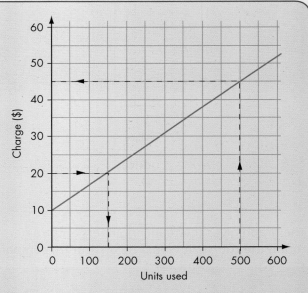

EXERCISE 14A

1 Mass can be measured in kilograms or pounds. This is a conversion graph between kilograms (kg) and pounds (lb).

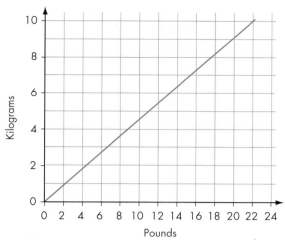

a Use the graph to make an approximate conversion of:
  i   18 lb to kilograms
  ii 5 lb to kilograms
  iii 4 kg to pounds
  iv 10 kg to pounds.

b Approximately how many pounds are equivalent to a total mass of 1 kg?

c Explain how you could use the graph to convert 48 lb to kilograms.

2 Distances can be measured in centimetres or inches. This is a conversion graph between inches (in) and centimetres (cm).

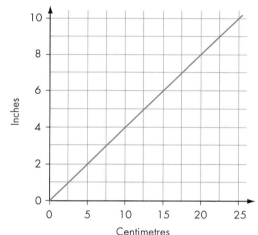

a Use the graph to make an approximate conversion of:
  i   4 in to centimetres
  ii 9 in to centimetres
  iii 5 cm to inches
  iv 22 cm to inches.

b Approximately how many centimetres are there in 1 inch?

c Explain how you could use the graph to convert 18 inches to centimetres.

3 This graph was produced to show approximately how much the British pound (£) is worth in Singapore dollars ($).

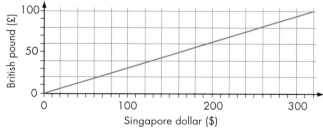

a Use the graph to make an approximate conversion of:
  i   £100 to Singapore dollars
  ii £30 to Singapore dollars
  iii $150 to British pounds
  iv $250 to British pounds.

b Approximately how many Singapore dollars would you get for £1?

c What would happen to the conversion line on the graph if the pound is worth fewer Singapore dollars?

Chapter 14: Graphs in practical situations

4 A hire firm hired out industrial heaters. They used this graph to approximate what the charges would be.

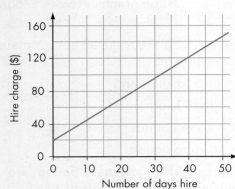

a Use the graph to find the approximate charge for hiring a heater for:
  i 40 days
  ii 25 days.
b Use the graph to find out how many days' hire you would get for a cost of:
  i $100
  ii $140.

5 A conference centre had this chart on the office wall so that the staff could see the approximate cost of a conference, based on the number of people attending it.

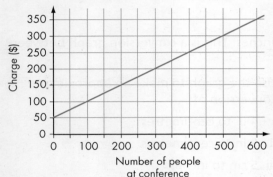

a Use the graph to find the approximate charge for:
  i 100 people
  ii 550 people.
b Use the graph to estimate how many people can attend a conference at the centre for a cost of:
  i $300
  ii $175.

6 A shopkeeper has to add sales tax to the prices of all goods he sells. He marked all goods with prices before tax was added and the sales assistant had to use this chart to convert these marked prices to selling prices.

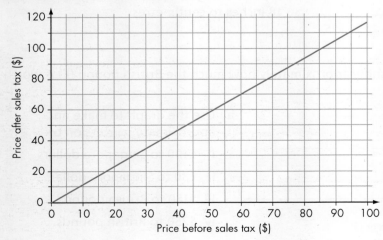

a Use the chart to find the selling price of goods marked:
  i $60
  ii $25.
b What was the marked price if you bought something for:
  i $100
  ii $45?

7 In Europe temperatures are measured in Celsius and in the USA they are measured in Fahrenheit. Here is a conversion graph for the two scales.

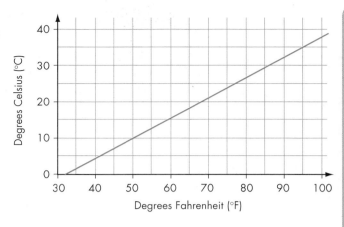

  a Use the graph to make an approximate conversion of:

  i 35°C to Fahrenheit

  ii 20°C to Fahrenheit

  iii 50°F to Celsius

  iv 90°F to Celsius.

  b Water freezes at 0°C. What temperature is this in Fahrenheit?

8 I lost my fuel bill, but while talking to my friends I found out that:

Bill, who had used 850 units, was charged $57.50

Wendy, who had used 320 units, was charged $31

Rhanni, who had used 540 units, was charged $42.

  a Plot the given information and draw a straight-line graph. Use a scale from 0 to 900 on the horizontal units axis, and from $0 to $60 on the vertical cost axis.

  b Use your graph to find what I will be charged for 700 units.

9 Distances can be measured in kilometres or miles.

80 kilometres is approximately the same distance as 50 miles.

  a Use this information to draw a conversion graph between kilometres and miles.

  b Use the graph to convert 30 miles into kilometres.

  c Use the graph to convert 25 kilometres into miles.

10 A candle gets shorter as it burns.

After 10 hours it has burnt down by 13 centimetres.

  a Show this information on a graph.

  b How much did the candle burn down in 7 hours?

  c How long did it take to burn down by 5 centimetres?

11 This table shows how far a snail has moved after different periods of time.

Time in minutes	5	15	30
Distance in centimetres	13	39	78

  a Draw a graph to show this information.

  b How long did the snail take to move 60 centimetres?

Chapter 14: Graphs in practical situations

14.2 Travel graphs

As the name suggests, a travel graph gives information about how far someone or something has travelled over a given time period.

Travel graphs are sometimes called **distance–time** graphs.

You read a travel graph in a similar way to the conversion graphs you have just done. But you can also find the **average speed** from a distance–time graph, using the formula:

$$\text{average speed} = \frac{\text{total distance travelled}}{\text{total time taken}}$$

Example 3

This distance–time graph represents a car journey from Murcia to Cartagena, a distance of 50 km, and back again.

a What can you say about points B, C and D?

b What can you say about the journey from D to F?

c Work out the average speed for each of the five stages of the journey.

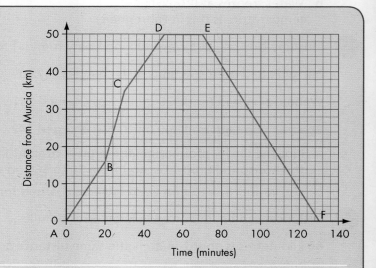

From the graph:

a B: After 20 minutes the car was 16 km away from Murcia.

C: After 30 minutes the car was 35 km away from Murcia.

D: After 50 minutes the car was 50 km away from Murcia, so at Cartagena.

b D–F: The car stayed at Cartagena for 20 minutes, and then took 60 minutes for the return journey.

c The average speeds over the five stages of the journey are worked out as follows.

A to B represents 16 km in 20 minutes.

20 minutes is $\frac{1}{3}$ of an hour, so multiply by 3 to give distance/hour.

Multiplying both numbers by 3 gives 48 km in 60 minutes, which is 48 km/h.

B to C represents 19 km in 10 minutes.

Multiplying both numbers by 6 gives 114 km in 60 minutes, which is 114 km/h.

C to D represents 15 km in 20 minutes.

Multiplying both numbers by 3 gives 45 km in 60 minutes, which is 45 km/h.

D to E represents a stop: speed is 0, no further distance travelled.

E to F represents the return journey of 50 km in 60 minutes, which is 50 km/h.

So, the return journey was at an average speed of 50 km/h.

EXERCISE 14B

1 Paulo was travelling in his car to a meeting. This distance–time graph illustrates his journey.

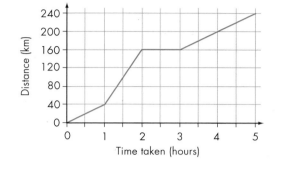

  a How long after he set off did he:

   i stop for his break

   ii set off after his break

   iii get to his meeting place?

  b At what average speed was he travelling:

   i over the first hour

   ii over the second hour

   iii for the last part of his journey?

  c The meeting was scheduled to start at 10.30 am.

   What is the latest time he should have left home?

Advice and Tips

If a part of a journey takes 30 minutes, just double the distance to get the average speed.

2 Farid was travelling by car in Europe on his holiday.

This distance–time graph illustrates his journey.

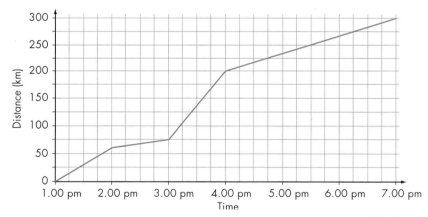

  a His greatest speed was on the motorway.

   i How far did he travel on the motorway?

   ii What was his average speed on the motorway?

  b i When did he travel the most slowly?

   ii What was his lowest average speed?

Chapter 14: Graphs in practical situations 243

3 A small bus set off from Auzio to pick up Mikel and his family. It then went on to pick up Mikel's parents and grandparents. It then travelled further, dropping them all off at a hotel. The bus then went on a further 10 km to pick up another party and it took them back to Auzio. This distance–time graph illustrates the journey.

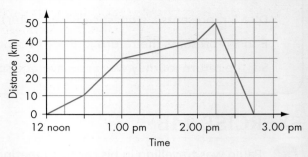

a How far from Auzio did Mikel's parents and grandparents live?

b How far from Auzio is the hotel at which they all stayed?

c What was the average speed of the bus on its way back to Auzio?

4 Reu and Yuto took part in a 5000 m race. It is illustrated in this graph.

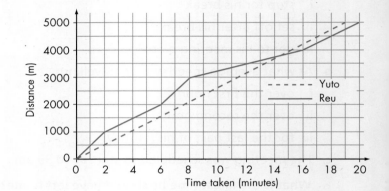

a Yuto ran a steady race. What is his average speed in:

  i metres per minute

  ii kilometres per hour?

b Reu ran in spurts. What was his highest average speed?

c Who won the race and by how much?

5 Three friends, Patrick, Araf and Sean, ran a 1000 m race. The race is illustrated on the distance–time graph.

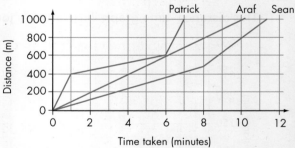

a Describe how each of them completed the race.

b i What is Araf's average speed in m/s?

  ii What is this speed in km/h?

6 A walker sets off at 0900 from point P to walk along a trail at a steady pace of 6 km per hour.

90 minutes later, a cyclist sets off from P on the same trail at a steady pace of 15 km per hour.

a Draw a graph to illustrate the journeys of the walker and cyclist.

b At what time did the cyclist overtake the walker?

Advice and Tips

Mark a grid with a horizontal axis as time from 0900–1300, and the vertical axis as distance from 0 to 24. Draw lines for both walker and cyclist. Remember that the cyclist doesn't start until 1030.

7 Three school friends set off from school at the same time, 1545. They all lived 12 km away from the school. The distance–time graph illustrates their journeys.

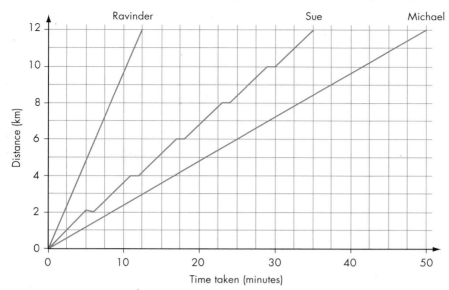

One of them went by bus, one cycled and one was taken by car.

a **i** Explain how you know that Sue used the bus.

  ii Who went by car?

b At what times did each of them get home?

c **i** When the bus was moving, it covered 2 km in 5 minutes. What is this speed in kilometres per hour?

  ii Overall, the bus covered 12 km in 35 minutes. What is this speed, in kilometres per hour?

  iii How many stops did the bus make before Sue got off?

8 A girl walks at a steady speed of 1.5 metres/second for 20 seconds.

She stops for 20 seconds.

Then she walks at 2 metres/second for 10 seconds.

a How far has she walked altogether?

b Show her journey on a distance–time graph.

c What is her average speed for the whole journey?

9 A car drives along a motorway for 2 hours at a steady speed of 100 km/hour.

The driver has a break of 30 minutes.

She then drives back to where she started.

The whole trip takes 5 hours.

a Draw a distance–time graph to show the journey.

b What is the average speed for the second half of the journey?

10 A man walks to the top of a high hill.

He starts at 0900 and by 1000 he has travelled 5 km.

From 1000 to 1200 he covers another 8 km.

He takes an hour to complete the final 2 km to the top of the hill.

a What time did he reach the top?
b How far did he walk?
c Illustrate the journey with a distance–time graph.
d Find the average speed for each stage of the walk and for the whole journey.

14.3 Speed–time graphs

This is the distance–time graph for a cyclist.

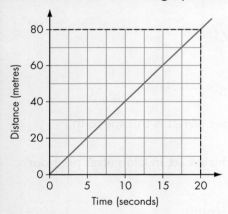

The graph is a straight line. This shows that she is moving at constant speed.

$$\text{Speed} = \frac{\text{distance}}{\text{time}} = \frac{80}{20} = 4 \text{ m/s}$$

The distance she travels in 20 seconds = 4 × 20 = 80 metres.

This is a **speed–time** graph of the same situation.

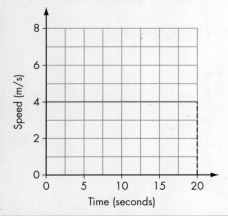

The horizontal line shows that she is moving at a **constant speed**.

The area between the line and the *x*-axis gives the **distance travelled**.

Area = 20 × 4 = 80 metres travelled in 20 seconds

Here is a different speed–time graph for the cyclist.

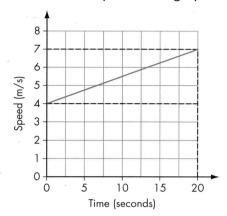

In this case the speed increases uniformly from 4 m/s to 7 m/s in 20 seconds.

The **rate of increase** of speed is the **acceleration**.

Acceleration = $\frac{7-4}{20}$ = 0.15 m/s²

As before, the area between the line and the *x*-axis gives the distance travelled.

Area = area of rectangle and area of triangle

= 20 × 4 + $\frac{1}{2}$ × 20 × 3 = 110 metres travelled

Note: if you know the formula for the area of a trapezium, you can use that here to give:

Area = $\frac{1}{2}$(4 + 7) × 20 = 110 metres, as above.

Example 4

This is a speed–time graph for a car over 15 seconds. Describe the car's journey.

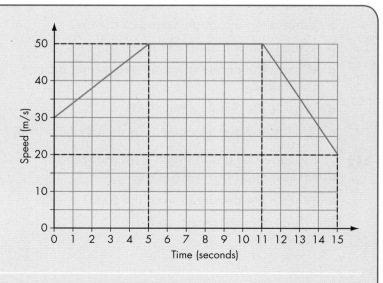

In the first five seconds the car accelerates steadily from 30 m/s to 50 m/s.

The acceleration is $\frac{50-30}{5}$ = 4 m/s².

For the next six seconds the car travels at a constant speed of 50 m/s.

For the last four seconds the car slows down steadily from 50 m/s to 20 m/s.

Because the speed is decreasing you can say the car is **decelerating**.

The **deceleration** is $\frac{50-20}{4} = 7.5$ m/s^2.

The area under the line gives the distance travelled.

Distance travelled = area of trapezium (from 0 to 5 seconds) + area of rectangle
(from 5 to 11 seconds) + area of trapezium (from 11 to 15 seconds).

= 200 + 300 + 140

= 640 metres.

EXERCISE 14C

1 The speed–time graph shows a car accelerating for 40 seconds and then decelerating.

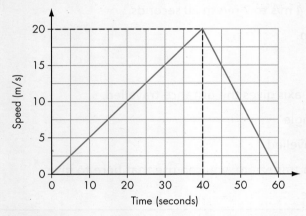

a What is the maximum speed of the car?

b What is the initial acceleration?

c What is the deceleration for the second part of the journey?

d What is the distance travelled?

e What is the average speed for the whole journey?

Advice and Tips

Average speed = $\frac{\text{distance travelled}}{\text{time taken}}$

2 The speed–time graph shows a train slowing down as it approaches a station.

a What is the deceleration of the train?

b What is the distance travelled?

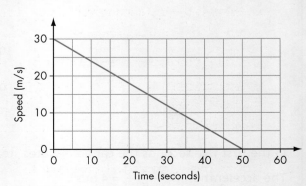

3 The speed–time graph shows the speed of a boat over thirty seconds.

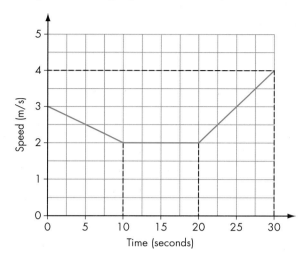

a What is the acceleration for the last ten seconds?
b What is the deceleration for the first ten seconds?
c What is the distance travelled?
d What is the average speed?

4 The graph illustrates a journey for a car.

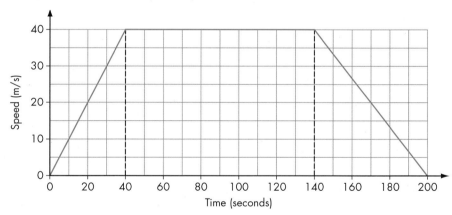

a What is the initial acceleration?
b What is the final deceleration?
c What is the distance travelled?
d What is the average speed?

5 This table shows the speed of a cyclist as he descends a hill.

Time (s)	0	5	10	15
Speed (m/s)	4	8		16

a Draw a speed–time graph for this journey.
b Find the acceleration of the cyclist.
c Find the distance travelled in the first ten seconds.

Chapter 14: Graphs in practical situations

6 A car accelerates from 10 m/s at a rate of 2 m/s² for 8 seconds.

    a What is the speed of the car after 8 seconds?

    b Draw a speed–time graph to illustrate this.

    c How far does the car travel in those 8 seconds?

7 This graph shows the journeys of two cars. The red car in this graph is travelling at a constant speed.

As the red car passes it the blue car starts from rest and accelerates for 15 seconds in the same direction as the red car.

The blue car then continues at a constant speed.

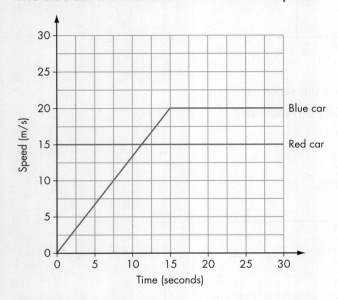

    a What is the initial acceleration of the blue car?

    b How far apart are the two cars after 30 seconds?

8 A car and a motorbike start at the same time and travel in the same direction.

Their speeds are shown on the speed–time graph.

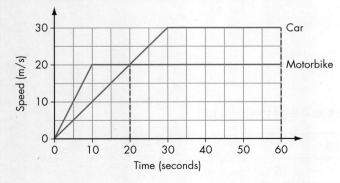

14.3 Speed–time graphs

Chapter 14 . Topic 4

   a What is the initial acceleration of the motorbike?
   b i When are they travelling at the same speed?
      ii How far apart are they at that time?
   c How far does the car travel when it is moving faster than the motorbike?

9 A truck decelerated uniformly from 20 m/s to rest. It travelled 150 m.

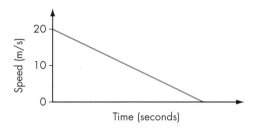

   a How long did it take to come to rest?
   b What was the rate of deceleration?

10 A cyclist started from rest and accelerated at 1.2 m/s² for 5 seconds.
   a What was the speed of the cyclist after 5 seconds?
   b Sketch a speed–time graph for the 5 seconds.
   c How far did the cyclist travel in the 5 seconds?

14.4 Curved graphs

Distance–time and speed–time graphs are not always straight lines. Often they are curves.

This graph shows the distance travelled by a runner in a 10-second interval.

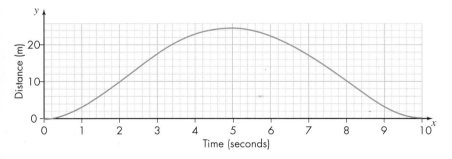

The runner goes 20 metres and then returns to the start.

To find the speed at any time, draw a tangent to the curve and find its gradient.

Example 5

For the runner described above, find the speed after 2 seconds.

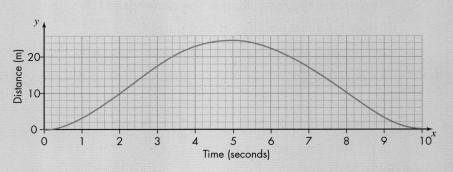

Draw a tangent at the point (2, 10)

This is a straight line with the same gradient as the curve at that point.

To find the gradient, draw a triangle underneath the tangent.

Gradient = $\frac{BC}{AB}$ = $\frac{31}{4}$ = 7.75

The speed is 7.75 m/s

Because this value is found by drawing a tangent on a graph, it will be an approximate answer.

For a speed–time graph that is a curve, you can find the acceleration in a similar way.

EXERCISE 14D

1. A car starts from rest and drives in a straight line.

   This graph shows the distance from the start for the first 5 seconds.

   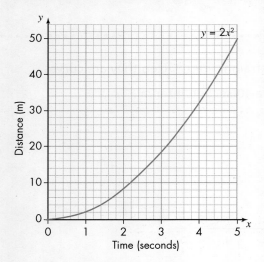

   a Use this table of values to make a copy of the graph.

Time (s)	0	1	2	3	4	5
Distance (m)	0	2	8	18	32	50

   b Draw a tangent at $x = 2$
   c Estimate the speed after 2 seconds.
   d Estimate the speed after 4 seconds.

2 A tennis ball is hit straight upwards.

This table shows the height of the ball.

Time (s)	0	1	2	3	4	5	6
Height (m)	0	25	40	45	40	25	0

a Use the table to draw a graph of the height of the ball.
b Draw a tangent to estimate the speed after 2 seconds.
c Draw a tangent to estimate the initial speed (when the time = 0).
d What is the speed after 3 seconds?
e Estimate the speed after 5 seconds.

3 A kite is blown vertically up and then vertically down to the ground.

Time (s)	0	1	2	3	4	5	6	7	8	9	10
Height (m)	20	27	32	35	36	35	32	27	20	11	0

a Draw a graph to show the height of the kite.
b Draw a tangent to estimate speed of the kite after 2 seconds.
c Estimate the speed after 7 seconds.
d When is the speed 0 m/s?
e Estimate the speed when the kite hits the ground.

4 This table shows the speed of an accelerating car.

Time (s)	0	1	2	3	4	5	6
Speed (m/s)	0	16	25	31	35	37	38

a Draw a graph to show the speed of the car.
b Draw a tangent to estimate the acceleration after 1 s.
c Estimate the acceleration after 2 s.
d Estimate the acceleration after 4 s.

5 This table gives the speed of a car.

Time (s)	0	10	20	30	40	50	60	70
Speed (m/s)	27	36	40	36	27	16	9	3

a Show the speed on a graph.
b Draw a tangent to estimate the acceleration after 10 seconds.
c When is the acceleration 0?
d Estimate the deceleration after 1 minute.

6 This table shows the speed of a car over an interval of 80 seconds.

Time (s)	0	10	20	30	40	50	60	70	80
Speed (m/s)	4	15	19	18	15	12	11	15	26

  a Draw a graph of the speed.
  b Find the acceleration after 10 s.
  c Find the deceleration after 40 s.
  d Find the acceleration after 60 s.
  e When is the acceleration 0?

Check your progress

Core
- I can interpret graphs in practical situations
- I can understand and use travel graphs that involve distance or speed
- I can find the speed from the gradient of a distance–time graph
- I can understand and use conversion graphs

Extended
- I understand the idea of a rate of change
- I can estimate speed from the gradient of the tangent to a distance–time graph
- I can estimate acceleration or deceleration using the gradient of the tangent to a speed–time graph

Chapter 15
Straight-line graphs

Topics	Level	Key words
1 Drawing straight-line graphs	CORE	quadrant, coordinates, negative coordinates, Cartesian coordinates, equation of line, Cartesian plane, straight-line graph
2 The equation $y = mx + c$	CORE	slope, gradient, intercept, coefficient
3 More about straight-line graphs	EXTENDED	
4 Solving equations graphically	CORE	
5 Parallel lines	CORE	parallel
6 Points and lines	EXTENDED	midpoint, line segment
7 Perpendicular lines	EXTENDED	perpendicular

In this chapter you will learn how to:

CORE	EXTENDED
• Demonstrate familiarity with Cartesian coordinates in two dimensions (C3.1 and E3.1) • Construct tables of values for functions of the form $ax + b$, $\pm x^2 + ax + b$, $(x \neq 0)$. where a and b are integer constants. (C2.11) • Draw and interpret such graphs. (C2.11) • Interpret and obtain the equation of a straight-line graph in the form $y = mx + c$. (C3.4 and E3.4) • Determine the equation of a straight line parallel to a given line. (C3.5 and E3.5) • Find the gradient of a straight line. (C3.2 and E3.2)	• Calculate the gradient of a straight line from the coordinates of two points on it. (E3.2) • Calculate the length and the coordinates of the midpoint of a straight line from the coordinates of its end points. (E3.3) • Find the gradient of a line perpendicular to a given line – find the gradient of parallel and perpendicular lines. (E3.6)

Why this chapter matters

The simplest type of graph is a straight-line graph. This shows that the value of one variable on the graph is always affected in a certain way by changes in the value of the other.

A graph gives you a good visual impression of the way two variables are related to each other.

This graph shows the results of an experiment that measured the voltage in an electrical circuit when different currents were flowing. The points are approximately in a straight line.

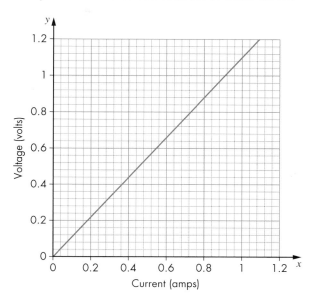

You could use this graph to find the voltage for any current you choose. For example, if you read up from 0.6 amps on the horizontal axis (the x-axis) to the graph line you find that the voltage on the vertical axis (the y-axis) is 0.66 volts and if the current is 0.8 amps the voltage is 0.88 volts.

You can see that there is a constant relationship between current and voltage: the current in amps is 1.1 × the voltage in volts. You can show this as an equation:

$y = 1.1x$ (voltage is on the y-axis and current is on the x-axis)

Then you can use this equation to find the voltage for any current you choose. You can also use it to find the current for any voltage. Once you know the relationship between variables, using the equation is quicker and easier than using the graph.

This connection between the geometry of graphs and the algebra of equations was made by a French mathematician, René Descartes in the 17th century CE. This picture shows him at work.

The coordinates in such graphs are called **Cartesian coordinates** (in Latin his name was *Cartesius*) and the xy grid they appear on is called the **Cartesian plane**.

The grid can be extended to include negative values as well. The graphs and equations shown on it are not always straight-line ones, as you will see in later chapters.

Chapter 15: Straight-line graphs

15.1 Drawing straight-line graphs

Using coordinates

A set of axes can form four sectors called **quadrants** but, so far, all the points you have read or plotted on graphs have been **coordinates** in the first quadrant (the top right section of a grid). The grid below shows you how to read and plot coordinates in all four quadrants and how to find the equations of vertical and horizontal lines. This involves using **negative coordinates**.

The coordinates of a point are given in the form (x, y), where x is the number along the x-axis and y is the number up the y-axis. These are sometimes called **Cartesian coordinates** after their inventor, René Descartes.

The grid is sometimes called the **Cartesian plane**.

The coordinates of the four points on the grid are:

A(2, 3)   B(–1, 2)   C(–3, –4)   D(1, –3)

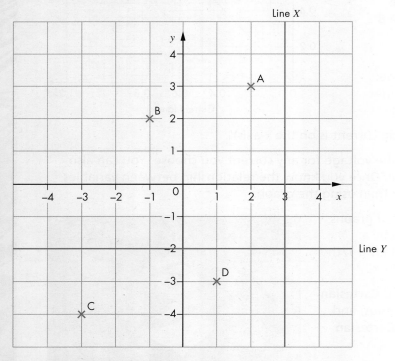

The x-coordinate of all the points on line X is 3.
So you can say the **equation of line** X is $x = 3$.

The y-coordinate of all the points on line Y is –2.
So you can say the equation of line Y is $y = -2$.

Note: The equation of the x-axis is $y = 0$ and the equation of the y-axis is $x = 0$.

15.1

Straight-line graphs

An equation of the form $y = mx + c$ where m and c are numbers will give a **straight-line graph**.

Example 1

Draw the graph of $y = 4x - 5$ for values of x from 0 to 5. This is usually written as $0 \leq x \leq 5$.

Choose three values for x: these should be the highest and lowest x-values and one in between.

Work out the y-values by substituting the x-values into the equation.

Keep a record of your calculations in a table, as shown below.

x	0	3	5
y			

When $x = 0$, $y = 4(0) - 5 = -5$
This gives the point $(0, -5)$.

When $x = 3$, $y = 4(3) - 5 = 7$
This gives the point $(3, 7)$.

When $x = 5$, $y = 4(5) - 5 = 15$
This gives the point $(5, 15)$.

Hence your table is:

x	0	3	5
y	−5	7	15

You now have to decide the extent (range) of the axes. You can find this out by looking at the coordinates that you have so far.

The smallest x-value is 0, the largest is 5. The smallest y-value is −5, the largest is 15.

Now draw the axes, plot the points and complete the graph.

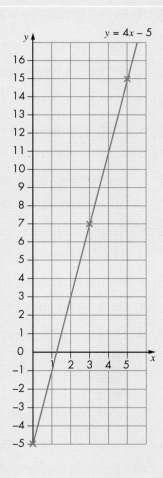

Chapter 15: Straight-line graphs

It is nearly always a good idea to choose 0 as one of the *x*-values. You will often be given the range for the *x*-values and a set of axes to draw your graph.

EXERCISE 15A

Read through these hints before drawing the straight-line graphs in this exercise.

- Use the highest and lowest values of *x* given in the range.
- Do not pick *x*-values that are too close together, such as 1 and 2. Try to space them out so that you can draw a more accurate graph.
- Always label your graph with its equation. This is particularly important when you are drawing two graphs on the same set of axes.
- Create a table of values.

1 Draw the graph of $y = 3x + 4$ for *x*-values from 0 to 5 ($0 \leq x \leq 5$).

2 Draw the graph of $y = 2x - 5$ for $0 \leq x \leq 5$.

3 Draw the graph of $y = \frac{x}{2} - 3$ for $0 \leq x \leq 10$.

4 Draw the graph of $y = 3x + 5$ for $-3 \leq x \leq 3$.

5 Draw the graph of $y = \frac{x}{3} + 4$ for $-6 \leq x \leq 6$.

Advice and Tips

Complete the table of values first, then you will know the extent of the *y*-axis.

6 a On the same set of axes, draw the graphs of $y = 3x - 2$ and $y = 2x + 1$ for $0 \leq x \leq 5$.

b At which point do the two lines intersect?

7 a On the same axes, draw the graphs of $y = 4x - 5$ and $y = 2x + 3$ for $0 \leq x \leq 5$.

b At which point do the two lines intersect?

8 a On the same axes, draw the graphs of $y = \frac{x}{3} - 1$ and $y = \frac{x}{2} - 2$ for $0 \leq x \leq 12$.

b At which point do the two lines intersect?

9 a On the same axes, draw the graphs of $y = 3x + 1$ and $y = 3x - 2$ for $0 \leq x \leq 4$.

b Do the two lines intersect? If not, why not?

10 a Copy and complete the table to draw the graph of $x + y = 5$ for $0 \leq x \leq 5$.

x	0	1	2	3	4	5
y	5		3		1	

b Now draw the graph of $x + y = 7$ for $0 \leq x \leq 7$.

11 A line has the equation $y = 1.5x + 3$.

Decide whether each of these points is on the line or not.

a (6, 12) b (0, 4.5)
c (2, 6)   d (10, 13)
e (−2, 0) f (−4, −6)

Chapter 15. Topic 2

12 This is a graph of $y = 2x - 7$.

Use the graph to solve these equations.

a $2x - 7 = 5$
b $2x - 7 = 0$
c $2x - 7 = -3$

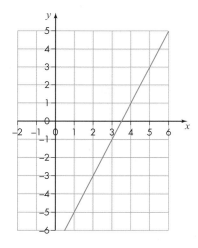

13 This is a graph of $y = 0.5x + 15$.

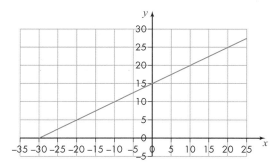

Use the graph to solve these equations.

a $0.5x + 15 = 25$ b $0.5x + 15 = 10$

15.2 The equation $y = mx + c$

Gradient

The **slope** of a line is called its **gradient**. The steeper the slope of the line, the larger the value of the gradient.

The graph shows the line with equation $y = 2x + 3$. You can measure the gradient of the line by drawing a right-angled triangle that has part of the line as its hypotenuse (sloping side). The gradient is then given by:

$$\text{gradient} = \frac{\text{distance measured up}}{\text{distance measured along}}$$
$$= \frac{6}{3}$$
$$= 2$$

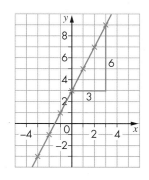

In fact, if the equation of the line is $y = mx + c$, then m is the gradient of the line.

Example 2

Show that the line with the equation $y = 0.5x - 1$ has a gradient of 0.5.

x	-2	0	2	4	6
y	-2	-1	0	1	2

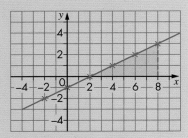

This is a graph of the line.

From the shaded triangle, you can see that the gradient is:

$\frac{3}{6} = 0.5$.

What does the 'c' in $y = mx + c$ represent?

From the graphs above you can see it is the value of y, where the line crosses the y-axis.

$y = 2x + 3$ passes through (0, 3).

$y = 0.5x - 1$ passes through (0, -1).

Note that a line that slopes downwards from left to right has a negative gradient.

This line has a gradient of -1.

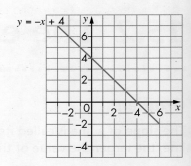

Summary

When a graph can be expressed in the form $y = mx + c$, the **coefficient** of x, which is m, is the **gradient**, and the constant term, which is c, is the **intercept** on the y-axis.

This means that if you know the gradient, m, of a line and its intercept, c, on the y-axis, you can write down the equation of the line immediately.

For example, if $m = 3$ and $c = -5$, the equation of the line is $y = 3x - 5$.

This gives a method of finding the equation of any line drawn on a pair of coordinate axes.

Example 3

Find the equation of the line shown in diagram **A**.

A 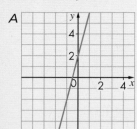 B 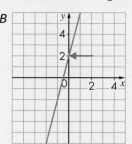 C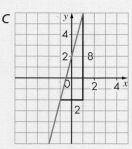

First, find where the graph crosses the *y*-axis (diagram **B**).

So $c = 2$

Next, measure the gradient of the line (diagram **C**).

y-step = 8

x-step = 2

gradient = $8 \div 2 = 4$

So $m = 4$

Finally, write down the equation of the line: $y = 4x + 2$.

EXERCISE 15B

1 You drew the graphs of the lines with these equations in Exercise 15A, questions **1–5**.

In each case state the gradient.

Then check from your drawing that you are correct.

a $y = 3x + 4$

b $y = 2x - 5$

c $y = \dfrac{x}{2} - 3$

d $y = 3x + 5$

e $y = \dfrac{x}{3} + 4$

2 Give the equation of each of these lines, all of which have positive gradients. (Each square represents one unit.)

a 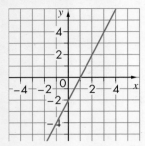 b 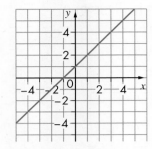 c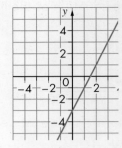

d 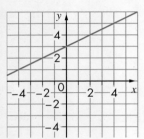 e 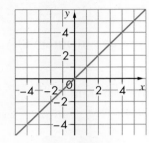 f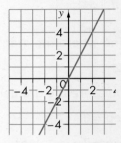

3 In each of these grids, there are two lines. (Each square represents one unit.)

a 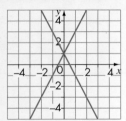 b 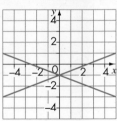 c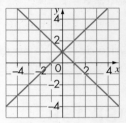

For each grid find the equation of each of the lines.

4 Give the equation of each of these lines, all of which have negative gradients. (Each square represents one unit.)

a 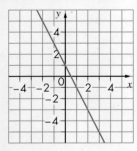 b 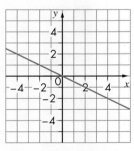 c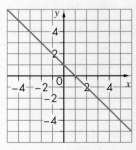

d 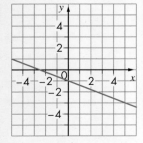 e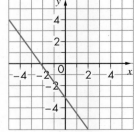

15.2 The equation $y = mx + c$

Chapter 15 . Topic 3

5 The line $y = 4x + c$ passes through (1, 7).
  a Find the value of c.
  b Where does the line cross the y-axis?

6 The line $y = mx - 6$ passes through (3, 6).
  a Find the value of m.
  b What is the gradient of the line?

7 Here are the equations of three lines.

   $y = 4x - 2$

   $y = 3x - 4$

   $y = 4x + 5$

   Which of these lines are parallel? Explain how you know.

8 Find the equation of a straight line that passes through the point (5, 3) and the origin.

9 a Draw a graph of the line $y = 0.5x + 3$.
  b Draw the reflection of the line $y = 0.5x + 3$ in the y-axis.
  c Find the equation of the line in part **b**.
  d Draw the reflection of the line $y = 0.5x + 3$ in the x-axis.
  e Find the equation of the line in part **d**.

15.3 More about straight-line graphs E

The equation of a straight line is not always written in the form $y = mx + c$

For example, $y = -2x + 3$ is the equation of a straight line with a gradient of -2 and an intercept on the y-axis of (3, 0)

By adding $2x$ to each side the equation can be written as $y + 2x = 3$ or $2x + y = 3$

By subtracting 3 from each side it can be written as $2x + y - 3 = 0$

Here are some more examples.

Equation	Alternatives
$y = -4x - 2$	$y + 4x = -2$ or $4x + y + 2 = 0$
$y = -5x + 1$	$y + 5x = 1$ or $5x + y = 1$
$y = \frac{3}{2}x + 1$	$2y = 3x + 2$
$y = 4x - 12$	$y = 4(x - 3)$

Chapter 15: Straight-line graphs **265**

Example 4

The equation of a straight line is $3x + 4y = 24$

a Find the gradient of the line.

b Find the intercept on the y-axis.

c Draw a graph of the line.

---

a Rewrite the equation.

Subtract $3x$ from both sides: $\quad 4y = -3x + 24$

Divide both sides by 4: $\quad y = -\dfrac{3}{4}x + 6$

The gradient is $-\dfrac{3}{4}$

b The intercept is (0, 6)

c You know that (0, 6) is on the line.

To find another point, substitute $y = 0$ into the original equation.

$3x + 0 = 24$

$3x = 24$

$x = 8 \quad$ The point (8, 0) is also on the line.

Draw the line by joining (0, 6) and (8, 0)

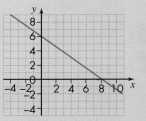

EXERCISE 15C

1 The equation of a straight line is $5x + 2y = 10$

 a Write the equation in the form $y = mx + c$

 b Find the gradient and the intercept on the y-axis.

 c Draw a graph of the line.

2 The equation of a straight line is $x + 3y = 15$

 a Write the equation in the form $y = mx + c$

 b Find the gradient and the intercept on the y-axis.

 c Draw a graph of the line.

3 The equation of a straight line is $x + 2y + 6 = 0$

 a Write the equation in the form $y = mx + c$

 b Find the gradient and the intercept on the y-axis.

 c Draw a graph of the line.

Chapter 15. Topic 4

4 Here are the equations of straight lines.

Write each one in the form $y = mx + c$

  a $x + y + 20 = 0$ b $y = 3(x + 5)$ c $7x + 10y = 30$
  d $x - 5y = 40$ e $2(y + 1) = 3x$ f $12 - 2y = x$

5 Find the gradient and the intercept on the y-axis for each of the lines in question **4**.

6 Find the gradient of each of these straight lines.

  a $4x + 4y = 15$ b $x = 3y + 6$ c $y = 6(x - 5)$
  d $5x + 10y = 28$ e $10x = 5y + 28$ f $28x - 5y = 10$

7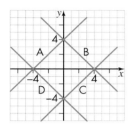

Match the equation to the line.

  a $x - y = 4$ b $y - x = 4$ c $x + y + 4 = 0$ d $x + y = 4 = 0$

8 Find the odd one out. Give a reason for your answer.

  a $2x + 3y = 12$ b $y = 4 - \frac{2}{3}x$ c $6y = -4x + 24$ d $3x - 2y = 12$ e $x = 6 - 1.5y$

9 Find the points where each of these straight lines cross the axes.

  a $3x + 5y = 45$ b $2y = 20 - x$ c $12x + 6y + 60 = 0$

15.4 Solving equations graphically

Here is a graph of $y = 0.73x - 1.15$

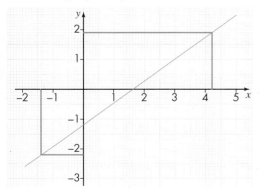

You can use this graph to solve equations.

Example 5

Use the graph to solve the equations:

a $0.73x - 1.15 = 0$

b $0.73x - 1.15 = -2.2$

c $0.73x - 2.65 = 0$

---

a Since the equation of the line is $y = 0.73x - 1.15$ then $0.73x - 1.15 = 0$ when the y coordinate is zero.

This is where the line crosses the x-axis.

Reading from the graph, $x = 1.6$

Because this is taken from a graph, it can only be correct to 1 d.p.

b To solve the equation $0.73x - 1.15 = -2.2$, find -2.2 go across the line and then up to the x-axis. The lines on the graph show this.

The solution is the x-coordinate. That is $x = 1.6$

c If $0.73x - 3.05 = 0$ then $0.73x - 1.15 - 1.9 = 0$

So $0.73x - 1.15 = 1.9$

Find 1.9 on the y-axis, go across to the line and down to the x-axis.

The solution is $x = 4.2$

EXERCISE 15D

1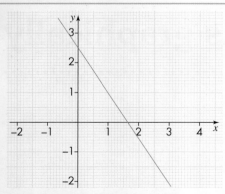

This is a graph of $y = 2.45 - 1.37x$

Use it to solve these equations:

a $2.45 - 1.37x = 0$

b $2.45 - 1.37x = 3$

c $2.45 = 1.37x - 1.5$

2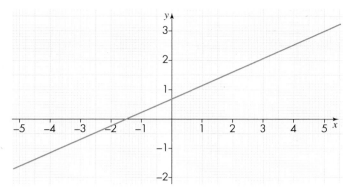

This is a graph of $y = 0.46x + 0.73$

Use the graph to solve these equations:

a $0.46x + 0.73 = 0$ b $0.46x + 0.73 = 2.5$ c $0.46x + 1.73 = 0$

3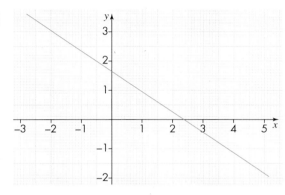

This is a graph of $2.7y + 1.9x = 4.42$

Use the graph to solve these equations:

a $1.9x = 4.42$ b $2.7y = 4.42$ c $2.7 + 1.9x = 4.42$

4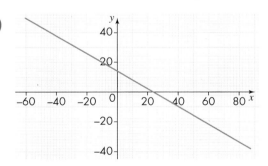

This is a graph of $y = 14.2 - 0.62x$

Use the graph to solve these equations:

a $14.2 - 0.62x = 0$ b $14.2 - 0.62x = 32$ c $44.2 - 0.62x = 0$

Chapter 15: Straight-line graphs

5

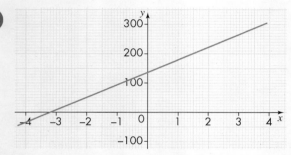

This is a graph of $y = 2.5 - 0.31x$ and $y = 0.42x - 0.6$

Use the graph to solve these equations:

a $0.42x - 0.6 = 2.3$

b $2.5 - 0.31x = 1.6$

c $0.42x - 0.6 = 2.5 - 0.31x$

6

This is a graph of $y = 43.8x + 136.2$

Use it to solve these equations:

a $43.8x + 36.2 = 0$

b $87.6x + 272.4 = 500$

c $438x + 1362 = 1500$

15.5 Parallel lines

Consider the line with the equation $y = 0.5x + 1$.

The gradient is 0.5. Any line **parallel** to it will have the same gradient.

Examples are $y = 0.5x + 3$ and $y = 0.5x - 4$. They all have 0.5 as the coefficient of x.

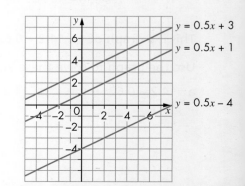

15.5

Example 6

Find the equation of the line parallel to $y = 4x$ that passes through (2, 5).

The line will have the equation $y = 4x + c$ for some value of c.

Substitute the coordinates of (2, 5) into the equation.

$5 = 4 \times 2 + c$

$5 = 8 + c$

$c = -3$

So the equation is $y = 4x - 3$.

EXERCISE 15E

1 A line has the equation $y = 2x + 6$.
  a Show that (3, 12) is on this line.
  b What is the gradient of the line?
  c Draw the graph of the line.
  d Find the equation of a parallel line passing through the origin.
  e Find the equation of a parallel line passing through (0, 3).

2 A line has the equation $y = \frac{1}{4}x - 1$.
  a Find the coordinates of the points where the line crosses the y-axis and the x-axis.
  b Find the equation of a parallel line through (8, 5).

3 A line has the equation $y = 8 - 2x$.
  a What is the gradient of this line?
  b Where does the line cross the x-axis?
  c Draw the line on a grid.
  d A parallel line passes through the origin. What is the equation?
  e Find the equation of a parallel line through (5, 4).

4 a The line with equation $y = 5x + k$ passes through (2, 11). Find the value of k.
  b A line that is parallel to the line in part **a** passes through (4, 11). Find the equation of this line.

5 The two lines in this graph are parallel.

Find the equation of the lower line.

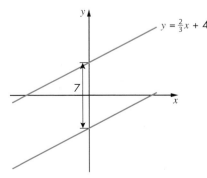

Chapter 15 . Topic 6

6 The three lines on this graph are parallel.

Find the equations of a and b.

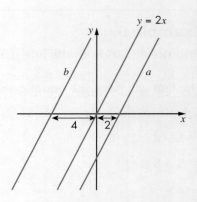

7 The equation of this line can be written as $y + 2x = 6$.

a Show that (1, 4) is on this line.
b Rearrange the formula in the form $y = mx + c$.
c What is the gradient of the line?
d Find the equation of a line that is parallel to this line, but passes through the origin.

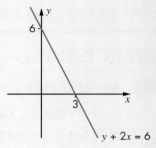

15.6 Points and lines

The equation of a line

If you know the coordinates of two points on a straight line you have enough information to find the equation of the line between them.

Consider the line through (–2, 2) and (4, 5).

Using the triangle shown:

$$\text{gradient} = \frac{\text{difference between } y\text{-coordinates}}{\text{difference between } x\text{-coordinates}}$$

$$= \frac{5 - 2}{4 - -2}$$

$$= \frac{3}{6}$$

$$= 0.5$$

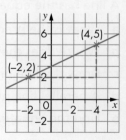

So the equation must be $y = 0.5x + c$ for some value of c.

Use the coordinates of either of the points to find c.

Using $x = 4$ and $y = 5$:
$5 = 0.5 \times 4 + c$
$5 = 2 + c$
$c = 3$

So the equation is $y = 0.5x + 3$.

15.6

Midpoints

The **midpoint** of a **line segment** is the same distance from each end.

To find the coordinates of the midpoint, add the coordinates of the end points and divide by 2.

In the example shown:

midpoint is $\left(\dfrac{-2+4}{2}, \dfrac{2+5}{2}\right) = (1, 3.5)$

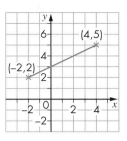

The distance between two points

You can use Pythagoras' theorem (see Chapter 25 section 1) to find the distance between two points.

The lengths of the sides of the right-angled triangle shown are the differences between the *x*-coordinates and the *y*-coordinates of the end points.

In this case the lengths are 3 and 6.

Distance between points = $\sqrt{3^2 + 6^2} = \sqrt{45} = 6.71$ to 2 decimal places.

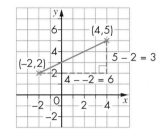

EXERCISE 15F

1 Find the gradient of the line through each pair of points.

   a (4, 0) and (6, 6)
   b (0, 3) and (8, 7)
   c (2, −2) and (4, 6)
   d (1, 5) and (5, 1)
   e (−4, 6) and (6, 1)
   f (−5, −3) and (4, 3)

   Advice and Tips

   It will help to plot them on a grid.

2 Find the equation of the line joining each pair of points.

   a (0, −3) and (4, 5)
   b (−4, 2) and (2, 5)
   c (−1, −6) and (2, 6)
   d (1, 5) and (4, −4)

3 Find the midpoints of the line segments joining the points in question **1**.

4 A is (−3, 5), B is (1, 1) and C is (5, 9).

   a Draw the triangle ABC on a coordinate grid.
   b Find the equation of the straight line through A and C.
   c Find the midpoint of AB.
   d Find the equation of the straight line through the midpoints of AC and BC.

Chapter 15: Straight-line graphs

5 Find the length of the line segment joining each pair of points.

  a (2, 2) and (6, 5)   **b** (−3, 2) and (9, 7)
  c (1, 5) and (7, −3)   **d** (−6, −4) and (9, 4)

6 A circle is drawn with its centre at (2, 1) and radius 5.

  Show that (−3, 1), (6, 4) and (5, −3) all lie on the circle.

7 Show that the triangle ABC in question **4** is an isosceles triangle.

15.7 Perpendicular lines

The lines in this diagram are **perpendicular** to each other.

This means that the angle between them is a right angle, 90°.

You can use the dotted triangles to work out the gradients.

The gradient of line A is $\frac{2}{3}$.

The gradient of line B is $-\frac{3}{2}$ or −1.5.

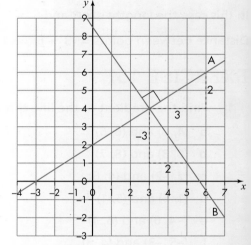

You can see that:

  the gradient of line B = −(the reciprocal of line B)

You can also see that the product of the gradients is −1:

  $\frac{2}{3} \times -\frac{3}{2} = -1$

The gradients of perpendicular lines usually have this property. The only exception is when the gradient of one of the lines is zero.

Example 7

Find the equation of the line that is perpendicular to $2x + y = 3$ and passes through the point (4, 6).

The given equation is equivalent to $y = -2x + 3$.

The gradient is −2.

The gradient of the line perpendicular to this line is $\frac{1}{2}$. (Find the reciprocal, change the sign).

The equation you want is $y = \frac{1}{2}x + c$ where c is a number.

Find the value of c by substituting the coordinates (4, 6) in the equation.

  $6 = \frac{1}{2} \times 4 + c$ ÷ $6 = 2 + c$ ÷ $c = 4$

The equation of the perpendicular line is $y = \frac{1}{2}x + 4$.

15.7

You can check the answer to example 5 by drawing a diagram.

The lines are perpendicular.

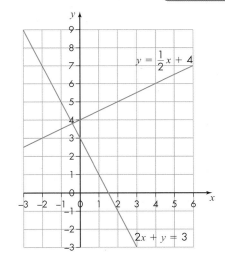

EXERCISE 15G

1 Work out the gradient of a line that is perpendicular to:

a line A

b line B

c line C.

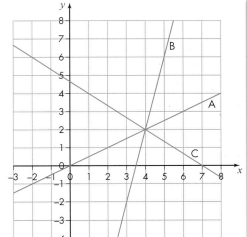

2 Here are the equations of two lines.

$y = 2 + 5x$ $5y + x = 6$

Show that the lines are perpendicular.

3 Work out the gradient of a line that is perpendicular to:

a $y = 3x$ b $y = \frac{1}{2}x - 15$ c $y = 3.5 - 0.4x$ d $y = 7.5x + 5.7$.

4 Work out the gradient of a line that is perpendicular to:

a $x + y + 5 = 0$ b $y + \frac{1}{2}x = 4$ c $3x + 4y = 5$ d $12y - 2x = 15$.

5 Find the equation of the line that is perpendicular to $y = 5x$ and passes through:

a the origin   b (0, 10)   c (10, 0).

6 Find the equation of the line that is perpendicular to $5x + 3y = 30$ and passes through $(-20, -10)$.

7 Find the equation of the line that is perpendicular to AB and passes through C.

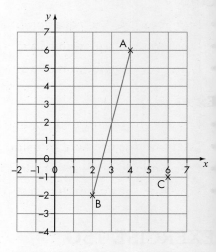

8 A is the point $(-4, 5)$ and B is the point $(6, 7)$.

The perpendicular bisector of AB is a line that is perpendicular to AB and passes through its midpoint.

Find the equation of the perpendicular bisector of AB.

9 This list includes the equations of two pairs of perpendicular lines.

a $3x + 5y = 7$
b $6x + 3y = 7$
c $3y - 5x = 7$
d $8x - 4y = 7$
e $4y + 2x = 7$

Which is the odd one out?

10 The centre of a circle is the point C(3, 2).

The circle passes through the point P(−1, 4).

Work out the equation of the tangent to the circle at P.

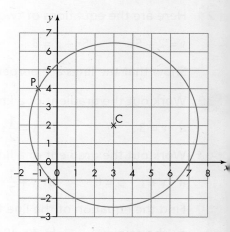

276   15.7 Perpendicular lines

15.7 Check your progress

Core

- I can interpret the equation of a straight-line graph in the form $y = mx + c$
- I can find the equation of a straight line from a graph
- I can find the equation of a straight line parallel to a given line

Extended

- I can interpret and find the equation of a straight-line graph given in different forms
- I can find the gradient of a line parallel to a given line
- I can find the gradient of a line perpendicular to a given line
- I can find the equation of a straight line passing through two points

Chapter 16
Graphs of functions

Topics	Level	Key words
1 Quadratic graphs	CORE	quadratic graph, quadratic equation, parabola
2 Turning points on a quadratic graph	EXTENDED	turning point
3 Reciprocal graphs	CORE	reciprocal
4 More graphs	EXTENDED	cubic, exponential functions, asymptotes
5 Exponential graphs	EXTENDED	exponential growth, exponential decay
6 Estimating gradients	EXTENDED	gradient, tangent

In this chapter you will learn how to:

CORE	EXTENDED
• Construct tables of values for functions of the form $\pm x^2 + ax + b$, $\frac{a}{x}$ ($x \neq 0$) where a and b are integer constants. (C2.11) • Draw and interpret such graphs. (C2.11) • Solve quadratic equations approximately by graphical methods. (C2.11)	• Construct tables of values and draw graphs for functions of the form ax^n (and simple sums of these) and functions of the form $ab^x + c$ where a and c are rational constants and $n = -2, -1, 0, 1, 2, 3$. (E2.11) • Solve associated equations approximately by graphical methods. (E2.11) • Draw and interpret graphs representing exponential growth and decay problems. (E2.11) • Estimate gradients of curves by drawing tangents. (E2.12)

Why this chapter matters

There are many curves that can be seen in everyday life. Did you know that all these curves can be represented mathematically?

Below are a few examples of simple curves that you may have noticed. Can you think of others?

In mathematics, curves can take many shapes. These can be demonstrated using a cone, as shown on the right and below. If you make a cone out of modelling clay, you can see this for yourself. As you look at these curves, try to think of where you have seen them in your own life.

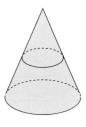

If you slice the cone parallel to the base, the shape you are left with is a circle.

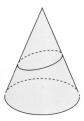

If you slice the cone at an angle to the base, the shape you are left with is an ellipse

If you slice the cone vertically, the shape you are left with is a hyperbola.

The curve that will be particularly important in this chapter is the parabola. Car headlights are shaped like parabolas.

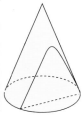

If you slice the cone parallel to its side, the shape you are left with is a parabola.

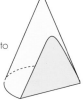

All parabolas can be represented by **quadratic graphs**. In this chapter you will look at how to use **quadratic equations** to draw graphs that have this kind of curve.

The suspension cables on this bridge are also parabolas.

16.1 Quadratic graphs

A graph with a ∪ or ∩ shape is a quadratic graph.

A **quadratic graph** has an equation that involves a 'squared' term, such as x^2.

All of the following are **quadratic equations** and each would produce a quadratic graph.

$y = x^2$

$y = x^2 + 5$

$y = x^2 - 3x$

$y = x^2 + 5x + 6$

$y = x^2 + 2x - 5$

Example 1

Draw the graph of $y = x^2$ for $-3 \leqslant x \leqslant 3$.

First make a table, as shown below.

x	−3	−2	−1	0	1	2	3
$y = x^2$	9	4	1	0	1	4	9

Now draw axes, with $-3 \leqslant x \leqslant 3$ and $0 \leqslant y \leqslant 9$, plot the points and join them to make a smooth curve.

This is the graph of $y = x^2$.
This type of graph is often referred to as a **parabola**.

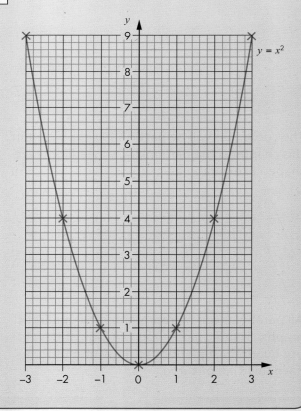

Here are some of the more common mistakes that you should try to avoid, when you are drawing a curve.

- When the points are too far apart, a curve tends to 'wobble'.

  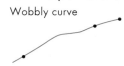
  Wobbly curve

- Drawing curves in small sections leads to 'feathering'.

  Feathering

- The place where a curve should turn smoothly is drawn 'flat'.

  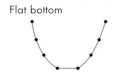
  Flat bottom

- A curve is drawn through a point that, clearly, has been incorrectly plotted.

  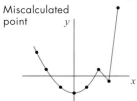
  Miscalculated point

A quadratic curve drawn correctly will *always* be a smooth curve.

Here are some tips that will make it easier for you to draw smooth, curved graphs.

- If you are *right-handed*, turn your piece of paper or your exercise book round so that you draw from left to right. Your hand may be steadier this way than if you try to draw from right to left or away from your body.

  If you are *left-handed*, you may find drawing from right to left the more accurate way.
- Move your pencil over the points as a practice run without drawing the curve.
- Do one continuous curve and only stop at a plotted point.
- Use a *sharp* pencil and do not press too heavily, so that you may easily rub out mistakes.

Example 2

a Draw the graph of $y = x^2 + 2x - 3$ for $-4 \leqslant x \leqslant 2$.

b Use your graph to find the value of y when $x = 1.6$.

c Use your graph to find solve the equation $x^2 + 2x - 3 = 1$.

---

a Draw a table like this, to help work out each step of the calculation.

x	−4	−3	−2	−1	0	1	2
x^2	16	9	4	1	0	1	4
$+2x$	−8	−6	−4	−2	0	2	4
-3	−3	−3	−3	−3	−3	−3	−3
$y = x^2 + 2x - 3$	5	0	−3	−4	−3	0	5

Generally, you do not need to work out all values in a table. If you use a calculator, you need only to work out the y-value. The other rows in the table are just working lines to break down the calculation.

b To find the corresponding y-value for any value of x, you start on the x-axis at that x-value, go up to the curve, across to the y-axis and read off the y-value. This procedure is marked on the graph with arrows.

So when $x = 1.6$, $y = 2.8$.

c If $x^2 + 2x - 3 = 1$, this means that $y = 1$. This time, start at 1 on the y-axis and read off the two x-values that correspond to a y-value of 1.

Again, this procedure is marked on the graph with arrows.

So when $y = 1$, $x = -3.2$ or $x = 1.2$.

There are two possible values of x that make $x^2 + 2x - 3 = 1$.

Notice that the graph is symmetrical. The equation of the line of symmetry is $x = -1$, and the lowest point is $(-1, -4)$.

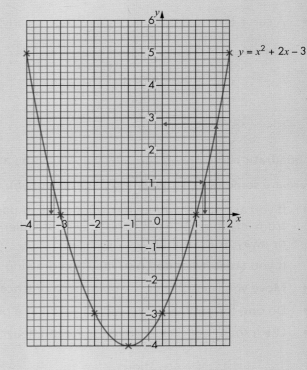

The solutions of an equation like $x^2 + 2x - 3 = 0$ are called the **roots** of the equation.

Look at the graph in Example 2. You can see that the roots of the equation $x^2 + 2x - 3 = 0$ are −3 and 1.

EXERCISE 16A

1 a Copy and complete the table for $y = x^2 + 2$.

x	−3	−2	−1	0	1	2	3
$y = x^2 + 2$	11			2		6	

b Draw a graph of $y = x^2 + 2$ for $-3 \leq x \leq 3$.

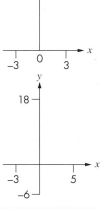

2 a Copy and complete the table for $y = x^2 - 3x$ for $-3 \leq x \leq 5$. Use your table to plot the graph.

x	−3	−2	−1	0	1	2	3	4	5
x^2						4			
$-3x$						−6			
y						−2			

b Use your graph to find the value of y when x = 3.5.

c What are the coordinates of the lowest point on the graph?

d What is the equation of the line of symmetry?

e Use your graph to solve the equation $x^2 - 3x = 5$.

> **Advice and Tips**
>
> You may find you do not need the second and third rows.

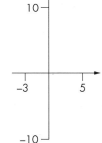

3 a Copy and complete the table for the graph of $y = x^2 - 2x - 8$ for $-3 \leq x \leq 5$. Use your table to plot the graph.

x	−3	−2	−1	0	1	2	3	4	5
y					−8				

b Find the roots of the equation $x^2 - 2x - 8 = 0$

c Use your graph to find the value of y when x = 0.5.

d Use your graph to solve the equation $x^2 - 2x - 8 = 3$.

4 a Copy and complete the table for $y = x^2 - 5x + 4$ for $-1 \leq x \leq 6$. Use your table to plot the graph.

x	−1	0	1	2	3	4	5	6
y		4			−2			

b What are the coordinates of the lowest point on the graph?

c What is the equation of the line of symmetry?

d Use your graph to find the value of y when x = −0.5.

e Find the roots of the equation $x^2 - 5x + 4 = 0$

f Use your graph to solve the equation $x^2 - 5x + 4 = 3$.

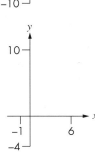

Chapter 16: Graphs of functions

5 **a** Copy and complete the table for $y = x^2 + 2x - 1$ for $-4 \leq x \leq 2$.
Use your table to plot the graph.

x	-4	-3	-2	-1	0	1	2
y	7						

b Use your graph to find the approximate values of the roots of the equation $x^2 - 6x + 3 = 0$
c Use your graph to find the approximate values of the roots of the equation $2x - 1 = 0$
d Use your graph to find the approximate values of the roots of the equation $x^2 + 2x = 2$

6 **a** Copy and complete the table to draw the graph of $y = 12 - x^2$ for $-4 \leq x \leq 4$.

x	-4	-3	-2	-1	0	1	2	3	4
y				11				3	

b Use your graph to find the y-value when $x = 1.5$.
c Use your graph to solve the equation $12 - x^2 = 0$.
d Use your graph to solve the equation $12 - x^2 = 7$.

7 **a** Copy and complete the table to draw the graph of $y = x^2 + 4x$ for $-5 \leq x \leq 2$.

x	-5	-4	-3	-2	-1	0	1	2
x^2	25			4			1	
+4x	-20			-8			4	
y	5			-4			5	

b Where does the graph cross the x-axis?
c Use your graph to find the y-value when $x = -2.5$.
d Use your graph to solve the equation $x^2 + 4x = 3$.

8 **a** Copy and complete the table to draw the graph of $y = x^2 - 6x + 3$ for $-1 \leq x \leq 7$.

x	-1	0	1	2	3	4	5	6	7
y	10			-5			-2		

b Where does the graph cross the x-axis?
c Use your graph to find the y-value when $x = 3.5$.
d Use your graph to solve the equation $x^2 - 6x + 3 = 5$.

9 $y = 5x - x^2$

a Copy and complete this table of values.

x	-1	0	1	2	3	4	5	6
y								

b Draw a graph of $y = 5x - x^2$ for $-1 \leq x \leq 6$.
c What is the highest point on the graph?
d What is the equation of the line of symmetry on the graph?
e Solve the equation $5x - x^2 = 2$.

16.1 Quadratic graphs

16.2 Turning points on a quadratic graph

Here is the equation of the quadratic graph from Example 2: $y = x^2 + 2x - 3$

In Section 13.5 you saw how to write expressions like $x^2 + 2x - 3$ in completed square form.

$y = x^2 + 2x - 3 = (x + 1)^2 - 1 - 3$

$= (x + 1)^2 - 4$

Now $(x + 1)^2 \geq 0$ and only equals 0 when $x = -1$

This means that $x^2 + 2x - 3$ has a minimum value of -4 when $x = -4$

This means that $(-1, -4)$ on the graph of $y = x^2 + 2x - 3$

It is called a **turning point**.

It is a point where the graph changes direction.

Advice and Tips

Look at Section 13.5 if you have forgotten how to do this.

Example 3

a Find the turning point of the graph of $y = x^2 - 3x - 3$

b Sketch the graph of $y = x^2 - 3x - 3$

a Write $x^2 - 3x - 3$ in completed square form.

$x^2 - 3x - 3 = (x - 1.5)^2 - 1.5^2 - 3$

$= (x - 1.5)^2 - 5.25$

$(x - 1.5)^2 = 0$ when $x = 1.5$ and so the turning point is $(1.5, -5.25)$

b To find a point on the curve, let $x = 0$ and then $y = -3$

Hence $(0, -3)$ is on the curve.

You can draw a sketch using $(1.5, -5.25)$ and $(0, -3)$

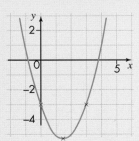

Notice that, by symmetry, $(3, -3)$ is also on the curve.

Chapter 16 . Topic 3

EXERCISE 16B

EXTENDED

1. The equation of a graph is $y = x^2 - 2x - 8$
   - a Write $x^2 - 2x - 8$ in completed square form.
   - b Find the coordinates of the turning point of the graph.
   - c Solve the equation $x^2 - 2x - 8 = 0$
   - d Hence sketch the graph of $y = x^2 - 2x - 8$

2. The equation of a graph is $y = x^2 + 10x + 21$
   - a Write $x^2 + 10x + 21$ in completed square form.
   - b Find the coordinates of the turning point of the graph.
   - c Find the roots of the equation $x^2 + 10x + 21 = 0$
   - d Sketch the graph of $y = x^2 + 10x + 21$

3.
   - a Write $x^2 - 7x + 10$ in completed square form.
   - b Find the coordinates of the turning point of the graph of $y = x^2 - 7x + 10$
   - c Solve the equation $x^2 - 7x + 10 = 0$
   - d Sketch the graph of $y = x^2 - 7x + 10$

4. The equation of a graph is $y = x^2 - 6x + 12$
   - a Find the intercept on the y-axis.
   - b Find the turning point of the graph.
   - c Sketch the graph of $y = x^2 - 6x + 12$
   - d Explain why the graph shows that the equation $x^2 - 6x + 12 = 0$ has no solution.

5.
   - a Find the turning point of the graph of $y = x^2 + 20x + 40$
   - b Sketch the graph of $y = x^2 + 20x + 40$

6. Find the smallest possible value of $x^2 - 25x + 100$

7. The equation of a curve is $y = x^2 + bx + c$ where b and c are constants.

   The intercept on the y-axis is (0, 14).

   The turning point of the graph is (5, −11)

   Find the values of b and c.

16.3 Reciprocal graphs

A **reciprocal** equation has the form $y = \dfrac{a}{x}$.

Examples of reciprocal equations are: $y = \dfrac{1}{x}, y = \dfrac{4}{x}, y = -\dfrac{3}{x}$

All reciprocal graphs have a similar shape and some symmetry properties.

Example 4

Complete the table to draw the graph of $y = \frac{1}{x}$ for $-4 \leq x \leq 4$.

x	−4	−3	−2	−1	0	1	2	3	4
y									

Values are rounded to two decimal places, as it is unlikely that you could plot a value more accurately than this. The completed table looks like this.

x	−4	−3	−2	−1	0	1	2	3	4
y	−0.25	−0.33	−0.5	−1	–	1	0.5	0.33	0.25

There is no value for $x = 0$ because 1/0 is undefined.

The graph plotted from these values is shown in **A**. This does not include much of the graph and does not show the properties of the reciprocal function. If you take *x*-values from −0.8 to 0.8 in steps of 0.2, you get the next table.

x	−0.8	−0.6	−0.4	−0.2	0.2	0.4	0.6	0.8
y	−1.25	−1.67	−2.5	−5	5	2.5	1.67	1.25

Plotting these points as well gives the graph in **B**.

A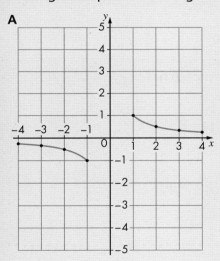

B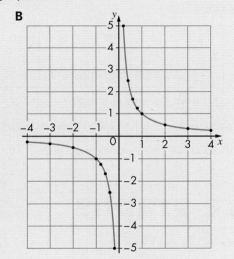

The graph in **B** shows these properties.

- The lines $y = x$ and $y = -x$ are lines of symmetry.
- The closer *x* gets to zero, the nearer the graph gets to the *y*-axis.
- As *x* increases, the graph gets closer to the *x*-axis.

The graph never actually touches the axes, it just gets closer and closer to them.

These properties are true for *all reciprocal graphs*.

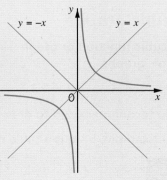

Chapter 16 . Topic 4

EXERCISE 16C

1 a Copy and complete the table to draw the graph of $y = \dfrac{2}{x}$ for $-4 \leqslant x \leqslant 4$.

x	0.2	0.4	0.5	0.8	1	1.5	2	3	4
y	10		4				1		0.5

   b Use your graph to find the y-value when $x = 2.5$.

   c Use your graph to solve the equation $\dfrac{2}{x} = 7$.

   d Use your graph to solve the equation $\dfrac{2}{x} = -1.25$.

2 a Copy and complete the table to draw the graph of $y = \dfrac{5}{x}$ for $-20 \leqslant x \leqslant 20$.

x	0.2	0.4	0.5	1	2	5	10	15	20
y	25		10		2.5				0.25

   b On the same axes, draw the line $y = x + 10$.

   c Use your graph to solve the equation $\dfrac{5}{x} = x + 10$.

3 Draw a graph of $y = \dfrac{-2}{x}$ for $-4 \leqslant x \leqslant 4$.

Use a table of values like the one in question **1**.

4 a Complete this table.

x	1	2	4	5	8	10	20
$\dfrac{20}{x}$							

   b Draw a graph of $y = \dfrac{20}{x}$ for $-20 \leqslant x \leqslant 20$.

   c On the same axes, draw the line with the equation $2x - 10$.

   d Use your graph to solve the equation $\dfrac{20}{x} = 2x - 10$.

16.4 More graphs

Cubic graphs

A **cubic** function or graph is one that contains a term in x^3. The following are examples of cubic graphs.

$y = x^3 \qquad y = x^3 + 3x \qquad y = x^3 + x^2 + x + 1$

The techniques used to draw them are exactly the same as those for quadratic and reciprocal graphs.

For example, here is a table of values and graph of $y = x^3 - x^2 - 4x + 4$

x	−3	−2.5	−2	−1.5	−1	−0.5	0	0.5	1	1.5	2	2.5	3
y	−20.00	−7.88	0.00	4.38	6.00	5.63	4.00	1.88	0.00	−0.88	0.00	3.38	10.00

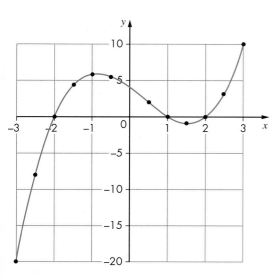

This graph has two turning points.

These have approximate coordinates of (−1, 6) and (1.5, −0.9).

Example 5

a Complete a table of values for $y = 2 + \dfrac{4}{x^2}$

x	−4	−3	−2	−1	0	1	2	3	4
y	2.25					6			

b Draw a graph of $y = 2 + \dfrac{4}{x^2}$

---

a

x	−4	−3	−2	−1	0	1	2	3	4
y	2.25	2.44	3	6	−	6	3	2.44	2.25

There is no value for $x = 0$

b To see what happens near $x = 0$ find more values of y.

If $y = 0.5$ or $−0.5$ then $x = 18$ so $(0.5, 18)$ and $(−0.5, 18)$ are on the curve.

If x is a large positive or negative value, the curve is close to the line $x = 2$

The line $x = 2$ is called an asymptote to the curve.

If y is large, the curve is close to the y-axis (the line $x = 0$).

The y-axis is also an asymptote.

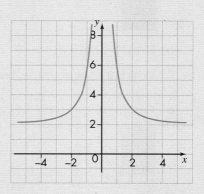

Chapter 16: Graphs of functions

EXERCISE 16D

1. Draw the graph of $y = x^3$ for $-2 \leq x \leq 2$.

2. a Complete the table to draw the graph of $y = 0.5x^3$ for $-2.5 \leq x \leq 2.5$.

x	-2.5	-2	-1.5	-1	-0.5	0	0.5	1	1.5	2	2.5
y			-1.69			0.00					7.81

   b Use your graph to solve the equation $0.5x^3 = 6$.

3. a Complete the table to draw the graph of $y = x^3 + 3$ for $-2.5 \leq x \leq 2.5$.

x	-2.5	-2	-1.5	-1	-0.5	0	0.5	1	1.5	2	2.5
y	-12.63		2.00			3.00	3.13			11.00	

   b Use your graph to solve the equation $x^3 + 3 = 0$.

4. a Complete the table to draw the graph of $y = x^3 - 2x + 5$ for $-2 \leq x \leq 2$.

x	-2	-1.5	-1	-0.5	0	0.5	1	1.5	2
y	1.00	4.63			5.00	4.13			

   b Use your graph to solve the equation $x^3 - 2x + 5 = 3$.
   c Use your graph to find the root of the equation $x^3 - 2x - 2 = 0$.
   d Use the graph to find the approximate coordinates of the turning points.

5. a Complete this table of values for $y = x^3 - x^2 - 6x$

x	-3	-2	-1	0	1	2	3	4
y		0			-6			

   b Draw a graph of $y = x^3 - x^2 - 6x$
   c Use your graph to solve the equation $x^3 - x^2 - 6x + 5 = 0$
   d Use your graph to find the coordinates of the turning points on the graph

6. a Complete this table of values for $y = \dfrac{20}{x^2}$. Give the values to 2 decimal places.

x	1	2	3	4	5
y			2.22		

   b Draw a graph of $y = \dfrac{20}{x^2}$ for $-5 \leq x \leq 5$.
   c What are the asymptotes of the curve?

7. a Complete this table of values for $y = \dfrac{100}{x^2} - 0.5x$.

x	4	5	6	7	8
y	4.25			-1.46	

   b Draw a graph of $y = \dfrac{100}{x^2} - 0.5x$ for $4 \leq x \leq 8$.
   c Use the graph to solve the equation $\dfrac{100}{x^2} - 0.5x = 0$.

16.4 More graphs

8 This is a sphere.

The diameter is x cm.

The volume is $\frac{1}{6}\pi x^3$ cm³.

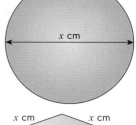

a Draw a graph of $y = \frac{1}{6}\pi x^3$ for $6 \leqslant x \leqslant 10$.

Use a scale of 2 cm to 1 unit on the x-axis and 2 cm to 100 units on the y-axis.

b Use your graph to find the diameter of a sphere with a volume of 300 cm³.

9 The volume of this cuboid is 1000 cm³.

The top is a square of side x cm.

The height is y cm.

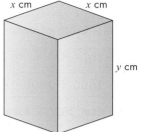

a Show that $y = \frac{1000}{x^2}$.

b Draw a graph of $y = \frac{1000}{x^2}$ for $5 \leqslant x \leqslant 10$.

Use a scale of 2 cm to 1 unit on the x-axis and 2 cm to 10 units on the y-axis

10 This is a cuboid. The end is a square of side x cm.

The length is 5 cm greater than the side of the square end.

The volume of the cuboid is $x^2(x + 5)$ cm³.

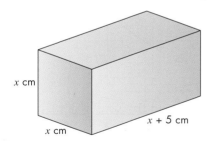

a Draw a graph of $y = x^2(x + 5)$ for $0 \leqslant x \leqslant 6$.

Use a scale of 2 cm to 1 unit on the x-axis and 2 cm to 100 units on the y-axis

b Use your graph to find the sides of the cuboid when the volume is 200 cm³.

11 a Complete this table of values of $y = x + 4/x$

x	−5	−4	−3	−2	−1	0	1	2	3	4	5
y		−5					5				5.8

b Draw the graph of $y = x + 4/x$

c Draw the line $y = x$ on your graph.

d Explain why $y = x$ is an asymptote to the curve.

e Find another asymptote.

Chapter 16 . Topic 5
16.5 Exponential graphs

There are 2000 monkeys in a forest. Scientists say that the numbers are increasing by 40% each year.

This is an example of **exponential growth**. The multiplier is 1.4.

After one year there will be 2000 × 1.4 = 2800 monkeys.

After two years there will be 2800 × 1.4 or 2000 × 1.4^2 = 3920 monkeys.

Advice and Tips

Ignore the decimal fraction, you need only consider whole monkeys.

After x years there will be 2000 × 1.4^x monkeys.

Here is a table of values.

x	0	1	2	3	4
y	2000	2800	3920	5488	7683

Check that 3920 × 1.4 = 5488 and 5488 × 1.4 = 7683.

A graph shows how the population of monkeys changes over time.

The graph shows that the population reaches 10 000 after nearly five years.

Graphs of exponential growth always have this shape.

As the value of x increases, the value of y increases and the graph gets steeper.

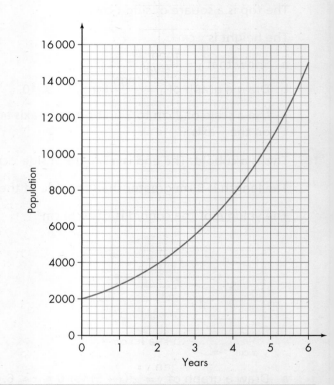

Example 6

Atmospheric pressure is measure in units called hectopascals (hPa).

It decreases as height above sea level increases.

At sea level atmospheric pressure is 1000 hPa.

It decreases by 12% for every kilometre increase in height.

a Draw a graph to show how atmospheric pressure changes, up to a height of six kilometres.

b At what height is the atmospheric pressure half the value at sea level?

a The multiplier for a decrease of 12% is 0.88.   100% − 12% = 88% = 0.88

At a height of x km the pressure is 1000×0.8^x hPa.

Height above sea level x (km)	Pressure y (hPa)
0	1000
1	1000 × 0.88 = 880
2	1000 × 0.88² = 774
3	1000 × 0.88³ = 681
4	681 × 0.88 = 600
5	600 × 0.88 = 528
6	464

This graph shows the values.

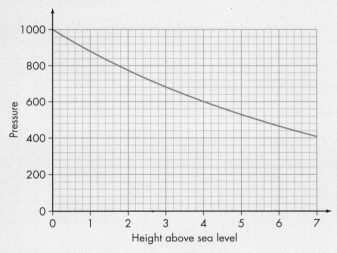

b Half the pressure at sea level is 500 mPa.

From the graph, when $y = 500$, $x = 5.4$.

The height is 5.4 km.

The graph in the example shows **exponential decay**.

As the value of x increases, the value of y decreases and the graph gets less steep.

EXERCISE 16E

1 A company buys some new machinery for $96 000.

The value of the machinery halves every year.

a Copy and complete this table.

Years	0	1	2	3	4
Value (thousand dollars)	96				

b Draw a graph to show how the value falls.

2 The price of a car is $25 000 when it is new.

The value falls by 20% per year.

a Show that the value after one year is $20 000.

b Copy and complete this table.

Years	0	1	2	3	4	5
Value ($)	25 000	20 000				

c Use the table to draw a graph showing the fall in value of the car.

d How long is it until the car is worth half its original value?

3 The population of a country in 2015 is 60 million.

The population increases by 15% every ten years.

a Copy and complete this table to show the predicted population.

Years	2015	2025	2035	2045	2055
Population (millions)	60	69			

b Draw a graph to show how the population changes.

c Estimate the year when the population will reach 100 million.

4 Mira has shares worth $500.

Their value increases by 30% each year.

a Copy and complete this table.

Years	0	1	2	3	4	5
Value ($)	500					1856

b Draw a graph to show how the value of the shares increases.

c How long does it take for the value of the shares to double?

5 There are 20 mice in a population.

Mice breed very quickly. The number of mice doubles every month.

   a Draw a graph to show how the number of mice increases over six months.

   b How long will it be until there are 200 mice?

6 Abram puts $400 in a bank.

After one year it is worth $480 dollars.

   a What is the annual percentage increase?

   b The value increases exponentially. Draw a graph to show how the value increases over four years.

   c Use the graph to find the value after 2.5 years.

7 Marcus buys a picture for $5000.

The value is decreasing exponentially by 10% a year.

   a Copy and complete this table to show the fall in value of the picture.

Years	0	1	2	3	4	5
Value ($)	5000		4050			2952

   b Draw a graph to show how the value falls over eight years.

   c How long does it take for the picture to halve in value?

8 This graph shows the rise in value of an investment.

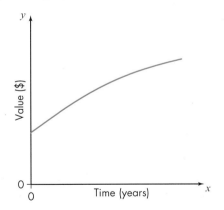

Mario says, 'The value increases all the time so this is exponential growth.'

Explain why Mario is not correct.

Chapter 16. Topic 6

9 The graph shows the number of bacteria in a population that is growing exponentially.

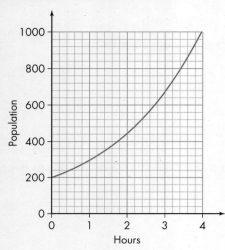

a What was the initial population?
b Work out the percentage growth of the population each hour.
c How long did the population take to triple in size?

10 Barak buys a motor bike for $1000.

The graph shows the fall in value of the bike.

a Show that the value halves every two years.
b Work out the percentage fall in the value of the bike every year.

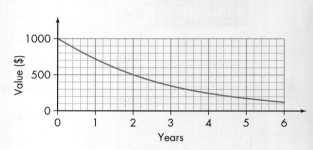

16.6 Estimating gradients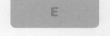

The **gradient** of a curve varies from point to point. At points on this curve to the left of A or to the right of B the gradient is positive.

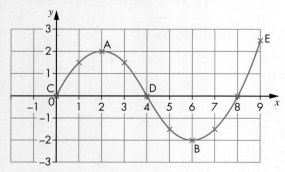

Between A and B the gradient is negative.

296   16.6 Estimating gradients

16.6

The curve is steepest near C, D or E and the gradient there will have the greatest magnitude.

You can estimate the gradient at any point on a curve by drawing a **tangent** at that point. This is a straight line that touches the curve at that point and has the same gradient. You can find the gradient of the tangent, which is a straight line, by drawing a triangle.

Example 7

Find the gradient at the point P with the coordinates (7, −1.5).

Draw a tangent at point P.

Draw a triangle and find the differences between the x-coordinates and y-coordinates.

$$\text{gradient} = \frac{\text{difference in } y}{\text{difference in } x} = \frac{3}{3} = 1$$

The gradient of the curve at P is 1.

Note that at A and B the gradient is 0.

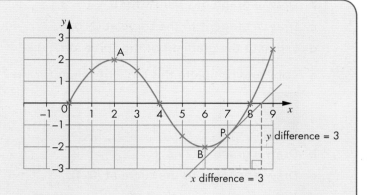

EXERCISE 16F

1. The straight line is a tangent to the curve at the point P with coordinates (2.5, 1).

   Find the gradient of the curve at P.

   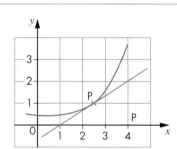

2. A and B have coordinates (−1, 1.5) and (3, 2).

   Tangents to the curve have been drawn at A and B.

   Calculate the gradient of the curve at A and at B.

   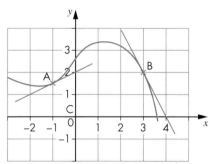

3. **a** Use this table of values to draw a graph of y for $0 \leq x \leq 4$.

x	0	1	2	3	4
y	2	1.5	2	3.5	6

   b Draw a tangent to the curve at the point (3, 3.5).

   c Estimate the gradient of the curve at the point (3, 3.5).

   d At which point is the gradient of the curve 0?

Chapter 16: Graphs of functions

4   a Copy and complete this table of values for $0.1x^3$.

x	0	0.5	1	1.5	2	2.5	3
$0.1x^3$		0.01	0.1		0.8		

   b Draw a graph of $y = 0.1x^3$ for $0 \leq x \leq 3$.
   c Draw a tangent to the curve at (2, 0.8).
   d Estimate the gradient of the curve at (2, 0.8).

5   a Copy and complete this table of values for $y = 2^x$.

x	−2	−1	0	1	2
y	0.25		1		

   b Draw a graph of $y = 2^x$ for $-2 \leq x \leq 2$.
   c Draw a tangent to the curve at (0, 1).
   d Estimate the gradient of the curve at (0, 1).

6   a Copy and complete this table of values for $\frac{5}{x}$.

x	2	3	4	5	6
$\frac{5}{x}$			1.25		0.83

   b Draw a graph of $y = \frac{5}{x}$ for $2 \leq x \leq 6$.
   c Estimate the gradient of the curve at (4, 1.25).

Check your progress

Core

- I can construct tables of values of functions of the form:
  - $+x^2 + ax + b$
  - a/x
- I can draw and interpret graphs of such functions
- I can solve quadratic equations approximately using a graph

Extended

- I can construct tables and draw graphs of functions of the form ax^n and simple sums of these, where $n = -2, -1, 0, 1, 2$ or 3
- I can construct tables and draw graphs of functions of the form
- I can solve approximately equations related to such graphs
- I can draw and interpret graphs representing exponential growth or decay
- I can recognise, sketch and interpret linear, quadratic, cubic, reciprocal and exponential graphs
- I can recognise turning points and asymptotes on a graph
- I can estimate the gradient of a curve by drawing a tangent

Chapter 17
Number sequences

Topics	Level	Key words
1 Patterns in number sequences	CORE	sequence, term, difference, consecutive
2 The nth term of a sequence	CORE	coefficient, linear sequence, nth term, quadratic sequence, cubic sequence
3 General rules from patterns	CORE	rule, patterns
4 Further sequences	EXTENDED	exponential sequence

In this chapter you will learn how to:

CORE	EXTENDED
Calculate: • Continue a given number sequence. (C2.7 and E2.7) • Recognise patterns in sequences and relationships between different sequences, including the term to term rule. (C2.7 and E2.7) • Find the nth term for linear sequences, and for simple quadratic and cubic sequences. (C2.7 and E2.7)	• Find the nth term of linear, quadratic, cubic and exponential sequences and simple combinations of these. (E2.7)

Why this chapter matters

Patterns often appear in numbers. Prime numbers, square numbers and multiples all form patterns. Mathematical patterns also appear in nature.

There are many mathematical patterns that appear in nature. The most famous of these is probably the **Fibonacci** series.

1   1   2   3   5   8   13   21   …

This is formed by adding the two previous terms to get the next term.

The sequence was discovered by the Italian, Leonardo Fibonacci, in 1202, when he was investigating the breeding patterns of rabbits!

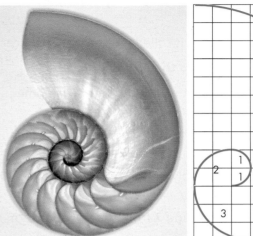
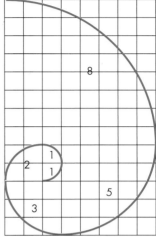

Since then, the pattern has been found in many other places in nature. The spirals found in a nautilus shell and in the seed heads of a sunflower plant also follow the Fibonacci series.

Fractals form another kind of pattern.

Fractals are geometric patterns that are continuously repeated on a smaller and smaller scale.

A good example of a fractal is this: start with an equilateral triangle and draw an equilateral triangle, a third the size of the original, on the middle of each side. Keep on repeating this and you will get an increasingly complex-looking shape.

The pattern shown here is called the Koch snowflake. It is named after the Swedish mathematician, Helge von Koch (1870–1924).

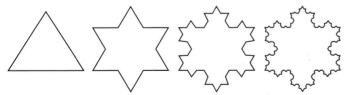

Fractals are commonly found in nature, a good example being the complex patterns found in plants, such as the leaves of a fern.

17.1 Patterns in number sequences

A number **sequence** is an ordered set of numbers with a rule for finding every number in the sequence. The rule that takes you from one number to the next could be a simple addition or multiplication, but often it is more tricky than that. You always need to look very carefully at the pattern of a sequence.

Each number in a sequence is called a **term** and is in a certain position in the sequence.

Look at these sequences and their rules.

3, 6, 12, 24, … doubling the previous term each time … 48, 96, … the term to term rule is 'multiply by 2'

2, 5, 8, 11, … adding 3 to the previous term each time … 14, 17, … the term to term rule is 'add 3'

1, 10, 100, 1000, … multiplying the previous term by 10 each time … 10 000, 100 000 the term to term rule is 'multiply by 10'

80, 73, 66, 59, … subtracting 7 from the previous term each time … 52, 45, … the term to term rule is 'subtract 7'

These are all quite straightforward once you have looked for the link from one term to the next (**consecutive** terms). This link is called the term to term rule.

Differences

For some sequences you need to look at the **differences** between consecutive terms to determine the pattern.

> **Example 1**
> Find the next two terms of the sequence 1, 3, 6, 10, 15, … .
>
> Looking at the differences between consecutive terms:
>
>   1 3 6 10 15
>   ↑ ↑ ↑ ↑
>   2 3 4 5
>
> So the sequence continues as follows.
>
>   1 3 6 10 15 21 28
>   ↑ ↑ ↑ ↑
>   2 3 4 5 +6 +7
>
> So the next two terms are 21 and 28.

The differences usually form a number sequence of their own, so you need to find the *sequence of the differences* before you can expand the original sequence.

Example 2

The nth term of a sequence is $3n + 1$, where $n = 1, 2, 3, 4, 5, 6, \ldots$.
Write down the first five terms of the sequence.

Substituting $n = 1, 2, 3, 4, 5$ in turn:

$(3 \times 1 + 1), (3 \times 2 + 1), (3 \times 3 + 1), (3 \times 4 + 1), (3 \times 5 + 1), \ldots$
    4   7   10   13   16

So the sequence is 4, 7, 10, 13, 16, … .

Notice that in Example 2 the difference between each term and the next is always 3, which is the **coefficient** of n (the number attached to n). Also, the constant term is the difference between the first term and the coefficient, that is, $4 - 3 = 1$.

EXERCISE 17B

1 Here are the nth terms of some sequences. Write down the first five terms of each sequence.

   a $2n + 1$ for $n = 1, 2, 3, 4, 5$
   b $3n - 2$ for $n = 1, 2, 3, 4, 5$
   c $5n + 2$ for $n = 1, 2, 3, 4, 5$
   d n^2 for $n = 1, 2, 3, 4, 5$
   e $n^2 + 3$ for $n = 1, 2, 3, 4, 5$
   f $20 - 2n$ for $n = 1, 2, 3, 4, 5$

2 Write down the first five terms of the sequence that has as its nth term:

   a $n + 3$
   b $3n - 1$
   c $5n - 2$
   d $n^2 - 1$
   e $4n + 5$
   f $45 - 3n$

   Advice and Tips

   Substitute numbers into the expressions until you can see how the sequence works.

3 The nth term of a sequence is $100 - 6n$.

   a Work out the first four terms.
   b Work out the first term that is less than zero.

4 Write down the first four terms of the sequence for which the nth term is:

   a n^2
   b $n^2 + 2$
   c $2n^2$
   d $4n^2 - 1$
   e $200 - n^2$
   f $\dfrac{n^2}{4}$

5 Write down the first four terms of a sequence for which the nth term is:

   a n^3
   b $n^3 + 1$
   c $n^3 - 2$
   d $3n^3 - 1$
   e $\dfrac{n^3}{2}$
   f $109 - n^3$

6 A haulage company uses this formula to calculate the cost of transporting n pallets.

   For $n \leqslant 5$, the cost will be $\$(40n + 50)$
   For $6 \leqslant n \leqslant 10$, the cost will be $\$(40n + 25)$
   For $n \geqslant 11$, the cost will be $\$40n$

   a How much will the company charge to transport 7 pallets?
   b How much will the company charge to transport 15 pallets?

c A company is charged $170 for transporting pallets. How many pallets did they transport?

  d Another haulage company uses the formula $50n$ to calculate the cost for transporting n pallets.

  At what value of n do the two companies charge the same amount?

7 The nth term of a sequence is $3n + 7$.
The nth term of another sequence is $4n - 2$.

These two series have several terms in common but only one term that is common and has the same position in the sequence.

Without writing out the sequences, show how you can tell, using the expressions for the nth term, that this is the 9th term.

Finding the nth term of a linear sequence

In a **linear sequence** the *difference* between one term and the next is always the same.

For example:

   2, 5, 8, 11, 14, … difference of 3

The nth term of this sequence is given by $3n - 1$.

Here is another linear sequence.

   5, 7, 9, 11, 13, … difference of 2

The nth term of this sequence is given by $2n + 3$.

So, you can see that the nth term of a linear sequence is *always* of the form $An + b$, where:

- A, the coefficient of n, is the difference between each term and the next term (**consecutive** term)
- b is the difference between the first term and A.

> ### Example 3
>
> From the sequence 5, 12, 19, 26, 33, … , find:
>
> **a** the nth term
>
> **b** the 50th term.
>
> ---
>
> **a** The difference between consecutive terms is 7. So the first part of the nth term is $7n$.
> Subtract the difference, 7, from the first term, 5, which gives $5 - 7 = -2$.
> So the nth term is given by $7n - 2$.
>
> **b** Now find the 50th term by substituting $n = 50$ into the rule, $7n - 2$.
>
>   50th term $= 7 \times 50 - 2 = 350 - 2$
>
> $= 348$

17.2

Quadratic and cubic sequences

The square numbers are:
  1, 4, 9, 16, 25, 36, …

The nth square number is n^2.

So, for example, the 15th square number is $15^2 = 225$.

The cube numbers are:
  1, 8, 27, 64, 125, …

The nth cube number is n^3.

So, for example, the 7th cube number is $7^3 = 7 \times 7 \times 7 = 343$.

Other sequences can be formed from square and cube numbers.

> **Example 4**
> Find the nth term of this sequence.
> 5, 8, 13, 20, 29, …
>
> The differences between terms are 3, 5, 7, 9, …
>
> These are not the same, so it is *not* a linear sequence.
>
> Compare the sequence with the square numbers.
>
> | 5 | 8 | 13 | 20 | 29 | … |
> |---|---|----|----|----|---|
> | 1 | 4 | 9 | 16 | 25 | … |
>
> You can see that each term is four more than a square number.
>
> The nth term of this sequence is $n^2 + 4$.

A sequence that is based on square numbers is called a **quadratic sequence**.

> **Example 5**
> Find the nth term of this sequence.
> 0, 7, 26, 63, 124
>
> The differences are not equal. It is not linear.
>
> Compare the terms with the square numbers and the cube numbers.
>
> | Sequence | 0 | 7 | 26 | 63 | 124 | … |
> |-----------------|---|---|----|----|-----|---|
> | Square numbers | 1 | 4 | 9 | 16 | 25 | … |
> | Cube numbers | 1 | 8 | 27 | 64 | 125 | … |
>
> Each term is one less than a cube number.
>
> The nth term of the sequence is $n^3 - 1$.

A sequence that is based on cube numbers is called a **cubic sequence**.

EXERCISE 17C

1 Find the next two terms and the nth term in each of these linear sequences.

- a 3, 5, 7, 9, 11, …
- b 5, 9, 13, 17, 21, …
- c 8, 13, 18, 23, 28, …
- d 2, 8, 14, 20, 26, …
- e 5, 8, 11, 14, 17, …
- f 2, 9, 16, 23, 30, …
- g 27, 25, 23, 21, 19, …
- h 42, 38, 34, 30, 26, …
- i 2, 5, 8, 11, 14, …
- j 2, 12, 22, 32, …
- k 8, 12, 16, 20, …
- l 4, 9, 14, 19, 24, …

Advice and Tips

Remember to look at the differences and the first term.

2 Find the nth term and the 50th term in each of these linear sequences.

- a 4, 7, 10, 13, 16, …
- b 7, 9, 11, 13, 15, …
- c 3, 8, 13, 18, 23, …
- d 1, 5, 9, 13, 17, …
- e 2, 10, 18, 26, …
- f 5, 6, 7, 8, 9, …
- g 6, 11, 16, 21, 26, …
- h 3, 11, 19, 27, 35, …
- i 1, 4, 7, 10, 13, …
- j 21, 24, 27, 30, 33, …
- k 12, 19, 26, 33, 40, …
- l 1, 9, 17, 25, 33, …

3 For each sequence **a** to **j**, find:

- i the nth term
- ii the 100th term.

- a 5, 9, 13, 17, 21, …
- b 3, 5, 7, 9, 11, 13, …
- c 4, 7, 10, 13, 16, …
- d 8, 10, 12, 14, 16, …
- e 9, 13, 17, 21, …
- f 6, 11, 16, 21, …
- g 0, 3, 6, 9, 12, …
- h 2, 8, 14, 20, 26, …
- i 7, 15, 23, 31, …
- j 25, 27, 29, 31, …

4 An online CD retail company uses this price chart. The company charges a standard basic price for a single CD, including postage and packing.

n	1	2	3	4	5	6	7	8	9	10	11	12	13	14	15
Charge ($)	10	18	26	34	42	49	57	65	73	81	88	96	104	112	120

- a Using the charges for 1 to 5 CDs, work out an expression for the nth term.
- b Using the charges for 6 to 10 CDs, work out an expression for the nth term.
- c Using the charges for 11 to 15 CDs, work out an expression for the nth term.
- d What is the basic charge for a CD?

5 Here are some quadratic sequences. Work out the nth term of each one.

- a 1, 4, 9, 16, 25, …
- b 3, 6, 11, 18, 27 …
- c 2, 8, 18, 32, 50, …
- d 0, 3, 8, 15, 24

17.2 The nth term of a sequence

6 Here are some cubic sequences. Work out the nth term of each one.

    a 1, 8, 27, 64, …   **b** 11, 18, 37, 74, …

    c 0.5, 4, 13.5, 32, …   **d** 10, 80, 270, 640, …

7 Work out the nth term of each of these sequences.

    a 6, 7, 8, 9, 10, …   **b** 6, 9, 14, 21, 30, …   **c** 6, 13, 32, 69, 130

    d 6, 11, 16, 21, 26, …   **e** 5, 20, 45, 80, 125, …   **f** 5, 40, 135, 320, 625, …

17.3 General rules from patterns

Many problem-solving situations that you are likely to meet involve number sequences. So you need to be able to formulate general **rules** from given number **patterns**.

Example 6

The diagram shows a pattern of squares building up.

a How many squares will there be in the nth pattern?

b Which pattern has 99 squares in it?

---

a First, build up a table for the patterns.

Pattern number	1	2	3	4	5
Number of squares	1	3	5	7	9

Looking at the **difference** between consecutive patterns, you can see it is always two squares. So, use $2n$.

Subtract the difference 2 from the first number, which gives $1 - 2 = -1$.

So the number of squares on the base of the nth pattern is $2n - 1$.

b Now find n when $2n - 1 = 99$:

    $2n - 1 = 99$

       $2n = 99 + 1 = 100$

        $n = 100 \div 2 = 50$

The pattern with 99 squares is the 50th.

When you are trying to find a general rule from a sequence of diagrams, always set up a table to connect the pattern number with the number of the variable (squares, matches, seats, etc.) for which you are trying to find the rule. Once you have set up the table, it is easy to find the nth term.

EXERCISE 17D

1 A pattern of squares is built up from matchsticks as shown.

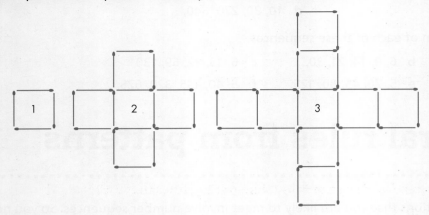

   a Draw the fourth diagram.

   b Copy and complete this table.

Pattern number	1	2	3	4
Number of squares	1	5	9	

   c How many squares are in the nth diagram?

   d How many squares are there in the 25th diagram?

   e With 200 squares, which is the biggest diagram that could be made?

Advice and Tips

Write out the number sequences to help you see the patterns.

2 A pattern of triangles is built up from matchsticks.

   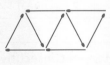

  1   2   3   4

   a Draw the fifth set of triangles in this pattern.

   b Copy and complete this table.

Pattern number	1	2	3	4	5
Number of matches	3				

   c How many matchsticks are needed for the nth set of triangles?

   d How many matchsticks are needed to make the 60th set of triangles?

   e If there are only 100 matchsticks, which is the largest set of triangles that could be made?

17.3 General rules from patterns

3 A conference centre had tables each of which could sit six people.

1

When put together, the tables could seat people as shown.

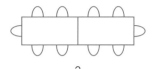   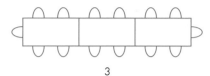
   2   3

a How many people could be seated at four tables put together this way?

b Copy and complete this table.

Number of tables	1	2	3	4
Number of seats	6			

c How many people could be seated at n tables put together in this way?

d At a conference, there were 50 people who wished to use the tables in this way. How many tables would they need?

4 Prepacked fencing units come in the shape shown, made of four pieces of wood.

1

When you put them together in stages to make a fence, you also need joining pieces, so the fence will start to build up as shown below.

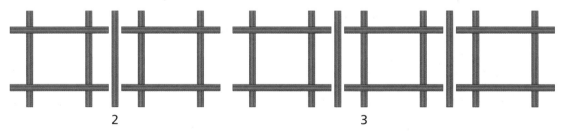

a How many pieces of wood would you have in a fence made up in:
   i five stages   **ii** n stages   **iii** 45 stages?

b I made a fence out of 124 pieces of wood. How many stages did I use?

5 Regular pentagons of side length 1 cm are joined together to make a pattern, as shown.

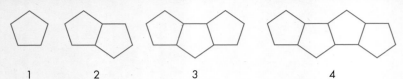

Copy this pattern and write down the perimeter of each shape. Put the results in a table.

a   What is the perimeter of patterns like this made from:

   i   six pentagons   ii n pentagons   iii   50 pentagons?

b   What is the largest number of pentagons that can be put together like this to have a perimeter less than 1000 cm?

6 Lamp-posts are put at the end of every 100 m stretch of a motorway, as shown.

 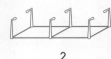 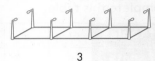

a   How many lamp-posts are needed for:

   i   four 100 m stretches
   ii n 100 m stretches
   iii 8 km of this motorway.

b   A new motorway is being built. The contractor has ordered 1598 lamp-posts. How long is this motorway?

7 A school dining hall had trapezium-shaped tables.
Each table could seat five people, as shown here.

When the tables were joined together, as shown below, the individual tables could not seat as many people.

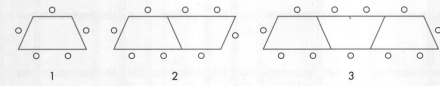

a   In this arrangement, how many could be seated if there were:

   i   four tables
   ii n tables
   iii 13 tables?

b   For an outside charity event, up to 200 people had to be seated.
How many tables arranged like this did they need?

8 When setting out tins to make a display of a certain height, you need to know how many tins to start with at the bottom.

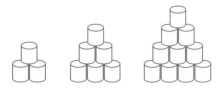

  a How many tins are needed on the bottom if you wish the display to be:
   i five tins high
   ii n tins high
   iii 18 tins high?

If the display is n tins high, the *total* number of tins is $T = \dfrac{n(n+1)}{2}$

  b Show that this formula gives the correct answer if $n = 2$ or $n = 3$.
  c How many tins will be needed to make a stack 10 tins high?

9 These are the instructions that were used to draw the patterns below.

For pattern 1, draw an equilateral triangle, mark the midpoints of each side and draw and colour in the equilateral triangle formed by these points.

For pattern 2, repeat this with the three white triangles remaining.

For pattern 3, repeat this with the nine white triangles.

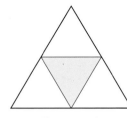
Pattern 1

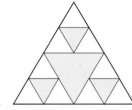
Pattern 2

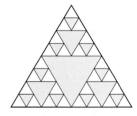
Pattern 3

The pattern is called a Sierpinski triangle and is one of the earliest examples of a fractal type pattern.

Pattern 4 is completed in the same way.

  a Copy and complete the table for patterns 1, 2 and 3.

Pattern	1	2	3	4
White triangles	3	9		
Coloured triangles	1	4		
Total	4			

  b Use the numbers in the first 3 columns of the table to help you complete column 4. Explain your method.

Chapter 17 . Topic 4

10 Thom is building three different patterns with matches.

He builds the patterns in steps.

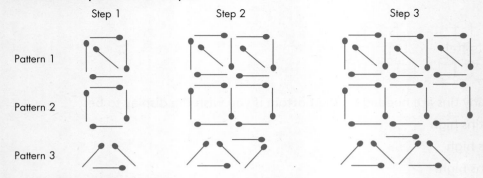

a Draw step 4 for each pattern.
b Complete this table for pattern 1.

Step	1	2	3	4
Number of matches	5			

c How many matches are needed for step n of pattern 1?
d How many matches are needed for step n of pattern 2?
e How many matches are needed for step n of pattern 3?
f What is the total number of matches needed for step n? Write your answer as simply as possible.

17.4 Further sequences

Look at these sequences.

  6, 24, 96, 384, 1536, …

  48, 24, 12, 6, 3, 1.5, …

In the first, you multiply each term by four to find the next one.

In the second, you multiply each term by 0.5 (or divide it by 2) to find the next term.

The nth term of the first sequence is 1.5×4^n. You can write this as $t_n = 1.5 \times 4^n$.

t_n is mathematical shorthand for the nth term.

Check: The first term is $1.5 \times 4 = 6$ (you can write this as $t_1 = 6$); the third term is $1.5 \times 4^3 = 1.5 \times 64 = 96$ ($t_3 = 6$).

The nth term of the second sequence is 96×0.5^n. You can write this as $t_n = 96 \times 0.5^n$.

Check: The first term is $96 \times 0.5 = 48$ ($t_1 = 48$); the third term is $96 \times 0.5^3 = 96 \times 0.125 = 12$ ($t_3 = 12$).

96×0.5^n can also be written as . Check that this is correct.

These sequences are called **exponential sequences**. In these sequences, n is an exponent or power.

17.4

> **Example 7**
>
> This is the start of an exponential sequence.
>
> 11, 12.1, 13.31, 14.641, ...
>
> Work out the nth term.
>
> The multiplier from one term to the next is always the same.
>
> $\frac{12.1}{11} = 1.1 \qquad \frac{13.31}{12.1} = 1.1 \qquad \frac{14.641}{13.31} = 1.1$
>
> The nth term is $a \times 1.1^n$ where a is a number.
>
> The first term is 11 so $a \times 1.1 = 11 \rightarrow a = \frac{11}{1.1} = 10$.
>
> The nth term is 10×1.1^n.

Sequences may be a combination of different types of sequence.

For example, look at this sequence.

3, 8, 15, 24, 35, ...

It is not linear, because the differences are not all the same.

It is not exponential because the multipliers are not all the same.

Compare the sequence with the square numbers.

n	1	2	3	4	5	...
Sequence	3	8	15	24	35	...
n^2	1	4	9	16	25	...
Difference between sequence and n^2	2	4	6	8	10	...

You can see that the difference is $2n$ each time.

The nth term is $n^2 + 2n$. You can write this as $t_n = n^2 + 2n$.

Check: If $n = 5$ then $n^2 + 2n = 25 + 10 = 35$ which is the fifth term ($t_5 = 35$).

EXERCISE 17E

1 Find the next term in each of these exponential sequences.

　a 12, 36, 144, ...　　　**b** 13, 39, 117, 351, ...　　　**c** 2, 10, 50, 250, ...

　d 20, 24, 28.8, 34.56, ...　**e** 4, 80, 1600, 32000, ...　**f** 240, 120, 60, 30, ...

　g 1000, 200, 40, 8, ...　**h** 162, 108, 72, 48, ...

2 Here are the nth terms of some sequences. In each case, find the first term (t_1) and the fourth term (t_4).

　a 6^n　　**b** 8×2^n　　**c** 5×3^n　　**d** 40×0.5^n　　**e** $81 \times \left(\frac{2}{3}\right)^n$

3. Work out the nth term of each of these sequences and write it in the form $t_n =$.

   a 50, 100, 200, 400, 800, …
   b 3, 6, 12, 24, 48, …
   c 6, 18, 54, 162, ….
   d 120, 60, 30, 15, …
   e 45, 40.5, 36.45, 32.805, …
   f 80, 100, 125, …

4. Selina has $5000 savings. She expects her savings to increase by 40% each year.

Years from now	0	1	2	3	4
Savings ($)	5000	7000	9800		

   a Work out the two numbers missing from the table.
   b Work out an expression for her savings after n years.

5. Find the nth term of each of these sequences.

   a 2, 5, 10, 17, 26, …   b 7, 10, 15, 22, 31, …   c 0, 2, 6, 12, 20, …
   d 3, 12, 27, 48, 75, …   e 1, 10, 25, 46, 73, …   f 4, 14, 30, 52, 80, …

6. These are the first four triangular numbers.

   1   3   6   10

   a Find the next two triangular numbers.
   b Find the nth term of this quadratic sequence.

      2, 6, 12, 20, 30, …

   c Use your answer from part **b** to find an expression for the nth triangular number.
   d Work out the 20th triangular number.
   e Show that 1275 is a triangular number.

7. Find the nth term of each of these sequences and write it in the form $t_n =$.

   a 0, 7, 26, 63, 124, …
   b 51, 58, 77, 114, 157, …
   c 2, 10, 30, 68, 130, …
   d 4, 14, 36, 76, 140, …

8. Here are some triangles made from oranges.

If these oranges are put in four layers they make a tetrahedron.

    a Find the number of oranges in a tetrahedron with four layers.

    The number of oranges in a tetrahedron with n layers is $\frac{1}{6}n(n+1)(n+2)$.

    b Check that this expression gives the correct answer with $n = 4$.

    c A shopkeeper has 100 oranges. He wants to display them in a tetrahedron. Work out the largest number of layers he can have.

    d Show how to use the formula to add the first 20 triangular numbers.

9 Here is a sequence.

    6, 16, 30, 48, 70, …

The nth term of this sequence is given by $t_n = an + bn^2$ where a and b are numbers.

    a By looking at the first term, show that $a + b = 6$.

    b Use the second term to find another equation involving a and b.

    c Solve the equation from parts **a** and **b** simultaneously to find the nth term of the sequence.

Check your progress

Core
- I can continue a given number sequence
- I can recognise patterns in sequences, including a term to term rule
- I can find and use the nth term of a linear sequence
- I can find and use the nth term of a simple quadratic or cubic sequence

Extended
- I understand subscript notation for the terms of a sequence
- I can find and use the nth term of an exponential sequence
- I can find and use the nth term of a simple combination of different types of sequences

Chapter 18
Indices

Topics	Level	Key words
1 Using indices	CORE	index, indices, power, power 1, power 0
2 Negative indices	CORE	negative index, reciprocal
3 Multiplying and dividing with indices	CORE	
4 Fractional indices	EXTENDED	

In this chapter you will learn how to:

CORE	EXTENDED
• Understand the meaning and rules of indices. (C1.7 and E1.7) • Use and interpret positive, negative and zero indices. (C2.4 and E2.4) • Use the rules of indices. (C2.4 and E2.4)	• Use and interpret fractional indices, e.g. solve $32^x = 2$. (E2.4)

Why this chapter matters

Indices are a useful way to write numbers. They show how different numbers are related to one another and they can make it easier to multiply or divide, or to compare the sizes of different numbers.

You probably already know about powers of numbers from Chapters 5 and 9.

You use find powers of 10 when you write numbers in standard form, such as 3.7×10^6 or 8.92×10^{-5}.

The first is 3 700 000 and the second is 0.000 08 92.

Using powers is a useful 'short cut'.
For example, the centre of the galaxy Andromeda is 24 000 000 000 000 000 000 km from our Sun.
It is much easier to write 24×10^{18}!

You will also find powers in squares and cubes such as 7^2 and 14^3.

Remember that $7^2 = 7 \times 7$ and $14^3 = 14 \times 14 \times 14$. This also saves time in writing out the numbers.

The small number that shows the power is called an index.

- In 7^2 the index is 2 and in 14^3 the index is 3.
- In 3.7×10^6 or 8.92×10^{-5} the indices are 6 and –5.

'Indices' is the plural of index. In the examples above 2, 3, 6 and –5 are indices.

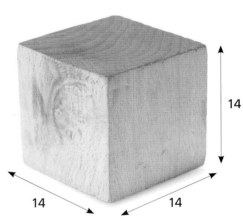

In this chapter you will discover more useful ways to use indices. You might be surprised to find you can write fractions using indices. You can probably see how 8 can be written as a power of two (2^3) but it is not so obvious how $\frac{1}{8}$ can also be written as a power of two.

Warning

The word index has a number of other meanings in English. For example, you will find an index at the back of this book. In this chapter the word is always used to mean a power.

Chapter 18. Topic 1

18.1 Using indices

An **index** is a convenient way of writing repetitive multiplications. The plural of index is **indices**.

The index tells you the number of times a number is multiplied by itself. For example:

$4^6 = 4 \times 4 \times 4 \times 4 \times 4 \times 4$ six lots of 4 multiplied together (call this '4 to the **power** 6')

$6^4 = 6 \times 6 \times 6 \times 6$ four lots of 6 multiplied together (call this '6 to the power 4')

$7^3 = 7 \times 7 \times 7$

$12^2 = 12 \times 12$

Example 1

a Write each of these numbers out in full.

   i 4^3 **ii** 6^2 **iii** 7^5 **iv** 12^4

b Write each multiplication as a power.

   i $3 \times 3 \times 3 \times 3 \times 3 \times 3 \times 3 \times 3$

   ii $13 \times 13 \times 13 \times 13 \times 13$

   iii $7 \times 7 \times 7 \times 7$

   iv $5 \times 5 \times 5 \times 5 \times 5 \times 5 \times 5$

---

a **i** $4^3 = 4 \times 4 \times 4$ **ii** $6^2 = 6 \times 6$

   iii $7^5 = 7 \times 7 \times 7 \times 7 \times 7$ **iv** $12^4 = 12 \times 12 \times 12 \times 12$

b **i** $3 \times 3 \times 3 \times 3 \times 3 \times 3 \times 3 \times 3 = 3^8$

   ii $13 \times 13 \times 13 \times 13 \times 13 = 13^5$

   iii $7 \times 7 \times 7 \times 7 = 7^4$

   iv $5 \times 5 \times 5 \times 5 \times 5 \times 5 \times 5 = 5^7$

Working out powers on your calculator

The power button on your calculator will probably look like this $\boxed{x^\blacksquare}$.

To work out 5^7 on your calculator use the power key.

$5^7 = \boxed{5}\ \boxed{x^\blacksquare}\ \boxed{7} = 78\,125$

Two special powers

Power 1

Any number to the power 1 is the same as the number itself. This is always true so normally you do not write the power 1.

For example: $5^1 = 5$ $32^1 = 32$ $(-8)^1 = -8$

18.1

Power zero

Any number to the power 0 is equal to 1.

For example:

$5^0 = 1 \qquad 32^0 = 1 \qquad (-8)^0 = 1$

You can check these results on your calculator.

EXERCISE 18A

1 Write these expressions using index notation. Do not work them out yet.

   a $2 \times 2 \times 2 \times 2$ **b** $3 \times 3 \times 3 \times 3 \times 3$

   c 7×7 **d** $5 \times 5 \times 5$

   e $10 \times 10 \times 10 \times 10 \times 10 \times 10 \times 10$ **f** $6 \times 6 \times 6 \times 6$

   g 4 **h** $1 \times 1 \times 1 \times 1 \times 1 \times 1 \times 1$

   i $0.5 \times 0.5 \times 0.5 \times 0.5$ **j** $100 \times 100 \times 100$

2 Write these power terms out in full. Do not work them out yet.

   a 3^4 **b** 9^3 **c** 6^2 **d** 10^5

   e 2^{10} **f** 8^1 **g** 0.1^3 **h** 2.5^2

   i 0.7^3 **j** 1000^2

3 Using the power key on your calculator (or another method), work out the values of the power terms in question **1**.

4 Using the power key on your calculator (or another method), work out the values of the power terms in question **2**.

5 A storage container is in the shape of a cube. The length of the container is 5 m.

To work out the volume of a cube, use the formula:

   volume = (length of edge)^3

Work out the total storage space in the container.

6 Write each number as a power of a different number.

The first one has been done for you.

   a $32 = 2^5$ **b** 100   **c** 8   **d** 25

7 Without using a calculator, work out the values of these power terms.

   a 2^0 **b** 4^1 **c** 5^0 **d** 1^9 **e** 1^{235}

8 The answers to question **7**, parts **d** and **e**, should tell you something special about powers of 1. What is it?

9 Write the answer to question **1**, part **j** as a power of 10.

10 Write the answer to question **2**, part **j** as a power of 10.

Chapter 18: Indices

Chapter 18 . Topic 2

11 Using your calculator, or otherwise, work out the values of these power terms.

a $(-1)^0$ b $(-1)^1$
c $(-1)^2$ d $(-1)^4$
e $(-1)^5$

12 Using your answers to question **11**, write down the answers to these power terms.

a $(-1)^8$ b $(-1)^{11}$
c $(-1)^{99}$ d $(-1)^{80}$
e $(-1)^{126}$

13 The number 16 777 216 is a power of 2.

It is also a power of 4, a power of 8 and a power of 16.

Write the number as a power of 4, a power of 8 and a power of 16.

18.2 Negative indices

A **negative index** is a convenient way of writing the **reciprocal** of a number or term. (That is, one divided by that number or term.) For example:

$$x^{-a} = \frac{1}{x^a}$$

Here are some other examples.

$5^{-2} = \frac{1}{5^2}$ $3^{-1} = \frac{1}{3}$ $5x^{-2} = \frac{5}{x^2}$

> **Example 2**
>
> Rewrite each number in the form 2^n.
>
> a 8   b $\frac{1}{4}$ c −32   d $-\frac{1}{64}$
>
> a $8 = 2 \times 2 \times 2 = 2^3$ b $\frac{1}{4} = \frac{1}{2^2} = 2^{-2}$
>
> c $-32 = -2^5$ d $-\frac{1}{64} = -\frac{1}{2^6} = -2^{-6}$

EXERCISE 18B

1 Write each of these in fraction form.

a 5^{-3} b 6^{-1} c 10^{-5} d 3^{-2} e 8^{-2}
f 9^{-1} g w^{-2} h t^{-1} i x^{-m} j $4m^{-3}$

2 Write each of these in negative index form.

a $\frac{1}{3^2}$ b $\frac{1}{5}$ c $\frac{1}{10^3}$ d $\frac{1}{m}$ e $\frac{1}{t^n}$

Advice and Tips

If you move a power from top to bottom, or vice versa its sign changes. A negative power means a reciprocal: it does not mean the answer is negative.

18.2

3 Change each expression into an index form of the type shown.

  a All of the form 2^n
   - **i** 16
   - **ii** $\frac{1}{2}$
   - **iii** $\frac{1}{16}$
   - **iv** -8

  b All of the form 10^n
   - **i** 1000
   - **ii** $\frac{1}{10}$
   - **iii** $\frac{1}{100}$
   - **iv** 1 million

  c All of the form 5^n
   - **i** 125
   - **ii** $\frac{1}{5}$
   - **iii** $\frac{1}{25}$
   - **iv** $\frac{1}{625}$

  d All of the form 3^n
   - **i** 9
   - **ii** $\frac{1}{27}$
   - **iii** $\frac{1}{81}$
   - **iv** -243

4 Rewrite each expression in fraction form.

  a $5x^{-3}$
  b $6t^{-1}$
  c $7m^{-2}$
  d $4q^{-4}$
  e $10y^{-5}$
  f $\frac{1}{2}x^{-3}$
  g $\frac{1}{2}m^{-1}$
  h $\frac{3}{4}t^{-4}$
  i $\frac{4}{5}y^{-3}$
  j $\frac{7}{8}x^{-5}$

5 Write each fraction in index form.

  a $\frac{7}{x^3}$
  b $\frac{10}{p}$
  c $\frac{5}{t^2}$
  d $\frac{8}{m^5}$
  e $\frac{3}{y}$

6 Find the value of each number.

  a $x = 5$
   - **i** x^2
   - **ii** x^{-3}
   - **iii** $4x^{-1}$

  b $t = 4$
   - **i** t^3
   - **ii** t^{-2}
   - **iii** $5t^{-4}$

  c $m = 2$
   - **i** m^3
   - **ii** m^{-5}
   - **iii** $9m^{-1}$

  d $w = 10$
   - **i** w^6
   - **ii** w^{-3}
   - **iii** $25w^{-2}$

7 Two different numbers can be written in the form 2^n.

The sum of the numbers is 40.

What is the difference of the numbers?

Chapter 18: Indices

Chapter 18. Topic 3

8 x and y are integers.

$x^2 - y^3 = 0$

Work out possible values of x and y.

9 You are given that $8^7 = 2\,097\,152$.

Write down the value of 8^{-7}.

10 Put these in order from smallest to largest:

$x^5 \qquad x^{-5} \qquad x^0$

a when x is greater than 1

b when x is between 0 and 1

c when $x = -10$.

11 $M = 3^{-8}$

Write the following in terms of M

a 3^{-9}

b 3^{-7}

c 3^{-5}

18.3 Multiplying and dividing with indices

When you *multiply* powers of the same number or variable, you *add* the indices. For example:

$3^4 \times 3^5 = 3^{(4+5)} = 3^9$

$2^3 \times 2^4 \times 2^5 = 2^{12}$

$10^4 \times 10^{-2} = 10^2$

$10^{-3} \times 10^{-1} = 10^{-4}$

$a^x \times a^y = a^{(x+y)}$

When you *divide* powers of the same number or variable, you *subtract* the indices. For example:

$a^4 \div a^3 = a^{(4-3)} = a^1 = a$

$b^4 \div b^7 = b^{-3}$

$10^4 \div 10^{-2} = 10^6$

$10^{-2} \div 10^{-4} = 10^2$

$a^x \div a^y = a^{(x-y)}$

18.3

When you *raise* a power to a further power, you *multiply* the indices. For example:

$(a^2)^3 = a^{2 \times 3} = a^6$

$(a^{-2})^4 = a^{-8}$

$(a^2)^6 = a^{12}$

$(a^x)^y = a^{xy}$

Here are some examples of different kinds of expressions that use powers.

$2a^2 \times 3a^4 = (2 \times 3) \times (a^2 \times a^4)$
$\qquad = 6 \times a^6 = 6a^6$

$4a^2b^3 \times 2ab^2 = (4 \times 2) \times (a^2 \times a) \times (b^3 \times b^2)$
$\qquad = 8a^3b^5$

$12a^5 \div 3a^2 = (12 \div 3) \times (a^5 \div a^2)$
$\qquad = 4a^3$

$(2a^2)^3 = (2)^3 \times (a^2)^3 = 8 \times a^6$
$\qquad = 8a^6$

EXERCISE 18C

1 Write these as single powers of 5.

  a $5^2 \times 5^2$ **b** 5×5^2 **c** $5^{-2} \times 5^4$
  d $5^6 \times 5^{-3}$ **e** $5^{-2} \times 5^{-3}$

2 Write these as single powers of 6.

  a $6^5 \div 6^2$ **b** $6^4 \div 6^4$ **c** $6^4 \div 6^{-2}$
  d $6^{-3} \div 6^4$ **e** $6^{-3} \div 6^{-5}$

3 Simplify these and write them as single powers of a.

  a $a^2 \times a$ **b** $a^3 \times a^2$ **c** $a^4 \times a^3$
  d $a^6 \div a^2$ **e** $a^3 \div a$ **f** $a^5 \div a^4$

4   **a** $a^x \times a^y = a^{10}$

  Write down a possible pair of values of x and y.

  b $a^x \div a^y = a^{10}$

  Write down a possible pair of values of x and y.

5 Write these as single powers of 4.

  a $(4^2)^3$ **b** $(4^3)^5$ **c** $(4^1)^6$
  d $(4^3)^{-2}$ **e** $(4^{-2})^{-3}$ **f** $(4^7)^0$

Chapter 18: Indices

6 Simplify these expressions.

a $2a^2 \times 3a^3$
b $3a^4 \times 3a^{-2}$
c $(2a^2)^3$
d $-2a^2 \times 3a^2$
e $-4a^3 \times -2a^5$
f $-2a^4 \times 5a^{-7}$

7 Simplify these expressions.

a $6a^3 \div 2a^2$
b $12a^5 \div 3a^2$
c $15a^5 \div 5a$
d $18a^{-2} \div 3a^{-1}$
e $24a^5 \div 6a^{-2}$
f $30a \div 6a^5$

Advice and Tips

Deal with numbers and indices separately and do not confuse the rules. For example: $12a^5 \div 4a^2$ = $(12 \div 4) \times (a^5 \div a^2)$

8 Simplify these expressions.

a $2a^2b^3 \times 4a^3b$
b $5a^2b^4 \times 2ab^{-3}$
c $6a^2b^3 \times 5a^{-4}b^{-5}$
d $12a^2b^4 \div 6ab$
e $24a^{-3}b^4 \div 3a^2b^{-3}$

9 Simplify these expressions.

a $\dfrac{6a^4b^3}{2ab}$

b $\dfrac{2a^2bc^2 \times 6abc^3}{4ab^2c}$

c $\dfrac{3abc \times 4a^3b^2c \times 6c^2}{9a^2bc}$

10 Write down **two** possible:

a multiplication questions with an answer of $12x^2y^5$
b division questions with an answer of $12x^2y^5$.

11 a, b and c are three different positive integers.

What is the smallest possible value of a^2b^3c?

12 $8^{12} = A$

Find the following in terms of A

a 8^{24}
b 8^{-12}
c 8^6
d 2^{12}

13 $x^n = y$

a Show that $x^{2n+1} = xy^2$
b Find a similar expression for x^{2n-1}

Chapter 18 . Topic 4

18.4 Fractional indices   E

Indices of the form $\frac{1}{n}$

Consider the problem $7^x \times 7^x = 7$. This can be written as:

$7^{(x+x)} = 7$

$7^{2x} = 7^1 \Rightarrow 2x = 1 \Rightarrow x = \frac{1}{2}$

If you now substitute $x = \frac{1}{2}$ back into the original equation, you see that:

$7^{\frac{1}{2}} \times 7^{\frac{1}{2}} = 7$

This makes $7^{\frac{1}{2}}$ the same as $\sqrt{7}$.

You can similarly show that $7^{\frac{1}{3}}$ is the same as $\sqrt[3]{7}$. And that, generally:

$x^{\frac{1}{n}} = \sqrt[n]{x}$ (nth root of x)

So in summary:

Power $\frac{1}{2}$ is the same as positive square root.

Power $\frac{1}{3}$ is the same as cube root.

Power $\frac{1}{n}$ is the same as nth root.

For example:

$49^{\frac{1}{2}} = \sqrt{49} = 7$

$8^{\frac{1}{3}} = \sqrt[3]{8} = 2$

$10000^{\frac{1}{4}} = \sqrt[4]{10000} = 10$

$36^{-\frac{1}{2}} = \frac{1}{\sqrt{36}} = \frac{1}{6}$

If you have an expression in the form $\left(\frac{a}{b}\right)^n$ you can calculate it as $\frac{a^n}{b^n}$ and then write it as a fraction.

> **Example 3**
>
> Write $\left(\frac{16}{25}\right)^{\frac{1}{2}}$ as a fraction.
>
> You can find the power of the numerator and denominator separately.
>
> $\left(\frac{16}{25}\right)^{\frac{1}{2}} = \frac{16^{\frac{1}{2}}}{25^{\frac{1}{2}}}$
>
> $= \frac{4}{5}$

Chapter 18: Indices

EXERCISE 18D

1 Evaluate each number.

a $25^{\frac{1}{2}}$ b $100^{\frac{1}{2}}$ c $64^{\frac{1}{2}}$ d $81^{\frac{1}{2}}$ e $625^{\frac{1}{2}}$
f $27^{\frac{1}{3}}$ g $64^{\frac{1}{3}}$ h $1000^{\frac{1}{3}}$ i $125^{\frac{1}{3}}$ j $512^{\frac{1}{3}}$
k $144^{\frac{1}{2}}$ l $400^{\frac{1}{2}}$ m $625^{\frac{1}{4}}$ n $81^{\frac{1}{4}}$ o $100\,000^{\frac{1}{5}}$
p $729^{\frac{1}{6}}$ q $32^{\frac{1}{5}}$ r $1024^{\frac{1}{10}}$ s $1296^{\frac{1}{4}}$ t $216^{\frac{1}{3}}$
u $16^{-\frac{1}{2}}$ v $8^{-\frac{1}{3}}$ w $81^{-\frac{1}{4}}$ x $3125^{-\frac{1}{5}}$ y $1\,000\,000^{-\frac{1}{6}}$

2 Evaluate each number.

a $\left(\frac{25}{36}\right)^{\frac{1}{2}}$ b $\left(\frac{100}{36}\right)^{\frac{1}{2}}$ c $\left(\frac{64}{81}\right)^{\frac{1}{2}}$ d $\left(\frac{81}{25}\right)^{\frac{1}{2}}$
e $\left(\frac{25}{64}\right)^{\frac{1}{2}}$ f $\left(\frac{27}{125}\right)^{\frac{1}{3}}$ g $\left(\frac{8}{512}\right)^{\frac{1}{3}}$ h $\left(\frac{1000}{64}\right)^{\frac{1}{3}}$
i $\left(\frac{64}{125}\right)^{\frac{1}{3}}$ j $\left(\frac{512}{343}\right)^{\frac{1}{3}}$

3 Use the general rule for raising a power to another power to prove that $x^{\frac{1}{n}}$ is equivalent to $\sqrt[n]{x}$.

4 Which of these is the odd one out?

$16^{-\frac{1}{4}}$ $64^{-\frac{1}{2}}$ $8^{-\frac{1}{3}}$

Show how you decided.

5 Imagine that you are the teacher.

Write down how you would teach the class that $27^{-\frac{1}{3}}$ is equal to $\frac{1}{3}$.

6 $x^{-\frac{2}{3}} = y^{\frac{1}{3}}$

Find values for x and y that make this equation work.

7 Solve these equations.

a $2^x = 8$ b $8^x = 2$ c $4^x = 1$ d $16^x = 4$
e $100^x = 10$ f $81^x = 3$ g $16^x = 2$ h $125^x = 5$
i $1000^x = 10$ j $400^x = 20$ k $512^x = 8$ l $128^x = 2$

Indices of the form $\frac{a}{b}$

Here are two examples of this form.

$t^{\frac{2}{3}} = t^{\frac{1}{3}} \times t^{\frac{1}{3}} = (\sqrt[3]{t})^2$ $81^{\frac{3}{4}} = (\sqrt[4]{81})^3 = 3^3 = 27$

If you have an expression of the form $\left(\frac{a}{b}\right)^{-n}$ you can invert it to calculate it as a fraction.

$\left(\frac{a}{b}\right)^{-n} = \left(\frac{b}{a}\right)^n$

18.4

Example 4

Evaluate each number.   **a** $16^{-\frac{1}{4}}$ **b** $32^{-\frac{4}{5}}$

When dealing with the negative index remember that it means reciprocal.

Work through problems like these one step at a time.

Step 1: Rewrite the calculation as a fraction by dealing with the negative power.

Step 2: Take the root of the base number given by the denominator of the fraction.

Step 3: Raise the result to the power given by the numerator of the fraction.

Step 4: Write out the answer as a fraction.

a Step 1: $16^{-\frac{1}{4}} = \left(\frac{1}{16}\right)^{\frac{1}{4}}$ Step 2: $16^{\frac{1}{4}} = \sqrt[4]{16} = 2$ Step 3: $2^1 = 2$ Step 4: $16^{-\frac{1}{4}} = \frac{1}{2}$

b Step 1: $32^{-\frac{4}{5}} = \left(\frac{1}{32}\right)^{\frac{4}{5}}$ Step 2: $32^{\frac{1}{5}} = \sqrt[5]{32} = 2$ Step 3: $2^4 = 16$ Step 4: $32^{-\frac{4}{5}} = \frac{1}{16}$

Example 5

Write $\left(\frac{8}{27}\right)^{-\frac{2}{3}}$ as a fraction.

$$\left(\frac{8}{27}\right)^{-\frac{2}{3}} = \left(\frac{27}{8}\right)^{\frac{2}{3}}$$

$$= \frac{27^{\frac{2}{3}}}{8^{\frac{2}{3}}}$$

$$= \frac{(\sqrt[3]{27})^2}{(\sqrt[3]{8})^2} = \frac{3^2}{2^2} = \frac{9}{4}$$

EXERCISE 18E

1 Evaluate these.

   a $32^{\frac{4}{5}}$ **b** $125^{\frac{2}{3}}$

   c $1296^{\frac{3}{4}}$ **d** $243^{\frac{4}{5}}$

2 Rewrite these in index form.

   a $\sqrt[3]{t^2}$ **b** $\sqrt[4]{m^3}$

   c $\sqrt[5]{k^2}$ **d** $\sqrt{x^3}$

3 Evaluate these.

   a $8^{\frac{2}{3}}$ **b** $27^{\frac{2}{3}}$

   c $16^{\frac{3}{2}}$ **d** $625^{\frac{5}{4}}$

4 Evaluate these.

a $25^{-\frac{1}{2}}$ b $36^{-\frac{1}{2}}$

c $16^{-\frac{1}{4}}$ d $81^{-\frac{1}{4}}$

e $16^{-\frac{1}{2}}$ f $8^{-\frac{1}{3}}$

g $32^{-\frac{1}{5}}$ h $27^{-\frac{1}{3}}$

5 Evaluate these.

a $25^{-\frac{3}{2}}$ b $36^{-\frac{3}{2}}$

c $16^{-\frac{3}{4}}$ d $81^{-\frac{3}{4}}$

e $64^{-\frac{4}{3}}$ f $8^{-\frac{2}{3}}$

g $32^{-\frac{2}{5}}$ h $27^{-\frac{2}{3}}$

6 Evaluate these.

a $100^{-\frac{5}{2}}$ b $144^{-\frac{1}{2}}$

c $125^{-\frac{2}{3}}$ d $9^{-\frac{3}{2}}$

e $4^{-\frac{5}{2}}$ f $64^{-\frac{5}{6}}$

g $27^{-\frac{4}{3}}$ h $169^{-\frac{1}{2}}$

7 Which of these is the odd one out?

$16^{-\frac{3}{4}}$ $64^{-\frac{1}{2}}$ $8^{-\frac{2}{3}}$

Show how you decided.

8 Imagine that you are the teacher.

Write down how you would teach the class that $27^{-\frac{2}{3}}$ is equal to $\frac{1}{9}$.

9 Write these as fractions.

a $\left(\frac{9}{4}\right)^{\frac{3}{2}}$ b $\left(\frac{27}{125}\right)^{\frac{2}{3}}$

c $\left(\frac{16}{9}\right)^{\frac{5}{2}}$ d $\left(\frac{4}{49}\right)^{\frac{3}{2}}$

e $\left(\frac{64}{27}\right)^{\frac{2}{3}}$ f $\left(\frac{16}{81}\right)^{\frac{3}{4}}$

g $\left(\frac{125}{64}\right)^{\frac{4}{3}}$ h $\left(\frac{64}{729}\right)^{\frac{5}{6}}$

10 Write these as fractions.

a $\left(\frac{3}{5}\right)^{-2}$ b $\left(\frac{4}{3}\right)^{-3}$

c $\left(\frac{9}{5}\right)^{-3}$ d $\left(\frac{2}{3}\right)^{-5}$

e $\left(\frac{9}{4}\right)^{-\frac{3}{2}}$ f $\left(\frac{4}{9}\right)^{-\frac{5}{2}}$

g $\left(\frac{8}{27}\right)^{-\frac{2}{3}}$ h $\left(\frac{49}{25}\right)^{-\frac{3}{2}}$

i $\left(\dfrac{125}{64}\right)^{-\frac{2}{3}}$ **j** $\left(\dfrac{25}{64}\right)^{-\frac{3}{2}}$

k $\left(\dfrac{16}{81}\right)^{-\frac{5}{4}}$ **l** $\left(\dfrac{2187}{128}\right)^{-\frac{5}{7}}$

11 Simplify these expressions.

a $x^{\frac{3}{2}} \times x^{\frac{5}{2}}$ **b** $x^{\frac{1}{2}} \times x^{-\frac{3}{2}}$

c $(8y^3)^{\frac{2}{3}}$ **d** $5x^{\frac{3}{2}} \div \frac{1}{2}x^{-\frac{1}{2}}$

e $4x^{\frac{1}{2}} \times 5x^{-\frac{3}{2}}$ **f** $\left(\dfrac{27}{y^3}\right)^{-\frac{1}{3}}$

12 Simplify these.

a $x^{\frac{1}{2}} \times x^{\frac{1}{2}}$ **b** $d^{-\frac{1}{2}} \times d^{-\frac{1}{2}}$ **c** $t^{\frac{1}{2}} \times t$

d $(x^{\frac{1}{2}})^4$ **e** $(y^2)^{\frac{1}{4}}$ **f** $a^{\frac{1}{2}} \times a^{\frac{3}{2}} \times a^2$

13 Simplify these.

a $x \div x^{\frac{1}{2}}$ **b** $y^{\frac{1}{2}} \div y^{\frac{1}{12}}$ **c** $a^{\frac{1}{3}} \times a^{\frac{4}{3}}$

d $t^{-\frac{1}{2}} \times t^{-\frac{3}{2}}$ **e** $\dfrac{1}{d^{-2}}$ **f** $\dfrac{k^{\frac{1}{2}} \times k^{\frac{3}{2}}}{k^2}$

14 $x^{\frac{2}{3}} = y$

Find $x^{\frac{3}{2}}$ in terms of y.

Check your progress

Core

- I understand the meaning of fractional, negative and zero indices
- I can use the rules of indices to simplify numerical expressions such as $2^{-3} \times 2^4$ or $(2^3)^2$

Chapter 19
Proportion

Topics	Level	Key words
1 Direct proportion	EXTENDED	direct proportion, constant of proportionality
2 Inverse proportion	EXTENDED	inverse proportion

In this chapter you will learn how to:

EXTENDED

- Express direct and inverse proportion in algebraic terms and use this form of expression to find unknown quantities. (E2.8)

Why this chapter matters

In many real-life situations, variables are connected by a rule or relationship. It may be that as one variable increases the other increases. Alternatively, it may be that as one variable increases the other decreases.

In this chapter you will look at how quantities vary when they are related in some way.

As this plant gets older it becomes taller.

As the storm increases the number of sunbathers decreases.

As this car gets older it is worth less (and eventually it is worthless!).

As more songs are downloaded, there is less money left on the voucher.

Try to think of other variables that are connected in this way.

Chapter 19: Proportion

Chapter 19. Topic 1

19.1 Direct proportion

There is direct proportion between two variables when one variable is a simple multiple of the other. That is, their ratio is a constant.

For example:

    1 kilogram = 2.2 pounds   There is a multiplying factor of 2.2 between kilograms and pounds.

    Area of a circle = πr^2 There is a multiplying factor of π between the area of a circle and the square of its radius.

Any question involving direct proportion usually requires you first to find this multiplying factor (called the **constant of proportionality**), then to use it to solve a problem.

The symbol for proportion is $\propto$.

So the statement 'Pay is directly proportional to time' can be mathematically written as:

    pay $\propto$ *time*

which implies that:

    pay = $k \times$ *time*

where k is the constant of proportionality.

There are four steps to be followed when you are using proportionality to solve problems.

Step 1: Set up the statement, using the proportionality symbol (you may use symbols to represent the variables).

Step 2: Set up the equation, using a constant of proportionality.

Step 3: Use given information to work out the value of the constant of proportionality.

Step 4: Substitute the value of the constant of proportionality into the equation and use this equation to find unknown values.

Example 1

The cost of an item is directly proportional to the time spent making it. An item taking 6 hours to make costs $30. Find:

a the cost of an item that takes 5 hours to make

b the length of time it takes to make an item costing $40.

Step 1: Let C dollars be the cost of making an item and t hours the time it takes.

    $C \propto t$

Step 2: Setting up the equation gives:

    $C = kt$

where k is the constant of proportionality.

Note that you can 'replace' the proportionality sign $\propto$ with $= k$ to obtain the proportionality equation.

Step 3: Since $C = 30$ when $t = 6$, then $30 = 6k$

$\Rightarrow \dfrac{30}{6} = k$

$\Rightarrow k = 5$

Step 4: So the formula is $C = 5t$.

a When $t = 5$ $\qquad C = 5 \times 5 = 25$

So the cost is $25.

b When $C = 40$ $\qquad 40 = 5 \times t$

$\Rightarrow \dfrac{40}{5} = t \Rightarrow t = 8$

So the time spent making the item is 8 hours.

EXERCISE 19A

For questions **1** to **4**, first find the value of k, the constant of proportionality, and then the formula connecting the variables.

1 T is directly proportional to M. If $T = 20$ when $M = 4$, find the value of:

   a T when $M = 3$
   b M when $T = 10$.

2 W is directly proportional to F. If $W = 45$ when $F = 3$, find the value of:

   a W when $F = 5$
   b F when $W = 90$.

3 Q varies directly with P. If $Q = 100$ when $P = 2$, find the value of:

   a Q when $P = 3$
   b P when $Q = 300$.

4 X varies directly with Y. If $X = 17.5$ when $Y = 7$, find the value of:

   a X when $Y = 9$
   b Y when $X = 30$.

5 The distance covered by a train is directly proportional to the time taken for the journey. The train travels 105 kilometres in 3 hours.

   a What distance will the train cover in 5 hours?
   b How much time will it take for the train to cover 280 kilometres?

6 The cost of fuel delivered to your door is directly proportional to the mass received. When 250 kg is delivered, it costs 47.50 dollars.

    a How much will it cost to have 350 kg delivered?

    b How much would be delivered if the cost were 33.25 dollars?

7 The number of children who can play safely in a playground is directly proportional to the area of the playground. A playground with an area of 210 m² is safe for 60 children.

    a How many children can safely play in a playground of area 154 m²?

    b A playgroup has 24 children. What is the smallest playground area in which they could safely play?

8 The number of spaces in a car park is directly proportional to the area of the car park.

    a A car park has 300 parking spaces in an area of 4500 m².

    It is decided to increase the area of the car park by 500 m² to make extra spaces.

    How many extra spaces will be made?

    b The old part of the car park is redesigned so that the original area has 10% more parking spaces.

    How many more spaces than in the original car park will there be altogether if the number of spaces in the new area is directly proportional to the number in the redesigned car park?

9 The number of passengers in a bus queue is directly proportional to the time that the person at the front of the queue has spent waiting.

Katya is the first to arrive at a bus stop. When she has been waiting 5 minutes the queue has 20 passengers.

A bus has room for 70 passengers.

How long had Katya been in the queue if the bus fills up from empty when it arrives and all passengers get on?

Direct proportions involving squares, cubes, square roots and cube roots

The process is the same as for a linear direct variation, as the next example shows.

Example 2

The cost of a circular badge is directly proportional to the square of its radius. The cost of a badge with a radius of 2 cm is $0.68. Find:

a the cost of a badge of radius 2.4 cm   **b** the radius of a badge costing $1.53.

Step 1: Let C be the cost in dollars and r the radius of a badge in centimetres.

$C \propto r^2$

Step 2: Setting up the equation gives:

$C = kr^2$

where k is the constant of proportionality.

Step 3: $C = 0.68$ when $r = 2$. So:

$0.68 = 4k$

$\Rightarrow \dfrac{0.68}{4} = k \Rightarrow k = 0.17$

Step 4: So the formula is $C = 0.17r^2$.

a When $r = 2.4$ $C = 0.17 \times 2.4^2 = 0.98$ to 2 decimal places.

Rounding gives the cost as $0.98.

b When $C = 1.53$ $1.53 = 0.17r^2$

$\Rightarrow \dfrac{1.53}{0.17} = 9 = r^2$

$\Rightarrow r = \sqrt{9} = 3$

Hence, the radius is 3 cm.

EXERCISE 19B

For questions **1** to **6**, first find k, the constant of proportionality, and then the formula connecting the variables.

1 T is directly proportional to x^2. If $T = 36$ when $x = 3$, find the value of:
  a T when $x = 5$
  b x when $T = 400$.

2 W is directly proportional to M^2. If $W = 12$ when $M = 2$, find the value of:
  a W when $M = 3$
  b M when $W = 75$.

3 E varies directly with $\sqrt{C}$. If $E = 40$ when $C = 25$, find the value of:
  a E when $C = 49$
  b C when $E = 10.4$.

4 X is directly proportional to $\sqrt{Y}$. If $X = 128$ when $Y = 16$, find the value of:
  a X when $Y = 36$
  b Y when $X = 48$.

5 P is directly proportional to f^3. If $P = 400$ when $f = 10$, find the value of:
  a P when $f = 4$
  b f when $P = 50$.

6 y is directly proportional to $\sqrt[3]{x}$. If $y = 100$ when $x = 125$, find the value of:
  a y when $x = 64$
  b x when $y = 40$.

7 The cost of serving tea and biscuits varies directly with the square root of the number of people at the buffet. It costs $25 to serve tea and biscuits to 100 people.
  a How much will it cost to serve tea and biscuits to 400 people?
  b For a cost of $37.50, how many people could be served tea and biscuits?

8 In an experiment, the temperature, in °C, varied directly with the square of the pressure, in atmospheres (atm). The temperature was 20 °C when the pressure was 5 atm.

   a What will the temperature be at 2 atm?

   b What will the pressure be at 80 °C?

9 The mass, in grams, of ball bearings varies directly with the cube of the radius, measured in millimetres. A ball bearing of radius 4 mm has a mass of 115.2 g.

   a What will be the mass of a ball bearing of radius 6 mm?

   b A ball bearing has a mass of 48.6 g. What is its radius?

10 The energy, in joules (J), of a particle varies directly with the square of its speed, in m/s. A particle moving at 20 m/s has 50 J of energy.

   a How much energy has a particle moving at 4 m/s?

   b At what speed is a particle moving if it has 200 J of energy?

11 The cost, in dollars, of a trip varies directly with the square root of the number of miles travelled. The cost of a 100-mile trip is 35 dollars.

   a What is the cost of a 500-mile trip (to the nearest dollar)?

   b What is the distance of a trip costing 70 dollars?

12 A sculptor is making statues.

The amount of clay used is directly proportional to the cube of the height of the statue.

A statue is 10 cm tall and uses 500 cm³ of clay.

How much clay will a similar statue use if it is twice as tall?

13 The cost of making different-sized machines is proportional to the time taken.

A small machine costs $100 and takes two hours to make.

How much will a large machine cost that takes 5 hours to build?

14 The sketch graphs show each of these proportion statements.

   a $y \propto x^2$ **b** $y \propto x$ **c** $y \propto \sqrt{x}$

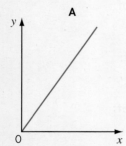
A

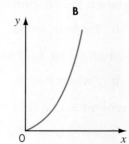
B

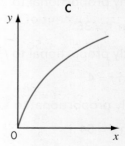
C

Match each statement to the correct sketch.

Chapter 19. Topic 2

15 Here are two tables.

Match each table to a graph in question **14**.

a

x	1	2	3
y	3	12	27

b

x	1	2	3
y	3	6	9

19.2 Inverse proportion

There is inverse proportion between two variables when one variable is directly proportional to the *reciprocal* of the other. That is, the product of the two variables is constant. So, as one variable increases, the other decreases.

For example, the faster you travel over a given distance, the less time it takes. So there is an inverse variation between speed and time. Speed is inversely proportional to time.

$S \propto \dfrac{1}{T}$ and so $S = \dfrac{k}{T}$

which can be written as $ST = k$.

Example 3

M is inversely proportional to R. If $M = 9$ when $R = 4$, find the value of:

a M when $R = 2$ **b** R when $M = 3$.

Step 1: $M \propto \dfrac{1}{R}$

Step 2: Setting up the equation gives:

$M = \dfrac{k}{R}$

where k is the **constant of proportionality**.

Step 3: $M = 9$ when $R = 4$. So $9 = \dfrac{k}{4}$

$\Rightarrow 9 \times 4 = k \Rightarrow k = 36$

Step 4: The formula is $M = \dfrac{36}{R}$

a When $R = 2$, then $M = \dfrac{36}{2} = 18$

b When $M = 3$, then $3 = \dfrac{36}{R} \Rightarrow 3R = 36 \Rightarrow R = 12$

EXERCISE 19C

For questions **1** to **6**, first find the formula connecting the variables.

1 T is inversely proportional to m. If $T = 6$ when $m = 2$, find the value of:

   a T when $m = 4$
   b m when $T = 4.8$.

2 W is inversely proportional to x. If $W = 5$ when $x = 12$, find the value of:

   a W when $x = 3$
   b x when $W = 10$.

3 Q varies inversely with $(5 - t)$. If $Q = 8$ when $t = 3$, find the value of:

   a Q when $t = 10$
   b t when $Q = 16$.

4 M varies inversely with t^2. If $M = 9$ when $t = 2$, find the value of:

   a M when $t = 3$
   b t when $M = 1.44$.

5 W is inversely proportional to $\sqrt{T}$. If $W = 6$ when $T = 16$, find the value of:

   a W when $T = 25$
   b T when $W = 2.4$.

6 y is inversely proportional to the cube of x. If $y = 4$ when $x = 2$, find the value of:

   a y when $x = 1$
   b x when $y = \frac{1}{2}$.

7 The grant available to a group of students was inversely proportional to the number of students. When 30 students needed a grant, they received $60 each.

   a What would the grant have been if 120 students had needed one?
   b If the grant had been $50 each, how many students would have received it?

8 While doing underwater tests in an ocean, scientists noticed that the temperature, in °C, was inversely proportional to the depth, in kilometres. When the temperature was 6 °C, the scientists were at a depth of 4 km.

   a What would the temperature have been at a depth of 8 km?
   b At what depth would they find the temperature at 2 °C?

9 A new engine had serious problems. The distance it went, in kilometres, without breaking down was inversely proportional to the square of its speed in metres per second (m/s). When the speed was 12 m/s, the engine lasted 3 km.

   a Find the distance covered before a breakdown, when the speed is 15 m/s.
   b On one test, the engine broke down after 6.75 km. What was the speed?

10 In a balloon it was noticed that the pressure, in atmospheres (atm), was inversely proportional to the square root of the height, in metres. When the balloon was at a height of 25 m, the pressure was 1.44 atm.

   a What was the pressure at a height of 9 m?
   b What would the height have been if the pressure was 0.72 atm?

11 The amount of waste from a firm, measured in tonnes per hour, is inversely proportional to the square root of the area of the filter beds that clean it, in square metres (m²). The firm produces 1.25 tonnes of waste per hour, with filter beds of size 0.16 m².

   a The filter beds used to be only 0.01 m². How much waste did the firm produce then?

   b How much waste could be produced if the filter beds were 0.75 m²?

12 Which statement is represented by the graph?
Give a reason for your answer.

A $y \propto x$

B $y \propto \dfrac{1}{x}$

C $y \propto \sqrt{x}$

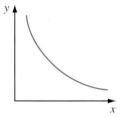

13 In the table, y is inversely proportional to the cube root of x.

Complete the table, leaving your answers as fractions.

x	8	27	
y	1		$\frac{1}{2}$

Check your progress

Extended

- I understand the terms direct and inverse proportion
- I can express direct and inverse proportion in algebraic terms
- I can use this form of expression to find unknown quantities

Chapter 20
Linear programming

Topics	Level	Key words
1 Graphical inequalities	EXTENDED	boundary, region, included, origin, dashed line, solid line
2 More than one inequality	EXTENDED	required
3 Linear programming	EXTENDED	linear programming

In this chapter you will learn how to:

EXTENDED

- Represent inequalities graphically. (E2.6)
- Use this representation in the solution of simple linear programming problems. (E2.6)

Why this chapter matters

The theory of linear programming has been used by many companies to reduce their costs and increase productivity.

The theory of linear programming was developed at the start of the Second World War in 1939.

It was used to work out ways to supply armaments as efficiently as possible. It was such a powerful tool that the British and Americans did not want the Germans to know about it, so it was not made public until 1947.

George Dantzig was one of the inventors of linear programming. He came late to a lecture at University one day and saw two problems written on the blackboard. He copied them, thinking they were the homework assignment.

He solved both problems, but apologised to the lecturer later as he found them a little harder than the usual homework, so he took a few days to solve them and was late handing them in.

The lecturer was astonished. The problems he had written on the board were not homework but examples of 'impossible problems'. Not any more!

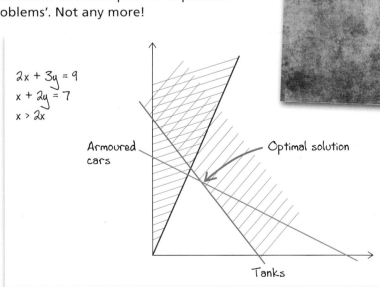

Chapter 20 . Topic 1
20.1 Graphical inequalities   E

A linear inequality can be plotted on a graph. The result is a **region** that lies on one side or the other of a straight line. You will recognise an inequality by the fact that it looks like an equation but instead of the equals sign it has an inequality sign: $<$, $>$, $\leq$, or $\geq$.

The following are examples of linear inequalities that can be represented on a graph.

$y < 3 \qquad x > 7 \qquad -3 \leq y < 5 \qquad y \geq 2x + 3 \qquad 2x + 3y < 6 \qquad y \leq x$

The method for graphing an inequality is to draw the **boundary** line that defines the inequality. This is found by replacing the inequality sign with an equals sign. When a strict inequality is stated ($<$ or $>$), the boundary line should be drawn as a *dashed* line to show that it is not included in the range of values. When $\leq$ or $\geq$ is used to state the inequality, the boundary line should be drawn as a *solid* line to show that the boundary is **included**.

After the boundary line has been drawn, shade the *unwanted region*.

To confirm on which side of the line the region lies, choose any point that is not on the boundary line and test it in the inequality. If it satisfies the inequality, that is the side required. If it doesn't, the other side is required.

> **Advice and Tips**
>
> You should shade the unwanted region of the graph, leaving the region that satisfies the inequality unshaded.

Work through the six inequalities in this example to see how the procedure is applied.

Example 1

Show each inequality on a graph.

a $y \leq 3$
b $x > 7$
c $-3 \leq y < 5$
d $y \leq 2x + 3$
e $2x + 3y < 6$
f $y \leq x$

a Draw the line $y = 3$. Since the inequality is stated as $\leq$, the line is *solid*. Test a point that is not on the line. The **origin** is always a good choice if possible, as 0 is easy to test.

Putting 0 into the inequality gives $0 \leq 3$. The inequality is satisfied and so the region containing the origin is the side we want.

Shade the region on the other side of the line.

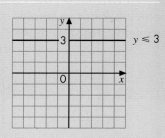

b Since the inequality is stated as $>$, the line is *dashed*. Draw the line $x = 7$.

Test the origin (0, 0), which gives $0 > 7$. This is not true, so you want the other side of the line from the origin.

Shade the unwanted region as shown.

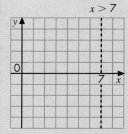

---

344   20.1 Graphical inequalities

c Draw the lines $y = -3$ (solid for $\leqslant$) and $y = 5$ (dashed for $<$).

Test a point that is not on either line, say (0, 0). Zero is between -3 and 5, so the required region lies between the lines.

Shade in the unwanted regions, outside these lines.

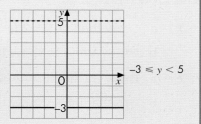

$-3 \leqslant y < 5$

d Draw the line $y = 2x + 3$. Since the inequality is stated as $\leqslant$, the line is solid.

Test a point that is not on the line, (0, 0). Putting these x- and y-values in the inequality gives $0 \leqslant 2(0) + 3$, which is true. So the region that includes the origin is what you want.

Shade in the unwanted region on the other side of the line.

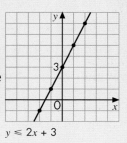

$y \leqslant 2x + 3$

e Draw the line $2x + 3y = 6$.

The easiest way is to find out where it crosses the axes.

If $x = 0$, $3y = 6 \Rightarrow y = 2$. Crosses y-axis at (0, 2).

If $y = 0$, $2x = 6 \Rightarrow x = 3$. Crosses x-axis at (3, 0).

Draw the line through these two points.

Since the inequality is stated as $<$, the line is dashed.

Test a point that is not on the line, say (0, 0). Is it true that $2(0) + 3(0) < 6$? The answer is yes, so the origin is in the region that you want.

Shade in the unwanted region on the other side of the line.

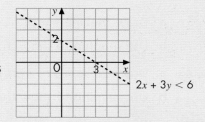

$2x + 3y < 6$

f Draw the line $y = x$. Since the inequality is stated as $\leqslant$, the line is solid.

This time the origin is on the line, so pick any other point, say (1, 3). Putting $x = 1$ and $y = 3$ in the inequality gives $3 \leqslant 1$. This is not true, so the point (1, 3) is not in the region you want.

Shade in the region that includes (1, 3), the unwanted region.

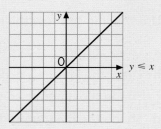

$y \leqslant x$

Chapter 20: Linear programming

EXERCISE 20A

EXTENDED

1 a Draw the line $x = 2$ (as a solid line).
   b Use shading to show the region defined by $x \leq 2$.

2 a Draw the line $y = -3$ (as a dashed line).
   b Use shading to show the region defined by $y > -3$.

3 a Draw the line $x = -2$ (as a solid line).
   b Draw the line $x = 1$ (as a solid line) on the same grid.
   c Use shading to show the region defined by $-2 \leq x \leq 1$.

4 a Draw the line $y = -1$ (as a dashed line).
   b Draw the line $y = 4$ (as a solid line) on the same grid.
   c Use shading to show the region defined by $-1 < y \leq 4$.

5 a On the same grid, draw the regions defined by these inequalities.
      i $-3 \leq x \leq 6$
      ii $-4 < y \leq 5$
   b Are these points in the region defined by both inequalities?
      i (2, 2)
      ii (1, 5)
      iii (−2, −4)

6 a Draw the line $y = 2x - 1$ (as a dashed line).
   b Use shading to show the region defined by $y < 2x - 1$.

7 a Draw the line $3x - 4y = 12$ (as a solid line).
   b Use shading to show the region defined by $3x - 4y \leq 12$.

8 a Draw the line $y = \frac{1}{2}x + 3$ (as a solid line).
   b Use shading to show the region defined by $y \geq \frac{1}{2}x + 3$.

9 Use shading to show the region defined by $y < -3$.

10 a Draw the line $y = 3x - 4$ (as a solid line).
   b Draw the line $x + y = 10$ (as a solid line) on the same diagram.
   c Shade the diagram so that the region defined by $y \geq 3x - 4$ is left *unshaded*.
   d Shade the diagram so that the region defined by $x + y \leq 10$ is left *unshaded*.
   e Are these points in the region defined by both inequalities?
      i (2, 1)
      ii (2, 2)
      iii (2, 3)

20.2 More than one inequality

When you have to show a region that satisfies more than one inequality, it is always clearer to *shade* the regions *not required*, so that the *required region* is left *blank*.

Example 2

a On the same grid, show the regions that represent each inequality by shading the unwanted regions.

  i $x > 2$

  ii $y \geqslant x$

  iii $x + y < 8$

b Are these points in the region that satisfies all three inequalities?

  i (3, 4)

  ii (2, 6)

  iii (3, 3)

---

a **i** 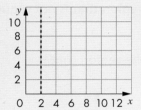 **ii** 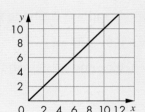 **iii**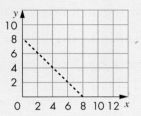

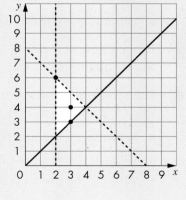

i The region $x > 2$ is shown unshaded in diagram **i**. The boundary line is $x = 2$ (dashed).

ii The region $y \geqslant x$ is shown unshaded in diagram **ii**. The boundary line is $y = x$ (solid).

iii The region $x + y < 8$ is shown unshaded in diagram **iii**.

The boundary line is $x + y = 8$ (dashed). The regions have first been drawn separately so that each may be clearly seen. The diagram on the right shows all three regions on the same grid. The white triangular area defines the region that satisfies all three inequalities.

b **i** The point (3, 4) is clearly within the region that satisfies all three inequalities.

  ii The point (2, 6) is on the boundary lines $x = 2$ and $x + y = 8$. As these are dashed lines, they are not included in the region defined by all three inequalities. So, the point (2, 6) is not in this region.

  iii The point (3, 3) is on the boundary line $y = x$. As this is a solid line, it is included in the region defined by all three inequalities. So, the point (3, 3) is included in this region.

EXERCISE 20B

1
  a Draw the line $y = x$ (as a solid line).
  b Draw the line $2x + 5y = 10$ (as a solid line) on the same diagram.
  c Draw the line $2x + y = 6$ (as a dashed line) on the same diagram.
  d Shade the diagram so that the region defined by $y \geq x$ is left *unshaded*.
  e Shade the diagram so that the region defined by $2x + 5y \geq 10$ is left *unshaded*.
  f Shade the diagram so that the region defined by $2x + y < 6$ is left *unshaded*.
  g Are these points in the region defined by these inequalities?
    i (1, 1)   **ii** (2, 2)   **iii** (1, 3)

Advice and Tips

Find the points where the line crosses each axis.

2
  a On the same grid, draw the regions defined by the following inequalities. (Shade the diagram so that the overlapping region is left blank.)
    i $y > x - 3$ **ii** $3y + 4x \leq 24$ **iii** $x \geq 2$
  b Are these points in the region defined by all three inequalities?
    i (1, 1)   **ii** (2, 2)   **iii** (3, 3)   **iv** (4, 4)

3
  a On a graph draw the lines $y = x$, $x + y = 8$ and $y = 2$.
  b Label the region R where $y \leq x$, $x + y \leq 8$ and $y \geq 2$. Shade the region that is *not* required.

4
  a On a graph draw the lines $y = x - 4$, $y = 0.5x$ and $y = -x$.
  b Show the region S where $y \geq x - 4$, $y \leq 0.5x$ and $y \geq -x$. Shade the region that is *not* required.
  c What is the largest y-coordinate of a point in S?
  d What is the smallest y-coordinate of a point in S?
  e What is the largest value of $x + y$ for a point in S?

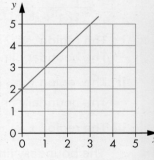

5 Explain how you would find which side of the line represents the inequality $y < x + 2$.

6 The region marked R is bounded by the lines $x + y = 3$, $y = \frac{1}{2}x + 3$ and $y = 5x - 15$.
  a What three inequalities are satisfied in region R?
  b What is the greatest value of $x + y$ in region R?
  c What is the greatest value of $x - y$ in region R?

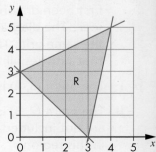

20.3 Linear programming

There are practical situations that give rise to inequalities that can be shown on a graph. Solving problems in this way is called **linear programming**.

> **Example 3**
> A boat trip costs $20 for adults and $10 for children.
>
> Suppose there are x adults and y children on the trip.
>
> **a** There must not be more than 10 people on the trip.
> Explain why $x + y \leq 10$.
>
> **b** The money taken for tickets must be at least $120.
> Explain why $2x + y \geq 12$.
>
> **c** Show the region satisfied by both these inequalities.
>
> **d** What is the smallest number of adults that should be carried?
>
> ---
>
> **a** The total number of passengers is 10.
>
> The total must be 10 or less so $x + y \leq 10$.
>
> **b** The total number of dollars taken is $20x + 10y$.
>
> This must be at least 120 so $20x + 10y \geq 120$.
>
> Divide both sides by 10 to get $2x + y \geq 12$.
>
> **c** Negative numbers on the axes are not needed.
>
> First draw the lines with equations $x + y = 10$ and $2x + y = 12$.
>
> The first crosses the axes at (10, 0) and (0, 10).
>
> The second crosses the axes at (6, 0) and (0, 12).
>
> Shade the regions that are *not* required.
>
> The solution must be in the unshaded region R.
>
>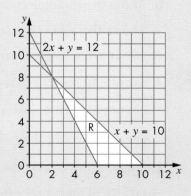
>
> **d** The smallest x-coordinate in this region is at (2, 8).
>
> x must be at least 2 so there must be at least 2 adults.

EXERCISE 20C

1 A man buys x cartons of milk and y bottles of water.

  a Explain what $x \geqslant 3$ means.
  b Write an inequality to show that he buys at least 2 bottles of water.
  c He cannot carry more than a total of 8 bottles and cartons. Write this as an inequality.
  d Draw a graph to show possible values of x and y.

2 There are x cars and y vans in a car park.

  a There are no more than 10 vehicles (cars and vans) in the car park.

  Write an inequality to show this.

  b Cars pay $2 and vans pay $5 to park. At least $30 was paid in total.

  Write an inequality to show this.

  c Show the two inequalities on a grid.
  d What is the largest possible number of cars?
  e What is the smallest possible number of vans?

3 x girls and y boys are at a children's party.

There are more girls than boys.

The number of girls is fewer than twice the number of boys.

There are fewer than 12 children at the party.

  a Write down three inequalities to represent each of the statements above.
  b Show the region indicated by these inequalities.

  Shade the areas that are not included.

  c What is the largest possible number of girls at the party?

4 A house is home to x cats and y dogs.

There are at least 6 animals.

There are no more than 12 animals.

There are more dogs than cats.

There are at least two cats.

  a Write down four inequalities from these sentences.
  b Show the region that gives possible values of x and y by shading the area that is not required.
  c What is the smallest possible number of dogs?
  d What is the largest possible number of cats?

5 At a concert, x people bought seats near the front for $15 and y people bought seats near the back for $10.

    a Ticket sales were at least $600. Show that $3x + 2y \geq 120$.

    b No more than 50 tickets were sold. Write an inequality for this.

    c Show the region which represents these inequalities on a grid.

    d What is the largest possible number of back seats sold?

6 A company has a fleet of coaches in two sizes. There are x large coaches and y small coaches.

    a Write an inequality for each of these statements.

        i The total number of coaches must not exceed 15.

        ii There must be at least 5 coaches in total.

        iii The number of small coaches must not be less than half the number of large coaches.

    b Explain in words the meaning of $y \leq x + 2$.

    c Label the region R to show the four inequalities in parts **a** and **b**. Shade the area which is not required.

    d If there are 6 large coaches, how many small coaches could there be? Give all the possible values.

7 In a football league a team score 3 points for a win, 1 point for a draw and none for a loss.

    a If a team wins x matches and draws y matches, write down an expression for the total number of points gained.

    b Wayne knows these facts about his team's performance so far this season.

        • They have gained at most 18 points.

        • They have more wins than draws.

        • They have at least two draws.

    Write an inequality, using x and y, for each of these three statements.

    c Show on a graph the region where the three inequalities are satisfied.

    d List all the possible numbers of wins and draws for Wayne's team.

Check your progress

Extended

- I can represent inequalities graphically
- I can use graphical representations to solve simple linear programming problems

Chapter 21
Functions

Topics	Level	Key words
1 Function notation	EXTENDED	function
2 Inverse functions	EXTENDED	inverse
3 Composite functions	EXTENDED	composite
4 More about composite functions	EXTENDED	

In this chapter you will learn how to:

EXTENDED

- Use function notation, e.g. f(x) = 3x − 5, f: x → 3x − 5 to describe simple functions. (E2.9)
- Find inverse functions f^{-1}(x). (E2.9)
- Form composite functions as defined by gf(x) = g(f(x)). (E2.9)

Why this chapter matters

Notation is important in mathematics. Try writing an equation or formula in words instead of symbols and you will see why notation makes things easier to understand.

If you drop a coin, how long does it take to reach the ground? That depends on the height you drop it from. We say that the time taken is a **function** of the height.

By dropping the coin from different heights and measuring the time it takes to fall, it would be possible to find a formula for the time in terms of the height.

Here are some other examples where one variable is a **function** of another.

- The cost of posting a parcel is a function of its mass.
- The time taken for a journey is a function of the distance travelled.
- The stopping distance of a car is a function of its speed.
- The cost of a second hand car is a function of its age.
- The time taken to download a computer file is a function of the size of the file.

The idea of an **inverse** occurs frequently in mathematics. It is not a difficult idea and it is a useful one. Putting a hat on and taking it off are inverse operations. Switching a light on and switching it off are also inverse operations.

Here are some examples of inverse operations in mathematics.

- Add 3 and subtract 3.
- Multiply by 5 and divide by 5.
- Rotate 90° clockwise and rotate 90° anticlockwise.
- Square a number and find the square root of the square number.

It is not always possible to find an inverse. Sadly the inverse of breaking a glass does not exist.

21.1 Function notation

E

You are familiar with equations written in x and y, such as $y = 3x - 4$ or $y = 2x^2 + 5x - 3$.

These equations are showing that y is a function of x. This means that the value of y depends on the value of x so that y changes when x changes.

Sometimes it is useful to use a different notation to show this. You could write the first equation above as $f(x) = 3x - 4$ and refer to it as 'function f'.

It is then easy to show the result of using different values for x. For example:

- 'the value of $f(x)$ when x is 5' can be written as $f(5)$.

  So $f(5) = 3 \times 5 - 4 = 11$

- $f(1)$ means 'the value of $f(x)$ when x is 1'

  So $f(1) = 3 \times 1 - 4 = -1$

  and $f(-1) = -7$

If there are different functions in the same problem they can be represented by different letters, for example:

$g(x) = 2x^2 + 5x - 3$ or 'function g'.

Sometimes instead of $f(x) = 3x - 4$ you may see $f: x \to 3x - 4$. These two forms mean exactly the same thing.

EXERCISE 21A

EXTENDED

1. $f(x) = 2x + 6$. Find the value of:
   - **a** $f(3)$
   - **b** $f(10)$
   - **c** $f(\frac{1}{2})$
   - **d** $f(-4)$
   - **e** $f(-1.5)$.

2. $g(x) = \dfrac{x^2 + 1}{2}$. Find the value of:
   - **a** $g(0)$
   - **b** $g(3)$
   - **c** $g(10)$
   - **d** $g(-2)$
   - **e** $g(-\frac{1}{2})$.

3. $f: x \to x^3 - 2x + 1$. Find the value of:
   - **a** $f(2)$
   - **b** $f(-2)$
   - **c** $f(100)$
   - **d** $f(0)$
   - **e** $f(\frac{1}{2})$.

4. $g: x \to 2^x$. Find the value of:
   - **a** $g(2)$
   - **b** $g(5)$
   - **c** $g(0)$
   - **d** $g(-1)$
   - **e** $g(-3)$.

5. $h(x) = \dfrac{x + 1}{x - 1}$. Find the value of:
   - **a** $h(2)$
   - **b** $h(3)$
   - **c** $h(-1)$
   - **d** $h(0)$
   - **e** $h(1\frac{1}{2})$.

6. $f: x \to 2x + 5$
   - **a** If $f(a) = 20$, what is the value of a?
   - **b** If $f(b) = 0$, what is the value of b?
   - **c** If $f(c) = -5$, what is the value of c?

Chapter 21 . Topic 2

7 $g(x) = \sqrt{x} + 3$

    a Find the value of g(33).
    b If $g(a) = 10$, find the value of a.
    c If $g(b) = 2.5$, find the value of b.

8 $f(x) = 2x - 8$ and $g(x) = 10 - x$.

    a What is the value of x for which $f(x) = g(x)$?
    b Sketch the graphs of $y = f(x)$ and $y = g(x)$. At what point do they cross?

9 $h: x \to \dfrac{12}{x} + 1$ $k: x \to 2^x - 1$

    a Find the value of h(6).
    b Find the value of k(−1).
    c Solve the equation $h(x) = k(3)$.
    d Solve the equation $k(x) = h(-12)$.

21.2 Inverse functions

Suppose $f(x) = 2x + 6$.

Then $f(1) = 8$, $f(3) = 12$ and $f(-4) = -2$.

The **inverse** of f is the function that has the opposite effect and 'undoes' f. You write the inverse of f as f^{-1}.

Since f above means 'multiply by 2 and then add 6', the inverse will be 'subtract 6 and then divide by 2':

$$f^{-1}(x) = \frac{x-6}{2}$$

So $f^{-1}(8) = 1$, $f^{-1}(12) = 3$ and $f^{-1}(-2) = -4$.

You can find the inverse by following these steps.

Step 1: Write $y = f(x)$

$$y = 2x + 6$$

Step 2: Rearrange to make x the subject. $y - 6 = 2x$

$$\frac{y-6}{2} = x$$

$$x = \frac{y-6}{2}$$

Step 3: Replace y by x in the result. $f^{-1}(x) = \dfrac{x-6}{2}$

EXERCISE 21B

EXTENDED

1 Find $f^{-1}(x)$ for each of these functions.

  a $f(x) = x + 7$
  b $f(x) = 8x$
  c $f(x) = \dfrac{x}{5}$
  d $f(x) = x - 3$

2 $f: x \to \dfrac{x}{2} + 6$. Find the value of:

  a $f(4)$
  b $f^{-1}(8)$
  c $f(-2)$
  d $f^{-1}(5)$.

3 Find $f^{-1}(x)$ for each of these functions.

  a $f(x) = \dfrac{x}{3} - 2$
  b $f(x) = 4(x - 5)$
  c $f(x) = \dfrac{x + 4}{5}$
  d $f(x) = \dfrac{(3x - 6)}{5}$
  e $f(x) = 3(\dfrac{x}{2} + 4)$
  f $f(x) = 4x^3$

4 $g(x) = \dfrac{2x + 5}{3}$. Find the value of:

  a $g^{-1}(3)$
  b $g^{-1}(2)$
  c $g^{-1}(0)$.

5 $f(x) = 10 - x$

  a Find an expression for $f^{-1}(x)$.
  b What do you notice about $f(x)$ and $f^{-1}(x)$?

6 Find $f^{-1}(x)$ for each of these functions.

  a $f(x) = \dfrac{8}{x}$
  b $f(x) = \dfrac{20}{x} - 1$
  c $f(x) = \dfrac{2}{x + 1}$

7 $f(x) = 2x - 4$

  a Find $f^{-1}(x)$.
  b On the same axes draw graphs of $y = f(x)$ and $y = f^{-1}(x)$.
  c Where do the lines cross?

8 $f(x) = \dfrac{x + 5}{2}$

  Solve the equation $f(x) = f^{-1}(x)$.

9 $f^{-1}(x) = 3x - 2$

  Find an expression for $f(x)$.

21.2 Inverse functions

21.3 Composite functions

Consider a function given by $f(x) = 2x$.

So f means 'double it'.

$$f(2) = 4$$
$$f(3) = 6$$
$$f(5) = 10$$
and $f(-3) = -6$

Now suppose you have another function, g, and $g(x) = x - 3$.

So g means 'subtract 3'.

$$g(4) = 1$$
$$g(6) = 3$$
$$g(10) = 7$$
and $g(-6) = -9$

Now put those side by side to make a **composite** function.

$f(2) = 4$ and $g(4) = 1$
$f(3) = 6$ and $g(6) = 3$
$f(5) = 10$ and $g(10) = 7$
$f(-3) = -6$ and $g(-6) = -9$

In words they mean:

Start with 2, double it, subtract 3, the answer is 1.

Start with 3, double it, subtract 3, the answer is 3.

Start with 5, double it, subtract 3, the answer is 7.

Start with −3, double it, subtract 3, the answer is −9.

You can write that in symbols as:

$$gf(2) = 1$$
$$gf(3) = 3$$
$$gf(5) = 7$$
$$gf(-3) = -9$$

Advice and Tips

gf means 'first f, then g'.

Example 1

$h(x) = x^2$ and $k(x) = x + 4$. Find:

a kh(3)

b kh(–2)

c hk(5)

---

a h means 'square it' and k means 'add 4'.

kh(3) means 'start with 3, square it, then add 4'.

So kh(3) = 3^2 + 4 = 9 + 4 = 13

b kh(–2) = $(-2)^2$ + 4 = 4 + 4 = 8

c hk(5) is the other way round.

It means 'start with 5, add 4, then square it'.

hk(5) = $(5 + 4)^2 = 9^2 = 81$

Look again at the functions f and g you considered earlier.

$f(x) = 2x$ and $g(x) = x - 3$

gf(x) means 'first double x, then subtract 3'. So you can write:

$gf(x) = 2x - 3$

What about fg(x)?

That means 'first subtract 3, then double it'. So you can write:

$fg(x) = 2(x - 3)$

Looking at example 1, $h(x) = x^2$ and $k(x) = x + 4$.

Then $kh(x) = x^2 + 4$

and $hk(x) = (x + 4)^2$

EXERCISE 21C

1 $s(x) = x + 4$ and $t(x) = \dfrac{x}{2}$

   a Find the value of s(2) and ts(2).   **b** Find the value of s(3) and ts(3).

   c Find the value of s(6) and ts(6).   **d** Find an expression for ts(x).

   e Find the value of t(2) and st(2).   **f** Find the value of t(3) and st(3).

   g Find the value of t(–10) and st(–10).   **h** Find an expression for st(x).

2 $c(x) = x^3$ and $d(x) = 2x$.

   a Find the value of d(3) and cd(3).   **b** Find the value of d(5) and cd(5).

   c Find an expression for cd(x).   **d** Find the value of c(4) and dc(4).

   e Find an expression for dc(x).

3 $r(x) = \sqrt{x}$ and $a(x) = 2x + 1$

  a Find the value of a(0), a(4) and a(12).   **b** Find the value of ra(0), ra(4) and ra(12).
  c Find an expression for ra(x).

4 $m(x) = 3x$

  a Find the value of m(2) and mm(2).   **b** Find the value of m(4) and mm(4).
  c Find an expression for mm(x).

5 $f(x) = 3x$ and $g(x) = x - 6$

  a Find an expression for fg(x).   **b** Find an expression for gf(x).

6 $a(x) = x + 4$ and $b(x) = x - 7$

  Show that ab(x) and ba(x) are identical.

21.4 More about composite functions

Suppose $f(x) = x^2 + 2$ and $g(x) = 2x - 3$.

Is it possible to find an expression for gf(x)?

Start with a value of x, say $x = 3$.

$\quad f(3) = 3^2 + 2 = 11$

Now apply g to that answer.

$\quad g(11) = 2 \times 11 - 3 = 19$

So gf(3) = 19

gf(x) means 'start with x, apply f, and then apply g to the answer.'

If you start with x and 'apply f' you get $x^2 + 2$.

Now take that answer and 'apply g', double it and subtract 3.

$\quad gf(x) = 2(x^2 + 2) - 3$

Now simplify that.

$\quad gf(x) = 2(x^2 + 2) - 3$
$\quad \quad \quad = 2x^2 + 4 - 3$
$\quad \quad \quad = 2x^2 + 1$

So $gf(x) = 2x^2 + 1$

Check this works when x is 3.

$\quad gf(3) = 2 \times 3^2 + 1$
$\quad \quad \quad = 18 + 1$
$\quad \quad \quad = 19$ as before.

EXERCISE 21D

1 $f(x) = \dfrac{x-3}{2}$ and $g(x) = 3x + 1$

Find the value of:

a fg(3)
b gf(3)
c fg(6)
d gf(6).

2 $f(x) = x^2 - x$ and $g(x) = \dfrac{x}{2} + 3$

Find the value of:

a fg(4)
b gf(4)
c fg(1)
d gf(1).

3 $f(x) = 2^x$ and $g(x) = 2x - 1$

Find the value of:

a gf(2)
b fg(2)
c ff(3)
d gg(6).

4 $f(x) = 3x + 1$ and $g(x) = 2x - 2$

a Find an expression for gf(x). Write your answer as simply as possible.
b Find an expression for fg(x). Write your answer as simply as possible.

5 In each case find an expression for fg(x). Write your answer as simply as possible.

a $f(x) = x^2$ and $g(x) = 3x + 4$
b $f(x) = 2x + 3$ and $g(x) = 3x - 4$
c $f(x) = \dfrac{x}{2} + 4$ and $g(x) = 4x - 2$
d $f(x) = 12 - x$ and $g(x) = 2x + 8$

6 $h(x) = 10 - x$ and $k(x) = 20 - x$

Find an expression for:

a hk(x)
b kh(x)
c kk(x).

7 $h(x) = x^2$ and $k(x) = \dfrac{12}{x}$

Find an expression for:

a $hh(x)$
b $hk(x)$
c $kh(x)$
d $kk(x)$.

8 $m(x) = x^2 + 2x$ and $n(x) = 2x - 1$

a Find the value of $mm(2)$.
b Show that $nn(x) = 4x - 3$.
c Show that $mn(x) = 4x^2 - 1$

9 $f(x) = \dfrac{1}{x - 4}$ and $g(x) = 3x + 1$.

a Find an expression for $fg(x)$.
b Find an expression for $gf(x)$, writing your answer as a single fraction.

10 $h(x) = \dfrac{x + 4}{2}$ and $k(x) = 3x - 5$

Find:

a $h^{-1}k(x)$
b $hh(x)$.

11 Suppose $f(x) = 0.5(x + 9)$.

a Show that $f(1) = 5$.
b Find the value of $f(5)$.
c Find $f(b)$ where b is the answer to part **b**.
d Continue in this way, using the last answer as the next value of x, to find the next six values.
e What is happening to the answers?

12 What happens in question 11 if you start with $f(25)$ instead of $f(1)$?

Check your progress

Extended

- I can use function notation to describe simple functions
- I can find inverse functions
- I can find composite functions

Chapter 22
Differentiation

Topics	Level	Key words
1 The gradient of a curve	EXTENDED	gradient
2 More complex curves	EXTENDED	derivative, differentiate, differentiation
3 Turning points	EXTENDED	turning point, maximum, minimum

In this chapter you will learn how to:

EXTENDED

- Differentiate integer powers of x and simple sums of these. (E2.13)
- Understand the idea of a derived function. (E2.13)
- Use the derivatives of functions of the form ax^n, and simple sums of not more than three of these. (E2.13)
- Determine gradients, and turning points (maxima and minima) by differentiation and relate these to graphs. (E2.13)
- Distinguish between maxima and minima by considering the general shape of the graph. (E2.13)

Why this chapter matters

When you look at a graph, it shows you how changes in one variable lead to changes in another. Differentiation helps us to find the rate at which this change happens.

When you look at a straight line graph you can easily calculate the gradient which tells you how quickly one variable changes compared to another. But when the line on a graph is curved, the gradient is changing all the time. Using differentiation you can calculate the gradient and so work out how quickly change is happening at any point.

For example:

- When you walk up a hill your height compared to ground level changes with respect to your position on the slope
- When you drive a car your position and your velocity vary with respect to time
- The brightness of a light bulb varies depending on the electric current flowing through it
- Hot drinks cool down and ice melts at different rates as time passes

We can use differentiation to understand how changes like this happen.

A form of differentiation was used in Ancient Greece and China. In the seventeenth century CE a German mathematician called Leibniz and an English scientist, Isaac Newton, started to develop the form of differentiation we use today.

Differentiation is now an essential tool in almost all areas of science, including engineering, physics, chemistry, biology, economics and computer science.

Chapter 22. Topic 1
22.1 The gradient of a curve

This is a graph of $y = x^2 - 3x + 4$.

What is the **gradient** at the point P(3, 4)?

In an earlier chapter you learnt how to find the gradient by drawing a tangent to the curve at P. It is hard to do this accurately and we need to draw the graph first.

We will now look at another method.

The point A(4, 8) is on the curve.

The gradient of the straight line PA is

$$\frac{\text{difference in } y \text{ - coordinates}}{\text{difference in } x \text{ - coordinates}} = \frac{8-4}{4-3} = 4$$

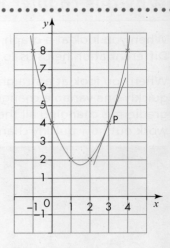

The line PA is steeper than the tangent so the gradient of the curve at P is less than 4.

Suppose A is a point on the curve closer to P.

If A is (3.5, 5.75) Then the gradient of PA is $\frac{5.75 - 4}{3.5 - 3} = 3.5$

If A is (3.1, 4.31) Then the gradient of PA is $\frac{4.31 - 4}{3.1 - 3} = 3.1$

If A is (3.01, 4.0301) Then the gradient of PA is $\frac{4.0301 - 4}{3.01 - 3} = 3.01$

As A gets closer to P, the gradient of PA approaches 3.

It is reasonable to assume that the gradient of the tangent at P(3, 4) is 3

This method could be used at any point on the curve and it will show the following:

The gradient at any point on the curve $y = x^2 - 3x + 4$ is $2x - 3$.

For example,   at (3, 4) the gradient is $2 \times 3 - 3 = 3$

                  at (4, 8) the gradient is $2 \times 4 - 3 = 5$

                  at (1, 2) the gradient is $2 \times 1 - 3 = -1$

Looking at the graph should convince you that these seem to be reasonable values.

The notation we use to represent the gradient of a curve is $\frac{dy}{dx}$ (read it as "dee y by dee x"). So the tangent to the curve on the graph shows that:

If $y = x^2 - 3x + 4$

$\frac{dy}{dx} = 2x - 3$

This is the case for all equations of this form. So the general result is:

If $y = ax^2 + bx + c$

then $\frac{dy}{dx} = 2ax + b$

We can use this to calculate the gradient at any point on a quadratic curve. We no longer need to draw the graph and this method can be used for any point on the curve.

Example 1

A curve has the equation $y = 0.5x^2 + 4x - 3$

a Find the gradient at $(0, -3)$ and at $(2, 7)$.

b Find the coordinates of the point where the gradient is 0.

a Using the general result above, if $y = 0.5x^2 + 4x - 3$,

then $\frac{dy}{dx} = 2 \times 0.5x + 4$

$\Rightarrow \frac{dy}{dx} = x + 4$

If $x = 0$, $\frac{dy}{dx} = 0 + 4 = 4$. The gradient at $(0, -3)$ is 4

If $x = 2$, $\frac{dy}{dx} = 2 + 4 = 6$. The gradient at $(2, 7)$ is 6

b If the gradient is 0, then $\frac{dy}{dx} = 0$

$\Rightarrow x + 4 = 0$

$\Rightarrow x = -4$

If $x = -4$, $y = 0.5 \times (-4)^2 + 4 \times (-4) - 3 = -11$

$\Rightarrow$ The gradient is 0 at $(-4, -11)$

EXERCISE 22A

1 A curve has the equation $y = x^2 - 2x$.

   a Copy and complete this table of values:

x	-2	-1	0	1	2	3	4
$x^2 - 2x$	8	3				3	8

   b Sketch the graph of $y = x^2 - 2x$.
   c Find $\frac{dy}{dx}$.
   d Find the gradient of the curve at $(3, 3)$.
   e Find the gradient of the curve at $(4, 8)$.
   f Find the gradient at two more points on the curve.
   g At what point on the graph is the gradient 0?
   h By looking at your graph, check that your answers to parts **d**, **e**, **f** and **g** seem sensible.

2 $y = x^2 - 6x + 15$

   a Find $\frac{dy}{dx}$.
   b Find the gradient at $(0, 15)$.
   c Find the gradient at $(5, 10)$.
   d Find the coordinates of the point where the gradient is 2.

Chapter 22: Differentiation

Chapter 22. Topic 2

EXTENDED

3 $y = 2x^2 - 10$

  a Find $\frac{dy}{dx}$.
  b Find the gradient at (2, −2).
  c Find the gradient at (−1, −9).
  d Find the point where the gradient is 12.

4 This is the graph of $y = 4x - x^2$

  a Find $\frac{dy}{dx}$.
  b Find the gradient at each point where the curve crosses the *x*-axis.
  c Where is the gradient equal to 2?
  d Where is the gradient equal to 1?

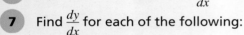

5 Find $\frac{dy}{dx}$ for each of the following:

  a $y = x^2 + x + 1$
  b $y = x^2 - 7x + 3$
  c $y = 4x^2 - x + 6$
  d $y = 0.3x^2 - 1.5x + 7.2$
  e $y = 6 - 2x + x^2$
  f $y = 10 + 3x - x^2$
  g $y = 2x + 5$
  h $y = 4$

6 If $y = (x + 4)(x - 2)$ what is $\frac{dy}{dx}$?

7 Find $\frac{dy}{dx}$ for each of the following:

  a $y = 2x(x + 1)$
  b $y = (x + 2)(x + 5)$
  c $y = (x + 3)(x - 3)$

> **Advice and Tips**
>
> First multiply out the brackets.

8 A curve has the equation $y = x^2 + 2x - 5$

  a Where does the curve cross the *y*-axis?
  b What is the gradient of the curve at that point?

22.2 More complex curves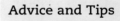

You have seen that if $y = x^2$ then $\frac{dy}{dx} = 2x$.

This table shows the value of $\frac{dy}{dx}$ for some other curves.

y	$\frac{dy}{dx}$
1	0
x	1
x^2	$2x$

→ The lines $y = 1$ has gradient 0
→ The line $y = x$ has gradient 1

x^3	$3x^2$
x^4	$4x^3$

There is a general pattern here:

If $y = x^n$ then $\frac{dy}{dx} = nx^{n-1}$

We call $\frac{dy}{dx}$ the derivative of y with respect to x.

If a is a constant and $y = ax^n$ then $\frac{dy}{dx} = anx^{n-1}$

For example, if $y = 5x^2$ then $\frac{dy}{dx} = 5 \times 2x = 10x$. If $y = 4x^3$ then $\frac{dy}{dx} = 12x^2$.

If $y = 6$ then $\frac{dy}{dx} = 0$. So the line with equation $y = 6$ is horizontal and has gradient 0.

Example 2

What is the gradient of the curve with equation $y = x^3 - 3x^2 + 4x + 7$ at the point $(2, -5)$?

$\frac{dy}{dx} = 3x^2 - 6x + 4$

If $x = 2$, $\frac{dy}{dx} = 3 \times 2^2 - 6 \times 2 + 4 = 4$

The gradient at $(2, -5)$ is 4.

The process of finding $\frac{dy}{dx}$ is called **differentiation**.

In example 2 we **differentiated** each term in turn:

$\left. \begin{array}{l} \text{Differentiate } x^3 \text{ to get } 3x^2 \\ \text{Differentiate } -3x^2 \text{ to get } -6x \\ \text{Differentiate } 4x \text{ to get } 4 \\ \text{Differentiate } 7 \text{ to get } 0 \end{array} \right\} \frac{dy}{dx} = 3x^2 - 6x + 4$

Note the minor change to the wording so that it now says "We call $\frac{dy}{dx}$ the derivative of y with respect to x"

EXERCISE 22B

1 The equation of a curve is $y = 2x^3$.
   a Find $\frac{dy}{dx}$.
   b Find the gradient of the curve at $(1, 2)$ and $(2, 16)$.

2 The equation of a curve is $y = x^3 - 6x^2 + 8x$.
   a Find $\frac{dy}{dx}$.
   b Show that $(0, 0)$, $(2, 0)$ and $(4, 0)$ are all on this curve.
   c Find the rate of change of y with respect to x at each point in part **b**.

3 This is a graph of $y = 0.5x^3 - 3x^2 + 4x$

a Find $\frac{dy}{dx}$

b Find the gradient at each point where the graph crosses the x-axis.

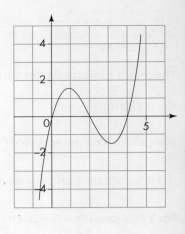

4 Find $\frac{dy}{dx}$ for each of the following:

a $y = 2x^4$

b $y = 2x^3 + 5x^2 - 8$

c $y = 5x^3 - 2x + 4$

d $y = 5 - x - \frac{2}{3}x^3$

e $y = 3x^3 + 5x - 7$

f $y = 10 - x^3$

g $y = x(x^3 - 1)$

h $y = 2x^3(x + 3)$

5 This is a sketch of the curve $y = x^4 - 4x^2$.

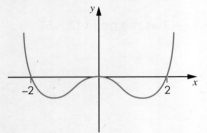

Not to scale

Find the gradient at the points where the curve meets the x-axis.

6 This is a graph of $y = x^4 - 2x^3 + 1$

a Show that the gradient at (0, 1) is 0

b Find the gradient at (−1, 4)

c Find the gradient at (2, 1)

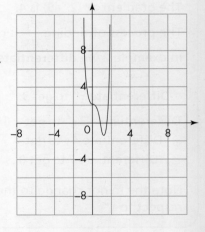

7 A curve has the equation $y = \frac{1}{3}x^3 - 5x + 4$.

Show that there are two points on the curve where the rate of change of y with respect to x is 4. Find the coordinates of the two points.

8 This is a graph of the curve $y = \frac{1}{8}x^3$ and the tangent at (2, 1)

Find the equation of the tangent at (2, 1)

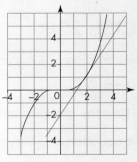

Chapter 22 . Topic 3

22.3

9 The equation of a curve is $y = x^3 - 2ax$ where a is a constant.

When $x = 2$ the gradient of the curve is -12

a Find the value of a

b Find the gradient when $x = 4$

22.3 Turning points E

A point where the gradient is zero is called a **turning point**.

A and B are turning points.

At any turning point $\frac{dy}{dx} = 0$

A is called a **maximum** point because it is higher than the points near it.

B is called a **minimum** point.

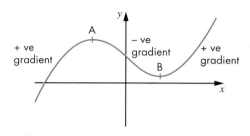

Example 3

Find the turning points of $y = x^3 - 12x + 4$ and state whether each is a maximum or a minimum point.

If $y = x^3 - 12x + 4$:

$\frac{dy}{dx} = 3x^2 - 12$

At a turning point $\frac{dy}{dx} = 0$:

$\Rightarrow 3x^2 - 12 = 0$

$\Rightarrow 3x^2 = 12$

$\Rightarrow x^2 = 4$

$\Rightarrow x = 2$ or -2

If $x = 2$, $y = 8 - 24 + 4 = -12$ $\Rightarrow$ $(2, -12)$ is a turning point

If $x = -2$, $y = -8 + 24 + 4 = 20$ $\Rightarrow$ $(-2, 20)$ is a turning point

A rough sketch makes it look likely that $(-2, 20)$ is a maximum point and $(2, -12)$ is a minimum point.

If you are not sure, check the gradient on each side of the point.

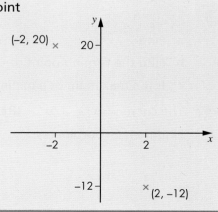

For (2, −12):

x	1.9	2	2.1
$\frac{dy}{dx}$	−1.17	0	1.23
gradient	negative	0	positive

This is $3 \times 2.1^2 - 12$

(2, −12) is a minimum point.

Because the gradient changes (from negative to zero to positive) as we move from left to right, (2, −12) must be a minimum point.

For (−2, 20):

x	−2.1	−2	−1.9
$\frac{dy}{dx}$	1.23	0	−1.17
gradient	positive	0	negative

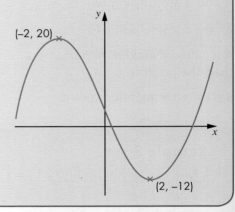

(−2, 20) is a maximum point.

Because the gradient changes (from positive to zero to negative) as we move from left to right, (−2, 20) must be a maximum point.

Here is a sketch of the curve.

EXERCISE 22C

1. $y = x^2 − 4x + 3$

   a Find $\frac{dy}{dx}$.

   b Show that the curve has one turning point and find its coordinates.

   c State whether it is a maximum or minimum point.

2. a Find the turning point of the curve $y = x^2 + 6x − 3$.

   b Is it a maximum or a minimum point?

3. $y = 1 + 5x − x^2$

   a Find $\frac{dy}{dx}$.

   b Find the turning point of the curve.

   c Is it a maximum or a minimum point?

4. The curve $y = x^3 + 1.5x^2 − 18x$ has two turning points.

   Find their x-coordinates.

5 This is a sketch of $y = x^3 - 3x^2$

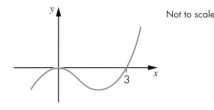

Not to scale

a Find $\frac{dy}{dx}$.

b Solve the equation $\frac{dy}{dx} = 0$.

c Find the coordinates of the two turning points shown on the graph.

6 $y = x^2 - 3x - 10$

a Show that the graph crosses the x-axis at $(-2, 0)$ and $(5, 0)$.

b Find $\frac{dy}{dx}$.

c Find the turning point of the curve. Is it a maximum or a minimum?

d The curve has a line of symmetry. What is the equation of this line?

7 The curve $y = 18x^2 - x^4$ has three turning points.

a Find their coordinates.

b Sketch the graph.

8 The equation of a curve is $y = 2x^3 - 6x + 4$.

a Find $\frac{dy}{dx}$.

b Find the turning points of the curve and state whether each one is a maximum or a minimum point.

9 A rectangle has a perimeter of 30 cm.

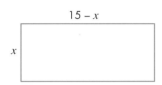

Advice and Tips

A is used instead of y here. Differentiate in the usual way.

a Explain why, if one side of the rectangle is x cm, the other will be $(15 - x)$ cm.

Suppose the area of the rectangle is A cm² then $A = x(15 - x)$.

b Find $\frac{dA}{dx}$.

c Find the turning point of the graph $A = x(15 - x)$.

d Is the turning point a maximum or a minimum?

e What do the answers to **c** and **d** tell you about the area of the rectangle?

Check your progress

Extended

- I understand the idea of a derivative of a function
- I can find the derivatives of functions of the form ax^n and simple sums of these where $n = 0, 1, 2, 3$ or 4
- I can apply differentiation to find gradients
- I can apply differentiation to find turning points
- I can discriminate between maximum points and minimum points

Examination questions: Algebra

Past paper questions reproduced by permission of Cambridge Assessment International Education. Other exam-style questions have been written by the authors.

PAPER 1

1 Simplify $\dfrac{r^6}{r^2}$. [1]

Cambridge International IGCSE Mathematics 0580 *Paper 11 Q3 Oct/Nov 2015*

2 a $s = 4t + 3u$
Calculate s when $t = 2.6$ and $u = -0.4$. [2]
b Solve $5x - 7 = 10$. [2]

Cambridge International IGCSE Mathematics 0580 *Paper 11 Q19 Oct/Nov 2015*

3 a Write down the next term in each of these sequences.
   i 5 9 13 17 ... [1]
   ii 3 6 12 24 ... [1]

b Here are the first four terms in a different sequence.
   2 7 12 17
   Find an expression for the nth term of this sequence. [2]

Cambridge International IGCSE Mathematics 0580 *Paper 11 Q22 Oct/Nov 2015*

4 a Maria travels by bus to the shopping mall.
She leaves home at 11 50 and arrives at the shopping mall at 12 17.
How many minutes does it take Maria to travel from home to the shopping mall? [1]

b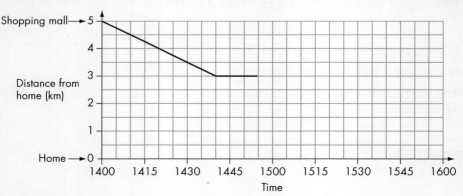

Maria walks home from the shopping mall.
The travel graph shows part of her journey.
   i Maria stops at her friend's house on the way home.
   How far from the shopping mall does her friend live? [1]

Examination questions: Algebra

    ii Maria leaves her friend's house at 14 55.

       She walks the rest of the way home at a constant speed of 4 km/h.

       Complete the travel graph. [2]

Cambridge International IGCSE Mathematics 0580 Paper 11 Q20 Oct/Nov 2015

5 Simplify.

$6uw^{-3} \times 4uw^6$ [2]

Cambridge International IGCSE Mathematics 0580 Paper 11 Q10 May/June 2015

6 a Factorise.

    $3w^2 - 2w$ [1]

  b Expand and simplify.

    $x(2x + 3) + 5(x - 7)$ [2]

Cambridge International IGCSE Mathematics 0580 Paper 11 Q13 May/June 2015

7 Solve the equation.

$\dfrac{n - 8}{2} = 11$ [2]

Cambridge International IGCSE Mathematics 0580 Paper 11 Q6 May/June 2014

8 Solve the simultaneous equations.

$5x + 2y = 16$

$3x - 4y = 7$ [3]

Cambridge International IGCSE Mathematics 0580 Paper 11 Q20 Oct/Nov 2014

9 a Find the value of $5x^2$ when $x = -4$. [2]

  b Make x the subject of the formula $y = 5x^2$. [2]

Cambridge International IGCSE Mathematics 0580 Paper 11 Q21 Oct/Nov 2014

10 a Simplify the expressions.

    i $p^3 \times p^7$ [1]

    ii $t^5 \div t^8$ [1]

  b $(h^3)^k = h^{12}$

    Find the value of k. [1]

Cambridge International IGCSE Mathematics 0580 Paper 11 Q16 May/June 2014

11 Rearrange this equation to make b the subject.

$a = \dfrac{b}{5} - 9$ [2]

Cambridge International IGCSE Mathematics 0580 Paper 11 Q9 May/June 2013

12 Here are the first four terms of a sequence.

    4   11   18   25

Write down an expression for the nth term. [2]

Cambridge International IGCSE Mathematics 0580 Paper 11 Q10 May/June 2013

Examination questions: Algebra

13 a Which **two** of these have the same value?

5^{-2} $\dfrac{2}{5}$ $\left(\dfrac{1}{2}\right)^2$ $\left(\dfrac{2}{5}\right)^2$ 0.2^2 [2]

b Simplify.

  i $a^6 \times a^3$ [1]

  ii $24b^{16} \div 6b^4$ [2]

Cambridge International IGCSE Mathematics 0580 *Paper 11 Q18 May/June 2013*

14 a Multiply out the brackets.

  $5(x + 3)$ [1]

b Factorise completely.

  $12xy - 3x^2$ [2]

c Solve.

  $5x - 24 = 51$ [2]

Cambridge International IGCSE Mathematics 0580 *Paper 11 Q19 May/June 2013*

Examination questions: Algebra

PAPER 3

1 a i Complete the table of values for $y = 8 - x^2$. [2]

x	−3	−2	−1	0	1	2	3
y	−1			8	7		−1

ii On the grid, draw the graph of $y = 8 - x^2$ for $-3 \leqslant x \leqslant 3$. [4]

iii Write down the equation of the line of symmetry of the graph. [1]

iv Use your graph to solve the equation $8 - x^2 = 0$. [2]

b i On the grid, plot the points (−2, 8) and (2.5, −1).

Draw a straight line through these points. [2]

ii Find the equation of your line in the form $y = mx + c$. [3]

iii Write down the co-ordinates of the point of intersection of your line with $y = 8 - x^2$. [1]

Cambridge International IGCSE Mathematics 0580 *Paper 31 Q6 Oct/Nov 2014*

2 a In 2001 Arnold was x years old.

Ken is **34 years younger** than Arnold.

i Complete the table, in terms of x, for Arnold's and Ken's ages. [3]

	2001	2013
Arnold's age	x	
Ken's age		

ii In 2013 Arnold is **three** times as old as Ken.

Write down an equation in x and solve it. [4]

Examination questions: Algebra

b Solve the simultaneous equations.

$3x + 2y = 18$

$2x - y = 19$ [3]

Cambridge International IGCSE Mathematics 0580 *Paper 31 Q10 May/June 2013*

3

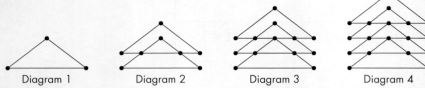

| Diagram 1 | Diagram 2 | Diagram 3 | Diagram 4 |

Diagrams 1 to 4 show a sequence of shapes made up of lines and dots at the intersections of lines.

a i Complete the table showing the number of dots in each diagram. [3]

Diagram	1	2	3	4	5	6
Dots	3	8	13			

  ii Write down the rule for continuing the sequence of dots. [1]

  iii Write down an expression, in terms of n, for the number of dots in Diagram n. [2]

  iv Find the number of dots in Diagram 15. [1]

b The dots are joined by sloping lines and horizontal lines.

  i Diagram 1 has 2 sloping lines and Diagram 2 has 6 sloping lines.

  Find the number of sloping lines in Diagrams 3 and 4. [2]

  ii Write down an expression, in terms of n, for the number of sloping lines in Diagram n. [2]

Cambridge International IGCSE Mathematics 0580 *Paper 31 Q9 Oct/Nov 2014*

4 a The cost, C, of hiring a meeting room for n people is calculated using the formula $C = 80 + 5n$.

  i Calculate C when $n = 12$. [2]

  ii Maria pays $230 to hire the meeting room.

  Work out the number of people at the meeting. [2]

  iii Make n the subject of the formula $C = 80 + 5n$. [2]

b Expand and simplify $2(3x + 4) - 3(2 - x)$. [2]

c Solve the simultaneous equations.

$3x + y = 13$

$2x + 3y = 18$ [3]

Cambridge International IGCSE Mathematics 0580 *Paper 31 Qa Oct/Nov 2012*

376 Examination questions: Algebra

Examination questions: Algebra

5 a Simplify $x^8 \div x^2$. [1]

b Simplify $\left(\dfrac{x^6}{27}\right)^{\frac{1}{3}}$. [2]

Cambridge International IGCSE Mathematics 0580 *Paper 21 Q11 Oct/Nov 2014*

6 Make x the subject of the formula.
$y = (x - 4)^2 + 6$ [3]

Cambridge International IGCSE Mathematics 0580 *Paper 21 Q7 May/June 2014*

Examination questions: Algebra

PAPER 2

1 Factorise completely.
   a $ax + ay + 3cx + 3cy$ [2]
   b $3a^2 - 12b^2$ [3]

   Cambridge International IGCSE Mathematics 0580 *Paper 21 Q9 Oct/Nov 2015*

2 Simplify.
   $$\frac{x^2 - 16}{x^2 - 3x - 4}$$ [4]

   Cambridge International IGCSE Mathematics 0580 *Paper 21 Q15 Oct/Nov 2015*

3 $f(x) = x^3$ $g(x) = 3x - 5$ $h(x) = 2x + 1$
   Work out
   a ff(2), [2]
   b gh(x) and simplify your answer, [2]
   c $h^{-1}(x)$, the inverse of h(x). [2]

   Cambridge International IGCSE Mathematics 0580 *Paper 21 Q21 Oct/Nov 2015*

4

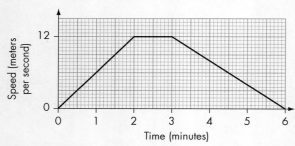

A tram leaves a station and accelerates for 2 **minutes** until it reaches a speed of 12 metres per second.
It continues at this speed for 1 minute.
It then decelerates for 3 minutes until it stops at the next station.
The diagram shows the speed-time graph for this journey.
Calculate the distance, in metres, between the two stations. [3]

Cambridge International IGCSE Mathematics 0580 *Paper 21 Q10 May/June 2015*

5 Find the nth term of each sequence.
   a 4, 8, 12, 16, 20, ... [1]
   b 11, 20, 35, 56, 83, ... [2]

   Cambridge International IGCSE Mathematics 0580 *Paper 21 Q11 May/June 2015*

Examination questions: Algebra

6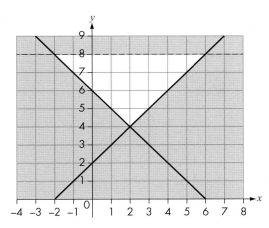

Write down the 3 inequalities which define the unshaded region. [4]

Cambridge International IGCSE Mathematics 0580 *Paper 21 Q15 May/June 2015*

7 Factorise completely.
   a $yp + yt + 2xp + 2xt$ [2]
   b $7(h + k)^2 - 21(h + k)$ [2]

Cambridge International IGCSE Mathematics 0580 *Paper 21 Q20 May/June 2015*

8 Solve the simultaneous equations.
$0.4x - 5y = 27$
$2x + 0.2y = 9$ [3]

Cambridge International IGCSE Mathematics 0580 *Paper 21 Q12 Oct/Nov 2014*

9 Write as a single fraction in its simplest form.
$$\frac{2}{x} - \frac{2}{x+1}$$ [3]

Cambridge International IGCSE Mathematics 0580 *Paper 21 Q8 May/June 2014*

10 Factorise completely.
   a $ax + ay + bx + by$ [2]
   b $3(x - 1)^2 + (x - 1)$ [2]

Cambridge International IGCSE Mathematics 0580 *Paper 21 Q10 May/June 2014*

11 Solve the inequality for positive integer values of x.
$$\frac{21 + x}{5} > x + 1$$ [4]

Cambridge International IGCSE Mathematics 0580 *Paper 21 Q15 May/June 2014*

12 **a** $(2^{24})^{\frac{1}{2}} = p^4$

Find the value of p. [2]

   b Simplify $\dfrac{q^2 + q^2}{q^{\frac{1}{4}} \times q^{\frac{1}{4}}}$. [3]

Cambridge International IGCSE Mathematics 0580 *Paper 21 Q16 may/June 2014*

Examination questions: Algebra

PAPER 4

1 a Calculate $2^{0.7}$. [1]

  b Find the value of x in each of the following.

   i $2x = 128$ [1]

   ii $2x \times 2^9 = 2^{13}$ [1]

   iii $2^9 \div 2^x = 4$ [1]

   iv $2^x = \sqrt[3]{2}$ [1]

  c i Complete this table of values for $y = 2^x$. [2]

x	−3	−2	−1	0	1	2	3
y	0.125		0.5		2	4	8

  ii On the grid, draw the graph of $y = 2^x$ for $-3 \leq x \leq 3$. [4]

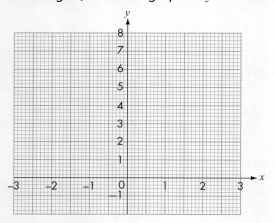

  iii Use your graph to solve $2^x = 5$. [1]

  iv Find the equation of the line joining the points (1, 2) and (3, 8). [3]

  v By drawing a suitable line on your graph, solve $2^x - 2 - x = 0$. [2]

Cambridge International IGCSE Mathematics 0580 Paper 41 Q2 Oct/Nov 2015

2 a Factorise $x^2 - 3x - 10$. [2]

  b i Show that $\dfrac{x+2}{x+1} + \dfrac{3}{x} = 3$ simplifies to $2x^2 - 2x - 3 = 0$. [3]

  ii Solve $2x^2 - 2x - 3 = 0$.

   Give your answers correct to 3 decimal places.

   Show all your working. [4]

  c Simplify $\dfrac{2x+3}{x+2} - \dfrac{x}{x+1}$. [4]

Cambridge International IGCSE Mathematics 0580 Paper 41 Q8 Oct/Nov 2015

Examination questions: Algebra

3 a $f(x) = 2x - 3$ $\quad g(x) = \dfrac{1}{x+1} + 2$ $\quad h(x) = 3^x$

    i Work out $f(4)$. [1]

    ii Work out $fh(-1)$. [2]

    iii Find $f^{-1}(x)$, the inverse of $f(x)$. [2]

    iv Find $ff(x)$ in its simplest form. [2]

    v Show that the equation $f(x) = g(x)$ simplifies to $2x^2 - 3x - 6 = 0$. [3]

    vi Solve the equation $2x^2 - 3x - 6 = 0$.

       Give your answers correct to 2 decimal places.

       Show all your working. [4]

b Simplify $\dfrac{x^2 - 3x + 2}{x^2 + 3x - 10}$. [4]

Cambridge International IGCSE Mathematics 0580 *Paper 41 Q10 May/June 2014*

4

Diagram 1   Diagram 2   Diagram 3   Diagram 4

The first four diagrams in a sequence are shown above.

The diagrams are drawn using white squares ☐ and grey squares ▨.

a Complete the columns in the table for Diagram 4 and Diagram n. [6]

Diagram	1	2	3	4	n
Number of white squares	12	20	28		
Number of grey squares	0	1	4		
Total number of squares	12	21	32		$(n+1)(n+5)$

b Work out the number of the diagram which has a total of 480 squares. [2]

c The **total** number of squares in the **first n diagrams** is $\dfrac{1}{3}n^3 + pn^2 + qn$.

    i Use $n = 1$ in this expression to show that $p + q = 11\dfrac{2}{3}$. [1]

    ii Use $n = 2$ in the expression to show that $4p + 2q = 30\dfrac{1}{3}$. [2]

    iii Find the values of p and q. [3]

Cambridge International IGCSE Mathematics 0580 *Paper 41 Q9 Oct/Nov 2014*

Chapter 23
Angle properties

Topics	Level	Key words
1 Angle facts	CORE	angles at a point, angles on straight lines, opposite angles, vertically opposite angles
2 Parallel lines	CORE	corresponding angles, alternate angles, allied angles, interior angles
3 Angles in a triangle	CORE	equilateral triangle, isosceles triangle, right-angled triangle
4 Angles in a quadrilateral	CORE	quadrilateral, parallelogram, rhombus, kite, trapezium
5 Regular polygons	CORE	polygon, regular polygon, external angles, pentagon, hexagon, octagon, square
6 Irregular polygons	EXTENDED	irregular polygon, heptagon, nonagon, decagon
7 Tangents and diameters	CORE	radius, tangent, point of contact, diameter, semi-circle
8 Angles in a circle	EXTENDED	arc, circumference, segment, subtended
9 Cyclic quadrilaterals	EXTENDED	cyclic quadrilateral, opposite segment, supplementary
10 Alternate segment theorem	EXTENDED	tangent, chord, major segment, minor segment, major arc, minor arc, alternate segment

In this chapter you will learn how to:

CORE	EXTENDED
• **Calculate unknown angles using the following geometrical properties:** (a) angles at a point (b) angles at a point on a straight line and intersecting straight lines (c) angles formed within parallel lines (d) angle properties of triangles and quadrilaterals (e) angle properties of regular polygons (f) angle in a semi-circle (g) angle between tangent and radius of a circle. (C4.7 and E4.7)	• **Use in addition the following geometrical properties:** (a) angle properties of irregular polygons (b) angle at the centre of a circle is twice the angle at the circumference (c) angles in the same segment are equal (d) angles in opposite segments are supplementary; cyclic quadrilaterals (e) alternate segment theorem. (E4.7)

Why this chapter matters

Angles describe an amount of turn around a point. It is important to be able to measure them and understand their properties.

In a regular polygon all the angles are the same size and all the sides are the same length. The shape of the polygon depends on how many angles and sides it has:

- a triangle has 3 sides and 3 angles
- a square has 4 sides and 4 angles
- a pentagon has 5 sides and 5 angles
- a hexagon has 6 sides and 6 angles.

The patterns below are made by putting together different regular polygons. How many different polygons can you see in them? Can you see any octagons (8-sided shapes) or decagons (10-sided shapes)?

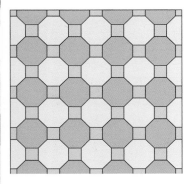
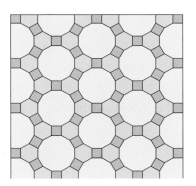
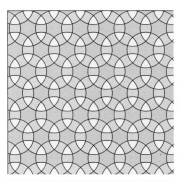

Some shapes fit together better than others because of the size of their angles.

Bees construct their hives from hexagon shapes, which can join together without gaps.

Squares and rectangles also fit together easily, which makes them an ideal shape to use in building.

In this chapter you will look at angles, the shapes they form, and their properties.

Chapter 23: Angle properties

Chapter 23 . Topic 1

23.1 Angle facts

Angles on a line

The **angles on a straight line** add up to 180°.

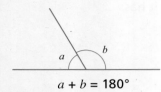

$a + b = 180°$

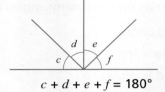

$c + d + e + f = 180°$

Draw an example for yourself (and measure all the angles) to show that the statement is true.

Angles at a point

The sum of the **angles at a point** is 360°. For example:

$a + b + c + d + e = 360°$

Again, check this for yourself by drawing an example and measuring the angles.

Sometimes equations can be used to solve angle problems.

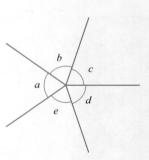

Opposite angles

Opposite angles are equal.

So $a = c$ and $b = d$.

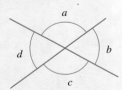

Sometimes opposite angles are called **vertically opposite angles**.

EXERCISE 23A

1 Calculate the value of x in each diagrams.

a

b

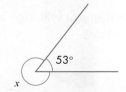

Advice and Tips

Never measure angles in questions like these. Diagrams in textbooks are not usually drawn accurately. Always calculate angles unless you are told to measure them.

384   23.1 Angle facts

c

d

e

f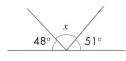

g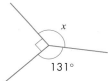

h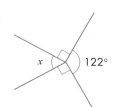

2 Write down the value of x in each of these diagrams.

a

b

c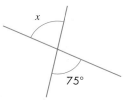

3 In the diagram, angle ABD is 45° and angle CBD is 125°.

Decide whether ABC is a straight line.

Write down how you decided.

4 Calculate the value of x in each of these diagrams.

a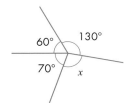

b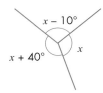

c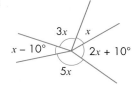

5 Calculate the value of x in each of these diagrams.

a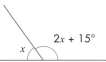

b

c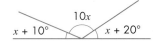

6 Calculate the value of x first and then calculate the value of y in each of these diagrams.

a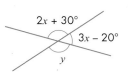

b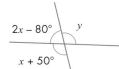

c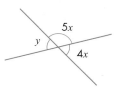

Chapter 23: Angle properties

Chapter 23 . Topic 2

7 Shalini has a collection of tiles.
They are all equilateral triangles and are all the same size.

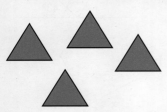

> **Advice and Tips**
>
> All the angles in an equilateral triangle are 60°.

She says that six of the tiles will fit together and leave no gaps.

Explain why Shalini is correct.

8 Work out the value of y in the diagram.

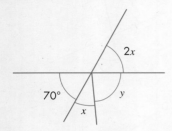

9 The ratio $a : b = 2 : 3$

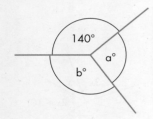

Work out the values of a and b.

23.2 Parallel lines

Angles in parallel lines

The arrowheads indicate that the lines are parallel and the line that crosses the parallel lines is called a transversal.

Notice that eight angles are formed.

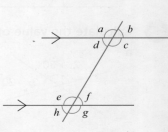

Angles like these 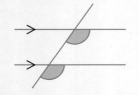 are called **corresponding angles** (Look for the letter F).

Corresponding angles are equal.

23.2

Angles like these 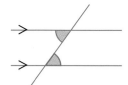 are called **alternate angles** (Look for the letter Z).

Alternate angles are equal.

Angles like these 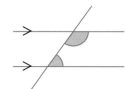 are called **allied angles** or **interior angles** (Look for the letter C).

Allied angles add to 180°.

EXERCISE 23B

1 State the sizes of the lettered angles in each diagram.

a 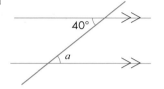   b 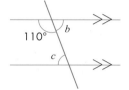   c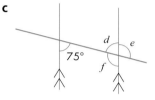

d 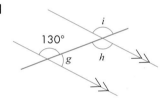   e 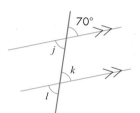   f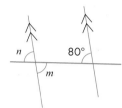

2 State the sizes of the lettered angles in each diagram.

a 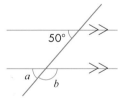   b 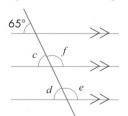   c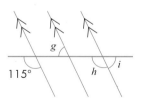

d 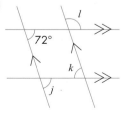   e 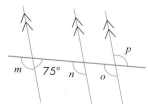   f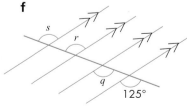

Chapter 23: Angle properties   **387**

3. State the sizes of the lettered angles in these diagrams.

   a

   b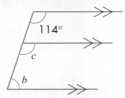

4. Calculate the values of x and y in these diagrams.

   a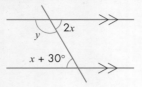

   b

   c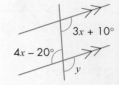

5. Calculate the values of x and y in these diagrams.

   a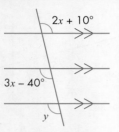

   b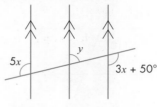

   c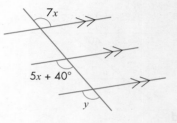

6. A company makes signs like the one below.

   It has one line of symmetry.

   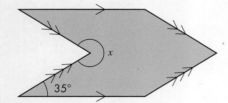

   Advice and Tips

   Draw the line of symmetry on the shape first.

   The company needs to know the size of the angle marked x on the diagram.

   Work out the size of the angle labelled x.

7. In the diagram, AE is parallel to BD.

   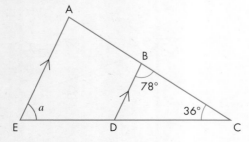

   Work out the size of the angle labelled a.

388 23.2 Parallel lines

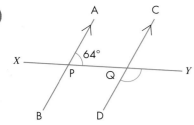

The line *XY* crosses the parallel lines *AB* and *CD* at *P* and *Q*.

a Work out the size of angle *DQY*. Give reasons for your answer.

b This is Vreni's solution.

> Angle *PQD* = 64° (corresponding angles)
>
> So angle *DQY* = 124° (angles on a line = 190°)

She has made a number of errors in her solution.

Write out the correct solution for the question.

 Use the diagram to prove that the three angles in a triangle add up to 180°.

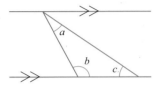

 Prove that $p + q + r = 180°$.

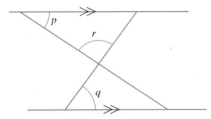

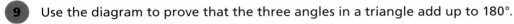
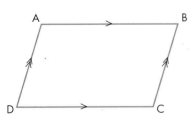

ABCD is a parallelogram.

AB is parallel to *CD* and *DA* is parallel to *CB*.

Prove that the opposing angles of the parallelogram are the same size.

Chapter 23: Angle properties

23.3 Angles in a triangle

The three angles in any triangle add up to 180°.

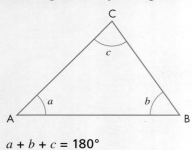

$a + b + c = 180°$

You can prove this by drawing a line through C parallel to AB.

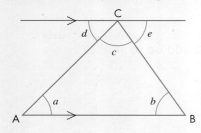

$a = d$ (alternate angles)
$b = e$ (alternate angles)
$d + e + c = 180°$ (angles on a straight line)

So $a + b + c = 180°$

Special triangles

Equilateral triangle

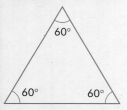

An **equilateral triangle** is a triangle with all its sides equal. Therefore, all three interior angles are 60°.

Isosceles triangle

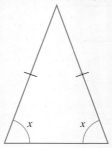

An **isosceles triangle** is a triangle with two equal sides and, therefore, with two equal interior angles (at the foot of the equal sides).

Notice how to mark the equal sides and equal angles.

Right-angled triangle

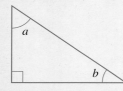

A **right-angled triangle** has an interior angle of 90°.
$a + b = 90°$

EXERCISE 23C

1 Find the size of the angle marked with a letter in each of these triangles.

a 　b 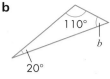　c 　d

e 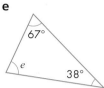　f 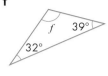　g 　h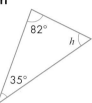

2 Do any of these sets of angles form the three angles of a triangle? Explain your answer.
  a 35°, 75°, 80°
  b 50°, 60°, 70°
  c 55°, 55°, 60°
  d 60°, 60°, 60°
  e 35°, 35°, 110°
  f 102°, 38°, 30°

3 In the triangle on the right, all the interior angles are the same.
  a What is the size of each angle?
  b What is the name of a special triangle like this?
  c What is special about the sides of this triangle?

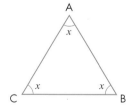

4 In the triangle on the right, two of the angles are the same.
  a Work out the size of the lettered angles.
  b What is the name of a special triangle like this?
  c What is special about the sides AC and AB of this triangle?

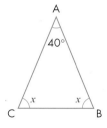

5 Find the size of the angle marked with a letter in each of these diagrams.

a 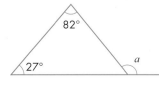　b 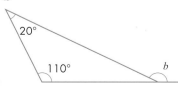　c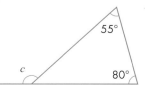

Chapter 23: Angle properties

Chapter 23 . Topic 4

6 What is the special name for triangle *DEF*?

Show all your working to explain your answer.

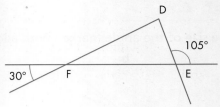

7 The diagram shows three intersecting straight lines.

Work out the values of *a*, *b* and *c*.

Give reasons for your answers.

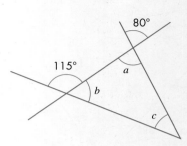

8 By using algebra, show that $x = a + b$.

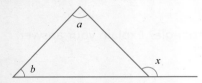

9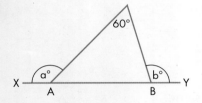

ABC is a triangle and *XABY* is a straight line.

Find a formula for *b* in terms of *a*.

23.4 Angles in a quadrilateral

The four angles in any **quadrilateral** add up to 360°.

$a + b + c + d = 360°$

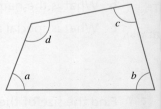

You can check this by dividing the quadrilateral into two triangles.

The six angles of the triangles are the same as the four angles of the quadrilateral.

Sum of angles of a quadrilateral = 180° + 180° = 360°

23.4

Example 1

Three angles of a quadrilateral are 125°, 130° and 60°.

Find the size of the fourth angle.

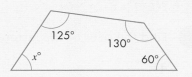

$125 + 130 + 60 + x = 360$

$\qquad 315 + x = 360$

$\qquad\qquad x = 360 - 315$

$\qquad\qquad x = 45$

So the fourth angle is 45°.

Special quadrilaterals

A **parallelogram** has opposite sides that are parallel.

Its opposite sides are equal. Its diagonals bisect each other.
Its opposite angles are equal: that is, angle A = angle C and
angle B = angle D

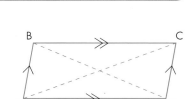

A **rhombus** is a parallelogram with all its sides equal.

Its diagonals bisect each other at right angles.
Its diagonals also bisect the angles at the vertices.

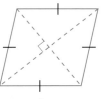

A **kite** is a quadrilateral with two pairs of equal adjacent sides.

Its longer diagonal bisects its shorter diagonal at right angles.
The opposite angles between the sides of different lengths are equal.

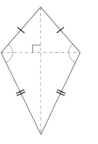

A **trapezium** has two parallel sides.

The sum of the interior angles at the ends of each non-parallel side is 180°.

angle A + angle D = 180° and angle B + angle C = 180°

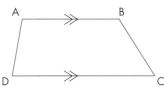

Example 2

Find the size of the angles marked x and y in this parallelogram.

$x = 55°$ (opposite angles are equal) and $y = 125°$ ($x + y = 180°$)

Chapter 23: Angle properties

EXERCISE 23D

1 Find the sizes of the lettered angles in these quadrilaterals.

a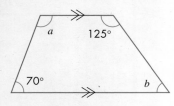
b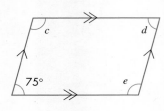
c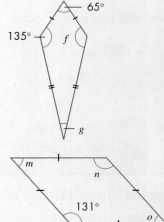

d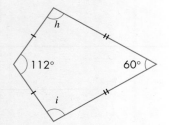
e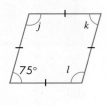
f (see diagram)

2 Calculate the values of x and y in each of these quadrilaterals.

a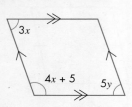
b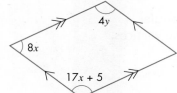
c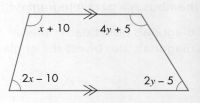

3 Find the value of x in each of these quadrilaterals and state what type of quadrilateral it could be. All angles are in degrees.

  a A quadrilateral with angles $x + 10$, $x + 20$, $2x + 20$, $2x + 10$
  b A quadrilateral with angles $x - 10$, $2x + 10$, $x - 10$, $2x + 10$
  c A quadrilateral with angles $x - 10$, $2x$, $5x - 10$, $5x - 10$
  d A quadrilateral with angles $4x + 10$, $5x - 10$, $3x + 30$, $2x + 50$

4 The diagram shows a parallelogram ABCD.

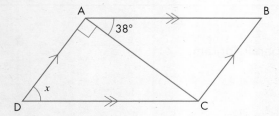

Work out the value of x, marked on the diagram.

394   23.4 Angles in a quadrilateral

Chapter 23 . Topic 5

5 Dani is making a kite and wants angle C to be half of angle A.

Work out the size of angles B and D.

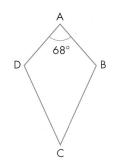

6 This quadrilateral is made from two isosceles triangles. They are both the same size.

Find the value of y in terms of x.

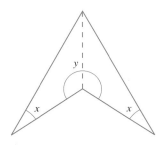

7 The four angles of a quadrilateral are in the ratio 1 : 2 : 3 : 4

Calculate the four angles.

23.5 Regular polygons

Regular polygons

Here are five regular polygons.

| Square | Pentagon | Hexagon | Octagon | Decagon |
| 4 sides | 5 sides | 6 sides | 8 sides | 10 sides |

A **polygon** is regular if all its interior angles are equal and all its sides are the same length.

A **square** is a regular four-sided shape that has an angle sum of 360°.

So, each angle is 360° ÷ 4 = 90°.

The angles of a regular polygon

Lines from the centre of a regular pentagon divide it into five isosceles triangles.

The angle at the centre is 360° so the angle of each isosceles triangle at the centre is:

$\qquad$ 360° ÷ 5 = 72°

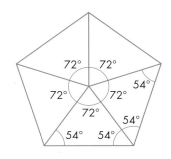

The other angles in each triangle are identical so each one is:

(180 − 72) ÷ 2 = 54°

So each interior angle of a regular pentagon is:

2 × 54° = 108°

There is also an **exterior angle** at each vertex. It is:

180 − 108 = 72°

Notice that 72° = 360° ÷ 5.

General result

If a regular polygon has n sides each exterior angle is $\frac{360°}{n}$.

If you want to find the interior angle of a regular polygon it may be easier to find the exterior angle like this first. Then subtract it from 180° to find the interior angle.

Regular polygon	Number of sides	Exterior angle	Interior angle
Equilateral triangle	3	120°	60°
Square	4	90°	90°
Pentagon	5	72°	108°
Hexagon	6	60°	120°
Octagon	8	45°	135°

Example 3

Calculate the size of the exterior and interior angle for a regular 12-sided polygon (a regular dodecagon).

Exterior angle = $\frac{360°}{12}$ = 30°

interior angle = 180° − 30° = 150°

EXERCISE 23E

1. Each diagram shows an interior angle of a regular polygon. For each polygon, answer these questions.

   i   What is the size of its exterior angle?
   ii How many sides does it have?
   iii What is the sum of its interior angles?

   a  135°

   b 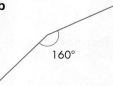 160°

   c 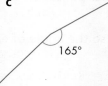 165°

   d  144°

23.5

2 Each diagram shows an exterior angle of a regular polygon. For each polygon, answer these questions.

Advice and Tips

Remember that the angle sum is calculated as (number of sides − 2) × 180°.

  i   What is the size of its interior angle?
  ii How many sides does it have?
  iii What is the sum of its interior angles?

a    b 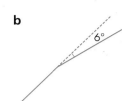   c    d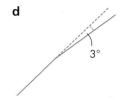

3 None of these angles can be the interior angle of a regular polygon. Explain why.

a b c 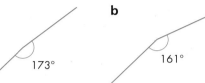   d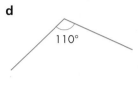

4 None of these angles can be the exterior angle of a regular polygon. Explain why.

a   b    c    d

5 Draw a sketch of a regular octagon and join each vertex to the centre.

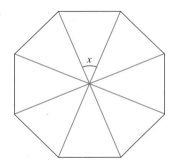

Calculate the value of the angle at the centre (marked x).

What connection does this have with the exterior angle?

Is this true for all regular polygons?

6 The diagram shows part of a regular polygon.

Each interior angle is 144°.

  a What is the size of each exterior angle of the polygon?
  b How many sides does the polygon have?

7   a Show that it is possible to draw a regular polygon with an interior angle of 170°.
       b Show that it is not possible to draw a regular polygon with an interior angle of 169°.

23.6 Irregular polygons

A polygon is regular if all its sides are the same length and all its angles are the same size. Any other polygon is **irregular**.

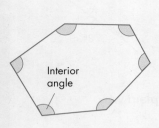
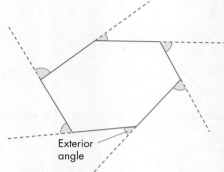

The *exterior* angles of *any* polygon add up to 360°.

Interior angles

You can find the sum of the interior angles of any polygon by splitting it into triangles.

Quadrilateral   Pentagon   Hexagon   Heptagon

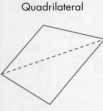  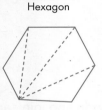 

Two triangles   Three triangles   Four triangles   Five triangles

Since you already know that the angles in a triangle add up to 180°, you find the sum of the interior angles in a polygon by multiplying the number of triangles in the polygon by 180°, as shown in this table.

Shape	Name	Sum of interior angles
4-sided	Quadrilateral	2 × 180° = 360°
5-sided	Pentagon	3 × 180° = 540°
6-sided	Hexagon	4 × 180° = 720°
7-sided	Heptagon	5 × 180° = 900°
8-sided	Octagon	6 × 180° = 1080°
9-sided	Nonagon	7 × 180° = 1260°
10-sided	Decagon	8 × 180° = 1440°

As you can see from the table, for an n-sided polygon, the sum of the interior angles, S, is given by the formula:

$$S = 180(n - 2)°$$

23.6

Exterior angles

As in regular polygons the sum of all the exterior angles in an irregular polygon is 360° but their sizes may not be the same.

The size of any specific exterior angle = 180° − the size of its adjacent interior angle.

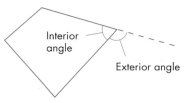

> **Example 4**
>
> Four angles of a pentagon are 100°.
>
> How big is the fifth angle?
>
> ---
>
> The interior angle sum of a pentagon is 3 × 180° = 540°.
>
> Four angles add up to 400°.
>
> The fifth angle must be 540 − 400 = 140°.

EXERCISE 23F

1 Calculate the sum of the interior angles of polygons with these numbers of sides.

    a 10 sides   **b** 15 sides

    c 100 sides   **d** 45 sides

2 Find the number of sides of polygons with these interior angle sums.

    a 1260°   **b** 2340°

    c 18 000°   **d** 8640°

3 Calculate the size of the lettered angles in each of these polygons.

a 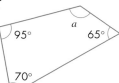   **b** 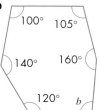   **c**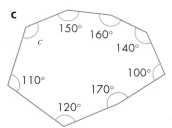

4 Find the value of x in each of these polygons.

a 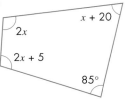   **b**    **c**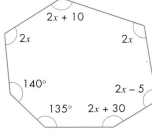

EXTENDED

Chapter 23. Topic 7

5 What is the name of the regular polygon in which the interior angles are twice its exterior angles?

6 Wesley measured all the interior angles in a polygon. He added them up to make 991°, but he had missed out one angle.

   a What type of polygon did Wesley measure?

   b What is the size of the missing angle?

7 **a** In the triangle ABC, angle A is 42° and angle B is 67°.

     i Calculate the value of angle C.

     ii What is the value of the exterior angle at C?

     iii What connects the exterior angle at C with the sum of the angles at A and B?

   b Prove that any exterior angle of a triangle is equal to the sum of the two opposite interior angles.

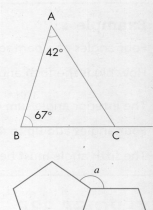

8 Two regular pentagons are placed together.

Work out the value of a.

9 A joiner is making tables so that the shape of each one is half a regular octagon, as shown in the diagram.

He needs to know the size of each angle on the top.

What are the sizes of the angles?

23.7 Tangents and diameters

A **tangent** is a straight line that touches a circle at one point only.

This point is called the **point of contact**.

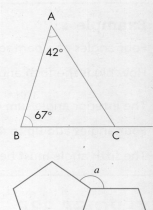

A tangent to a circle is perpendicular to the **radius** drawn to the point of contact.

The radius OX is perpendicular to the tangent AB.

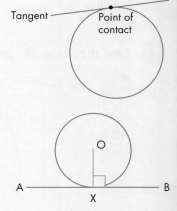

The **diameter** of a circle divides it into two **semi-circles**.

Every angle at the circumference of a semi-circle that is subtended by the diameter of the semi-circle is a right angle.

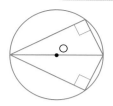

Example 5

O is the centre of a circle. AP and BP are tangents.

Calculate the angle at P.

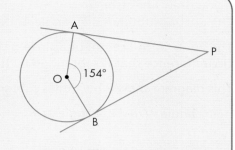

The angles at A and B are right angles.

OAPB is a quadrilateral so the interior angles add up to 360°.

Angle P is 360 − (90 + 90 + 154) = 26°.

EXERCISE 23G

1. In each diagram, TP and TQ are tangents to a circle with centre O. Find each value of x.

   a 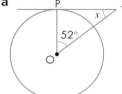 b 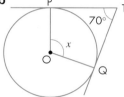 c 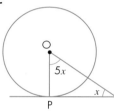 d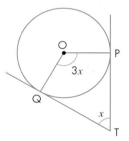

2. Each diagram shows a tangent to a circle with centre O. Find the value of x and y in each case.

   a 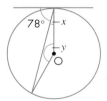 b 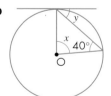 c 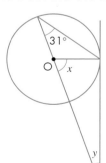 d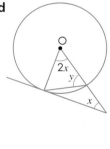

3. In the diagram, O is the centre of the circle and AB is a tangent to the circle at C.

   Explain why triangle BCD is isosceles

   Give reasons to justify your answer.

   Advice and Tips

   Look for isosceles triangles

   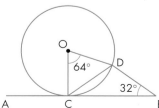

23.8 Angles in a circle

Here are two more theorems you need to know about angles in circles.

If you draw lines from each end of an **arc** to the centre of a circle they form an angle at the centre. The arc has **subtended** an angle at the centre.

The angle at the centre of a circle is twice the angle at the circumference that is subtended by the same arc.

This diagram shows the angles subtended by arc AB.

angle $AOB = 2 \times$ angle ACB

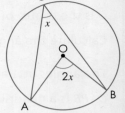

Angles subtended at the circumference in the same **segment** of a circle are equal.

Points C_1, C_2, C_3 and C_4 on the circumference are subtended by the same arc AB.

So angle AC_1B = angle AC_2B = angle AC_3B = angle AC_4B

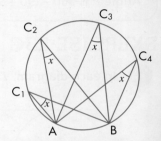

Example 6

O is the centre of each circle. Find the sizes of angles marked a and b in each circle.

a
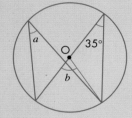

b
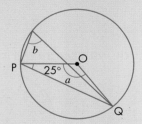

a $a = 35°$ (angles in same segment)

$b = 2 \times 35°$ (angle at centre = twice angle at circumference)

$= 70°$

b With $OP = OQ$, triangle OPQ is isosceles and the sum of the angles in this triangle = $180°$

So $a + (2 \times 25°) = 180°$

$a = 180° - (2 \times 25°)$

$= 130°$

$b = 130° \div 2$ (angle at centre = twice angle at circumference)

$= 65°$

23.8

Example 7

O is the centre of the circle. PQR is a straight line.

Find the size of the angle labelled a.

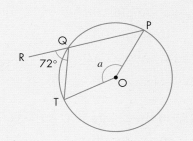

angle $PQT = 180° − 72° = 108°$ (angles on straight line)

The reflex angle angle $POT = 2 × 108°$
(angle at centre = twice angle at circumference)

$\qquad = 216°$

$a + 216° = 360°$ (sum of angles around a point)

$\qquad a = 360° − 216°$

$\qquad a = 144°$

EXERCISE 23H

1 Find the value of x in each of these circles with centre O.

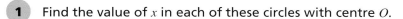

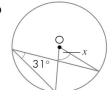
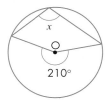
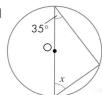

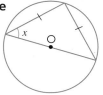
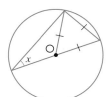
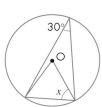
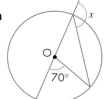

2 Find the value of x in each of these circles with centre O.

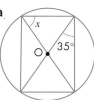
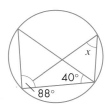

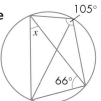
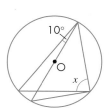
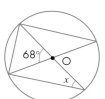
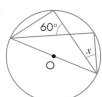

Chapter 23: Angle properties

3 In the diagram, *O* is the centre of the circle. Calculate these angles.

a Angle *ADB*

b Angle *DBA*

c Angle *CAD*

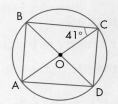

4 In the diagram, *O* is the centre of the circle. Calculate these angles.

a Angle *EDF*

b Angle *DEG*

c Angle *EGF*

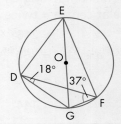

5 In the diagram *XY* is a diameter of the circle and angle *AZX* is *a*.

Ben says that the value of *a* is 50°.

Give reasons to explain why he is wrong.

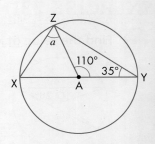

6 Find the values of *x* and *y* in each of these circles. *O* is the centre where shown.

a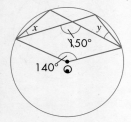

b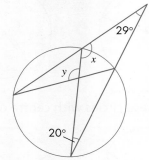

c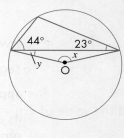

d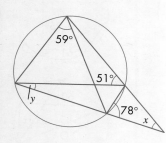

e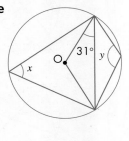

f

404 23.8 Angles in a circle

Chapter 23. Topic 9

7 In the diagram, O is the centre and AD is the diameter of the circle.

Find the value of x.

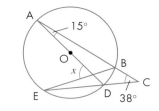

8 In the diagram, O is the centre of the circle and angle CBD is x.

Show that the reflex angle AOC is $2x$, giving reasons to explain your answer.

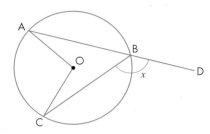

9 A, B, C and D are points on the circumference of a circle with centre O. Angle ABO is $x°$ and angle CBO is $y°$.

  a State the value of angle BAO.

  b State the value of angle AOD.

  c Prove that the angle subtended by the chord AC at the centre of a circle is twice the angle subtended at the circumference.

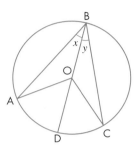

23.9 Cyclic quadrilaterals

There are *two* segments between points P and Q.

    a is the angle in one segment.

    b is the angle in the **opposite segment**.

Angles in opposite segments add up to 180°.
These angles (a and b) are **supplementary**.

  Proof: $a + b = \frac{1}{2}c + \frac{1}{2}d = \frac{1}{2}(c + d) = \frac{1}{2} \times 360 = 180°$

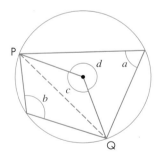

A quadrilateral whose four vertices lie on the circumference of a circle is called a **cyclic quadrilateral**.

The sum of the opposite angles of a cyclic quadrilateral is 180°.

  $a + c = 180°$ and $b + d = 180°$

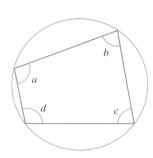

EXERCISE 23I

1 Find the sizes of the lettered angles in each of these circles.

a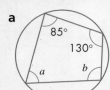
b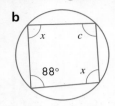
c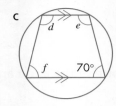
d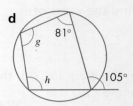

e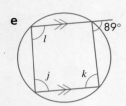
f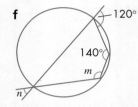
g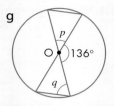
h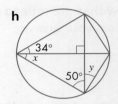

2 Find the values of x and y in each of these circles. Where shown, O marks the centre of the circle.

a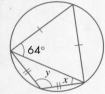
b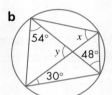
c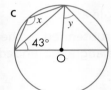
d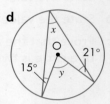

e
f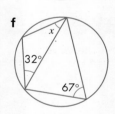
g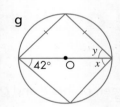
h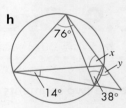

3 Find the values of x and y in each of these circles. Where shown, O marks the centre of the circle.

a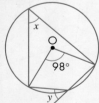
b
c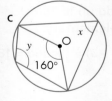
d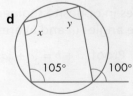

4 Find the values of x and y in each of these circles.

a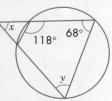
b
c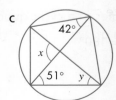
d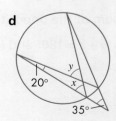

23.9 Cyclic quadrilaterals

5 Find the values of the angles labelled with letters in each of these circles with centre O.

a 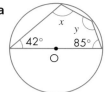   b 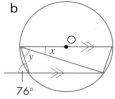   c 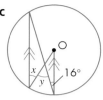   d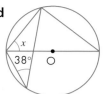

6 The cyclic quadrilateral PQRT has angle ROQ equal to 38° where O is the centre of the circle. POT is a diameter and parallel to QR. Calculate these angles.

   a ROT

   b QRT

   c QPT

7 In the diagram, O is the centre of the circle.

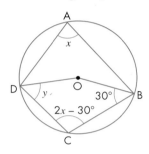

   a Explain why $3x - 30° = 180°$.

   b Work out the size of angle CDO, labelled y on the diagram.
   Give reasons in your working.

8 ABCD is a cyclic quadrilateral within a circle centre O and angle AOC is $2x°$.

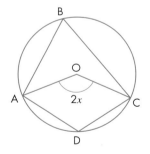

   a Write down the size of angle ABC.

   b Write down the size of the reflex angle AOC.

   c Prove that the sum of a pair of opposite angles of a cyclic quadrilateral is 180°.

Chapter 23: Angle properties

Chapter 23 . Topic 10

EXTENDED

9 In the diagram, ABCE is a parallelogram.

Prove that angle AED = angle ADE.

Give reasons in your working.

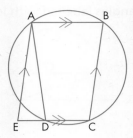

10 Two circles touch at D.

ADE and CDG are straight lines.

Explain why angles ABC and EFG must be equal in size.

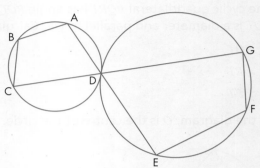

23.10 Alternate segment theorem   E

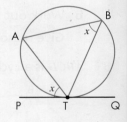

PTQ is the tangent to a circle at T. The segment containing angle TBA is known as the **alternate segment** of angle PTA, because it is on the other side of the chord AT from angle PTA.

The angle between a tangent and a chord through the point of contact is equal to the angle in the alternate segment.

angle PTA = angle TBA

Example 8

In the diagram, find a angle ATS and b angle TSR.

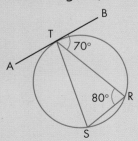

a angle ATS = 80° (angle in alternate segment)

b angle TSR = 70° (angle in alternate segment)

EXERCISE 23J

1 Find the size of each lettered angle.

a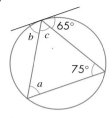

b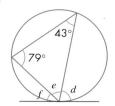

c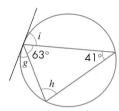

d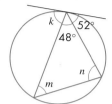

2 In each diagram, find the size of each lettered angle.

a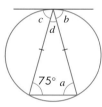

b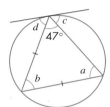

c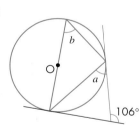

d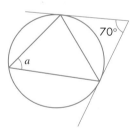

3 In each diagram, find the value of x.

a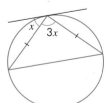

b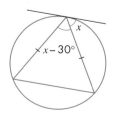

4 *ATB* is a tangent to each circle with centre *O*. Find the size of each lettered angle.

a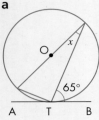

b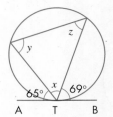

c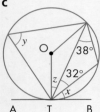

d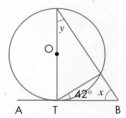

5 In the diagram, *O* is the centre of the circle.

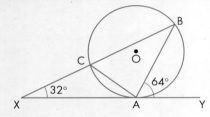

XY is a tangent to the circle at *A*.

BCX is a straight line.

Show that triangle *ACX* is isosceles.

Give reasons to justify your answer.

6 *AB* and *AC* are tangents to the circle at *X* and *Y*.

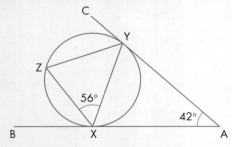

Work out the size of angle *XYZ*.

Give reasons to justify your answer.

410   23.10 Alternate segment theorem

7 *PT* is a tangent to a circle with centre *O*.
AB are points on the circumference. Angle *PBA* is $x°$.

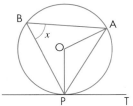

  a Write down the value of angle *AOP* in terms of x.
  b Calculate the angle *OPA* in terms of x.
  c Prove that the angle *APT* is equal to the angle *PBA*.

Check your progress

Core
- I can calculate unknown angles using the following geometrical properties:
  - angles at a point
  - angles on a straight line
  - angles formed within parallel lines
  - angle properties of triangles
  - angle properties of quadrilaterals
  - angle properties of regular polygons
  - an angle in a semi-circle
  - the angle between a tangent and a radius of a circle

Extended
- I can calculate unknown angles using the following geometrical properties:
  - angle properties of irregular polygons
  - angles in the same segment are equal
  - the angle at the centre of a circle is twice the angle at the circumference
  - angle properties of a cyclic quadrilateral
  - the alternate segment theorem

Chapter 24
Geometrical terms and relationships

Topics	Level	Key words
1 Measuring and drawing angles	CORE	right angles, acute angles, obtuse angles, reflex angles, perpendicular, protractor
2 Bearings	CORE	bearing, three-figure bearing
3 Nets	CORE	net, cube, cuboid, prism, pyramid, vertex
4 Congruent shapes	CORE	congruent
5 Congruent triangles	EXTENDED	congruent
6 Similar shapes	CORE	similar, enlargement, linear scale factor, corresponding angles, corresponding sides
7 Areas of similar triangles	EXTENDED	area scale factor
8 Areas and volumes of similar shapes	EXTENDED	solid shapes, volume scale factor

In this chapter you will learn how to:

CORE	EXTENDED
• Use and interpret the geometrical terms: point, line, parallel, bearing, right angle, acute, obtuse and reflex angles, perpendicular, similarity and congruence. (C4.1 and E4.1) • Calculate lengths of similar figures. (C4.4 and E4.4) • Use and interpret vocabulary of triangles, quadrilaterals, circles, polygons and simple solid figures including nets. (C4.1 and E4.1) • Calculate lengths of similar figures. (C4.4 and E4.4) • Interpret and use three-figure bearings measured clockwise from the North (that is 000°–360°). (C6.1 and E6.1) • Recognise congruent shapes. (C4.5)	• Use the relationships between areas of similar triangles, with corresponding results for similar figures. (E4.4) • Extend to volumes and surface areas of similar solids. (E4.4) • Use the basic congruence criteria for triangles (SSS, ASA, SAS, RHS). (E4.5)

Why this chapter matters

Thales of Miletus (624–547 BCE) was a Greek philosopher. Mathematicians believe he was the first person to use similar triangles to find the height of tall objects.

Thales discovered that, at a particular time of day, the height of an object and the length of its shadow were the same. He used this to calculate the height of the Egyptian pyramids.

You can apply the geometry of triangles to calculate heights. You can do this with an instrument called a clinometer. This means you can find the heights of trees, buildings and towers, mountains and other objects which are difficult to measure physically.

Astronomers use the geometry of triangles to measure the distance to nearby stars. They use the Earth's journey in its orbit around the Sun.

They measure the angle of the star twice, from the same point on Earth, but at opposite ends of its orbit. This gives them angle measurements at a known distance apart and from this triangle they can calculate the distance to the star.

Telescopes and binoculars also use the geometry of triangles.

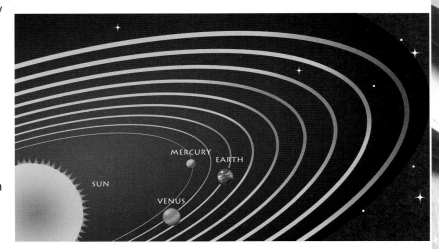

Chapter 24: Geometrical terms and relationships

Chapter 24 . Topic 1
24.1 Measuring and drawing angles

A whole turn is divided into 360° or four **right angles** of 90° each.

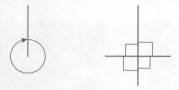

An **acute angle** is less than one right angle.

An **obtuse angle** is between one and two right angles (90° and 180°).

A **reflex angle** is between two and four right angles (180° and 360°).

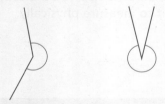

Two lines are **perpendicular** if the angle between them is 90°.

When you are using a **protractor**, it is important that you:
- place the centre of the protractor *exactly* on the corner (vertex) of the angle
- lay the baseline of the protractor *exactly* along one side of the angle.

You must follow these two steps to obtain an accurate value for the angle you are measuring.

You should already have discovered how easy it is to measure acute angles and obtuse angles, using the common semicircular protractor.

Example 1

Measure the angles ABC, DEF and reflex GHI in the diagrams below.

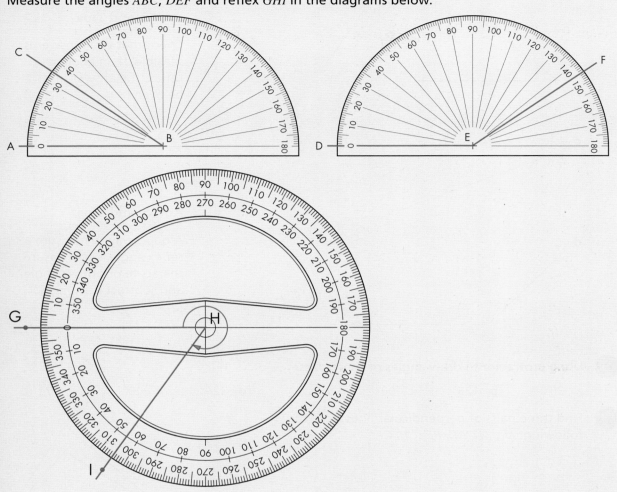

Acute angle ABC is 35° and obtuse angle DEF is 145°.

To measure reflex angles, such as angle GHI, it is easier to use a circular protractor if you have one.

Note the notation for angles.

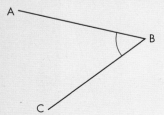

Angle ABC, or ∠ABC, means the angle at B between the lines AB and BC.

Reflex angle GHI is 305°.

EXERCISE 24A

1 Use a protractor to measure the size of each marked angle.

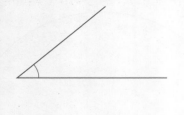

Advice and Tips

Check that your answer is sensible. If an angle is reflex, the answer must be over 180°.

Advice and Tips

A good way to measure a reflex angle is to measure the acute or obtuse angle and subtract it from 360°.

2 Use a protractor to draw angles of these sizes.

    a 30°   **b** 125°   **c** 90°   **d** 212°   **e** 324°   **f** 19°   **g** 171°

3 Find three pairs of perpendicular lines from the following: *AC*, *AD*, *AE*, *BE*, *CE*, *CF*.

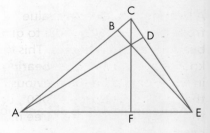

4 It is only safe to climb this ladder if the angle between the ground and the ladder is between 72° and 78°.

Is it safe for Oliver to climb the ladder?

Chapter 24. Topic 2

5 An obtuse angle is 10° more than an acute angle.
Write down a possible value for the size of the obtuse angle.

6 Use a ruler and a protractor to draw these triangles accurately. Then measure the unmarked angle in each one.

a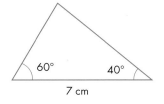

b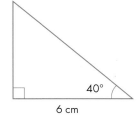

c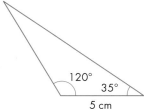

24.2 Bearings

The **bearing** of a point B from a point A is the angle through which you turn *clockwise* as you change direction from *due north* to the direction of B.

For example, in this diagram the bearing of B from A is 060°.

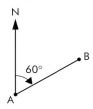

A bearing can have any value from 0° to 360°. It is usual to give all bearings as three figures. This is known as a **three-figure bearing**. So, in the example on the previous page, the bearing is written as 060°, using three figures. Here are three more examples.

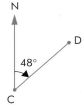
D is on a bearing of 048° from C

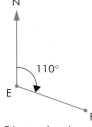
F is on a bearing of 110° from E

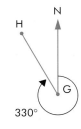
H is on a bearing of 330° from G

There are eight bearings that you should know. They are shown in the diagram.

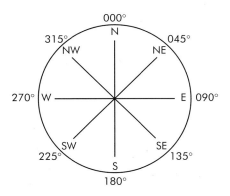

Chapter 24: Geometrical terms and relationships

Example 2

A, B and C are three towns.

Write down the bearing of B from A and the bearing of C from A.

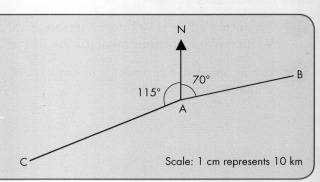

The bearing of B from A is 070°.

The bearing of C from A is 360° − 115° = 245°.

EXERCISE 24B

1. Look at this map. By measuring angles, find the bearings of:

   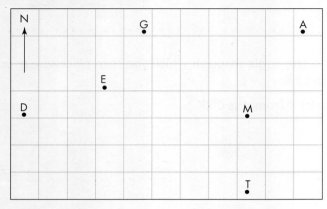

   a T from D
   b D from E
   c M from D
   d G from A
   e M from G
   f T from M.

2. Draw sketches to illustrate these situations.

   a C is on a bearing of 170° from H.
   b B is on a bearing of 310° from W.

3. A is due north from C. B is due east from A. B is on a bearing of 045° from C. Sketch the layout of the three points, A, B and C.

4. The Captain decided to sail his ship around the four sides of a square kilometre.

   a Assuming he started sailing due north, write down the next three bearings he would use in order to complete the square in a clockwise direction.

   b Assuming he started sailing on a bearing of 090°, write down the next three bearings he would use in order to complete the square in an anticlockwise direction.

5 The map shows a boat journey around an island, starting and finishing at S. On the map, 1 centimetre represents 10 kilometres. Measure the distance and bearing of each leg of the journey.

Copy and complete this table.

Leg	Actual distance	Bearing
1		
2		
3		
4		
5		

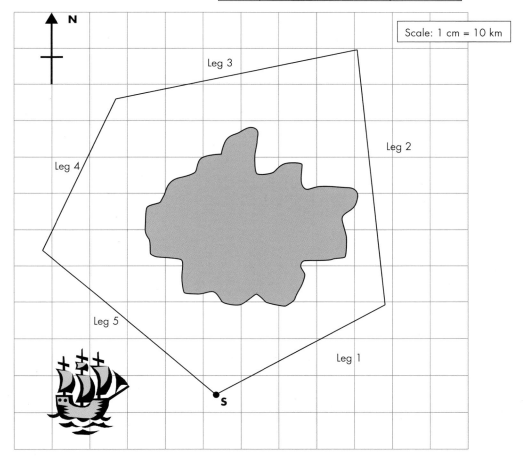

6 The diagram shows a port P and two harbours X and Y on the coast.

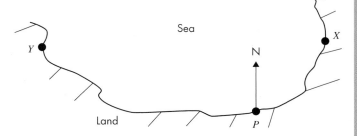

  a A fishing boat sails to X from P.

  What is the three-figure bearing of X from P?

  b A yacht sails to Y from P.

  What is the three-figure bearing of Y from P?

Chapter 24 . Topic 3

7 Draw diagrams to solve these problems.

  a The three-figure bearing of A from B is 070°. Work out the three-figure bearing of B from A.

  b The three-figure bearing of P from Q is 145°. Work out the three-figure bearing of Q from P.

  c The three-figure bearing of X from Y is 324°. Work out the three-figure bearing of Y from X.

8 The diagram shows the position of Kim's house H and the college C.

Scale: 1 cm represents 200 m

  a Use the diagram to work out the actual distance from Kim's house to the college.

  b Measure and write down the three-figure bearing of the college from Kim's house.

  c The supermarket S is 600 m from Kim's house on a bearing of 150°.

  Mark the position of S on a copy of the diagram.

9 Chen is flying a plane on a bearing of 072°.

He is told to fly due south towards an airport.

Through what angle does he need to turn?

10 A, B and C are three villages in a bay.

They lie on the vertices of a square.

The bearing of B from A is 030°.

Work out the bearing of A from C.

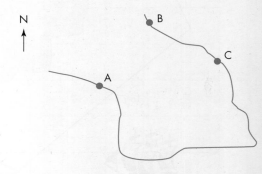

24.3 Nets

You should know these solid shapes:

A **cube** has six square faces.

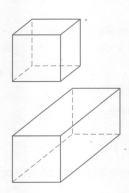

A **cuboid** has rectangular faces.

A **prism** has a uniform cross-section.

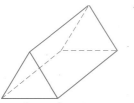

Triangular prism   Hexagonal prism

A **pyramid** has a polygon-shaped base. The other faces are triangles and meet at the **vertex**.

Square-based pyramid   Pentagon-based pyramid

Each of these solid shapes can be made from a **net**.

A net is a flat shape which can be cut out and folded to make a solid shape.

Example 3

Sketch the net for:

a cube

b square-based pyramid.

a This is a sketch of a net for a cube.

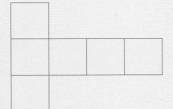

b This is a sketch of a net for a square-based pyramid.

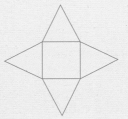

EXERCISE 24C

1 Draw, on squared paper, an accurate net for each of these cuboids.

a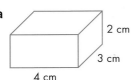

b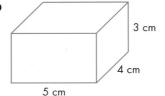

c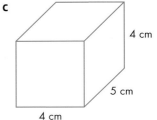

2 Bashira is making an open box from card.

This is a sketch of the box.

Bashira has a piece of card that measures 15 cm by 21 cm. Can she make the box from this card?

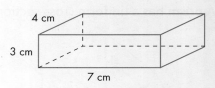

3 The shape on the right is a triangular prism. Its ends are isosceles triangles and its other faces are rectangles. Draw an accurate net for this prism. Use squared paper.

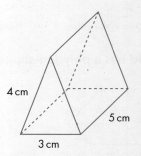

4 Draw the nets of these shapes.

a

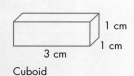

Cuboid

b

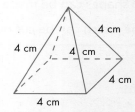

Square-based pyramid

c

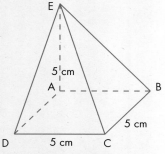

Square-based pyramid, with point E directly above point A

d

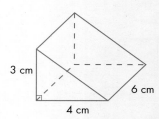

Right-angled triangular prism

5 Here is a net for a cube.

How many different nets can you draw for a cube?

Advice and Tips

There are 11 altogether. How many can you find?

6 Which of these are nets for a square-based pyramid?

a

b

c

24.4 Congruent shapes

Two-dimensional shapes that are exactly the same size and shape as each other are **congruent**. For example, although they are in different positions, the triangles below are all congruent, because they are all exactly the same size and shape.

Congruent shapes fit exactly on top of each other. So, one way to see whether shapes are congruent is to trace one of them and check that it covers the other shapes exactly. For some of the shapes, you may have to turn your tracing paper over.

Example 4

Which of these shapes is not congruent to the others?

Trace shape **a** and check whether it fits exactly on top of the others.

You should find that shape **b** is not congruent to the others.

EXERCISE 24D

1. State whether the shapes in each pair, **a** to **f** are congruent or not.

   a   b   c

   d   e    f

Chapter 24. Topic 5

2 Which figure in each group, **a** to **c**, is not congruent to the other two?

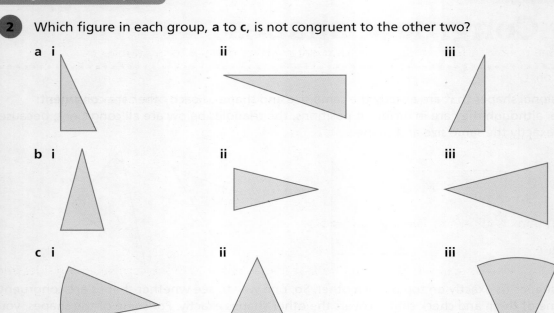

3 Draw a square *PQRS*. Draw in the diagonals *PR* and *QS*. Which triangles are congruent to each other?

4 Draw a rectangle *EFGH*. Draw in the diagonals *EG* and *FH*. Which triangles are congruent to each other?

5 Draw a parallelogram *ABCD*. Draw in the diagonals *AC* and *BD*. Which triangles are congruent to each other?

6 Draw an isosceles triangle *ABC* where *AB* = *AC*. Draw the line from *A* to the midpoint of *BC*. Which triangles are congruent to each other?

24.5 Congruent triangles

Two shapes are **congruent** if they are exactly the same size and shape.

For example, these triangles are all congruent.

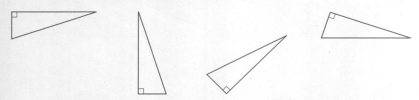

Notice that the triangles can be differently oriented (reflected or rotated).

24.5

Conditions for congruent triangles

Any **one** of the following four conditions is sufficient for two triangles to be congruent.

- **Condition 1**

  All three sides of one triangle are equal to the corresponding sides of the other triangle.

  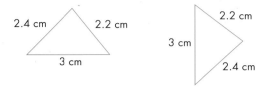

  This condition is known as SSS (side, side, side).

- **Condition 2**

  Two sides and the angle between them of one triangle are equal to the corresponding sides and angle of the other triangle.

  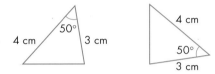

  This condition is known as SAS (side, angle, side).

- **Condition 3**

  Two angles and a side of one triangle are equal to the corresponding angles and side of the other triangle.

  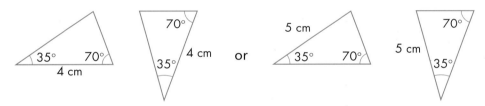

  This condition is known as ASA (angle, side, angle) or AAS (angle, angle, side).

- **Condition 4**

  Both triangles have a right angle, an equal hypotenuse and another equal side.

  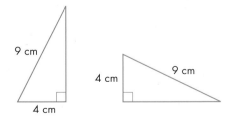

  This condition is known as RHS (right angle, hypotenuse, side).

  Note that SSA (or ASS) does not show congruency.

Notation

Once you have shown that triangle *ABC* is congruent to triangle *PQR* by one of the above conditions, it means that:

angle *A* = angle *P*   *AB* = *PQ*
angle *B* = angle *Q*   *BC* = *QR*
angle *C* = angle *R*   *AC* = *PR*

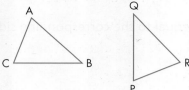

In other words, the points *ABC* correspond exactly to the points *PQR* in that order. Triangle *ABC* is congruent to triangle *PQR* can be written as △*ABC* ≡ △*PQR*.

Example 5

ABCD is a kite. Show that triangle *ABC* is congruent to triangle *ADC*.

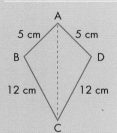

AB = *AD*

BC = *CD*

AC is common

So △*ABC* ≡ △*ADC* (SSS)

EXERCISE 24E

1 The triangles in each pair are congruent. State the condition that shows that the triangles are congruent.

a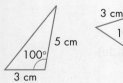

b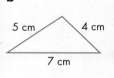

c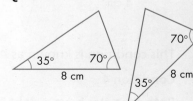

d

e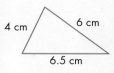

f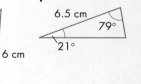

2 The triangles in each pair are congruent. State the condition that shows that the triangles are congruent and say which points correspond to which.

   a *ABC* where *AB* = 8 cm, *BC* = 9 cm, *AC* = 7.4 cm

   PQR where *PQ* = 9 cm, *QR* = 7.4 cm, *PR* = 8 cm

Chapter 24. Topic 6

   b ABC where $AB = 5$ cm, $BC = 6$ cm, angle $B = 35°$
      PQR where $PQ = 6$ cm, $QR = 50$ mm, angle $Q = 35°$

3 Triangle ABC is congruent to triangle PQR, angle $A = 60°$, angle $B = 80°$ and $AB = 5$ cm. Find these:
   a angle P **b** angle Q **c** angle R **d** PQ

4 $ABCD$ is congruent to $PQRS$, angle $A = 110°$, angle $B = 55°$, angle $C = 85°$ and $RS = 4$ cm. Find these:
   a angle P **b** angle Q **c** angle R **d** angle S **e** CD

5 Draw a rectangle, $EFGH$. Draw in the diagonal EG. Prove that triangle EFG is congruent to triangle EHG.

6 Draw an isosceles triangle (ABC) where $AB = AC$. Draw the line from A to X, the midpoint of BC. Prove that triangle ABX is congruent to triangle ACX.

7 In the diagram, $ABCD$ and $DEFG$ are squares.

Use congruent triangles to prove that $AE = CG$.

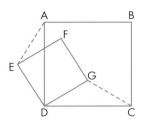

8 Jez says that these two triangles are congruent because two angles and a side are the same.

Explain why he is wrong.

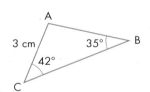 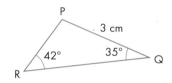

24.6 Similar shapes

Two shapes are **similar** if one is an **enlargement** of the other.

Corresponding angles of similar shapes are equal.

These shapes are similar. Their corresponding angles are equal.

Advice and Tips

In this section, corresponding angles are those angles that appear in the same position on both shapes. Corresponding sides are those sides that appear in the same position on both shapes.

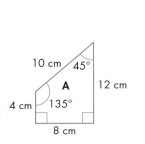 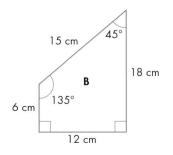

The ratio of their **corresponding sides** is 2 : 3.

The **linear scale factor** is $\frac{3}{2}$ because the lengths in shape B are $\frac{3}{2}$ of the corresponding lengths in shape A.

Chapter 24: Geometrical terms and relationships   **427**

Example 6

These two shapes are similar.

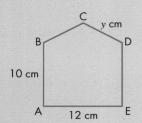 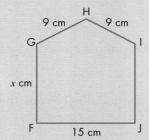

Find the values of x and y.

Look for two corresponding sides where you know the lengths.

AE corresponds to FJ and the lengths are 12 cm and 15 cm.

The scale factor is $\frac{15}{12} = 1.25$.

To find the value of x (GF) from the length given for BA:

$x = 1.25 \times AB$

$ = 1.25 \times 10$

$ = 12.5$

$x = 12.5$ cm

To find the value of y (CD) from the length of HI:

$HI = 1.25 \times y$

$9 = 1.25 \times y$

$y = \frac{9}{1.25}$

$ = 7.2$

$y = 7.2$ cm

EXERCISE 24F

1. These diagrams are drawn to scale. What is the linear scale factor of the enlargement in each case?

   a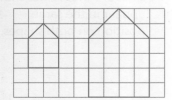

   b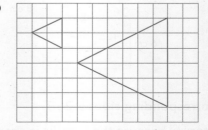

428   24.6 Similar shapes

2 Are the shapes in each pair similar? If so, give the scale factor. If not, give a reason.

a

b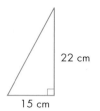

3
a Explain why these triangles are similar.
b Give the ratio of the sides.
c Which angle corresponds to angle C?
d Which side corresponds to side QP?

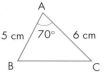 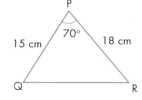

4
a Explain why these triangles are similar.
b Which angle corresponds to angle A?
c Which side corresponds to side AC?

 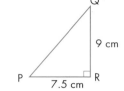

5
a Explain why triangle ABC is similar to triangle AQR.
b Which angle corresponds to the angle at B?
c Which side of triangle AQR corresponds to side AC of triangle ABC?
Your answers to question **4** may help you.

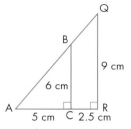

6 In the diagrams **a** to **d**, each pair of shapes are similar but not drawn to scale.

a Find the value of x.

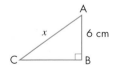 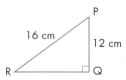

b Find the values of x and y.

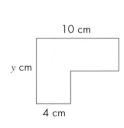 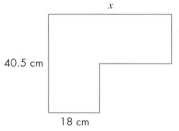

Chapter 24: Geometrical terms and relationships **429**

Chapter 24 . Topic 7

c Find the values of x and y.

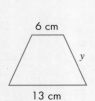
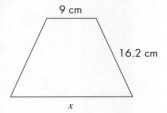

d Calculate the length of QR.

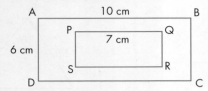

7 a Explain why all squares are similar.

b Are all rectangles similar? Explain your answer.

8 Sean is standing next to a tree.

His height is 1.6 m and he casts a shadow that has a length of 2.4 m.

The tree casts a shadow that has a length of 7.8 m.

Use what you know about similar triangles to work out the height of the tree, h.

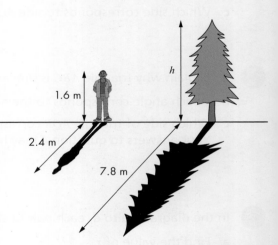

24.7 Areas of similar triangles   E

If two triangles have the same angles then they are similar.

Triangles ABC and DEF are similar.

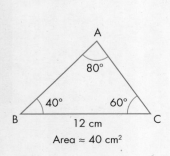
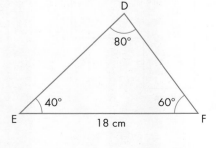

430 24.7 Areas of similar triangles

The linear scale factor is $\frac{18}{12} = 1.5$. The **area scale factor** is $1.5^2 = 2.25$.

If the area of triangle ABC is 40 cm² then the area of DEF is $40 \times 2.25 = 90$ cm².

If the linear scale factor is k, then the area scale factor is k^2.

EXERCISE 24G

1 These triangles are similar.

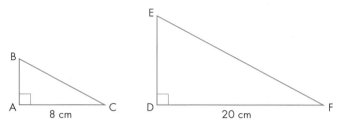

  a   What is the linear scale factor?
  b   The area of triangle ABC is 20 cm².
      Calculate the area of triangle DEF.

2 These are equilateral triangles.

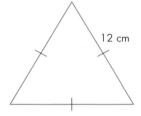

  a   Explain why they are similar.
  b   The area of the larger one is 60 cm² (to 1 significant figure)
      Calculate the area of the smaller one.

3 The area of triangle ABC is 7 cm².

Calculate the area of triangle ADE.

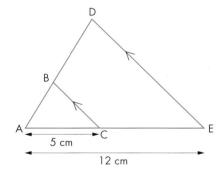

4 The area of triangle ABD is 25 cm².

Calculate the area of the trapezium $CBDE$.

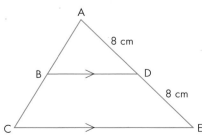

Chapter 24: Geometrical terms and relationships

5 P is enlarged by a linear scale factor of 1.2 to make Q.

Q is enlarged by a linear scale factor of 1.2 to make R.

The area of Q is 100 cm².

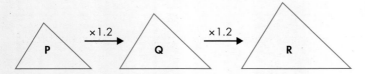

  a What is the area of R?
  b What is the area of P?

6 The area scale factor also applies to other similar shapes.

A and B are regular pentagons.

  a Explain why they are similar.
  b Calculate the area of B.

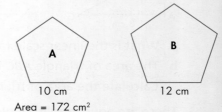

7

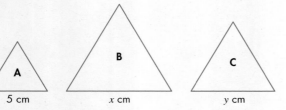

A, B and C are equilateral triangles.

B is four times the area of A.

  a What is the linear scale factor?
  b What is the value of x?
  c C is twice the area of A.
  Calculate the value of y.

8 These shapes are similar.

The smaller one has an area of 210 cm².

Find the area of the larger one.

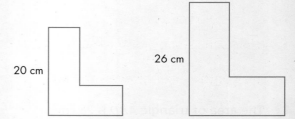

9 A photocopier has a setting for enlargements with a scale factor of 1.41%.

What is special about this particular value?

10 All circles are similar.

If a circle with a diameter of 8 cm has an area of 50.3 cm², what is the area of a circle with diameter 6 cm?

24.8 Areas and volumes of similar shapes E

You saw that if two shapes are similar and the linear scale factor is k then the area scale factor is k^2.

Two **solid shapes** are similar if corresponding lengths are in the same ratio and corresponding angles are equal. In that case the **volume scale factor** is k^3.

Generally, the relationship between similar shapes can be expressed as:

Length ratio $x : y$ Area ratio $x^2 : y^2$ Volume ratio $x^3 : y^3$

Example 7

A model yacht is made to a scale of $\frac{1}{20}$ of the size of the real yacht. The area of the sail of the model is 150 cm². What is the area of the sail of the real yacht?

At first, it may appear that you do not have enough information to solve this problem, but you can do it like this.

Linear scale factor = 1 : 20

Area scale factor = 1 : 400 (square of the linear scale factor)

Area of real sail   = 400 × area of model sail

= 400 × 150 cm²

= 60 000 cm² = 6 m²

Example 8

A bottle has a base radius of 4 cm, a height of 15 cm and a capacity of 650 cm³. A similar bottle has a base radius of 3 cm.

a What is the length ratio?

b What is the volume ratio?

c What is the volume of the smaller bottle?

a The length ratio is given by the ratio of the two radii, that is 4 : 3.

b The volume ratio is therefore $4^3 : 3^3 = 64 : 27$.

c Let v be the volume of the smaller bottle. Then the volume ratio is:

$$\frac{\text{volume of smaller bottle}}{\text{volume of larger bottle}} = \frac{v}{650} = \frac{27}{64}$$

$$\Rightarrow v = \frac{27 \times 650}{64} = 274 \text{ cm}^3 \text{ (3 significant figures)}$$

Chapter 24: Geometrical terms and relationships

EXERCISE 24H

1 The length ratio between two similar solids is 2 : 5.
   a What is the area ratio between the solids?
   b What is the volume ratio between the solids?

2 The length ratio between two similar solids is 4 : 7.
   a What is the area ratio between the solids?
   b What is the volume ratio between the solids?

3 Copy and complete this table.

Linear scale factor	Linear ratio	Linear fraction	Area scale factor	Volume scale factor
2	1 : 2	$\frac{2}{1}$		
3				
$\frac{1}{4}$	4 : 1	$\frac{1}{4}$		$\frac{1}{64}$
			25	
				$\frac{1}{1000}$

4 A shape has an area of 15 cm². What is the area of a similar shape with lengths that are three times the corresponding lengths of the first shape?

5 A toy brick has a surface area of 14 cm². What would be the surface area of a similar toy brick with lengths that are:
   a twice the corresponding lengths of the first brick
   b three times the corresponding lengths of the first brick?

6 A rug has an area of 12 m². What area would be covered by rugs with lengths that are:
   a twice the corresponding lengths of the first rug
   b half the corresponding lengths of the first rug?

7 A brick has a volume of 300 cm³. What would be the volume of a similar brick whose lengths are:
   a twice the corresponding lengths of the first brick
   b three times the corresponding lengths of the first brick?

8 A tin of paint, 6 cm high, holds a half a litre of paint. How much paint would go into a similar tin which is 12 cm high?

9 A model statue is 10 cm high and has a volume of 100 cm³. The real statue is 2.4 m high. What is the volume of the real statue? Give your answer in m³.

10 A small tin of paint costs $0.75. What is the cost of a larger similar tin with height twice that of the smaller tin? Assume that the cost is based only on the volume of paint in the tin.

11 A small box of width 2 cm has a volume of 10 cm³. What is the width of a similar box with a volume of 80 cm³?

12 A cinema sells popcorn in two different-sized tubs that are similar in shape.

Show that it is true that the big tub is better value.

13 The diameters of two ball bearings are given below.

Work out:

a the ratio of their radii
b the ratio of their surface areas
c the ratio of their volumes.

14 Cuboid A is similar to cuboid B.

The length of cuboid A is 10 cm and the length of cuboid B is 5 cm.

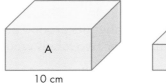

The volume of cuboid A is 720 cm³.

Zainab says that the volume of cuboid B must be 360 cm³.

Explain why she is wrong.

More complex problems on area and volume ratios

In some problems involving similar shapes, the length ratio is not given, so you need to start with the area ratio or the volume ratio. Then you will need to find the length ratio in order to proceed with the solution.

> ### Example 9
> A manufacturer makes a range of clown hats that are all similar in shape. The smallest hat is 8 cm tall and uses 180 cm² of card. What will be the height of a hat made from 300 cm² of card?
>
> The area ratio is 180 : 300
>
> Therefore, the length ratio is $\sqrt{180} : \sqrt{300}$ (Do not calculate these yet.)
>
> Let the height of the larger hat be H, then:
>
> $$\frac{H}{8} = \frac{\sqrt{300}}{\sqrt{180}} = \sqrt{\frac{300}{180}}$$
>
> $\Rightarrow H = 8 \times \sqrt{\frac{300}{180}} = 10.3$ cm (1 decimal place)

Chapter 24: Geometrical terms and relationships

Example 10

Two similar tins hold respectively 1.5 litres and 2.5 litres of paint. The area of the label on the smaller tin is 85 cm². What is the area of the label on the larger tin?

The volume ratio is 1.5 : 2.5

Therefore, the length ratio is $\sqrt[3]{1.5} : \sqrt[3]{2.5}$ (Do not calculate these yet.)

So the area ratio is $(\sqrt[3]{1.5})^2 : (\sqrt[3]{2.5})^2$

Let the area of the label on the larger tin be A, then:

$$\frac{A}{85} = \frac{(\sqrt[3]{2.5})^2}{(\sqrt[3]{1.5})^2} = \left(\sqrt[3]{\frac{2.5}{1.5}}\right)^2 \Rightarrow A = 85 \times \left(\sqrt[3]{\frac{2.5}{1.5}}\right)^2 = 119 \text{ cm}^2 \text{ (3 significant figures)}$$

EXERCISE 24I

1. A firm produces three sizes of similar-shaped labels for its products. Their areas are 150 cm², 250 cm² and 400 cm².

   The 250 cm² label fits around a can of height 8 cm. Find the heights of similar cans around which the other two labels would fit.

2. A firm makes similar boxes in three different sizes: small, medium and large. These are the areas of their lids.

   Small: 30 cm²   Medium: 50 cm²   Large: 75 cm²

   The medium box is 5.5 cm high. Find the heights of the other two sizes.

3. A cone of height 8 cm can be made from a piece of card with an area of 140 cm². What is the height of a similar cone made from a similar piece of card with an area of 200 cm²?

4. It takes 5.6 litres of paint to paint a chimney that is 3 m high. What is the tallest similar chimney that can be painted with 8 litres of paint?

5. A piece of card, 1200 cm² in area, will make a tube 13 cm long. What is the length of a similar tube made from a similar piece of card with an area of 500 cm²?

6. If a television screen of area 220 cm² has a diagonal length of 21 cm, what is the diagonal length of a similar screen of area 350 cm²?

7. There are two similar bronze statues. One has a mass of 300 g, the other has a mass of 2 kg. The height of the smaller statue is 9 cm.

   What is the height of the larger statue?

8. These are the sizes of the labels around three similar cans.

   Small can: 24 cm²   Medium can: 46 cm²   Large can: 78 cm²

   The medium size can is 6 cm tall with a mass of 380 g. Calculate:

   a the heights of the other two sizes   b the masses of the other two sizes.

9 A statue has a mass of 840 kg. A similar statue was made out of the same material but is only two-fifths the height of the first one. What was the mass of the smaller statue?

10 A wooden model stands on a base of area 12 cm². A similar wooden model stands on a base of area 7.5 cm².

Calculate the mass of the smaller model if the larger one has a mass of 3.5 kg.

11 Stefan fills two similar jugs with orange juice.

The first jug holds 1.5 litres of juice and has a base diameter of 8 cm.

The second jug holds 2 litres of juice. Work out the base diameter of the second jug.

12 The total surface areas of two similar cuboids are 500 cm² and 800 cm².

If the width of one of the cuboids is 10 cm, calculate the two possible widths for the other cuboid.

13 The volumes of two similar cylinders are 256 cm³ and 864 cm³.

Which of these is the ratio of their surface areas?

   a   2 : 3   b   4 : 9   c   8 : 27

Check your progress

Core

- I can measure and draw angles
- I can understand and use geometrical terms including right angle, acute angle, obtuse angle, reflex angle and perpendicular lines
- I can understand and use the terms similar and congruent
- I know and can use the names of polygons and solid figures including nets
- I can calculate lengths in similar figures
- I can interpret and use three-figure bearings

Extended

- I can use the relationship between the area of similar triangles and similar figures
- I can use the relationship between the volumes of similar figures
- I can use the relationship between the surface areas of similar figures
- I can use the congruence criteria for triangles

Chapter 25
Geometrical constructions

Topics	Level	Key words
1 Constructing shapes	CORE	construct, ruler, protractor, compasses, set square
2 Scale drawings	CORE	scale drawing

In this chapter you will learn how to:

CORE

- Measure and draw lines and angles. (C4.2 and E4.2)
- Construct other simple geometrical figures from given data using protractors and set squares as necessary. (C4.2 and E4.2)
- Read and make scale drawings. (C4.3 and E4.3)

Why this chapter matters

Engineers, town planners, architects, surveyors, builders and computer designers all need to work with great precision. Some projects involve working with very big lengths or distances; others involve very tiny ones. So how do they manage to work with these difficult measures?

The answer is that they all work with drawings drawn to scale. This allows them to represent lengths they cannot easily measure with standard equipment.

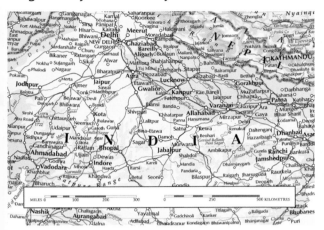

In a scale drawing, one length is used to represent another.

For example, a map cannot be drawn to the same size as the area it represents. The measurements are scaled down, to make a map of a size that can be conveniently used by drivers, tourists and walkers.

Architects use scale drawings to show views of a planned house from different directions.

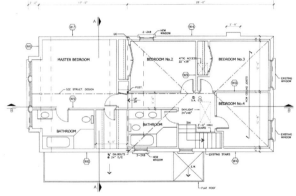

In computer design, people who design microchips need to scale up their drawings, as the dimensions they work with are so small. Use the internet to research more about how scale drawings are used.

Chapter 25: Geometrical constructions 439

25.1 Constructing shapes

You need to be able to draw a triangle when you are given the lengths of its sides.

You need a ruler and a pair of compasses to do this.

Use a sharp pencil and do not use a pen.

Leave any construction lines on your drawing. Do not erase them.

Example 1

Construct a triangle with sides that are 5 cm, 4 cm and 6 cm long.

Step 1: Draw the longest side as the base. In this case, the base will be 6 cm, which you draw along a ruler. (The diagrams in this example are drawn at half-size.)

───────────

Step 2: Draw the second longest side, in this case the 5 cm side. Open the compasses to a radius of 5 cm (the length of the side), place the point on one end of the 6 cm line and draw a short faint arc, as shown here.

Step 3: Draw the shortest side, in this case the 4 cm side. Open the compasses to a radius of 4 cm, place the point on the other end of the 6 cm line and draw a second short faint arc to intersect the first arc, as shown here.

Step 4: Complete the triangle by joining each end of the base line to the point where the two arcs intersect.

Note: The arcs are construction lines and so you must leave them in, to show how you constructed the triangle.

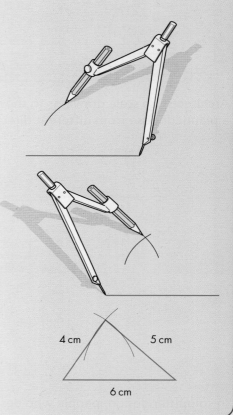

EXERCISE 25A

1 Draw each triangle accurately and measure the sides and angles not given in the diagram.

Advice and Tips

Always make a sketch if one is not given in the question.

a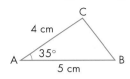

b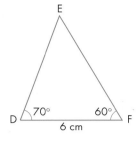

c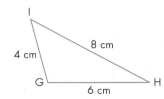

d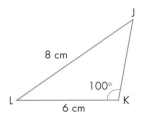

e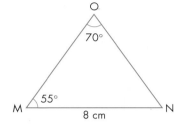

f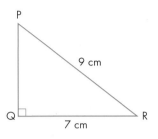

2 a Draw a triangle ABC, with $AB = 7$ cm, $BC = 6$ cm and $AC = 5$ cm.

b Measure the sizes of angle ABC, angle BCA and angle CAB.

Advice and Tips

Sketch the triangle first.

3 Draw an isosceles triangle that has two sides of length 7 cm and the included angle of 50°. Measure the length of the base of the triangle.

4 Make an accurate drawing of this quadrilateral.

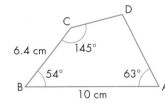

5 Make an accurate drawing of this trapezium.

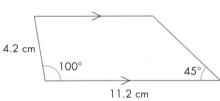

6 Construct an equilateral triangle of side length 5 cm. Measure the height of the triangle.

7 Construct a parallelogram with sides of length 5 cm and 8 cm and with an angle of 120° between them. Measure the height of the parallelogram.

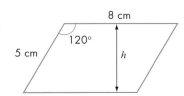

Chapter 25: Geometrical constructions   **441**

Chapter 25 . Topic 2

8 A rope has 12 equally-spaced knots. It can be laid out to give a triangle, like this.

It will always be a right-angled triangle.

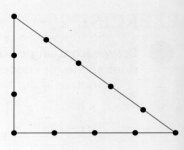

Here are two examples of such ropes.

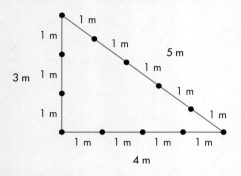
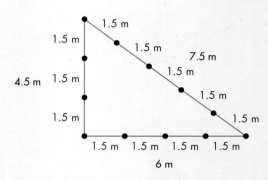

a Show, by constructing both of the above triangles (use a scale of 1 cm : 1 m), that each is a right-angled triangle.

b Choose a different triangle that you think might also be right-angled. Use the same knotted-rope idea to check.

9 Construct the triangle with the largest area that has a total perimeter of 12 cm.

10 Anil says that, as long as he knows all three angles of a triangle, he can draw it. Explain why Anil is wrong.

25.2 Scale drawings

A **scale drawing** is an accurate representation of a real object.

Scale drawings are usually smaller in size than the original objects. However, in certain cases, they are enlargements. Examples of these are drawings of miniature electronic circuits and very small watch movements.

You will generally be given the scale being used, for example, '1 cm represents 20 m'.

Example 2

The diagram shows the front of a kennel. 1 cm on the diagram represents a measurement of 30 cm. Find:

a the actual width of the front

b the actual height of the doorway.

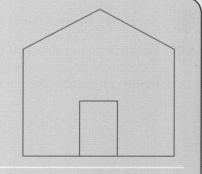

a The actual width of the front is:

4 cm × 30 = 120 cm

b The actual height of the doorway is:

1.5 cm × 30 = 45 cm

EXERCISE 25B

1 Look at this plan of a garden.

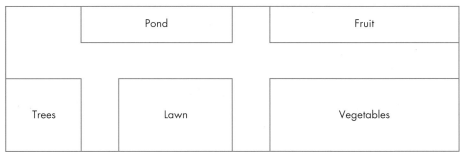

Scale: 1 cm represents 10 m

   a State the actual dimensions of each plot of the garden.

   b Calculate the actual area of each plot.

2 Below is a plan for a computer mouse mat.

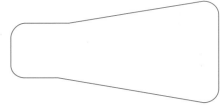

Scale: 1 cm represents 6 cm

Advice and Tips

Remember to check the scale.

   a How long is the actual mouse mat?

   b How wide is the narrowest part of the mouse mat?

3 Below is a scale plan of the top of Ahmed's desk, in which 1 cm represents 10 cm.

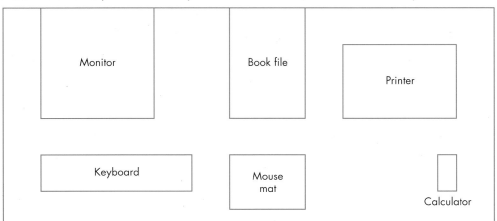

What are the actual dimensions of each of these objects?

a monitor
b keyboard
c mouse mat
d book file
e printer
f calculator

4 The diagram shows a sketch of a garden.

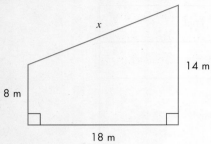

a Make an accurate scale drawing of the garden.

Use a scale of 1 cm to represent 2 m.

b Marie wants to plant flowers along the side marked x on the diagram. The flowers need to be planted 0.5 m apart. Use your scale drawing to work out how many plants she needs.

5 Look at the map below, which is drawn to a scale of 1 cm representing 2 km. Towns are shown with letters.

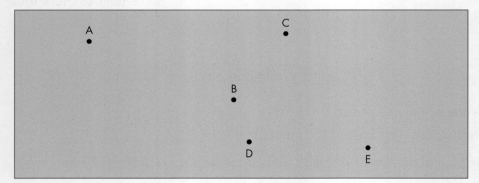

State these actual distances, correct to the nearest tenth of a kilometre.

a A to B
b B to C
c C to D
d D to E
e E to B
f B to D

444 25.2 Scale drawings

25.2

6 This sketch shows the outline of a car park.

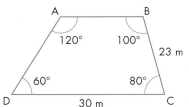

a Make a scale drawing in which 1 cm represents 5 m.

b What is the length of AB, in metres?

7 This map is drawn to a scale of 1 cm to 200 km.

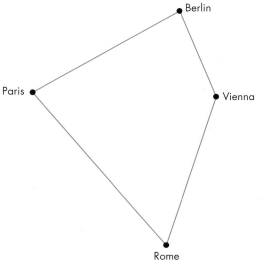

Find the distances between:

a Paris and Berlin

b Paris and Rome

c Rome and Vienna.

8 This is a scale drawing of the Great Beijing Wheel in China.

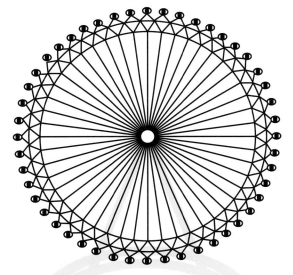

The height of the wheel is 210 m.

Which of these is the correct scale?

a 1 cm represents 30 cm

b 1 cm represents 7 m

c 1 cm represents 30 m

d 1 cm represents 300 m

Check your progress

Core

- I can measure and draw lines and angles
- I can read and make scale drawings
- I can draw a triangle with a ruler and a pair of compasses when I know the lengths of the sides

Chapter 26
Trigonometry

Topics	Level	Key words
1 Pythagoras' theorem	CORE	hypotenuse, Pythagoras' theorem
2 Trigonometric ratios	CORE	ratio, sine, cosine, tangent, opposite side, adjacent side
3 Calculating angles	CORE	inverse
4 Using sine, cosine and tangent functions	CORE	
5 Which ratio to use	CORE	
6 Applications of trigonometric ratios	EXTENDED	angle of elevation, angle of depression
7 Problems in three dimensions	EXTENDED	
8 Sine and cosine of obtuse angles	EXTENDED	obtuse angle
9 The sine rule and the cosine rule	EXTENDED	sine rule, cosine rule, included angle
10 Using sine to find the area of a triangle	EXTENDED	area sine rule
11 Sine, cosine and tangent of any angle	EXTENDED	

In this chapter you will learn how to:

CORE	EXTENDED
• Apply Pythagoras' theorem and the sine, cosine and tangent ratios for acute angles to the calculation of a side or of an angle of a right-angled triangle. (C6.2 and E6.2)	• Solve trigonometrical problems in two dimensions involving angles of elevation and depression. (E6.2) • Extend sine and cosine values to angles between 90° and 180°. (E6.2) • Recognise, sketch and interpret graphs of simple trigonometric functions. Graph and know the properties of trigonometric functions. Solve simple trigonometric equations for values between 0° and 360°. (E6.3) • Solve problems using the sine and cosine rules for any triangle and the formula area of triangle = $\frac{1}{2}ab\sin C$. (E6.4) • Solve simple trigonometrical problems in three dimensions including angle between a line and a plane. (E6.5)

Why this chapter matters

How can you find the height of a mountain?

How do you draw an accurate map?

How can computers take an image and make it rotate so that you can view it from different directions?

How do Global Positioning Systems (GPS) work?

How can music be produced electronically?

The answer is by using the angles and sides of triangles and the connections between them. This important branch of mathematics is called trigonometry and is used in science, engineering, electronics and everyday life. This chapter gives a brief introduction to trigonometry.

The first major book of trigonometry was a written by an astronomer called Ptolemy, who lived in Alexandria, Egypt, over 1800 years ago.

It has tables of numbers, called 'trigonometric ratios', used in making calculations about the positions of stars and planets.

Trigonometry also helped Ptolemy to make a map of the world he knew. Today you no longer need to look up tables of values of trigonometric ratios because they are programmed into calculators and computers.

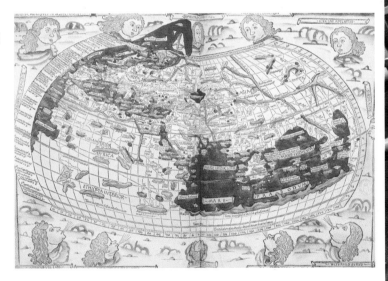

In the 19th century the French mathematician Jean Fourier showed how all musical sounds can be broken down into a combination of tones that can be described by trigonometry. His work makes it possible to imitate the sound of any instrument electronically.

Chapter 26: Trigonometry   **447**

26.1 Pythagoras' theorem

Pythagoras, who was a philosopher as well as a mathematician, was born in 580BCE in Greece. He later moved to Italy, where he established the Pythagorean Brotherhood, which was a secret society devoted to politics, mathematics and astronomy.

This is his famous theorem.

Consider squares being drawn on each side of a right-angled triangle, with sides 3 cm, 4 cm and 5 cm.

The longest side is called the **hypotenuse** and is always opposite the right angle.

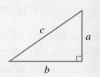

Pythagoras' theorem can then be stated as:

> For any right-angled triangle, the area of the square drawn on the hypotenuse is equal to the sum of the areas of the squares drawn on the other two sides.

The usual description is:

> In any right-angled triangle, the square of the hypotenuse is equal to the sum of the squares of the other two sides.

Pythagoras' theorem is more usually written as a formula:

$$c^2 = a^2 + b^2$$

Remember that Pythagoras' theorem can only be used in right-angled triangles.

Finding the hypotenuse

Example 1

Find the length of the hypotenuse, marked x on the diagram.

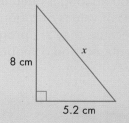

Using Pythagoras' theorem gives:

$x^2 = 8^2 + 5.2^2$ cm²

$= 64 + 27.04$ cm²

$= 91.04$ cm²

So $x = \sqrt{91.04} = 9.5$ cm (1 decimal place)

EXERCISE 26A

For each of the triangles in questions **1** to **9**, calculate the length of the hypotenuse, x, giving your answers to 1 decimal place.

1

2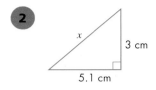

Advice and Tips

In these examples you are finding the hypotenuse. You need to add the squares of the two short sides in every case.

3 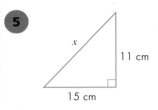   **4**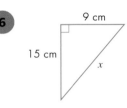

5 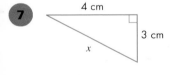   **6**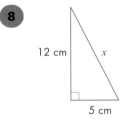

7 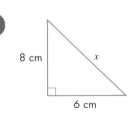   **8**

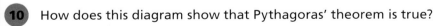

9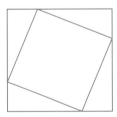

10 How does this diagram show that Pythagoras' theorem is true?

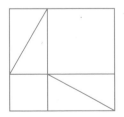

By rearranging the formula for Pythagoras' theorem, you can calculate the length of one of the shorter sides.

$$c^2 = a^2 + b^2$$

So, $a^2 = c^2 - b^2$ or $b^2 = c^2 - a^2$

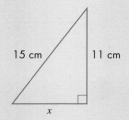

Example 2

Find the length of the side labelled x.

x is one of the shorter sides.

So using Pythagoras' theorem gives:

$x^2 = 15^2 - 11^2$ cm^2

$= 225 - 121$ cm^2

$= 104$ cm^2

So $x = \sqrt{104} = 10.2$ cm (1 decimal place)

EXERCISE 26B

1. For each of these triangles, calculate the length of the side labelled x, giving your answers to 1 decimal place.

   a

   b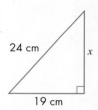

   c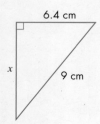

   d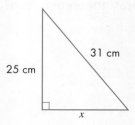

Advice and Tips

In these examples you are finding a short side. You need to subtract the square of the other short side from the square of the hypotenuse in every case.

2 For each of these triangles, calculate the length labelled x, giving your answers to 1 decimal place.

a 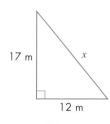   b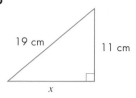

Advice and Tips

These examples are a mixture. Make sure you combine the squares of the sides correctly.

c   d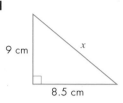

3 For each of these triangles, calculate the length marked x.

a 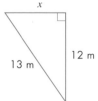   b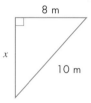

c 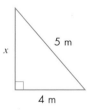   d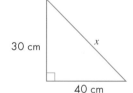

4 In question **3** you found sets of three numbers which satisfy $a^2 + b^2 = c^2$.

Can you find any more?

5 Calculate the value of x.

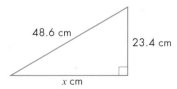

Chapter 26: Trigonometry

26.2 Trigonometric ratios

In trigonometry you will use three important **ratios** to calculate sides and angles: **sine, cosine** and **tangent**. These ratios are defined in terms of the sides of a right-angled triangle and an angle. The angle is often written as θ.

In a right-angled triangle:

- the side opposite the right angle is called the hypotenuse and is the longest side
- the side opposite the angle θ is called the **opposite side**
- the other side next to both the right angle and the angle θ is called the **adjacent side**.

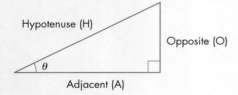

The sine, cosine and tangent ratios for θ are defined as:

$$\text{sine } \theta = \frac{\text{Opposite}}{\text{Hypotenuse}} \qquad \text{cosine } \theta = \frac{\text{Adjacent}}{\text{Hypotenuse}} \qquad \text{tangent } \theta = \frac{\text{Opposite}}{\text{Adjacent}}$$

These ratios are usually abbreviated as:

$$\sin \theta = \frac{O}{H} \qquad \cos \theta = \frac{A}{H} \qquad \tan \theta = \frac{O}{A}$$

These abbreviated forms are also used on calculator keys.

Using your calculator

You will need to use a calculator to find trigonometric ratios.

Different calculators work in different ways, so make sure you know how to use your model.

Angles are not always measured in degrees. Sometimes radians or grads are used instead. You do not need to learn about those in your IGCSE course. Calculators can be set to operate in any of these three units, so make sure your calculator is operating in degrees.

Use your calculator to find the sine of 60 degrees.

You will probably press the keys sin 6 0 = in that order, but it might be different on your calculator.

The answer should be 0.8660…

3 cos 57° is a short way of writing 3 × cos 57°.

On most calculators you do not need to use the × button and you can just press the keys in the way it is written: 3 cos 5 7 =

Check to see whether your calculator works this way.

The answer should be 1.63.

26.2

Example 3

Find the value of 5.6 sin 30°.

This means 5.6 × sine of 30 degrees.

Remember that you may not need to press the × button.

5.6 sin 30° = 2.8

EXERCISE 26C

1 Find these values, rounding your answers to 3 significant figures.

   a sin 43°   **b** sin 56°   **c** sin 67.2°   **d** sin 90°

2 Find these values, rounding your answers to 3 significant figures.

   a cos 43°   **b** cos 56°   **c** cos 67.2°   **d** cos 90°

3 **a** **i** What is sin 35°?   **ii** What is cos 55°?

   b **i** What is sin 12°?   **ii** What is cos 78°?

   c **i** What is cos 67°?   **ii** What is sin 23°?

   d What connects the values in parts **a**, **b** and **c**?

   e Copy and complete these sentences.

      i sin 15° is the same as cos …

      ii cos 82° is the same as sin …

      iii sin x is the same as cos …

4 Use your calculator to work out the value of each ratio.

   a tan 43°   **b** tan 56°   **c** tan 67.2°   **d** tan 90°

   e tan 45°   **f** tan 20°   **g** tan 22°   **h** tan 0°

5 What is so different about tan compared with both sin and cos?

6 Use your calculator to work out the value of each ratio.

   a 4 sin 63°   **b** 7 tan 52°   **c** 5 tan 80°   **d** 9 cos 8°

7 Use your calculator to work out the values of the these ratios.

   a $\dfrac{5}{\sin 63°}$ **b** $\dfrac{6}{\cos 32°}$ **c** $\dfrac{3}{\tan 64°}$ **d** $\dfrac{7}{\tan 42°}$

8 Calculate sin x, cos x, and tan x for each triangle. Leave your answers as fractions.

   a   **b**   **c**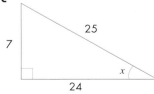

Chapter 26: Trigonometry   453

26.3 Calculating angles

What angle has a cosine of 0.6? You can use a calculator to find out.

'The angle with a cosine of 0.6' is written as $\cos^{-1} 0.6$ and is called the '**inverse** cosine of 0.6'.

Find out where $\cos^{-1}$ is on your calculator.

You will probably find it on the same key as cos, but you will need to press SHIFT or INV or 2ndF first.

Look to see if $\cos^{-1}$ is written above the cos key.

Check that $\cos^{-1} 0.6 = 53.1301... = 53.1°$ (1 decimal place)

Check that $\cos 53.1° = 0.600$ (3 decimal places)

Check that you can find the inverse sine and the inverse tangent in the same way.

> **Example 4**
> What angle has a sine of $\frac{3}{8}$?
>
> You need to find $\sin^{-1} \frac{3}{8}$.
>
> You could use the fraction button on your calculator or you could calculate $\sin^{-1} (3 \div 8)$.
>
> If you use the fraction key you may not need brackets, or your calculator may put them in automatically.
>
> Try to do it in both of these ways and then use whichever method you prefer.
>
> The answer should be 22.0°.

EXERCISE 26D

Use your calculator to find the answers to these questions. Give your answers to 1 decimal place.

1 What angles have these sines?
   a 0.5   b 0.785   c 0.64   d 0.877   e 0.999   f 0.707

2 What angles have these cosines?
   a 0.5   b 0.64   c 0.999   d 0.707   e 0.2   f 0.7

3 What angles have these tangents?
   a 0.6   b 0.38   c 0.895   d 1.05   e 2.67   f 4.38

4 What happens when you try to find the angle with a sine of 1.2? What is the largest value of sine you can put into your calculator without getting an error when you ask for the inverse sine? What is the smallest?

5 a i What angle has a sine of 0.3? (Keep the answer in your calculator memory.)
       ii What angle has a cosine of 0.3?
       iii Add the two accurate answers of parts **i** and **ii** together.
   b Will you always get the same answer to this question, whatever number you start with?

26.4 Using sine, cosine and tangent functions

Sine function

Remember $\sin \theta = \dfrac{\text{Opposite}}{\text{Hypotenuse}}$

You can use the **sine** ratio to calculate the lengths of sides and angles in right-angled triangles.

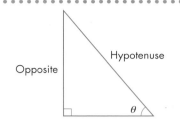

Example 5
Find the size of angle θ, given that the opposite side is 7 cm and the hypotenuse is 10 cm.

Draw a diagram. (This is an essential step.)

From the information given, use sine.

$\sin \theta = \dfrac{O}{H} = \dfrac{7}{10} = 0.7$

What angle has a sine of 0.7?

To find out, use the inverse sine function on your calculator.

$\sin^{-1} 0.7 = 44.4°$ (1 decimal place)

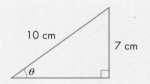

Example 6
Find the length of the side marked a in this triangle.

Side a is the opposite side, with 12 cm as the hypotenuse, so use sine.

$\sin \theta \ = \dfrac{O}{H}$

$\sin 35° \ = \dfrac{a}{12}$

So $a = 12 \sin 35° = 6.88$ cm (3 significant figures)

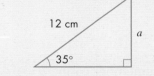

Example 7
Find the length of the hypotenuse, h, in this triangle.

Note that although the angle is in the other corner, the opposite side is again given. So use sine.

$\sin \theta \ = \dfrac{O}{H}$

$\sin 52° \ = \dfrac{8}{h}$

So $h = \dfrac{8}{\sin 52°} = 10.2$ cm (3 significant figures)

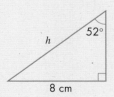

Chapter 26: Trigonometry

EXERCISE 26E

1 Find the size of the angle marked *x* in each of these triangles.

a
b
c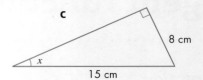

2 Find the length of the side marked *x* in each of these triangles.

a
b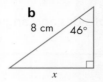
c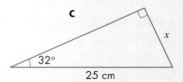

3 Find the length of the side marked *x* in each of these triangles.

a
b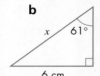
c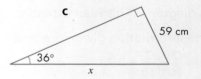

4 Find the length of the side marked *x* in each of these triangles.

a
b
c
d

Cosine function

Remember cosine $\theta = \dfrac{\text{Adjacent}}{\text{Hypotenuse}}$

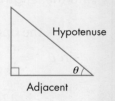

You can use the **cosine** ratio to calculate the lengths of sides and angles in right-angled triangles.

Example 8

Find the size of angle θ, given that the adjacent side is 5 cm and the hypotenuse is 12 cm.

Draw a diagram. (This is an essential step.)

From the information given, use cosine.

$\cos \theta = \dfrac{A}{H} = \dfrac{5}{12}$

What angle has a cosine of $\dfrac{5}{12}$?

To find out, use the inverse cosine function on your calculator.

$\cos^{-1} = 65.4°$ (1 decimal place)

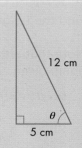

26.4

Example 9

Find the length of the hypotenuse, h, in this triangle.

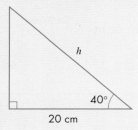

The adjacent side is given. So use cosine.

$\cos \theta = \dfrac{A}{H}$

$\cos 40° = \dfrac{20}{h}$

So $h = \dfrac{20}{\cos 40°} = 26.1$ cm (3 significant figures)

EXERCISE 26F

1 Find the size of the angle marked x in each of these triangles.

a
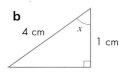

b

c
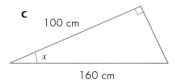

2 Find the length of the side marked x in each of these triangles.

a
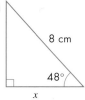

b
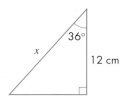

c

d

3 Find the value of x in each of these triangles.

a
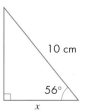

b
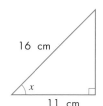

c
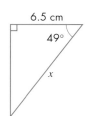

d
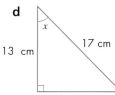

Chapter 26: Trigonometry

Tangent function

Remember tangent $\theta = \dfrac{\text{Opposite}}{\text{Adjacent}}$

You can use the **tangent** ratio to calculate the lengths of sides and angles in right-angled triangles.

Example 10

Find the length of the side marked x in this triangle.

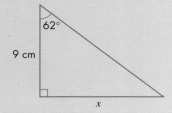

The side marked x is the opposite side, with 9 cm as the adjacent side, so use tangent.

$\tan \theta = \dfrac{O}{A}$

$\tan 62° = \dfrac{x}{9}$

So $x = 9 \tan 62° = 16.9$ cm (3 significant figures)

EXERCISE 26G

1 Find the size of the angle marked x in each of these triangles.

a 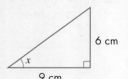   b 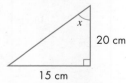   c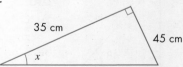

2 Find the length of the side marked x in each of these triangles.

a 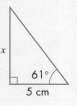   b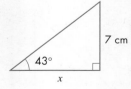

c 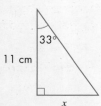   d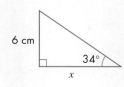

Chapter 26 . Topic 5

3 Find the value of x in each of these triangles.

a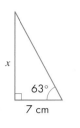

b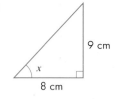

c

d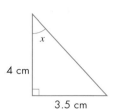

26.5 Which ratio to use

The difficulty with any trigonometric problem is knowing which ratio to use to solve it.

These examples show you how to decide which ratio you need in any given situation.

> **Example 11**
>
> Find the length of the side marked x in this triangle.
>
> **Step 1** Identify what information is given and what needs to be found. Namely, x is opposite the angle and 16 cm is the hypotenuse.
>
> **Step 2** Decide which ratio to use. Only one ratio uses opposite and hypotenuse: **sine**.
>
> **Step 3** Remember $\sin \theta = \dfrac{O}{H}$
>
> **Step 4** Put in the numbers and letters: $\sin 37° = \dfrac{x}{16}$
>
> **Step 5** Rearrange the equation and work out the answer:
> $x = 16 \sin 37° = 9.629040371$ cm
>
> **Step 6** Give the answer to an appropriate degree of accuracy:
> $x = 9.63$ cm (3 significant figures)

There is no need to write down every step as in Example 11. Step 1 can be done by marking the triangle. Steps 2 and 3 can be done in your head. Steps 4 to 6 are what you write down.

Remember that you must always show evidence of your working. Any reasonable attempt at identifying the sides and using a ratio will probably show that you undertand the method, but only if the fraction is the right way round.

The next examples are set out in a way that requires the *minimum* amount of working but gets *maximum* results.

Chapter 26: Trigonometry   459

Example 12

Find the length of the side marked x in this triangle.

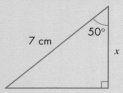

Mark on the triangle the side you know (H) and the side you want to find (A).

Recognise it is a **cosine** problem because you have A and H.

So $\cos 50° = \dfrac{x}{7}$

$x = 7 \cos 50° = 4.50$ cm (3 significant figures)

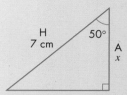

Example 13

Find the size of the angle marked x in this triangle.

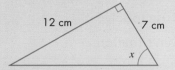

Mark on the triangle the sides you know.

Recognise it is a **tangent** problem because you have O and A.

So $\tan x = \dfrac{12}{7}$

$x = \tan^{-1} \dfrac{12}{7} = 59.7°$ (1 decimal place)

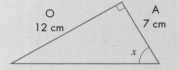

EXERCISE 26H

1 Find the length marked x in each of these triangles.

a
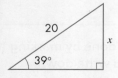

b
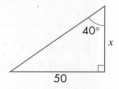

c
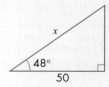

d
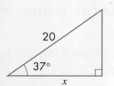

e
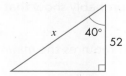

f
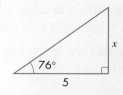

2 Find the size of the angle marked x in each of these triangles.

a 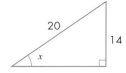 b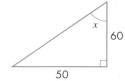

c 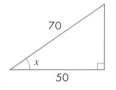 d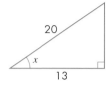

e 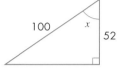 f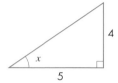

3 Find the value of the angle or length marked x in each of these triangles.

a 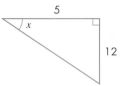 b  c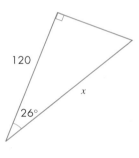

d 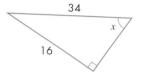 e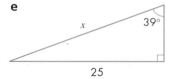

f 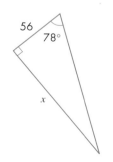 g 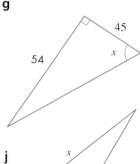 h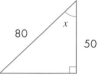

i  j

Chapter 26: Trigonometry

Chapter 26 . Topic 6

4 a How does this diagram show that $\tan \theta = \dfrac{\sin \theta}{\cos \theta}$?

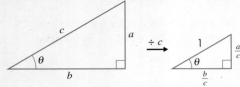

b How does the diagram show that $(\sin \theta)^2 + (\cos \theta)^2 = 1$?

c Choose a value for θ and check the two results in parts **a** and **b** are true.

26.6 Applications of trigonometric ratios E

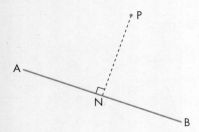

P is a point and AB is a straight line.

The shortest distance from P to AB is the length of PN where PN is perpendicular to AB.

The angle PNA is a right angle.

Example 14

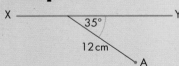

Find the shortest distance from the point A to the line XY.

Draw a right-angled triangle.

You want to find x.

$\sin 35° = \dfrac{x}{12}$ and so $x = 12 \sin 35° = 6.88$ cm (to 3 s.f.)

---

462   26.6 Applications of trigonometric ratios

26.6

When you look *up* at an aircraft in the sky, the angle through which your line of sight turns, from looking straight ahead (the horizontal), is called the **angle of elevation**.

When you are standing on a high point and look *down* at a boat, the angle through which your line of sight turns, from looking straight ahead (the horizontal), is called the **angle of depression**.

Example 15

From the top of a vertical cliff, 100 m high, Ali sees a boat out at sea. The angle of depression from Ali to the boat is 42°. How far from the base of the cliff is the boat?

The diagram of the situation is shown in figure **i**.

From this, you get the triangle shown in figure **ii**.

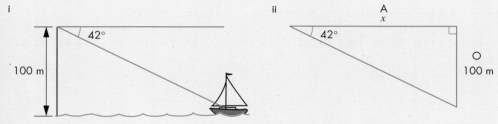

From figure **ii**, you see that this is a tangent problem.

So $\tan 42° = \dfrac{100}{x}$

$x = \dfrac{100}{\tan 42°}$

$= 111$ m (3 significant figures)

EXERCISE 26I

In these questions, give any answers involving angles to the nearest degree.

 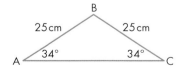

ABC is an isosceles triangle.

Find the shortest distance from **B** to **AC**.

2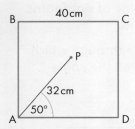

ABCD is a square. Each side is 40 cm.

P is a point inside the square.

Calculate the distance from P to

a AD

b AB

c CD

3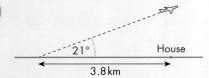

A plane takes off and rises at an angle of 21° to the ground.

It flies over a house 3.8 km away from the point where the plane took off.

Find the closest distance between the plane and the house. Give your answer in kilometres rounded to 2 d.p.

4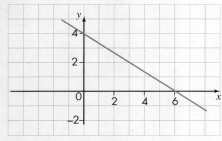

This is a graph of the line with equation $2x + 3y = 12$

Find the shortest distance from the origin to the line.

5 Erik sees an aircraft in the sky. The aircraft is at a horizontal distance of 25 km from Erik. The angle of elevation is 22°.

How high is the aircraft?

6 An aircraft is flying at an altitude of 4000 m and is 10 km from the airport. If a passenger can see the airport, what is the angle of depression?

7 A man standing 200 m from the base of a television transmitter looks at the top of it and notices that the angle of elevation of the top is 65°.

How high is the tower?

8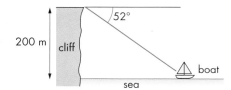

a From the top of a vertical cliff, 200 m high, a boat has an angle of depression of 52°. How far from the base of the cliff is the boat?

b The boat now sails away from the cliff so that the distance is doubled. Does that mean that the angle of depression is halved?

Give a reason for your answer.

9 From a boat, the angle of elevation of the foot of a lighthouse on the edge of a cliff is 34°.

a If the cliff is 150 m high, how far from the base of the cliff is the boat?

b If the lighthouse is 50 m high, what would be the angle of elevation of the top of the lighthouse from the boat?

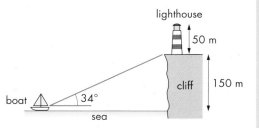

10 A bird flies from the top of a 12 m tall tree, at an angle of depression of 34°, to catch a worm on the ground.

a How far does the bird actually fly?

b How far was the worm from the base of the tree?

11 Sunil wants to work out the height of a building. He stands about 50 m away from the building. The angle of elevation from Sunil to the top of the building is about 15°. How tall is the building?

12 The top of a ski run is 100 m above the finishing line. The run is 300 m long. What is the angle of depression of the ski run?

13 Nessie and Cara are standing on opposite sides of a tree.

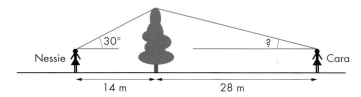

Nessie is 14 m away and the angle of elevation of the top of the tree is 30°.

Cara is 28 m away. She says the angle of elevation for her must be 15° because she is twice as far away.

Is she correct?

What do you think the angle of elevation is?

26.7 Problems in three dimensions

To find the value of an angle or side in a three-dimensional figure you need to find a right-angled triangle in the figure. This triangle also has to include two known values that you can use in the calculation.

You must redraw this triangle separately as a plain, right-angled triangle. Add the known values and the unknown value you want to find. Then use the trigonometric ratios and Pythagoras' theorem to solve the problem.

Example 16

A, B and C are three points at ground level. They are in the same horizontal plane. C is 50 km east of B. B is north of A. C is on a bearing of 050° from A.

An aircraft, flying east, passes over B and over C at the same height. When it passes over B, the angle of elevation from A is 12°. Find the angle of elevation of the aircraft from A when it is over C.

First, draw a diagram containing all the known information.

Next, use the right-angled triangle ABC to calculate AB and AC.

$AB = \dfrac{50}{\tan 50°} = 41.95$ km   (4 significant figures)

$AC = \dfrac{50}{\sin 50°} = 65.27$ km   (4 significant figures)

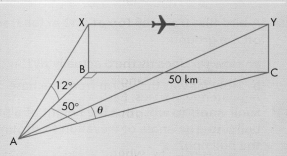

Then use the right-angled triangle ABX to calculate BX, and hence CY.

BX = 41.95 tan 12° = 8.917 km   (4 significant figures)

Finally, use the right-angled triangle ACY to calculate the required angle of elevation, θ.

$\tan \theta = \dfrac{8.917}{65.27} = 0.1366$

$\Rightarrow \theta = \tan^{-1} 0.1366 = 7.8°$ (1 decimal place)

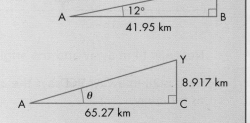

Always write down intermediate working values to at least 4 significant figures, or use the answer on your calculator display to avoid inaccuracies in the final answer.

EXERCISE 26J

1

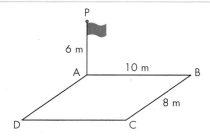

A vertical flagpole AP stands at the corner of a rectangular courtyard ABCD.

Calculate the angle of elevation of P from C.

2 The diagram shows a pyramid. The base is a horizontal rectangle ABCD, 20 cm by 15 cm. The length of each sloping edge is 24 cm. The apex, V, is over the centre of the rectangular base. Calculate:

a the length of the diagonal AC
b the size of the angle VAC
c the height of the pyramid.

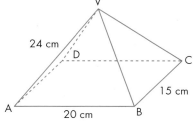

3 The diagram shows the roof of a building. The base ABCD is a horizontal rectangle 7 m by 4 m. The triangular ends are equilateral triangles. Each side of the roof is an isosceles trapezium. The length of the top of the roof, EF, is 5 m. Calculate:

a the length EM, where M is the midpoint of AB
b the size of angle EBC
c the size of the angle between EM and the base ABCD.

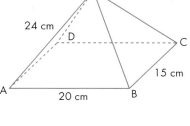

4 ABCD is a vertical rectangular plane. EDC is a horizontal triangular plane. Angle CDE = 90°, AB = 10 cm, BC = 4 cm and ED = 9 cm. Calculate:

a angle AED
b angle DEC
c EC
d angle BEC.

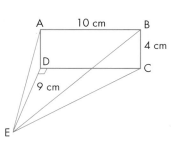

5 In the diagram, XABCD is a pyramid with a rectangular base.

Revina says that the angle between the edge XD and the base ABCD is 56.3°.

Work out the correct answer to show that Revina is wrong.

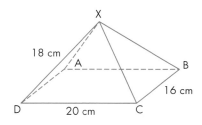

26.8 Sine and cosine of obtuse angles E

So far you have only used sines and cosines in right-angled triangles. A calculator also gives the sine and cosine of **obtuse angles**. Check that sin 115° = 0.906 and cos 115° = −0.423.

You cannot have a right-angled triangle with an obtuse angle so how can you calculate sin and cos of an obtuse angle?

Imagine a rod OP lying on the x-axis as shown.

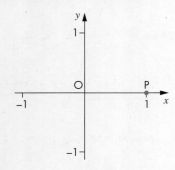

It rotates anticlockwise about the origin O.

ONP is a right-angled triangle.

The hypotenuse OP = 1

If the angle that OP makes with the x-axis is θ:

The adjacent side ON = OP × cos θ = 1 × cos θ = cos θ

So the x-coordinate of P is cos θ

Similarly:

The opposite side NP = OP × sin θ = 1 × sin θ = sin θ

So the y-coordinate of P is sin θ
Therefore the coordinates of P are (cos θ, sin θ).

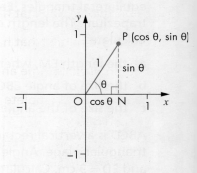

Imagine that OP continues to rotate so that angle θ becomes obtuse.
You can still define cos θ and sin θ as the coordinates of P.
You can see from the diagram that the x-coordinate for P must now be negative.
The y-coordinate is still positive.

For example, if θ = 115°, the diagram looks like this:

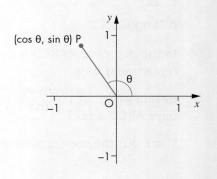

You can see that the angle adjoining θ on the x-axis must have the same sine as θ. When adjacent angles add up to 180° their sines are the same. So:

If angle θ is obtuse then $\sin \theta = \sin (180° - \theta)$

Their cosines have the same numeric value but the cosine of the obtuse angle is negative:

If angle θ is obtuse $\cos \theta = -\cos (180° - \theta)$

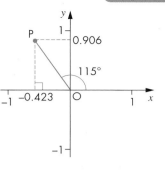

For example, if $\theta = 115°$ then $180° - \theta = 65°$
Sin 115° = sin 65° = 0.906
Cos 115° = − cos 65° = − 0.423

EXERCISE 26K

1 **a** Copy and complete this table.

Angle	10°	30°	50°	85°	90°	95°	130°	150°	170°
Sine		0.5			1				

b Draw a graph of $y = \sin x$ for $0 \leq x \leq 180°$.

c Describe the symmetry of the graph.

d Find two examples from the table to show that if two angles add up to 180 degrees they have the same sine.

2 Find two angles that have a sine of 0.5.

3 Find two angles that have a sine of 0.72.

4 x is an obtuse angle and $\sin x = 0.84$. Find the value of x.

5 **a** Copy and complete this table.

Angle	15°	35°	60°	80°	90°	100°	120°	145°	165°
Cosine									

b Draw a graph of $y = \cos x$ for $0 \leq x \leq 180°$.

c Describe the symmetry of the graph.

6 Find the size of each of these angles.

   a $\cos^{-1} 0.85$ **b** $\cos^{-1}(-0.85)$
   c $\cos^{-1}(-0.5)$ **d** $\cos^{-1} 0$
   e $\cos^{-1} 0.125$ **f** $\cos^{-1}(-0.125)$

7 Solve these equations where $0 \leq x \leq 180°$. Give your answers to the nearest degree. There may be more than one solution.

   a $\cos x = 0.6$ **b** $\cos x = -0.25$
   c $\sin x = \dfrac{3}{4}$ **d** $\sin x = 1$
   e $\cos x = 0$ **f** $\sin x = 0.95$
   g $\sin x = 2$ **h** $\sin x = \cos x$

26.9 The sine rule and the cosine rule

Any triangle has six measurements: three sides and three angles. To find any unknown angles or sides), you need to know at least three of the measurements. Any combination of three measurements – except that of all three angles – is enough to work out the rest.

When you need to find the value of a side or an angle in a triangle that contains no right angle, you can use one of two rules, depending on what you know about the triangle. These are the **sine rule** and the **cosine rule**.

The sine rule

Take a triangle ABC and draw the perpendicular from A to the opposite side BC.

From right-angled triangle ADB, $h = c \sin B$

From right-angled triangle ADC, $h = b \sin C$

Therefore,

$$c \sin B = b \sin C$$

which can be rearranged to give:

$$\frac{c}{\sin C} = \frac{b}{\sin B}$$

By drawing a perpendicular from each of the other two vertices to the opposite side (or by algebraic symmetry), you see that:

$$\frac{a}{\sin A} = \frac{c}{\sin C} \quad \text{and that} \quad \frac{a}{\sin A} = \frac{b}{\sin B}$$

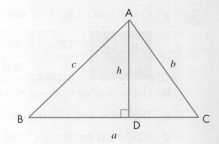

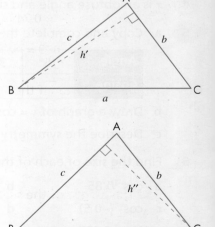

These are usually combined in the form:

$$\frac{a}{\sin A} = \frac{b}{\sin B} = \frac{c}{\sin C}$$

which can be inverted to give:

$$\frac{\sin A}{a} = \frac{\sin B}{b} = \frac{\sin C}{c}$$

Remember, when using the sine rule: take each side in turn, divide it by the sine of the angle opposite and then equate the results.

Note:

- When you are calculating a *side*, use the rule with the *sides on top*.
- When you are calculating an *angle*, use the rule with the *sines on top*.

Example 17

In triangle ABC, find the value of x.

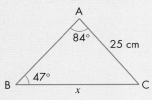

Use the sine rule with sides on top, which gives:

$$\frac{x}{\sin 84°} = \frac{25}{\sin 47°}$$

$$\Rightarrow x = \frac{25 \sin 84°}{\sin 47°} = 34.0 \text{ cm} \quad \text{(3 significant figures)}$$

Example 18

In triangle ABC, find the value of the acute angle x.

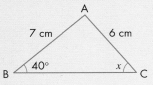

Use the sine rule with sines on top, which gives:

$$\frac{\sin x}{7} = \frac{\sin 40°}{6}$$

$$\Rightarrow \sin x = \frac{7 \sin 40°}{6} = 0.7499$$

$$\Rightarrow x = \sin^{-1} 0.7499 = 48.6° \quad \text{(3 significant figures)}$$

The sine rule works even if the triangle has an obtuse angle, because you can find the sine of an obtuse angle.

EXERCISE 26L

1 Find the length of the side labelled x in each of these triangles.

a

b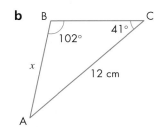

c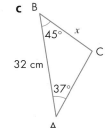

...e size of the angle labelled x in each of these triangles.

a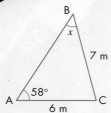
b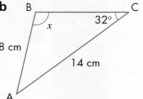
c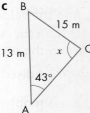

3 To find the height of a tower standing on a small hill, Maria made some measurements (see diagram).

From a point B, the angle of elevation of C is 20°, the angle of elevation of A is 50° and the distance BC is 25 m.

a Calculate these angles.
  i ABC
  ii BAC
b Using the sine rule and triangle ABC, calculate the height h of the tower.

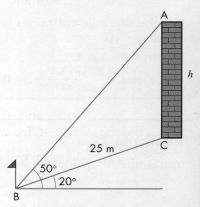

4 Use the information on this sketch to calculate the width, w, of the river.

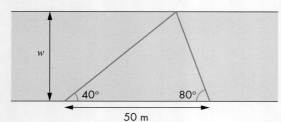

5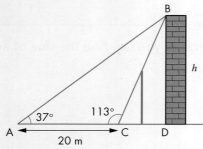

An old building is unsafe and is protected by a fence. A company is going to demolish the building and has to work out its height BD, marked h on the diagram.

Calculate the value of h, using the given information.

6 A mass is hung from a horizontal beam using two strings. The shorter string is 2.5 m long and makes an angle of 71° with the horizontal. The longer string makes an angle of 43° with the horizontal. What is the length of the longer string?

7 A rescue helicopter is based at an airfield at A.

It is sent out to rescue a man from a mountain at M, due north of A.

The helicopter then flies on a bearing of 145° to a hospital at H, as shown on the diagram.

Calculate the direct distance from the mountain to the hospital.

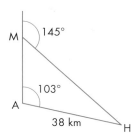

8 Triangle ABC has an obtuse angle at B.

Calculate the size of angle ABC.

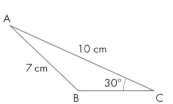

The cosine rule

Take the triangle, shown on the right, where D is the foot of the perpendicular to BC from A. The length of BD is x.

Using Pythagoras' theorem on triangle BDA:

$h^2 = c^2 - x^2$

Using Pythagoras' theorem on triangle ADC:

$h^2 = b^2 - (a - x)^2$

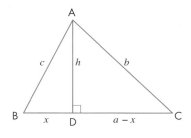

Therefore,

$c^2 - x^2 = b^2 - (a - x)^2$

$c^2 - x^2 = b^2 - a^2 + 2ax - x^2$

$\Rightarrow c^2 = b^2 - a^2 + 2ax$

From triangle BDA, $x = c \cos B$.

So:

$c^2 = b^2 - a^2 + 2ac \cos B$

Rearranging gives:

$b^2 = a^2 + c^2 - 2ac \cos B$

By algebraic symmetry:

$a^2 = b^2 + c^2 - 2bc \cos A$ and $c^2 = a^2 + b^2 - 2ab \cos C$

This is the cosine rule, which can be best remembered from the diagram on the right, where:

$a^2 = b^2 + c^2 - 2bc \cos A$

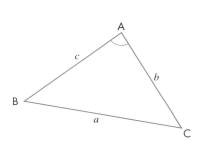

Note the symmetry of the rule and how the rule works using two adjacent sides and the angle between them (the **included angle**).

The formula can be rearranged to find any of the three angles.

$\cos A = \dfrac{b^2 + c^2 - a^2}{2bc}$ $\cos B = \dfrac{a^2 + c^2 - b^2}{2ac}$ $\cos C = \dfrac{a^2 + b^2 - c^2}{2ab}$

Example 19

Find the value of x in this triangle.

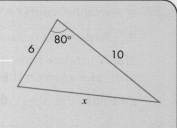

By the cosine rule:

$x^2 = 6^2 + 10^2 - 2 \times 6 \times 10 \times \cos 80°$

$x^2 = 115.16$

$\Rightarrow x = 10.7$ (3 significant figures)

Example 20

Find the value of x in this triangle.

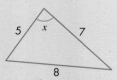

By the cosine rule:

$\cos x = \dfrac{5^2 + 7^2 - 8^2}{2 \times 5 \times 7} = 0.1428$

$\Rightarrow x = 81.8°$ (3 significant figures)

It is possible to find the cosine of an angle that is greater than 90°. For example, $\cos 120° = -0.5$.

EXERCISE 26M

1 Find the length of the side marked x in each of these triangles.

a 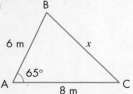 **b** 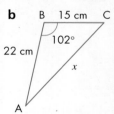 **c**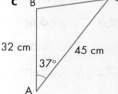

2 Find the angle x in each of these triangles.

a 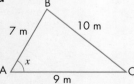 **b** 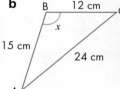 **c**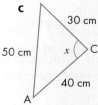

   d Explain the significance of the answer to part **c**.

3 In triangle ABC, AB = 5 cm, BC = 6 cm and angle ABC = 55°. Find AC.

4 A triangle has two sides of length 40 cm and an angle of 110°. Work out the length of the third side of the triangle.

5 The diagram shows a trapezium ABCD. AB = 6.7 cm, AD = 7.2 cm, CB = 9.3 cm and angle DAB = 100°.

Calculate:

a the length DB b angle DBA
c angle DBC d the length DC

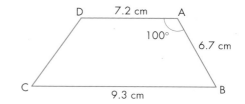

6 A ship sails from a port on a bearing of 050° for 50 km then turns on a bearing of 150° for 40 km. A crewman is taken ill, so the ship drops anchor. What course and distance should a rescue helicopter from the port fly to reach the ship in the shortest possible time?

7 The three sides of a triangle are given as $3a$, $5a$ and $7a$. Calculate the smallest angle in the triangle.

8 Two ships, X and Y, leave a port at 9 am.

Ship X travels at an average speed of 20 km/h on a bearing of 075° from the port.

Ship Y travels at an average speed of 25 km/h on a bearing of 130° from the port.

Calculate the distance between the two ships at 11 am.

9 Calculate the size of the largest angle in the triangle ABC.

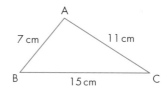

Choosing the correct rule

When finding the unknown sides and angles in a triangle, there are several situations that can occur.

Two sides and the included angle

1 Use the cosine rule to find the third side.

2 Use the sine rule to find either of the other angles.

3 Use the sum of the angles in a triangle to find the third angle.

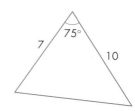

Two angles and a side

1 Use the sum of the angles in a triangle to find the third angle.

2, 3 Use the sine rule to find the other two sides.

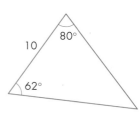

Three sides

1 Use the cosine rule to find one angle.

2 Use the sine rule to find another angle.

3 Use the sum of the angles in a triangle to find the third angle.

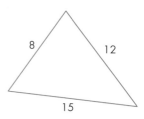

Chapter 26: Trigonometry

EXERCISE 26N

1 Find the length or angle x in each of these triangles.

a 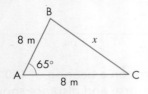   b 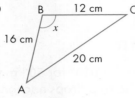   c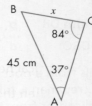

d 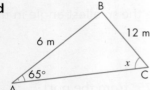   e 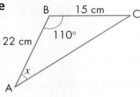   f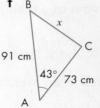

g 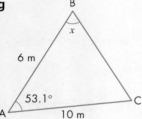   h 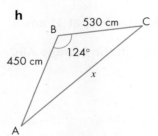   i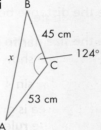

2 The hands of a clock have lengths 3 cm and 5 cm. Find the distance between the tips of the hands at 4 o'clock.

3 A helicopter is seen hovering at a point which is in the same vertical plane as two towns, X and F, which are on the same level. Its distances from X and F are 8.5 km and 12 km respectively. The angle of elevation of the helicopter when observed from F is 43°. Calculate the distance between the two towns.

4 Triangle ABC has sides with lengths a, b and c, as shown in the diagram.

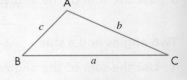

  a What can you say about the angle BAC, if $b^2 + c^2 - a^2 = 0$?
  b What can you say about the angle BAC, if $b^2 + c^2 - a^2 \geq 0$?
  c What can you say about the angle BAC, if $b^2 + c^2 - a^2 \leq 0$?

5 The diagram shows a sketch of a field ABCD.

A farmer wants to put a new fence round the perimeter of the field.

Calculate the perimeter of the field.

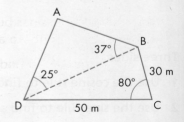

26.9 The sine rule and the cosine rule

26.10 Using sine to find the area of a triangle E

In triangle ABC, the vertical height is BD and the base is AC.

Let $BD = h$ and $AC = b$, then the **area** of the triangle is given by:

$\frac{1}{2} \times AC \times BD = \frac{1}{2}bh$

However, in triangle BCD:

$h = BC \sin C = a \sin C$

where $BC = a$.

Substituting into $\frac{1}{2}bh$ gives:

$\frac{1}{2}b \times (a \sin C) = \frac{1}{2}ab \sin C$

as the area of the triangle.

By taking the perpendicular from A to its opposite side BC, and the perpendicular from C to its opposite side AB, you can show that the area of the triangle is also given by:

$\frac{1}{2}ac \sin B$ and $\frac{1}{2}bc \sin A$

Note the pattern: the area is given by the product of two sides multiplied by the sine of the included angle. This is the **area sine rule**. Starting from any of the three forms, you can use the sine rule to establish the other two.

Example 21
Find the area of triangle ABC.

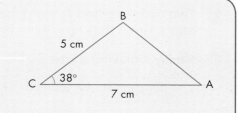

Area $= \frac{1}{2}ab \sin C$

Area $= \frac{1}{2} \times 5 \times 7 \times \sin 38°$

$= 10.8 \text{ cm}^2$ (3 significant figures)

Example 22
Find the area of triangle ABC.

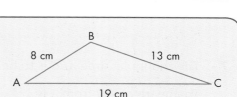

You have all three sides but no angle. So first you must find an angle in order to apply the area sine rule.

Use the cosine rule to find angle C.

$\cos C = \dfrac{a^2 + b^2 - c^2}{2ab} = \dfrac{13^2 + 19^2 - 8^2}{2 \times 13 \times 19} = 0.9433 \quad \Rightarrow C = \cos^{-1} 0.9433 = 19.4°$

(Keep the exact value in your calculator memory.)

Chapter 26: Trigonometry

Now apply the area sine rule.

$\frac{1}{2}ab \sin C = \frac{1}{2} \times 13 \times 19 \times \sin 19.4°$

$= 41.0 \text{ cm}^2$ (3 significant figures)

EXERCISE 26O

1 Find the area of each of these triangles.

    a Triangle *ABC* with *BC* = 7 cm, *AC* = 8 cm and angle *ACB* = 59°

    b Triangle *ABC* with angle *BAC* = 86°, *AC* = 6.7 cm and *AB* = 8 cm

    c Triangle *PQR* with *QR* = 27 cm, *PR* = 19 cm and angle *QRP* = 109°

    d Triangle *XYZ* with *XY* = 231 cm, *XZ* = 191 cm and angle *YXZ* = 73°

    e Triangle *LMN* with *LN* = 63 cm, *LM* = 39 cm and angle *NLM* = 85°

2 The area of triangle *ABC* is 27 cm². If *BC* = 14 cm and angle *BCA* = 115°, find the length of *AC*.

3 The area of triangle *LMN* is 113 cm², *LM* = 16 cm and *MN* = 21 cm. Angle *LMN* is acute. Calculate these angles.

    a Angle *LMN*

    b Angle *MNL*

4 A board is in the shape of a triangle with sides 60 cm, 70 cm and 80 cm. Find the area of the board.

5 Two circles, centres *P* and *Q*, have radii of 6 cm and 7 cm respectively. The circles intersect at *X* and *Y*. Given that *PQ* = 9 cm, find the area of triangle *PXQ*.

6 Sanjay is making a kite.

The diagram shows a sketch of his kite.

Calculate the area of the material required to make the kite.

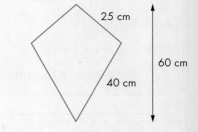

7 The triangular area *ABC* is to be made a national park.

The bearing of *B* from *A* is 324°.

The bearing of *C* from *A* is 42°.

Calculate the area of the park.

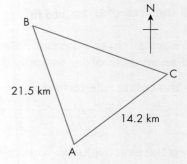

Chapter 26 . Topic 11

26 . 11

8 The diagram shows the dimensions of a four-sided field.

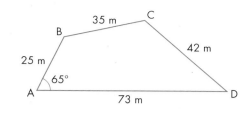

  a Show that the length of the diagonal BD is 66 metres to the nearest metre.

  b Calculate the size of angle C.

  c Calculate the area of the field.

9

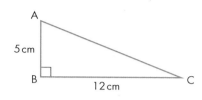
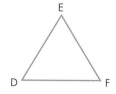

Triangle ABC is right-angled.

Triangle DEF is isosceles.

They have the same perimeter.

Calculate the area of triangle DEF.

26.11 Sine, cosine and tangent of any angle E

In Section 26.8 you used a rod of length 1 on a coordinate grid to find the sine and cosine of obtuse angles.

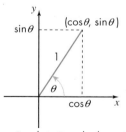
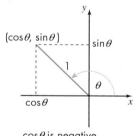

$\cos\theta$ and $\sin\theta$ are both positive

$\cos\theta$ is negative
$\sin\theta$ is positive

You can use the same idea to find the sine and cosine of a reflex angle. That is an angle bigger than 180°.

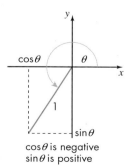
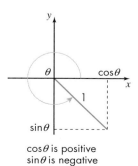

$\cos\theta$ is negative
$\sin\theta$ is positive

$\cos\theta$ is positive
$\sin\theta$ is negative

Chapter 26: Trigonometry   **479**

These graphs show the sine and cosine of any angle between 0° and 360°

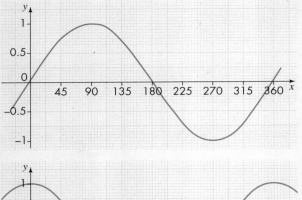

$y = \sin x°$

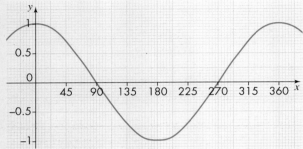

$y = \cos x°$

Notice the symmetry of the graphs.

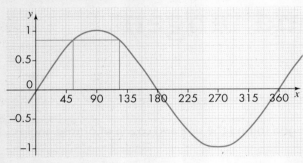

$y = \sin x°$

180 − 55 = 125 and so 55° and 125° have the same sine.

sin 55° = sin 125° = 0.819 to 3 d.p.

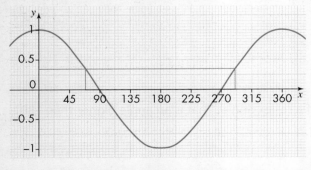

$y = \cos x°$

360 − 70 = 290 and so 70° and 290° have the same cosine.

cos 70° = cos 290° = 0.342 to 3 d.p.

26.11

Here are some more examples of the symmetry of the sine graph:

- 10° and 180 − 10 = 170° have the same sine.
- 50° and 180 − 50 = 130° have the same sine.
- 205° and 360 − 15 = 345° have the same sine.
- sin 30° = 0.5 and sin (180 + 30)° = sin 210° = −0.5

Here are some more examples of the symmetry of the cosine graph:

- 10° and 360 − 10 = 350° have the same cosine.
- 125° and 360 − 125 = 235° have the same cosine.
- cos 70° = 0.342 and cos (180 − 70)° = cos 110° = −0.342

You need to remember the shapes of these graphs and their symmetries.

Example 23

Solve these equations, giving your answers to the nearest degree.

a $\sin x = 0.3$ $\quad 0° \leqslant x \leqslant 360°$

b $\sin w = 0.3$ $\quad 0° \leqslant w \leqslant 360°$

c $2 \cos y + 1 = 0$ $\quad 0° \leqslant y \leqslant 360°$

a A calculator gives $\sin^{-1} 0.3 = 17°$. This is one answer.

Look at the symmetry of the sine graph.

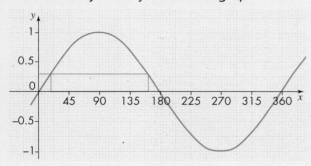

Another answer is 180 − 17 = 163°

There are two possible answers, $x = 17°$ or $163°$

b A calculator gives $\sin^{-1} (−0.3) = −17°$ but this is outside the interval $0° \leqslant w \leqslant 360°$

−17° is a *clockwise* turn of 17°. It is the same as an *anticlockwise* turn of 360 − 17 = 343°

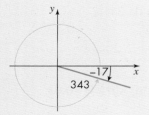

One solution is 343°.

Chapter 26: Trigonometry

To find another, look at the symmetry of the sine graph.

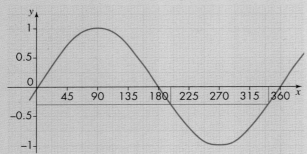

Another solution is 180 + 17 = 197°.

There are two possible answers, w = 197° or 343°

c Rearrange the equation $2 \cos y = -1$

  Divide by 2 $\cos y = -0.5$

A calculator gives $\cos^{-1}(-0.5) = 120°$ and this is one answer.

Look at the symmetry of the cosine graph.

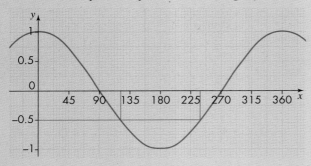

Another answer is 360 − 120 = 240°

There are two possible answers, y = 120° or 240°

EXERCISE 26P

1 Find an angle between 0° and 360° that has the same sine as:

   a 80°   **b** 146°   **c** 215°   **d** 306°

2 Find an angle between 0° and 360° that has the same cosine as:

   a 10°   **b** 125°   **c** 208°   **d** 311°

3 Solve these equations when $0° \leq x \leq 360°$

   Give your answers to the nearest degree.

   a $\sin x = 0.45$ **b** $\sin x = 0.83$ **c** $\sin x = -0.45$ **d** $\sin x = -0.83$

4 Solve these equations when $0° \leq x \leq 360°$

Give your answers to the nearest degree.

   a $\cos x = 0.8$ **b** $\cos x = -0.23$ **c** $\cos x = -0.92$ **d** $\cos x = 0.087$

5 Solve these equations when $0° \leq x \leq 360°$

   a $\sin x = 0.5$ **b** $\sin x = \dfrac{\sqrt{3}}{2}$ **c** $\sin x = -\dfrac{1}{\sqrt{2}}$ **d** $\sin x = -1$

6 Solve these equations when $0° \leq x \leq 360°$

   a $\cos x = -0.5$ **b** $\cos x = \dfrac{\sqrt{3}}{2}$ **c** $\cos x = \dfrac{1}{\sqrt{2}}$ **d** $\cos x = 0$

7 Solve these equations when $0° \leq x \leq 360°$

Round your answers to 1 d.p.

   a $\cos x = 0.05$ **b** $\sin x = 0.812$ **c** $\cos x + 0.65 = 0$ **d** $\sin x + 0.9 = 0.3$

8 Solve these equations when $0° \leq x \leq 360°$

Round your answers to 1 d.p.

   a $3 \sin x = 2$ **b** $5 \cos x - 4 = 0$ **c** $7 \sin x + 5 = 0$

9 Two different obtuse angles have the same sine. Find the sum of the two angles.

10 Solve the equation $4(\sin x)^2 = 1$ $0° \leq x \leq 360°$

11 Solve the equation $2(\cos x)^2 = 1$ $0° \leq x \leq 360°$

The tangent of any angle

In a right-angled triangle $\tan \theta = \dfrac{\text{Opposite side}}{\text{Adjacent side}}$

You know that in a right-angled triangle $\tan \theta = \dfrac{\text{Opposite side}}{\text{Adjacent side}}$

You can see from this diagram that $\tan \theta = \dfrac{\sin \theta}{\cos \theta}$

Here is a graph of $y = \tan x$ where $0° \leq x \leq 90°$

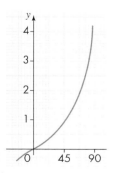

Notice that $\tan 45° = 0$

As the angle gets close to 90° the tangent becomes very large.

$\tan 90°$ is not defined because $\cos 90° = 0$

Chapter 26: Trigonometry

You can use $\tan \theta = \dfrac{\sin \theta}{\cos \theta}$ to define the tangent of any angle, not just acute ones.

Here is a graph of $y = \tan x$ where $0° \leqslant x \leqslant 180°$

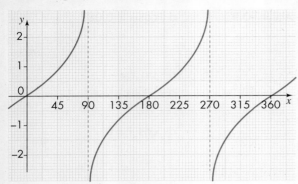

The graph is in three separate branches.

The lines $x = 90$ and $x = 270$ are asymptotes. The graph gets closer and closer to those lines as y gets larger.

The section from 180° to 360° is identical to the section from 0° to 180°

This means that tan 10° = tan 190° and tan 20° = tan 200° and so on.

> **Example 24**
>
> Solve these equations for $0° \leqslant x \leqslant 360°$. Give your answers to 1 d.p.
>
> **a** tan $x = 2$ **b** 4 tan $x + 3 = 0$
>
> ---
>
> **a** A calculator gives $\tan^{-1} 2 = 63.4$ so this is one answer.
>
> From the symmetry of the graph, another answer is 180 + 63.4 = 243.4°
>
> The two answers are $x = 63.4°$ and 243.4°
>
> **b** Rearrange as tan $x = -0.75$
>
> A calculator gives $\tan^{-1} (-0.75) = -36.9°$ but this is outside the interval $0° \leqslant x \leqslant 360°$.
>
> However −36.9° is a clockwise turn of 36.9°
>
> It is the same as an anticlockwise turn of 360 − 36.9 = 323.1°
>
> One answer is 323.1°
>
> By the symmetry of the tangent graph, another answer is 323.1 − 180 = 143.1°
>
> The two answers are $x = 143.1°$ and 323.1°

EXERCISE 26Q

1 Find the angles between 0° and 360° that have the same tangent as

   a 35°   **b** 85°   **c** 118°   **d** 200°   **e** 243°   **f** 337°

2 Solve these equations when $0° \leqslant x \leqslant 360°$.

   a tan $x = 1$ **b** tan $x = -1$ **c** tan $x = \sqrt{3}$ **d** tan $x = -\sqrt{3}$

3 Solve these equations when $0° \leq x \leq 360°$. Give your answers to 1 d.p.

   a $\tan x = 0.2$ **b** $\tan x = 5$ **c** $\tan x = -0.35$ **d** $\tan x = -4.17$

4 The sine of an angle is 0.6 and the cosine of the angle is -0.4

   Find the tangent of the angle.

5 Solve these equations when $0° \leq x \leq 360°$. Give your answers to 1 d.p.

   a $8 \tan x = 3$ **b** $3 \tan x = 8$ **c** $25 \tan x + 18 = 0$ **d** $9 \tan x + 40 = 0$

6 Solve these equations $(\tan x)^2 = 1$ and $0° \leq x \leq 360°$.

   a $(\tan x)^2 = 1$ **b** $(\tan x)^2 = 3$ **c** $3(\tan x)^2 = 1$

7 Solve the equation $(\tan x)^2 + \tan x - 12 = 0$ and $0° \leq x \leq 360°$.

8 The difference between two angles is 180°. What can you say about

   a the tangents of the angles?

   b the sines of the angles?

   c the cosines of the angles?

Check your progress

Core

- I can use Pythagoras' theorem the calculate a side of a right-angled triangle
- I can use sine, cosine and tangent ratios to calculate a side or an angle of a right-angled triangle

Extended

- I can solve trigonometric problems in two dimensions involving angles of elevation or depression
- I know that the perpendicular distance from a point to a line is the shortest distance to the line
- I can extend the sine, cosine and tangent values to angles between 90° and 360°
- I know the shapes and properties of graphs of trigonometric functions
- I can find the shortest distance from a point to a line
- I can solve simple trigonometric equations
- I can solve problems using the sine rule or the cosine rule for any triangle
- I can use the formula area of a triangle = $\frac{1}{2} ab \sin C$
- I can solve simple trigonometric problems in three dimensions, including the angle between a line and a plane

Chapter 27
Mensuration

Topics	Level	Key words
1 Perimeter and area of a rectangle	CORE	length, width, perimeter, area
2 Area of a triangle	CORE	base, perpendicular height
3 Area of a parallelogram	CORE	parallelogram
4 Area of a trapezium	CORE	trapezium
5 Circumference and area of a circle	CORE	circumference, diameter, radius, π
6 Surface area and volume of a cuboid	CORE	volume, cuboid, surface area, litre
7 Volume and surface area of a prism	CORE	cross-section, prism
8 Volume and surface area of a cylinder	CORE	curved surface, cylinder
9 Sectors and arcs: 1	CORE	arc, sector, subtended
10 Sectors and arcs: 2	EXTENDED	
11 Volume of a pyramid	CORE	pyramid, vertical height, vertex
12 Volume and surface area of a cone	CORE	cone, vertical height, slant height
13 Volume and surface area of a sphere	CORE	sphere

In this chapter you will learn how to:

CORE	EXTENDED
Carry out the following calculations involving multiples of π where appropriate: • the perimeter and area of a rectangle, triangle, parallelogram and trapezium and compound shapes derived from these (C5.2 and E5.2) • the circumference and area of a circle (C5.3 and E5.3) • the volume of a cuboid, prism and cylinder (C5.4 and E5.4) • the surface area of a cuboid, prism and a cylinder (C5.4 and E5.4) • the areas and volumes of compound shapes (C5.5 and E5.5) • the arc length and sector area as fractions of the circumference and area of a circle (C5.3 and E5.3) • the surface area and volume of a sphere, pyramid and cone (given formulae for the sphere, pyramid and cone). (C5.4 and E5.4)	Solve problems involving: • the arc length and sector area with more complicated angle fractions. (E5.3)

Why this chapter matters

People have always needed to measure areas and volumes.

In everyday life, you will, for instance, need to find the area to work out how many tiles to buy to cover a floor; or you will need to find the volume to see how much water is needed to fill a swimming pool. You can do this quickly using formulae.

Measuring the world

From earliest times, farmers have wanted to know the area of their fields to see how many crops they could grow or animals they could support. When land is bought and sold, the cost depends on the area.

Volumes are important too. Volumes tell you how much space there is inside any structure. Whether it is a house, barn, aeroplane, car or office, the volume is important. In some countries there are regulations about the number of people who can use an office, based on the volume of the room.

Volumes of containers for liquids also need to be measured. Think, for example, of a car fuel tank, the water tank in a building, or a reservoir. It is important to be able to calculate the capacity of all these things.

So how do you measure areas and volumes? In this chapter, you will learn formulae that can be used to calculate areas and volumes of different shapes, based on a few measurements.

Many of these formulae were first worked out thousands of years ago. They are still in use today because they are important in everyday life.

The process of calculating areas and volumes using formulae is called **mensuration**.

Chapter 27: Mensuration

27.1 Perimeter and area of a rectangle

The **perimeter** of a rectangle is the total distance around the outside.

perimeter = $l + w + l + w$
= $2(l + w)$

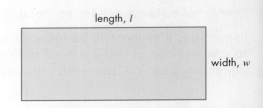

The **area** of a rectangle is **length × width**.

area = lw

Example 1

Calculate the area and perimeter of this rectangle.

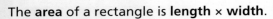

Area of rectangle = length × width
= 11 cm × 4 cm
= 44 cm^2

Perimeter = 2 × 11 + 2 × 4
= 30 cm

Some two-dimensional shapes are made up of two or more rectangles.

These shapes can be split into simpler shapes, which makes it easy to calculate their areas.

Example 2

Find the area and perimeter of the shape shown on the right.

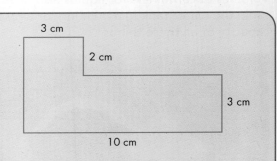

First, split the shape into two rectangles, A and B and find the missing lengths.

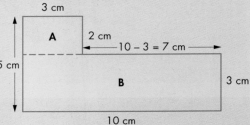

The perimeter is
3 + 2 + 7 + 3 + 10 + 5 = 30 cm
area of A = 2 × 3 = 6 cm^2
area of B = 10 × 3 = 30 cm^2

The area of the shape is:
area of A + area of B = 6 + 30
= 36 cm^2

EXERCISE 27A

1 Calculate the area and the perimeter for each of these rectangles.

a   7 cm, 5 cm

b 11 cm, 3 cm

c 15 cm, 3 cm

d 10 cm, 7 cm

e 8 cm, 7 cm

f 5 cm, 2 cm

2 Calculate the area and the perimeter for each of these rectangles.

a 8.2 cm, 6.5 cm

b 11.8 cm, 7.2 cm

3 A rectangular field is 150 m long and 45 m wide.

Fencing is needed to go all the way around the field.

The fencing is sold in 10-metre long pieces.

How many pieces are needed?

4 A soccer pitch is 160 m long and 70 m wide.

a Before a game, the players have to run about 1500 m to help them loosen up. How many times will they need to run round the perimeter of the pitch to do this?

b The groundsman waters the pitch at the rate of 100 m² per minute. How long will it take him to water the whole pitch?

5 What is the perimeter of a square with an area of 100 cm²?

6 Which rectangle has the largest area? Which has the largest perimeter?

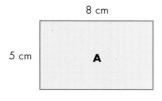 A: 8 cm × 5 cm

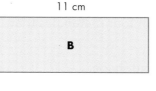 B: 11 cm × 4 cm

 C: 7 cm × 6 cm

Explain your answers.

Chapter 27: Mensuration

7 Doubling the length and width of a rectangle doubles the area of the rectangle.

Is this statement:
- always true
- sometimes true
- never true?

Explain your answer.

Advice and Tips

Draw some diagrams with different lengths and widths.

8 Calculate the perimeter and area of each of these compound shapes below.

a

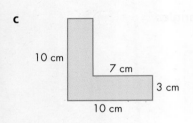

b

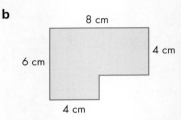

Advice and Tips

- First, split the compound shape into rectangles.
- Then, calculate the area of each rectangle
- Finally, add together the areas of the rectangles.

c

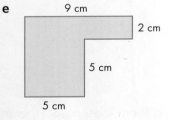

d

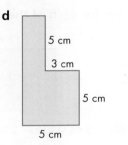

Advice and Tips

Be careful to work out the length and width of each separate rectangle. You will usually have to add or subtract lengths to find some of these.

e

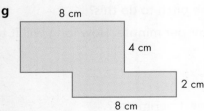

f

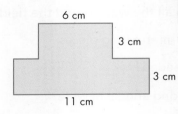

g

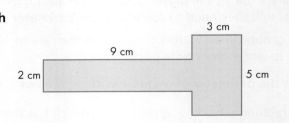

h

9 This compound shape is made from four rectangles that are all the same size.

Work out the area of the compound shape.

490 27.1 Perimeter and area of a rectangle

Chapter 27 . Topic 2

10 This shape is made from five squares that are all the same size.

It has an area of 80 cm².

Work out the perimeter of the shape.

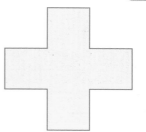

27.2 Area of a triangle

The area of any triangle is given by the formula:

area = ½ × **base** × **perpendicular height**

As an algebraic formula, this is written as:

$A = \frac{1}{2}bh$

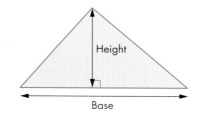

This diagram shows why the area for a triangle is half the area of a rectangle with the same base and height.

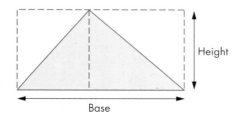

Example 3

Calculate the area of this triangle.

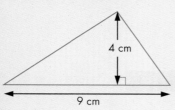

Area = ½ × 9 cm × 4 cm

= ½ × 36 cm²

= 18 cm²

Example 4

Calculate the area of the shape shown below.

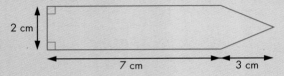

This shape can be split into a rectangle (R) and a triangle (T).

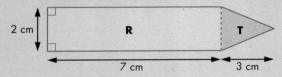

Area of the shape = area of R + area of T

$$= 7 \times 2 + \frac{1}{2} \times 2 \times 3$$

$$= 14 + 3$$

$$= 17 \text{ cm}^2$$

EXERCISE 27B

1. Calculate the area of each of these triangles.

   a   b 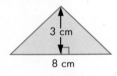   c

   d 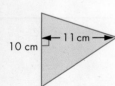   e 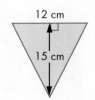   f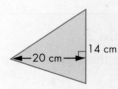

2. Copy and complete the table for triangles **a** to **f**.

	Base	Perpendicular height	Area
a	8 cm	7 cm	
b		9 cm	36 cm²
c		5 cm	10 cm²
d	4 cm		6 cm²
e	6 cm		21 cm²
f	8 cm	11 cm	

27.2

3 This regular hexagon has an area of 48 cm².

What is the area of the square that surrounds the hexagon?

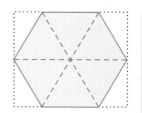

4 Find the area of each of these shapes.

a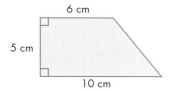

Advice and Tips

Refer to Example 4 on how to find the area of a compound shape.

b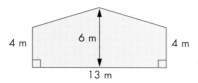

c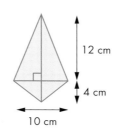

5 Find the area of each shaded shape.

a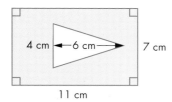

Advice and Tips

Find the area of the outer shape and subtract the area of the inner shape.

b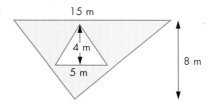

6 Write down the dimensions of two different-sized triangles that have the same area of 50 cm².

7 Which triangle is the odd one out? Give a reason for your answer.

a 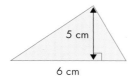   b 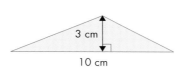   c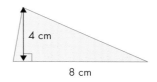

Chapter 27: Mensuration

Chapter 27 . Topic 3
27.3 Area of a parallelogram

A **parallelogram** can be changed into a rectangle by moving a triangle.

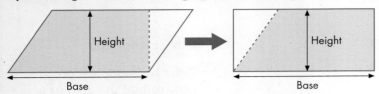

This shows that the area of the parallelogram is the area of a rectangle with the same base and height. The formula is:

area of a parallelogram = base × height

As an algebraic formula, this is written as:

$A = bh$

Example 5
Find the area of this parallelogram.

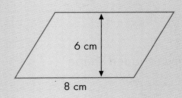

Area = 8 cm × 6 cm
     = 48 cm²

EXERCISE 27C

1. Calculate the area of each parallelogram below.

   a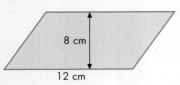

   b

   c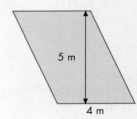

   d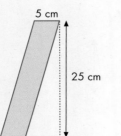

   e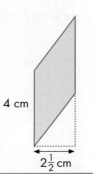

   f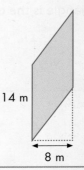

2 Sandeep says that the area of this parallelogram is 30 cm².

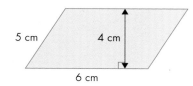

Is she correct? Give a reason for your answer.

3 This shape is made from four parallelograms that are all the same size. The area of the shape is 120 cm².

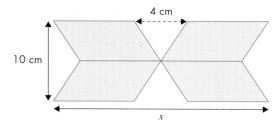

Work out the length marked x on the diagram.

4 This logo, made from two identical parallelograms, is cut from a sheet of card.

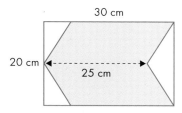

a Calculate the area of the logo.

b How many logos can be cut from a sheet of card that measures 1 m by 1 m?

27.4 Area of a trapezium

You can calculate the area of a **trapezium** by finding the average of the lengths of its parallel sides and multiplying this by the perpendicular height between them.

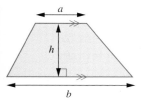

The area of a trapezium is given by this formula:

$A = \frac{1}{2}(a + b)h$

Example 6

Find the area of the trapezium ABCD.

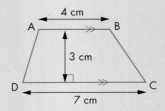

Area = $\frac{1}{2}(4 + 7) \times 3$

= $\frac{1}{2} \times 11 \times 3$

= 16.5 cm²

EXERCISE 27D

1. Copy and complete the table for the trapezia a–g.

	Parallel side 1	Parallel side 2	Perpendicular height	Area
a	8 cm	4 cm	5 cm	
b	10 cm	12 cm	7 cm	
c	7 cm	5 cm	4 cm	
d	5 cm	9 cm	6 cm	
e	3 cm	13 cm	5 cm	
f	4 cm	10 cm		42 cm²
g	7 cm	8 cm		22.5 cm²

Advice and Tips

Trapezia is the plural of trapezium.

2. Calculate the perimeter and the area of each trapezium.

a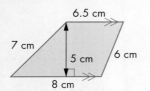

b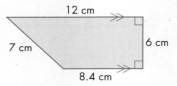

c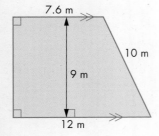

Advice and Tips

Make sure you use the right measurement for the height. Sometimes you might be told the slant side length, which is not used for the area.

3. A trapezium has an area of 25 cm². Its vertical height is 5 cm. Work out a possible pair of lengths for the two parallel sides.

4 Which of these shapes has the largest area?

a 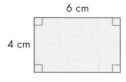 b 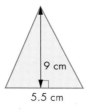 c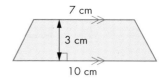

5 Which of these shapes has the smallest area?

a 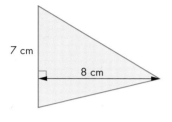 b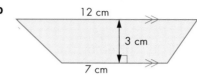

c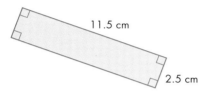

6 Which of these is the area of this trapezium?

a 45 cm²

b 65 cm²

c 70 cm²

You must show your workings.

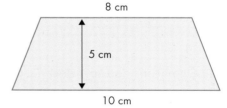

7 Work out the value of *a* so that the square and the trapezium have the same area.

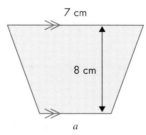

8 The side of a ramp is a trapezium, as shown in the diagram. Calculate its area, giving your answer in square metres.

Advice and Tips

Change the height into metres first.

27.5 Circumference and area of a circle

The perimeter of a circle is called the **circumference**.

You can calculate the circumference, C, of a circle by multiplying its **diameter**, d, by π.

The value of π is found on all scientific calculators, with $\pi = 3.141\,592\,654$, but if it is not on your calculator, then take $\pi = 3.142$.

The circumference of a circle is given by the formula:

circumference = $\pi \times$ diameter or $C = \pi d$

As the diameter is twice the **radius**, r, this formula can also be written as $C = 2\pi r$.

> ### Example 7
> Calculate the circumference of the circle with a diameter of 4 cm.
>
>
>
> Use the formula:
>
> $C = \pi d$
>
> $\quad = \pi \times 4$
>
> $\quad = 12.6$ cm (rounded to 1 decimal place)
>
> **Remember** The length of the radius of a circle is half the length of its diameter. So, when you are given a radius, in order to find a circumference you must first double the radius to get the diameter.

> ### Example 8
> Calculate the diameter of a circle that has a circumference of 40 cm.
>
> $C = \pi \times d$
>
> $40 = \pi \times d$
>
> $d = \dfrac{40}{\pi} = 12.7$ cm (rounded to 1 decimal place)
>
>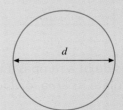

The area, A, of a circle is given by the formula:

area = $\pi \times$ radius^2 or $A = \pi \times r \times r$ or $A = \pi r^2$

Remember This formula uses the radius of a circle. So, when you are given the diameter of a circle, you must *halve* it to get the radius.

27.5

Example 9

Calculate the area of a circle with a diameter of 12 cm. Give your answer as a multiple of π.

First, halve the diameter to find the radius:

radius = 12 ÷ 2 = 6 cm

Then, find the area:

area = πr^2

= $\pi \times 6^2$

= 36 π cm²

EXERCISE 27E

1 Calculate the circumference and area of each circle. Give your answers as a multiple of π.

a 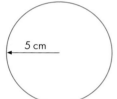 b   c   d

2 Calculate the circumference and area of each circle. Give your answers to 1 decimal place.

a   b   c   d

3 Calculate:

  i the circumference

  ii the area of each of these circles. Give your answers as a multiple of π and also to 1 decimal place.

a   b  c   d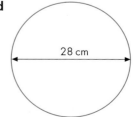

Chapter 27: Mensuration **499**

4. A circle has a circumference of 60 cm.

   a Calculate the diameter of the circle to 1 decimal place.

   b What is the radius of the circle to 1 decimal place?

   c Calculate the area of the circle to 1 decimal place.

5. Calculate the area of a circle with a circumference of 110 cm.

6. The circumference of a circle is 40π cm.

   Find **a** the radius of the circle **b** the area of the circle

7. Calculate the area of the shaded part of each of these diagrams. Give your answer as a multiple of π.

   Advice and Tips

   In each diagram, subtract the area of the small circle from the area of the large circle.

   a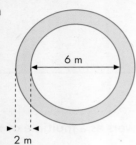
   6 m
   2 m

   b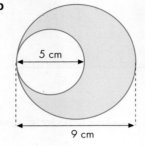
   5 cm
   9 cm

   c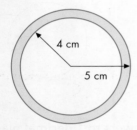
   4 cm
   5 cm

8. The diagram shows a circular photograph frame.

   Work out the area of the frame. Give your answer as a multiple of π.

   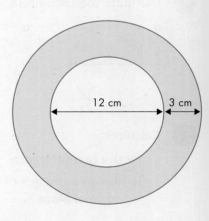
   12 cm 3 cm

9. A square has sides of length a and a circle has radius r.

   The area of the square is equal to the area of the circle.

   Show that $r = \dfrac{a}{\sqrt{\pi}}$

   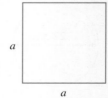
   a

   r

27.5 Circumference and area of a circle

10 A circle fits exactly inside a square of sides 10 cm.

Calculate the area of the shaded region. Give your answer to 1 decimal place.

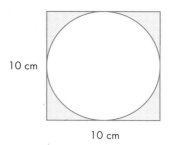

27.6 Surface area and volume of a cuboid

A cuboid is a box shape, all six faces of which are rectangles.

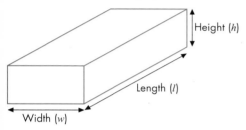

Every day you will see many examples of **cuboids**, such as food packets, smart phone – and even this book.

The **volume** of a cuboid is given by the formula:

volume = length × width × height *or* $V = l \times w \times h$ *or* $V = lwh$

You can claculate the **surface area** of a cuboid by finding the total area of the six faces, which are rectangles. Notice that each pair of opposite rectangles have the same area. So, from the diagram above:

area of top and bottom rectangles = 2 × length × width = $2lw$

area of front and back rectangles = 2 × height × width = $2hw$

area of two side rectangles   = 2 × height × length = $2hl$

Hence, the surface area of a cuboid is given by the formula:

surface area = $A = 2lw + 2hw + 2hl$

Example 10

Calculate the volume and surface area of this cuboid.

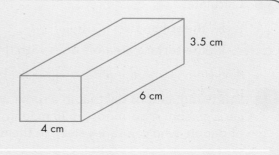

Chapter 27: Mensuration **501**

Volume = $V = lwh = 6 \times 4 \times 3.5 = 84$ cm^3

Surface area = $A = 2lw + 2hw + 2hl$
$= (2 \times 6 \times 4) + (2 \times 3.5 \times 4) + (2 \times 3.5 \times 6)$
$= 48 + 28 + 42 = 118$ cm^2

Note:
1 cm^3 = 1000 mm^3 and 1 m^3 = 1 000 000 cm^3
1000 cm^3 = 1 litre
1 m^3 = 1000 litres

EXERCISE 27F

1. Find **i** the volume and **ii** the surface area of each of these cuboids.

   a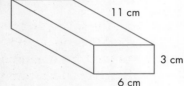

   b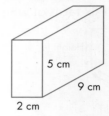

   c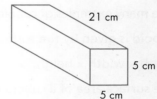

   d

2. Find the capacity of a fish-tank with dimensions: length 40 cm, width 30 cm and height 20 cm. Give your answer in litres.

3. Find the volume of each cuboid
   a The area of the base is 40 cm^2 and the height is 4 cm.
   b The base has one side 10 cm and the other side 2 cm longer, and the height is 4 cm.
   c The area of the top is 25 cm^2 and the depth is 6 cm.

4. Calculate:
   i the volume
   ii the surface area of cubes with these edge lengths.
   a 4 cm   **b** 7 cm   **c** 10 mm   **d** 5 m   **e** 12 m

5. Safety regulations say that in a room where people sleep there should be at least 12 m^3 for each person. A dormitory is 20 m long, 13 m wide and 4 m high. What is the greatest number of people who can safely sleep in the dormitory?

6 Copy and complete the table for cuboids **a** to **e**.

	Length	Width	Height	Volume
a	8 cm	5 cm	4.5 cm	
b	12 cm	8 cm		480 cm³
c	9 cm		5 cm	270 cm³
d		7 cm	3.5 cm	245 cm³
e	7.5 cm	5.4 cm	2 cm	

7 A tank contains 32 000 litres of water. The base of the tank measures 6.5 m by 3.1 m. Find the depth of water in the tank. Give your answer to one decimal place.

8 A room contains 168 m³ of air. The height of the room is 3.5 m. What is the area of the floor?

9 What are the dimensions of cubes with these volumes?

    a 27 cm³   **b** 125 m³   **c** 8 mm³   **d** 1.728 m³

10 Calculate the volume of each of these shapes.

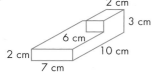

Advice and Tips

Split the solid into two separate cuboids and work out the dimensions of each of them from the information given.

11 A cuboid has volume of 125 cm³ and a total surface area of 160 cm².

Is it possible that this cuboid is a cube? Give a reason for your answer.

12 The volume of a cube is N cm³. The area of the cube is N cm².

    a How long is each side of the cube?   **b** What is the value of N?

27.7 Volume and surface area of a prism

A **prism** is a three-dimensional shape that has the same **cross-section** running all the way through it.

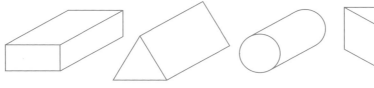

Name:	Cuboid	Triangular prism	Cylinder	Cuboid	Hexagonal prism
Cross-section:	Rectangle	Triangle	Circle	Square	Hexagon

Chapter 27: Mensuration

The volume of a prism is found by multiplying the area of its cross-section by the length of the prism (or height if the prism is stood on end).

That is, volume of prism = area of cross-section × length
or $V = Al$

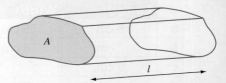

Example 11

Find the volume of the triangular prism.

The area of the triangular cross-section $A = \dfrac{5 \times 7}{2}$

$= 17.5 \text{ cm}^2$

The volume is the area of its cross-section × length = Al

$= 17.5 \times 9$

$= 157.5 \text{ cm}^3$

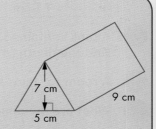

EXERCISE 27G

1 For each prism shown:

   i calculate the area of the cross-section   **ii** calculate the volume.

   a 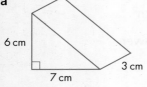   **b** 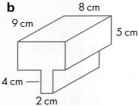   **c**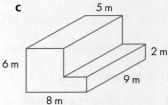

2 Calculate the volume of each of these prisms.

   a 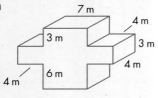   **b** 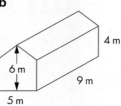   **c**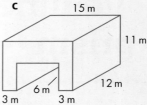

3 A swimming pool is 10 m wide and 25 m long.

It is 1.2 m deep at one end and 2.2 m deep at the other end. The floor slopes uniformly from one end to the other.

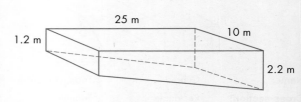

   a Explain why the shape of the pool is a prism.

   b The pool is filled with water at a rate of 2 m³ per minute. How long will it take to fill the pool?

27.7

4 A building is in the shape of a prism with the dimensions shown in the diagram. Calculate the volume of air (in litres) inside the building.

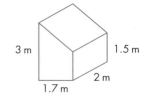

5 Each of these prisms has a uniform cross-section in the shape of a right-angled triangle.

  a Find the volume of each prism.   **b** Find the total surface area of each prism.

  i 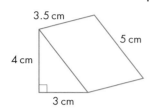   **ii**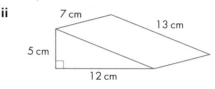

6 The top and bottom of the container shown here are the same size, both consisting of a rectangle, 4 cm by 9 cm, with a semi-circle at each end. The depth is 3 cm. Find the volume of the container.

7 In 2009 the sculptor Anish Kapoor exhibited a work called *Svayambh*. It was a block of red wax in the shape of a prism.

The cross-section was in the shape of an arched entrance.

It was 8 m long and weighed 30 tonnes. It slowly travelled through the galleries on a track.

Calculate the volume of wax used.

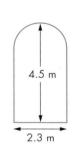

8 A horse trough is in the shape of a semi-circular prism, as shown.

What volume of water will the trough hold when it is filled to the top? Give your answer in litres.

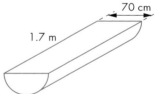

9 The dimensions of the cross-section of a girder (in the shape of a prism), 2 m in length, are shown on the diagram. The girder is made of iron. 1 cm³ of iron weighs 79 g.

What is the mass of the girder?

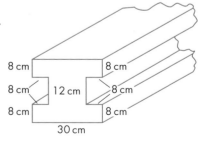

10 Calculate the volume of this prism.

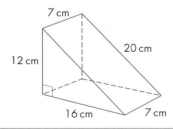

Chapter 27: Mensuration   **505**

27.8 Volume and surface area of a cylinder

Volume

Since a **cylinder** is an example of a prism, you can calculate its **volume** by multiplying the area of one of its circular ends by the height.

That is,

$$\text{volume} = \pi r^2 h$$

where r is the radius of the cylinder and h is its height or length.

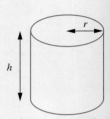

> **Example 12**
>
> What is the volume of a cylinder with a radius of 5 cm and a height of 12 cm?
>
> Volume = area of circular base × height
> $= \pi r^2 h = \pi \times 5^2 \times 12 \text{ cm}^3 = 942 \text{ cm}^3$ (3 significant figures)

Surface area

The total surface area of a cylinder is made up of the area of its **curved surface** plus the area of its two circular ends.

The curved surface area, when opened out, is a rectangle with length equal to the circumference of the circular end.

curved surface area = circumference of end × height of cylinder

$= 2\pi rh \quad \text{or} \quad \pi dh$

area of one end $= \pi r^2$

Therefore, total surface area $= 2\pi rh + 2\pi r^2 \quad \text{or} \quad \pi dh + 2\pi r^2$

> **Example 13**
>
> What is the total surface area of a cylinder with a radius of 15 cm and a height of 2.5 m?
>
> First, you must change the dimensions to a *common unit*. Use centimetres in this case.
>
> Total surface area $= \pi dh + 2\pi r^2$
>
> $= \pi \times 30 \times 250 + 2 \times \pi \times 15^2 \text{ cm}^2$
>
> $= 23\,562 + 1414 \text{ cm}^2 = 24\,976 \text{ cm}^2 = 25\,000 \text{ cm}^2$ (3 significant figures)

EXERCISE 27H

1 For the cylinders below find:

   i the volume

   ii the total surface area.

Give your answers as a multiple of π and also to 3 significant figures.

a   b   c   d

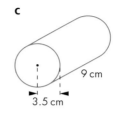
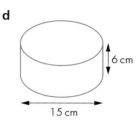

2 For each of these cylinder dimensions find:

   i the volume

   ii the curved surface area.

Give your answers in terms of π.

   a Base radius 3 cm and height 8 cm

   b Base diameter 8 cm and height 7 cm

   c Base diameter 12 cm and height 5 cm

   d Base radius of 10 m and length 6 m

3 A solid cylinder has a diameter of 8.4 cm and a height of 12.0 cm. Calculate the volume of the cylinder.

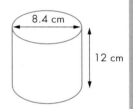

4 A cylindrical food can has a height of 10.5 cm and a diameter of 7.4 cm.

What can you say about the size of the paper label around the can?

5 A cylindrical container is 65 cm in diameter. Water is poured into the container until it is 1 m deep.

How much water is in the container? Give your answer in litres.

6 A drinks manufacturer plans a new drink in a can. The quantity in each can must be 330 ml.

Suggest a suitable height and diameter for the can. (You might like to look at the dimensions of a real drinks can.)

7 Wire is commonly made by putting hot metal through a hole in a plate.

What length of wire of diameter 1 mm can be made from a 1 cm cube of metal?

Chapter 27 . Topic 9

8 The engine size of a car is measured in litres. This tells you the total volume of its cylinders. Cylinders with the same volume can be long and thin or short and thick.

In a racing car, the diameter of a cylinder is twice its length. Suggest possible dimensions for a 0.4 litre racing car cylinder.

27.9 Sectors and arcs: 1

A **sector** is part of a circle.

It is formed by two radii and part of the circumference.

The part of the circumference is called an **arc**.

O is the angle of the sector.

The sector is a fraction of the circle.

You can use the angle to find the fraction.

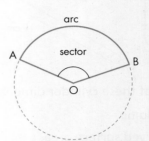

Example 14

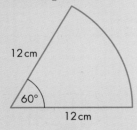

This is a sector. Find

a the arc length.   **b** the sector area.

Leave π in your answers.

a The sector is a fraction of a circle.

The fraction is $\frac{60}{360} = \frac{1}{6}$

The diameter of the circle is 24 cm.

The circumference of the circle is $\pi \times 24 = 24\pi$ cm

The arc length is $\frac{1}{6}$ of $24\pi = 24\pi \div 6 = 4\pi$ cm

b The area of the circle is $\pi \times 12^2 = 144\pi$ cm²

The sector area is $\frac{1}{6}$ of $144\pi = 144\pi \div 6 = 24\pi$ cm²

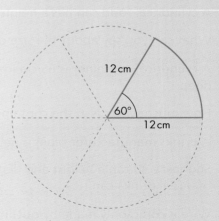

EXERCISE 27I

1 In this question leave π in your answers.

  a Find the circumference of this circle.

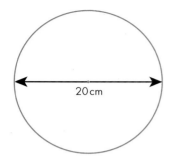

  b Find the arc length for each of these sectors.

  i  **ii**   **iii**

2 Find the area of each of the shapes in question 1. Leave π in your answers.

3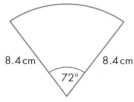

  a What fraction of a circle is this sector?
  b Find the arc length in cm. Round your answer to 1 d.p.
  c Find the area of the sector in cm². Round your answer to 1 d.p.

4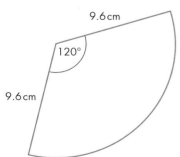

  a Find the area of this sector in cm². Round your answer to 1 d.p.
  b Find the length of the arc in cm. Round your answer to 1 d.p.
  c Find the perimeter of the shape.

Chapter 27: Mensuration

Chapter 27 . Topic 10

5

This is a semicircle.

a Find the area.

b Find the perimeter.

6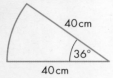

This is a sector.

a Show that the length of the arc is 8π cm.

b Find the area of the sector.

7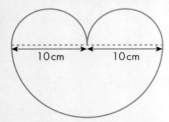

This shape is made from 3 semicircles.

a Find the perimeter of the shape. Leave π in your answer.

b Find the area of the shape. Leave π in your answer.

27.10 Sectors and arcs: 2

In the last section the sectors were all simple fractions of a circle.

You can always use the angle of the sector to find the arc length and sector area, even if it is not a simple fraction.

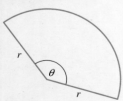

If the angle of the sector is θ, then:

arc length = $\dfrac{\theta}{360}$ × circumference = $\dfrac{\theta}{360} \times 2\pi r$

and sector area = $\dfrac{\theta}{360}$ × circle area = $\dfrac{\theta}{360} \times \pi r^2$

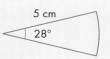

Example 15

Calculate the arc length and the area of the sector in the diagram.

The sector angle is 28° and the radius is 5 cm. Therefore:

arc length = $\frac{28}{360} \times \pi \times 2 \times 5$ = 2.4 cm (1 decimal place)

sector area = $\frac{28}{360} \times \pi \times 5^2$ = 6.1 cm² (1 decimal place)

EXERCISE 27J

1. For each of these sectors, calculate: **i** the arc length **ii** the sector area.

   a   b   c   d

2. For this sector work out: **a** the arc length **b** the sector area. Give your answers in terms of π.

   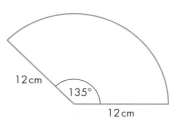

3. Calculate the total perimeter of each of these sectors.

   a   b

4. Calculate the area of each of these sectors.

   a   b

5. O is the centre of a circle of radius 12.5 cm.

   Calculate the length of the arc ACB.

   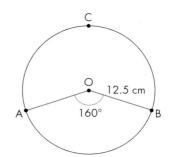

Chapter 27: Mensuration **511**

Chapter 27 . Topic 11

EXTENDED

6 The diagram shows quarter of a circle. Calculate the area of the shaded shape, giving your answer in terms of π.

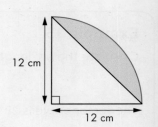

7 ABCD is a square of side length 8 cm. APC and AQC are arcs of the circles with centres D and B. Calculate the area of the shaded part, giving your answer in terms of π.

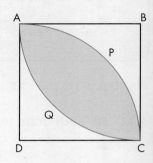

8 Find:
  a the perimeter
  b the area
of this shape.

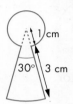

27.11 Volume of a pyramid

A **pyramid** is a three-dimensional shape with a base from which triangular faces rise to a common **vertex**. The base can be any polygon, but is usually a triangle, a rectangle or a square.

The volume of a pyramid is given by:

volume = $\frac{1}{3}$ × base area × **vertical height**

$V = \frac{1}{3}Ah$

where A is the base area and h is the vertical height.

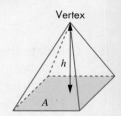

Example 16

Calculate the volume of the pyramid.

Base area = 5 × 4 = 20 cm²

Volume = $\frac{1}{3}$ × 20 × 6 = 40 cm³

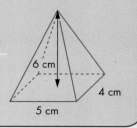

Example 17

A pyramid, with a square base of side 8 cm, has a volume of 320 cm³. What is the vertical height of the pyramid?

Let h be the vertical height of the pyramid. Then:

volume = $\frac{1}{3} \times 64 \times h$ = 320 cm³

$\frac{64h}{3}$ = 320 cm³

$h = \frac{960}{64}$ cm

h = 15 cm

EXERCISE 27K

1 Calculate the volume of each of these pyramids, all with rectangular bases.

a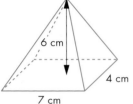

b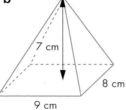

c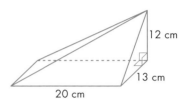

d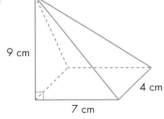

e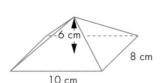

2 Calculate the volume of a pyramid that has a square base of side 9 cm and a vertical height of 10 cm.

3 Suppose you have six pyramids which have a height that is half the side of the square base.

a Explain how they can fit together to make a cube.

b How does this show that the formula for the volume of a pyramid is correct?

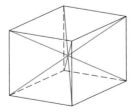

4 The glass pyramid outside the Louvre Museum in Paris was built in the 1980s. It is 20.6 m tall and the base is a square of side 35 m.

Suppose that instead of a pyramid, the building was a cuboid with the same square base, a flat roof and the same volume. How high would it have been?

Chapter 27 . Topic 12

5 Calculate the volume of each of these shapes.

a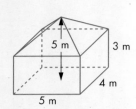

b

c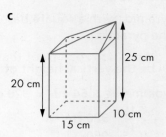

6 A crystal is in the form of two square-based pyramids joined at their bases (see diagram).

The crystal has a mass of 31.5 g.

What is the mass of 1 cm³ of the substance?

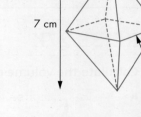

7 A pyramid has a square base of side 6.4 cm. Its volume is 81.3 cm³.

Calculate the height of the pyramid.

8 A pyramid has the same volume as a cube of side 10.0 cm.

The height of the pyramid is the same as the side of the square base.

Calculate the height of the pyramid.

9 The pyramid in the diagram has its top 5 cm cut off as shown. The shape that is left is called a frustum. Calculate the volume of the frustum.

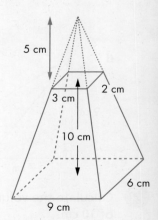

27.12 Volume and surface area of a cone

A **cone** can be treated as a pyramid with a circular base. Therefore, the formula for the volume of a cone is the same as that for a pyramid.

volume = $\frac{1}{3}$ × base area × vertical height

$V = \frac{1}{3}\pi r^2 h$

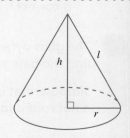

where r is the radius of the base and h is the **vertical height** of the cone.
The area of the curved surface of a cone is given by:

area of curved surface = π × radius × slant height
$$S = \pi r l$$

where l is the **slant height** of the cone.
So the total surface area of a cone is given by the are of the curved surface plus the area of its circular base.
$$A = \pi r l + \pi r^2$$

> **Example 18**
>
> For the cone in the diagram, calculate:
> **a** its volume   **b** its total surface area.
> Give your answers in terms of π.
>
>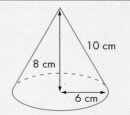
>
> **a** The volume is given by $V = \frac{1}{3}\pi r^2 h$
> $\qquad = \frac{1}{3} \times \pi \times 36 \times 8 = 96\pi$ cm³
>
> **b** The total surface area is given by $A = \pi r l + \pi r^2$
> $\qquad = \pi \times 6 \times 10 + \pi \times 36 = 96\pi$ cm²

EXERCISE 27L

1 For each cone, calculate:

    i its volume   **ii** its total surface area.

Give your answers to 3 significant figures.

a 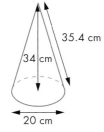   **b** 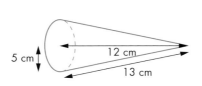   **c**

2 Find the total surface area of a cone with base radius 3 cm and slant height 5 cm. Give your answer in terms of π.

3 Calculate the volume of each of these shapes. Give your answers in terms of π.

a 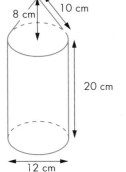   **b**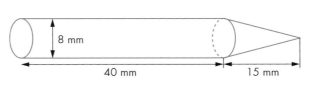

Chapter 27: Mensuration   **515**

4 You could work with a partner on this question.

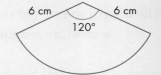

A sector of a circle, as in the diagram, can be made into a cone (without a base) by sticking the two straight edges together.

  a What would be the diameter of the base of the cone in this case?
  b What is the diameter if the angle is changed to 180°?
  c Investigate other angles.

5 A cone has the dimensions shown in the diagram.

Calculate the total surface area, leaving your answer in terms of π.

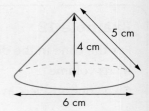

6 The slant height of a cone is equal to the base diameter. Show that the area of the curved surface is twice the area of the base.

7 The model shown on the right is made from aluminium.

What is the mass of the model, given that the density of aluminium is 2.7 g/cm³? (This means that 1 cm³ of aluminium has a mass of 2.7 g.)

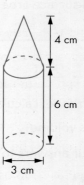

8 A container in the shape of a cone, base radius 10 cm and vertical height 19 cm, is full of water. The water is poured into an empty cylinder of radius 15 cm. How high is the water in the cylinder?

27.13 Volume and surface area of a sphere

The volume of a **sphere**, radius r, is given by:

$$V = \tfrac{4}{3}\pi r^3$$

Its surface area is given by:

$$A = 4\pi r^2$$

Example 19

A sphere has a radius of 8 cm. Calculate:

a its volume

b its surface area.

a The volume is given by:

$V = \frac{4}{3}\pi r^3$

$= \frac{4}{3} \times \pi \times 8^3 = \frac{2048}{3} \times \pi$

$= 2140 \text{ cm}^3$ (3 significant figures)

b The surface area is given by:

$A = 4\pi r^2$

$= 4 \times \pi \times 8^2 = 256 \times \pi$

$= 804 \text{ cm}^2$ (3 significant figures)

EXERCISE 27M

1 Calculate the volume and surface area of each of these spheres. Give your answers in terms of π.

   a Radius 3 cm   **b** Radius 6 cm   **c** Diameter 20 cm

2 Calculate the volume and the surface area of a sphere with a diameter of 50 cm.

3 A sphere fits exactly into an open cubical box of side 25 cm. Calculate:

   a the surface area of the sphere   **b** the volume of the sphere.

4 A metal sphere of radius 15 cm is melted down and recast into a solid cylinder of radius 6 cm. Calculate the height of the cylinder.

5 Lead has a density of 11.35 g/cm³. Calculate the maximum number of lead spheres of radius 1.5 mm that can be made from 1 kg of lead.

6 A sphere has a radius of 5.0 cm. A cone has a base radius of 8.0 cm. The sphere and the cone have the same volume. Calculate the height of the cone.

7 A sphere of diameter 10 cm is carved out of a wooden block in the shape of a cube of side 10 cm. What percentage of the wood is wasted?

Check your progress

Core

- I can carry out calculations involving the perimeter and area of a rectangle, triangle, parallelogram and trapezium
- I can calculate the areas of compound two-dimensional shapes
- I can calculate the circumference and area of a circle, leaving π in the answer if required
- I can calculate the arc length and area of sectors that are a simple fraction of a circle
- I can calculate the volume and surface area of a cuboid, prism, cylinder, sphere, pyramid and cone
- I can calculate the volumes of compound three-dimensional shapes

Extended

- I can calculate the arc length and area of sectors

Chapter 28
Symmetry

Topics	Level	Key words
1 Lines of symmetry	CORE	line of symmetry, mirror line
2 Rotational symmetry	CORE	rotational symmetry, order of rotational symmetry
3 Symmetry of special two-dimensional shapes	CORE	rectangle, square, parallelogram, kite, rhombus, trapezium, isosceles triangle, equilateral triangle
4 Symmetry of three-dimensional shapes	EXTENDED	plane of symmetry, axis of symmetry, cuboid, prism, pyramid, cone, cylinder, reflection
5 Symmetry in circles	EXTENDED	centre, chord, tangent, perpendicular bisector

In this chapter you will learn how to:

CORE	EXTENDED

• Recognise rotational and line symmetry (including order of rotational symmetry) in two dimensions. (C4.6 and E4.6)
• Recognise properties of triangles, quadrilaterals and circles directly related to their symmetries. (C4.6 and E3.5)
	• Recognise symmetry properties of the prism (including cylinder) and the pyramid (including cone). (E4.6)
	• Use the following symmetry properties of circles:
	   – equal chords are equidistant from the centre
	   – the perpendicular bisector of a chord passes through the centre
	   – tangents from an external point are equal in length. (E4.6)

Why this chapter matters

If you look carefully, you will be able to spot symmetry all around you. It is present in the natural world and in objects made by humans. But, does it have a purpose and why do we need it?

Symmetry in nature, art and literature

Symmetry is everywhere you look in nature. Plants and animals have symmetrical body shapes and patterns. For example, if you divide a leaf in half, you will see that one half is the same shape as the other half.

Where is the symmetry in this butterfly, star fish and peacock? What effect does this symmetry have?

Pegasus.

This painting by a Dutch artist called M.C. Escher (1898–1972) uses line symmetry and rotational symmetry. Why do you think Escher used symmetry in his paintings?

Every tiger has its own unique pattern of stripes. These appear on the tiger's skin as well as its fur. What purpose do these symmetrical stripes serve?

Can you identify symmetry in the face of the tiger?

Symmetry in structures

St Peter's Basilica, Rome.

St Peter's Basilica, in the Vatican City in Rome, was started in 1506 and completed in 1626. It is a very symmetrical structure – see if you can identify all the symmetry that is present.

Why do you think that the designers of this building used symmetry?

These examples show some of the uses of symmetry in the world. Now think about where symmetry occurs in your own life – how important is it to you?

28.1 Lines of symmetry

Many two-dimensional shapes have one or more lines of symmetry.

A **line of symmetry** is a line that can be drawn through a shape so that what can be seen on one side of the line is the mirror image of what is on the other side. This is why a line of symmetry is sometimes called a **mirror line**.

It is also the line along which a shape can be folded exactly onto itself.

Advice and Tips

Remember you can use tracing paper to check for symmetry. For line symmetry, use it to find the mirror line. To check for rotational symmetry, trace the shape and turn your tracing around, over the shape.

Example 1

Find the number of lines of symmetry for this cross.

There are four altogether.

EXERCISE 28A

1. Copy these shapes and draw on the lines of symmetry for each one. If it will help you, use tracing paper or a mirror to check your results.

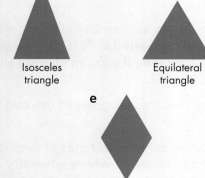

a Isosceles triangle
b Equilateral triangle
c Square
d Parallelogram

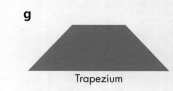

e Rhombus
f Kite
g Trapezium

2 a Find the number of lines of symmetry for each of these regular polygons.

i
Regular pentagon

ii
Regular hexagon

iii
Regular octagon

b How many lines of symmetry do you think a regular decagon has? (A decagon is a ten-sided polygon.)

3 Write down the number of lines of symmetry for each of these flags.

Austria   Canada   Iceland   Switzerland   Greece

4 These road signs all have lines of symmetry. Copy them and draw on the lines of symmetry for each one.

5 The animal and plant kingdoms are full of symmetry. Four examples are given below. State the number of lines of symmetry for each one.

a   b   c   d

Can you find other examples? Find suitable pictures, copy them and state the number of lines of symmetry each one has.

6 Copy this diagram.

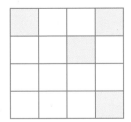

On your copy, shade in four more squares so that the diagram has four lines of symmetry.

28.2 Rotational symmetry

A two-dimensional shape has **rotational symmetry** if it can be rotated about a point to look exactly the same in a new position.

The **order of rotational symmetry** is the number of different positions in which the shape looks the same when it is rotated 360° about the point (that is, one complete turn).

The easiest way to find the order of rotational symmetry for any shape is to trace it and count the number of times that the shape looks the same as you turn the tracing paper through one complete turn.

Example 2

Find the order of rotational symmetry for this shape.

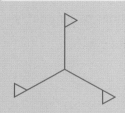

First, hold the tracing paper on top of the shape and trace the shape. Then rotate the tracing paper and count how many times the tracing matches the original shape in one complete turn.

You will find three different positions.

So, the order of rotational symmetry for the shape is 3.

EXERCISE 28B

1 Copy these shapes and write below each one the order of rotational symmetry. If it will help you, use tracing paper.

a
Square

b
Rectangle

c
Parallelogram

d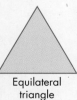
Equilateral triangle

e
Regular hexagon

Advice and Tips

Remember a shape with rotational symmetry of order 1 has no rotational symmetry.

2 Find the order of rotational symmetry for each of these shapes.

a   b   c   d   e

Chapter 28 . Topic 3

3 These are Greek capital letters. Write down the order of rotational symmetry for each one.

a Φ b H c Z d Θ e Ξ

4 Here is a star pattern.

Inside the star there are two patterns that have rotational symmetry.

a What is the order of rotational symmetry of the whole star?

b What is the order of rotational symmetry of the two patterns inside the star?

5 Copy the grid on the right. On your copy, shade in four squares so that the shape has rotational symmetry of order 2.

28.3 Symmetry of special two-dimensional shapes

Some three and four-sided shapes have special names such as **isosceles triangle** or **parallelogram**. You need to know the symmetry properties of these shapes.

For example:

- An isosceles triangle has one line of symmetry and no rotational symmetry.

- A parallelogram has no lines of symmetry and rotational symmetry of order 2.

EXERCISE 28C

1. Draw diagrams to show all the lines of symmetry on:
   a a **rectangle**
   b a **kite**
   c a **square**
   d an **equilateral triangle**
   e a **rhombus**.

2. a Which shape in question **1** has no rotational symmetry?
   b Find the order of rotational symmetry for each of the others.

3. a What do you call a triangle with one line of symmetry?
   b Can you draw a triangle with exactly two lines of symmetry?

4. What is the name for:
   a a quadrilateral with no lines of symmetry and rotational symmetry of order 2
   b a quadrilateral with rotational symmetry of order 4?

5. a Name two different quadrilaterals that have two lines of symmetry and rotational symmetry of order 2.
   b Can you draw a quadrilateral that has two lines of symmetry but no rotational symmetry?

6. The dotted line is a line of symmetry.
   a Which angles *must* be equal?
   b Which sides *must* be equal?
   c What is the name of this shape?

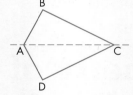

7. a What is the special name for a line of symmetry of a circle?
   b How many lines of symmetry does a circle have?
   c What is the order of rotational symmetry of a circle?

8. This shape has rotational symmetry of order 2.

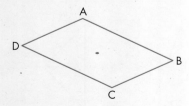

   a Which angles *must* be equal?
   b Which sides *must* be equal?
   c What is the name of this shape?

9. If a **trapezium** has a line of symmetry, what can you say about its angles?

28.4 Symmetry of three-dimensional shapes

Three-dimensional shapes can be symmetrical in two ways.

A **cuboid** has a **plane of symmetry** that divides it into two halves. One is the **reflection** of the other half in a mirror on the plane of symmetry.

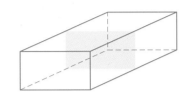

A cuboid has an **axis of symmetry**. It has rotational symmetry of order 2 about this axis.

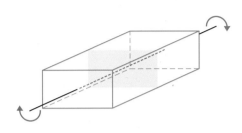

EXERCISE 28D

1 A cuboid has three planes of symmetry. Draw diagrams to show them.

2 A cuboid has three axes of symmetry.

    a Show them on a diagram.

    b What is the order of rotational symmetry about each axis?

3 This is an equilateral triangular **prism**.
It has *four* planes of symmetry.
One is shown in the diagram.

    Draw diagrams to show the other three.

4 The diagram shows an equilateral prism.
AB and CD are axes of symmetry.

    a What is the order of rotational symmetry about each one?

    b The shape has two more axes of symmetry. Show them on a diagram.

5 The diagram shows an axis of symmetry for a square-based **pyramid**.

    a What is the order of rotational symmetry about the axis?

    b Does the pyramid have any other axes of symmetry?

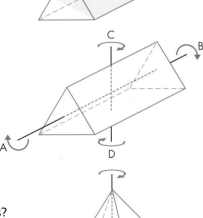

Chapter 28 . Topic 5

6 Draw diagrams to show the planes of symmetry of a square-based pyramid.

How many are there?

7 a Draw a diagram to show a plane of symmetry for a **cylinder**.

b One axis of symmetry for a cylinder is shown in this diagram.

Draw a diagram to show a different one.

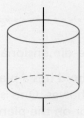

8 a How many axes of rotational symmetry does a **cone** have?

b How many planes of symmetry does a cone have?

9 The end of a prism is a regular hexagon.

a Show that the prism has seven planes of symmetry.

b How many axes of symmetry does it have?

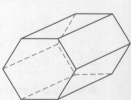

28.5 Symmetry in circles   E

Here are some facts about circles it is useful to know:

1 If two **chords** are the same length, they are the same distance from the centre.

If O is the centre of the circle and AB and CD are equal in length, then OM = ON.

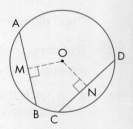

2 The **perpendicular bisector** of a chord passes through the centre of the circle.

In this diagram the **centre** must be on the broken line.

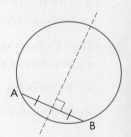

28.5

3 The two **tangents** from the point to a circle are equal in length.

PS and PT are the same length.

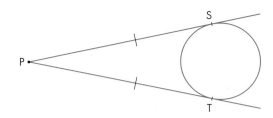

EXERCISE 28E

1 **a** Draw a circle. Use a pair of compasses.
   b Draw two chords, AB and CD.
   c Use compasses and a ruler to construct the perpendicular bisector of AB.
   d Construct the perpendicular bisector of CD.
   e You should find that the perpendicular bisectors cross at the centre of the circle. Explain why this is the case.
   f You could draw a circle by drawing round a circular object such as a plate or food can. How could you use chords to find the centre of the circle?

2 AB and CD are two chords of a circle that are the same length. O is the centre of the circle.

   a What sort of triangle is AOB? Give a reason for your answer.
   b Explain why triangles AOB and COD are congruent.
   c If angle OAB is 65°, find the angle COD.

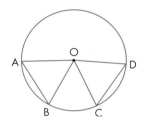

3 O is the centre of a circle. EF is a chord.
   M is the mid-point of EF.

   a Show that triangles EOM and FOM are congruent.
   b Explain why angle EMO is a right angle.
   c If angle MOF is 72°, find angle MEO.

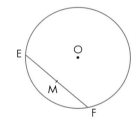

4 O is the centre of a circle.
   PX and PY are tangents.

   a Explain why angle PXO must be 90°.
   b Show that triangles XPO and YPO are congruent.
   c Angle XPO = 17°. Calculate angle XOY.

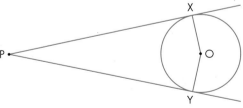

5 O is the centre of the circle.
   AB is a chord 6 cm long.
   Angle ACO is 90°.
   OC = 4 cm.

   Calculate the radius of the circle.

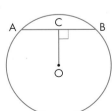

Chapter 28: Symmetry **527**

6 O is the centre of the circle.

XZ and YZ are tangents.

Calculate the value of a.

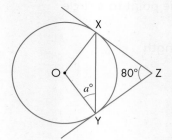

Check your progress

Core
- I can recognise lines of symmetry
- I can recognise rotational symmetry and order of rotational symmetry in two dimensions

Extended
- I can recognise symmetry properties of prisms, cylinders, pyramids and cones
- I can use these symmetry properties of a circle:
  - equal chords are the same distance from the centre
  - the perpendicular bisector of a chord passes through the centre
  - tangents from an external point are the same length

Chapter 29
Vectors

Topics	Level	Key words
1 Introduction to vectors	CORE	magnitude, direction, vector, column vector, scalar, coordinate grid
2 Using vectors	EXTENDED	position vector
3 The magnitude of a vector	EXTENDED	magnitude, Pythagoras' theorem

In this chapter you will learn how to:

CORE	EXTENDED
• Describe a translation by using a vector represented by eg: $\begin{pmatrix} x \\ y \end{pmatrix}$, $\overrightarrow{AB}$ or **a**. (C7.1 and E7.1) • Add and subtract vectors. (C7.1 and E7.1) • Multiply a vector by a scalar. (C7.1 and E7.1)	• Calculate the magnitude of a vector $\begin{pmatrix} x \\ y \end{pmatrix}$ as $\sqrt{x^2 + y^2}$. (E7.3) • Represent vectors by directed line segments. (E7.3) • Use the sum and difference of two vectors to express given vectors in terms of two coplanar vectors. (E7.3) • Use position vectors. (E7.3)

Why this chapter matters

Vectors are used to represent any quantity that has both magnitude and direction. The velocity of a speeding car is its direction and its speed. Velocity is a vector.

To understand how a force acts on an object, you need to know the magnitude of the force and the direction in which it moves – that is, its vector.

In science, vectors are used to describe displacement, acceleration and momentum.

But are vectors used in real life? Yes! Here are some examples.

In the 1950s, a group of talented Brazilian footballers invented the swerving free kick. By kicking the ball in just the right place, they managed to make it curl around the wall of defending players and go into the goal. When a ball is in flight, it is acted upon by various forces which can be described by vectors.

Formula One teams always employ physicists and mathematicians to help them build the perfect racing car. Since vectors describe movements and forces, they are used as the basis of a car's design.

Pilots have to consider wind speed and direction when they plan to land an aircraft at an airport. Vectors are an important part of the computerised landing system.

Vectors play a key role in the design of aircraft wings, where an upward force or lift is needed to enable the aircraft to fly.

Vectors are used extensively in computer graphics. Software for animations uses the mathematics of vectors.

Chapter 29: Vectors 531

29.1 Introduction to vectors

A **vector** is something that has both **magnitude** and **direction** and can be represented by an arrow.

Examples are velocity, acceleration, force and momentum.

Vectors can be written down in several ways:

$\overrightarrow{AB}$ – giving the start and end points with an arrow over the top.

a – as a lower-case letter printed in **bold**. When you are writing vectors **a**, **b** … by hand you cannot show them in bold type. Instead write them with a line underneath as <u>a</u>, <u>b</u> … to show that they stand for a vector and not a number.

When vectors are drawn on a **coordinate grid** they can be represented by two numbers in brackets in a **column**.

The top number shows how far the line moves from one side to the other between its start and end points, and the bottom number shows how far it moves up or down.

For example, on this grid:

$\overrightarrow{AB} = \begin{pmatrix} 2 \\ 3 \end{pmatrix}$ means move 2 right and 3 upwards to get to point B from point A.

If a line moves either downwards or left the coordinate is negative. So:

$\overrightarrow{BC} = \begin{pmatrix} 2 \\ -2 \end{pmatrix}$ means move 2 right and 2 downwards to get to point C from point B.

$\overrightarrow{DA} = \begin{pmatrix} -4 \\ 1 \end{pmatrix}$ means move 4 left and 1 upwards to get to point A from point D.

$\overrightarrow{DC} = \begin{pmatrix} 0 \\ 2 \end{pmatrix}$ means move neither left or right, just upwards to get to point C from point D.

Notice that the line joining A and B can be written as AB and BA and these both have the same magnitude (in this case length).

But the vectors $\overrightarrow{AB}$ and $\overrightarrow{BA}$ are not the same because their directions are different:

$\overrightarrow{AB} = \begin{pmatrix} 2 \\ 3 \end{pmatrix}$ and $\overrightarrow{BA} = \begin{pmatrix} -2 \\ -3 \end{pmatrix}$

Therefore $\overrightarrow{BA} = -\overrightarrow{AB}$

> **Advice and Tips**
>
> Do not forget the arrow to indicate a vector.

The vectors on this grid show that if $\mathbf{a} = \begin{pmatrix} 1 \\ 2 \end{pmatrix}$ then $-\mathbf{a} = \begin{pmatrix} -1 \\ -2 \end{pmatrix}$.

−**a** is a vector with the same length (magnitude) but acting in the opposite direction as the vector **a**.

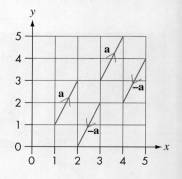

29.1

Adding, subtracting and multiplying vectors

Vectors are added together by placing them end to end.

On this grid **a** + **b** = **c**

or $\begin{pmatrix} 3 \\ 2 \end{pmatrix} + \begin{pmatrix} 1 \\ -3 \end{pmatrix} = \begin{pmatrix} 4 \\ -1 \end{pmatrix}$

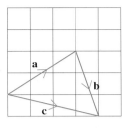

Notice that you add the top figures together (3 + 1 = 4) and the bottom figures together (2 + −3 = −1).

Vectors can be subtracted too.

a − **b** = **d**

a − **b** is the same as **a** + (−**b**)

$\begin{pmatrix} 3 \\ 2 \end{pmatrix} - \begin{pmatrix} 1 \\ -3 \end{pmatrix} = \begin{pmatrix} 3 \\ 2 \end{pmatrix} + \begin{pmatrix} -1 \\ 3 \end{pmatrix} = \begin{pmatrix} 2 \\ 5 \end{pmatrix}$

Notice that 3 − 1 = 2 and 2 − −3 = 5.

Vectors can be multiplied by a number.

If $\mathbf{a} = \begin{pmatrix} 3 \\ 2 \end{pmatrix}$

$3\mathbf{a} = 3 \times \begin{pmatrix} 3 \\ 2 \end{pmatrix} = \begin{pmatrix} 9 \\ 6 \end{pmatrix}$

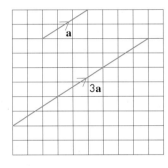

Notice that 3 × 3 = 9 and 3 × 2 = 6.

Here 3 is called a **scalar**, to distinguish it from a vector.

If k is a scalar, then $k \begin{pmatrix} x \\ y \end{pmatrix} = \begin{pmatrix} kx \\ ky \end{pmatrix}$.

EXERCISE 29A

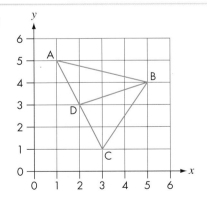

    a Write these as column vectors.

        i $\overrightarrow{AB}$ **ii** $\overrightarrow{DB}$ **iii** $\overrightarrow{CB}$

        iv $\overrightarrow{CA}$ **v** $\overrightarrow{AC}$ **vi** $\overrightarrow{DA}$

    b Show that $\overrightarrow{AD} = \overrightarrow{DC}$. What does this tell you about the position of D on line AC?

Chapter 29: Vectors

2 F has the coordinates (4, 2), G has the coordinates (2, 6), M is the midpoint of FG and O is the origin.

   a Mark F, G and M on a coordinate grid.

   b Write these as column vectors.

   i $\vec{FG}$ **ii** $\vec{GF}$ **iii** $\vec{OM}$ **iv** $\vec{MG}$ **v** $\vec{GO}$

3 A has the coordinates (3, 4).

$\vec{AP} = \begin{pmatrix} 2 \\ 1 \end{pmatrix}$, $\vec{AQ} = \begin{pmatrix} 1 \\ -3 \end{pmatrix}$, $\vec{AR} = \begin{pmatrix} -2 \\ 0 \end{pmatrix}$

Mark A, P, Q and R on a coordinate grid.

4 $\mathbf{a} = \begin{pmatrix} 2 \\ 3 \end{pmatrix}$ and $\mathbf{b} = \begin{pmatrix} 1 \\ -4 \end{pmatrix}$

Draw diagrams to show these vectors.

   a a + b
   b −b
   c a − b
   d b + a
   e b − a
   f 2b

5 $\mathbf{e} = \begin{pmatrix} -3 \\ -2 \end{pmatrix}$ and $\mathbf{f} = \begin{pmatrix} 2 \\ 4 \end{pmatrix}$

Work out these vectors.

   a e + f
   b 3f
   c e − f
   d f − e
   e 4e
   f 2f + e

6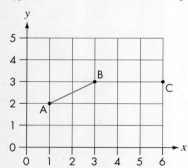

Copy this diagram.

   a $\vec{AE} = 2\vec{AB}$. Mark E on the grid.

   b $\vec{CD} = -2\vec{AB}$. Mark D on the grid.

   c $\vec{AB} = 2\vec{AM}$. Mark M on the grid.

   d $\vec{CN} = 2\vec{CB}$. Mark N on the grid.

   e If $\vec{DC} = k\vec{AM}$, what is the value of k?

Chapter 29. Topic 2

29.2 Using vectors **E**

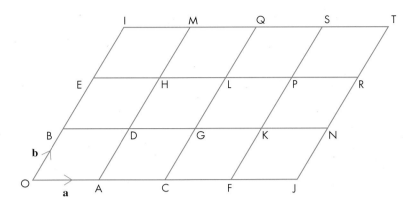

This grid is made of identical parallelograms.

O is the origin.

The **position vector** of A is

$\overrightarrow{OA} = \mathbf{a}$

The **position vector** of B is

$\overrightarrow{OB} = \mathbf{b}$

You can give the position vectors of the other points in terms of **a** and **b**. For example:

The position vector of G = $\overrightarrow{OG}$ = 2**a** + **b**

The position vector of S = $\overrightarrow{OS}$ = 3**a** + 3**b**

You can write other vectors in terms of **a** and **b**. For example:

$\overrightarrow{CL}$ = 2**b**

$\overrightarrow{CP} = \overrightarrow{CL} + \overrightarrow{LP}$ = 2**b** + **a** or **a** + 2**b**

$\overrightarrow{CH} = \overrightarrow{CL} + \overrightarrow{LH}$ = 2**b** + −**a** = 2**b** − **a**

Example 1

a Using the grid above, write down these vectors in terms of **a** and **b**.

   i $\overrightarrow{BH}$ ii $\overrightarrow{HP}$ iii $\overrightarrow{GT}$

   iv $\overrightarrow{TI}$ v $\overrightarrow{FH}$ vi $\overrightarrow{BQ}$

b What is the relationship between the following vectors?

   i $\overrightarrow{BH}$ and $\overrightarrow{GT}$ ii $\overrightarrow{BQ}$ and $\overrightarrow{GT}$ iii $\overrightarrow{HP}$ and $\overrightarrow{TI}$

c Show that B, H and Q lie on the same straight line.

---

a i **a** + **b**   ii 2**a**   iii 2**a** + 2**b**   iv −4**a**   v −2**a** + 2**b**   vi 2**a** + 2**b**

b i $\overrightarrow{BH}$ and $\overrightarrow{GT}$ are parallel and $\overrightarrow{GT}$ is twice the length of $\overrightarrow{BH}$.

   ii $\overrightarrow{BQ}$ and $\overrightarrow{GT}$ are equal.

   iii $\overrightarrow{HP}$ and $\overrightarrow{TI}$ are in opposite directions and $\overrightarrow{TI}$ is twice the length of $\overrightarrow{HP}$.

c $\overrightarrow{BH}$ and $\overrightarrow{BQ}$ are parallel and start at the same point B. Therefore, B, H and Q must lie on the same straight line.

EXERCISE 29B

1 On this grid O is the origin, $\overrightarrow{OA}$ is **a** and $\overrightarrow{OB}$ is **b**.

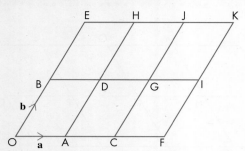

    a Name three other vectors equivalent to **a**.
    b Name three other vectors equivalent to **b**.
    c Name three vectors equivalent to −**a**.
    d Name three vectors equivalent to −**b**.

2 Using the same grid as in question **1**, write these vectors in terms of **a** and **b**.

    a the position vector of C
    b the position vector of E
    c the position vector of K
    d $\overrightarrow{OH}$
    e $\overrightarrow{AG}$
    f $\overrightarrow{AK}$
    g $\overrightarrow{BK}$

3 On the grid in question **1**, there are three vectors equivalent to $\overrightarrow{OG}$. Name all three.

4 On the grid in question **1**, there are three vectors that are three times the magnitude of $\overrightarrow{OA}$ and act in the same direction. Name all three.

5 Copy this grid.

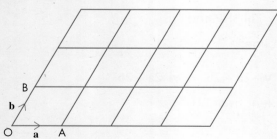

On your copy, mark the points C to G to show these vectors.

    a $\overrightarrow{OC} = 2\mathbf{a} + 3\mathbf{b}$ **b** $\overrightarrow{OD} = 2\mathbf{a} + \mathbf{b}$ **c** $\overrightarrow{OE} = 4\mathbf{a}$
    d $\overrightarrow{OF} = 4\mathbf{a} + 2\mathbf{b}$ **e** $\overrightarrow{OG} = \frac{1}{2}\mathbf{a} + 2\mathbf{b}$

6 On this grid, $\vec{OA}$ is **a** and $\vec{OB}$ is **b**.

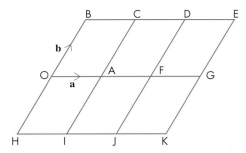

Write down vectors in terms of **a** and **b**.

a $\vec{OH}$ b $\vec{OK}$
c $\vec{OJ}$ d $\vec{OI}$
e $\vec{OC}$ f $\vec{CO}$
g $\vec{AK}$ h $\vec{DI}$
i $\vec{JE}$ j $\vec{AB}$
k $\vec{CK}$ l $\vec{DK}$

7 **a** On the grid in question **6**, there are two vectors that are twice the size of $\vec{AB}$ and act in the opposite direction. Name both of them.

b On the grid in question **6**, there are three vectors that are three times the size of $\vec{OA}$ and act in the opposite direction. Name all three.

8 Copy this grid.

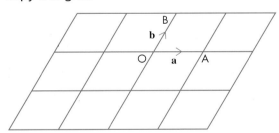

On your copy, mark the points C to P to show these vectors.

a $\vec{OC} = 2\mathbf{a} - \mathbf{b}$ b $\vec{OD} = 2\mathbf{a} + \mathbf{b}$
c $\vec{OE} = \mathbf{a} - 2\mathbf{b}$ d $\vec{OF} = \mathbf{b} - 2\mathbf{a}$
e $\vec{OG} = -\mathbf{a}$ f $\vec{OH} = -\mathbf{a} - 2\mathbf{b}$
g $\vec{OI} = 2\mathbf{a} - 2\mathbf{b}$ h $\vec{OJ} = -\mathbf{a} + \mathbf{b}$
i $\vec{OK} = -\mathbf{a} - \mathbf{b}$ j $\vec{OM} = -\mathbf{a} - \tfrac{3}{2}\mathbf{b}$
k $\vec{ON} = -\tfrac{1}{2}\mathbf{a} - 2\mathbf{b}$ l $\vec{OP} = \tfrac{3}{2}\mathbf{a} - \tfrac{3}{2}\mathbf{b}$

9 The diagram shows two sets of parallel lines. O is the origin.

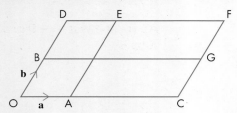

$\overrightarrow{OA}$ = a and $\overrightarrow{OB}$ = b
$\overrightarrow{OC}$ = 3$\overrightarrow{OA}$ and $\overrightarrow{OD}$ = 2$\overrightarrow{OB}$

a Write down these vectors in terms of **a** and **b**.
  i $\overrightarrow{OF}$ **ii** $\overrightarrow{OG}$ **iii** $\overrightarrow{EG}$ **iv** $\overrightarrow{CE}$

b Write down two vectors that can be written as 3**a** − **b**.

10 This grid shows the vectors $\overrightarrow{OA}$ = **a** and $\overrightarrow{OB}$ = **b**. O is the origin.

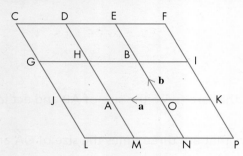

Write down the position vectors of these points.

a G
b F
c The midpoint of DH
d The centre of OAHB
e The centre of DCGH
f The centre of AMLJ

11 The diagram shows the vectors $\overrightarrow{OA}$ = **a** and $\overrightarrow{OB}$ = **b**. M is the midpoint of AB.

a i Work out the vector $\overrightarrow{AB}$ in terms of **a** and **b**.
  ii Work out the vector $\overrightarrow{AM}$.
  iii Explain why $\overrightarrow{OM} = \overrightarrow{OA} + \overrightarrow{AM}$.
  iv Using your answers to parts **ii** and **iii**, work out $\overrightarrow{OM}$ in terms of **a** and **b**.

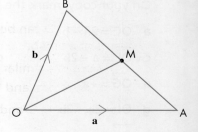

b Copy the diagram and show on it the vector $\overrightarrow{OC}$ which is equal to **a** + **b**.
c Describe in geometrical terms the position of M in relation to O, A, B and C.

538   29.2 Using vectors

12 The diagram shows the vectors $\overrightarrow{OA} = \mathbf{a}$ and $\overrightarrow{OB} = \mathbf{b}$.
The point C divides the line AB in the ratio 1:2.

Advice and Tips

AC is $\frac{1}{3}$ the distance from A to B.

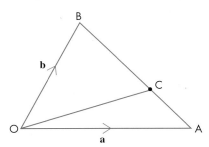

a i Work out the vector $\overrightarrow{AB}$.
   ii Work out the vector $\overrightarrow{AC}$.
   iii Work out the vector $\overrightarrow{OC}$ in terms of **a** and **b**.

b If C now divides the line AB in the ratio 1:3, write down the vector that represents $\overrightarrow{OC}$.

Advice and Tips

AC is now $\frac{1}{4}$ the distance from A to B.

13 The diagram shows the vectors $\overrightarrow{OA} = \mathbf{a}$ and $\overrightarrow{OB} = \mathbf{b}$.

Advice and Tips

OC is $\frac{2}{3}$ the distance from O to B.

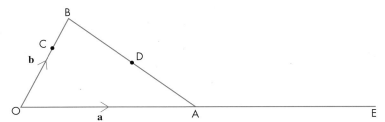

The point C divides OB in the ratio 2:1.

The point E is such that $\overrightarrow{OE} = 2\overrightarrow{OA}$.

D is the midpoint of AB.

a Write down (or work out) these vectors in terms of **a** and **b**.
   i $\overrightarrow{OC}$
   ii $\overrightarrow{OD}$
   iii $\overrightarrow{CO}$

b The vector $\overrightarrow{CD}$ can be written as $\overrightarrow{CD} = \overrightarrow{CO} + \overrightarrow{OD}$. Use this fact to work out $\overrightarrow{CD}$ in terms of **a** and **b**.

c Write down a similar rule to that in part **b** for the vector $\overrightarrow{DE}$. Use this rule to work out $\overrightarrow{DE}$ in terms of **a** and **b**.

d Explain why C, D and E lie on the same straight line.

Chapter 29: Vectors

29.3 The magnitude of a vector

The size or **magnitude** of a vector is represented by two vertical lines which stand for 'magnitude of' or 'length of', eg $|\overrightarrow{AB}|$ or $|\mathbf{a}|$.

If a vector is drawn on a rectangular coordinate grid you can use **Pythagoras' theorem** to calculate the magnitude.

For example:

if $\overrightarrow{AB} = \begin{pmatrix} 4 \\ -3 \end{pmatrix}$

then it can form the hypotenuse of a triangle with sides of lengths 3 and 4 as shown on the grid.

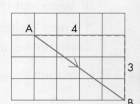

The square of the hypotenuse is equal to the sum of the squares of the other two sides so:

$|\overrightarrow{AB}| = \sqrt{4^2 + 3^2}$

$= \sqrt{25}$

$= 5$

In general, if $\mathbf{a} = \begin{pmatrix} x \\ y \end{pmatrix}$ then $|\mathbf{a}| = \sqrt{x^2 + y^2}$.

Example 2

If A has a coordinates (3, −2) and B has coordinates (−3, 5), calculate $|\overrightarrow{AB}|$.

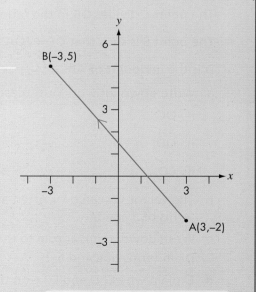

$\overrightarrow{AB} = \begin{pmatrix} -6 \\ 7 \end{pmatrix}$

$|\overrightarrow{AB}| = \sqrt{(-6)^2 + 7^2}$

$= \sqrt{36 + 49}$

$= \sqrt{85}$

$= 9.22$ to 2 decimal places

Advice and Tips

Remember that $(-6)^2 = 6^2$

EXERCISE 29C

1 O is the origin.

P, Q and R have position vectors $\begin{pmatrix} 3 \\ 5 \end{pmatrix}, \begin{pmatrix} 6 \\ -2 \end{pmatrix}, \begin{pmatrix} 0 \\ -4 \end{pmatrix}$.

a Show O, P, Q and R on a diagram.
b Calculate $|\overrightarrow{OP}|$, $|\overrightarrow{OQ}|$ and $|\overrightarrow{OR}|$. You can leave square root signs in your answers.
c Calculate $|\overrightarrow{PQ}|$.
d Calculate $|\overrightarrow{QR}|$.

2 $\mathbf{a} = \begin{pmatrix} 6 \\ 8 \end{pmatrix}$ and $\mathbf{b} = \begin{pmatrix} 5 \\ -12 \end{pmatrix}$

a Calculate $|\mathbf{a}|$ and $|\mathbf{b}|$.
b Calculate $\mathbf{a} + \mathbf{b}$.
c Calculate $|\mathbf{a} + \mathbf{b}|$.
d Is it true that $|\mathbf{a} + \mathbf{b}| = |\mathbf{a}| + |\mathbf{b}|$? Give a reason for your answer.
e Calculate $|\mathbf{a} - \mathbf{b}|$.
f Calculate $|\mathbf{b} - \mathbf{a}|$.
g Is it always true that $|\mathbf{a} - \mathbf{b}| = |\mathbf{b} - \mathbf{a}|$? Give a reason for your answer.

3 A, B, C and D have position vectors $\begin{pmatrix} 4 \\ 4 \end{pmatrix}, \begin{pmatrix} 10 \\ 12 \end{pmatrix}, \begin{pmatrix} -4 \\ 10 \end{pmatrix}$ and $\begin{pmatrix} 4 \\ -6 \end{pmatrix}$.

a Calculate $|\overrightarrow{AB}|$, $|\overrightarrow{AC}|$ and $|\overrightarrow{AD}|$.
b Explain why B, C and D must lie on a circle with centre A and state the radius of the circle.

4 $\mathbf{c} = \begin{pmatrix} 1 \\ 4 \end{pmatrix}$ and $\mathbf{d} = \begin{pmatrix} -2 \\ 5 \end{pmatrix}$

Calculate:

a $|\mathbf{c}|$
b $|3\mathbf{d}|$
c $|2\mathbf{c} + \mathbf{d}|$
d $|4\mathbf{c} - 2\mathbf{d}|$.

Check your progress

Core

- I can describe a translation with a vector of the form $\begin{pmatrix} x \\ y \end{pmatrix}$
- I can add and subtract vectors
- I can multiply a vector by a scalar

Extended

- I can calculate the magnitude of a vector
- I can represent vectors by line segments
- I can use sums and difference to express one vector in terms of two others
- I can use position vectors

Chapter 30
Transformations

Topics	Level	Key words
1 Translations	CORE	transformation, translation, vector
2 Reflections: 1	CORE	reflection, object, image, mirror line
3 Reflections: 2	EXTENDED	
4 Rotations: 1	CORE	rotation, centre of rotation, angle of rotation, clockwise, anticlockwise
5 Rotations: 2	EXTENDED	
6 Enlargements: 1	CORE	scale factor, enlargement, centre of enlargement, ray methods, coordinate method
7 Enlargements: 2	EXTENDED	negative enlargement
8 Combined transformations	EXTENDED	

In this chapter you will learn how to:

CORE	EXTENDED
• Reflect simple plane figures in horizontal or vertical lines. (C7.2 and E7.2) • Construct given translations and enlargements (with positive and fractional scale factors) of simple plane figures. (C7.2) • Recognise and describe reflections, rotations, translations and enlargements (with positive and fractional scale factors). (C7.2) • Rotate simple plane figures about the origin, vertices or midpoints of edges of the figures, through multiples of 90°. (C7.2)	• Construct given translations and enlargements (including positive, negative and fractional scale factors) of simple plane figures. (E7.2) • Recognise and describe reflections, rotations, translations and enlargements (including positive, negative and fractional scale factors). (E7.2) • Rotate simple plane figures through multiples of 90°. (E7.2)

Why this chapter matters

How many sides does a strip of paper have? Two or one?

Take a strip of paper about 20 cm by 2 cm.

How many sides does it have? Easy! You can see that this has two sides, a topside and an underside. If you were to draw a line along one side of the strip, you would have one side with a line 20 cm long on it and one side blank.

Now mark the ends A and B, put a single twist in the strip of paper and tape (or glue) the two ends together, as shown.

How many sides does this strip of paper have now?

Take a pen and draw a line on the paper, starting at any point you like. Continue the line along the length of the paper – you will eventually come back to your starting point. Your strip has only one side now! There is no blank side.

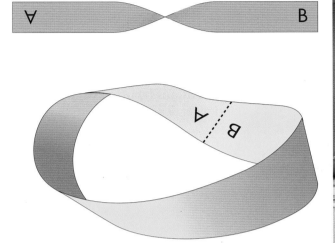

You have transformed a two-sided piece of paper into a one-sided piece of paper.

This shape is called a Möbius strip. It is named after August Ferdinand Möbius, a 19th-century German mathematician. Möbius caused a revolution in geometry.

Möbius strips have a number of applications that use its property of one-sidedness, including conveyor belts in industry and in vacuum cleaners.

The Möbius strip has become the universal symbol of recycling. The symbol was created in 1970 by Gary Anderson at the University of Southern California, as part of a contest sponsored by a paper company.

The Möbius strip is a form of transformation. In this chapter, you will look at some other transformations of shapes.

Chapter 30: Transformations 543

30.1 Translations

A **transformation** changes the position or the size of a shape.

There are four basic ways of changing the position and size of two-dimensional shapes: a **translation**, a reflection, a rotation or an enlargement. All of these transformations, except enlargement, keep shapes congruent.

A translation is the 'movement' of a shape from one place to another without reflecting it or rotating it. It is sometimes called a glide, since the shape appears to glide from one place to another. Every point in the shape moves in the same direction and through the same distance.

You can describe translations by using **vectors**. A vector is represented by the combination of a horizontal shift and a vertical shift.

Example 1

Use vectors to describe the translations of these triangles.

a A to B
b B to C
c C to D
d D to A

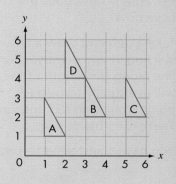

a The vector describing the translation from A to B is $\begin{pmatrix} 2 \\ 1 \end{pmatrix}$.

b The vector describing the translation from B to C is $\begin{pmatrix} 2 \\ 0 \end{pmatrix}$.

c The vector describing the translation from C to D is $\begin{pmatrix} -3 \\ 2 \end{pmatrix}$.

d The vector describing the translation from D to A is $\begin{pmatrix} -1 \\ -3 \end{pmatrix}$.

Note:

- The top number in the vector describes the horizontal movement. To the right +, to the left −.
- The bottom number in the vector describes the vertical movement. Upwards +, downwards −.
- These vectors are also called *direction vectors*.

EXERCISE 30A

1 Use vectors to describe these translations of the shapes on the grid below.

a i A to B
   ii A to C
   iii A to D
b i B to E
   ii B to F
   iii B to G
c i C to A
   ii C to E
   iii C to G
d i G to D
   ii F to G
   iii G to E

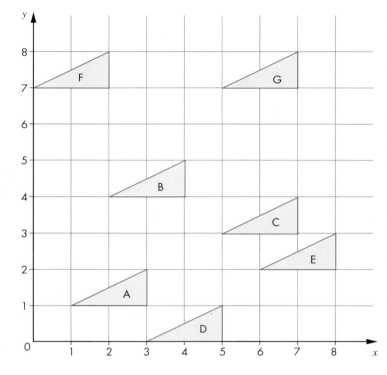

2 a Draw a set of coordinate axes and on it the triangle with coordinates A(1, 1), B(2, 1) and C(1, 3).

b Draw the image of ABC after a translation with vector $\begin{pmatrix} 2 \\ 3 \end{pmatrix}$. Label this triangle P.

c Draw the image of ABC after a translation with vector $\begin{pmatrix} -1 \\ 2 \end{pmatrix}$. Label this triangle Q.

d Draw the image of ABC after a translation with vector $\begin{pmatrix} 3 \\ -2 \end{pmatrix}$. Label this triangle R.

e Draw the image of ABC after a translation with vector $\begin{pmatrix} -2 \\ -4 \end{pmatrix}$. Label this triangle S.

3 Using your diagram from question **2**, use vectors to describe the translation that will move

a P to Q   b Q to R   c R to S   d S to P
e R to P   f S to Q   g R to Q   h P to S.

4 If a translation is given by:

$$\begin{pmatrix} x \\ y \end{pmatrix}$$

describe the translation that would take the image back to the original position.

Chapter 30: Transformations

Chapter 30 . Topic 2

5 A boat travels between three jetties X, Y and Z on a lake. Its journeys are described by direction vectors, with distance in kilometres.

The direction vector from X to Y is $\begin{pmatrix} 3 \\ -1 \end{pmatrix}$ and the direction vector from Y to Z is $\begin{pmatrix} -2 \\ -3 \end{pmatrix}$.

Using centimetre-squared paper and a scale of 1 cm : 1 km, draw a diagram to show journeys between X, Y and Z. Work out the direction vector for the journey from Z to X.

30.2 Reflections: 1

A **reflection** transforms a shape so that it becomes a mirror image of itself.

Example 2
Reflect this shape in the line provided.

The reflected shape looks like this.

Note: The reflection of each point in the original shape, called the **object**, is perpendicular to the mirror line. So if you 'fold' the whole diagram along the **mirror line**, the object will coincide with its reflection, called its **image**.

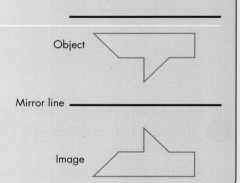

EXERCISE 30B

1 Copy the diagram.

On your copy, draw the reflection of the given triangle in each of these lines.

a $x = 2$
b $x = -1$
c $x = 3$
d $y = 2$
e $y = -1$
f y-axis

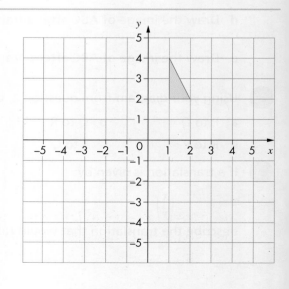

30.2

2
  a Draw a pair of axes.
     Label the x-axis from −5 to 5 and the y-axis from −5 to 5.
  b Draw the triangle with coordinates A(1, 1), B(3, 1), C(4, 5).
  c Reflect the triangle ABC in the x-axis. Label the image P.
  d Reflect triangle P in the y-axis. Label the image Q.
  e Reflect triangle Q in the x-axis. Label the image R.
  f Describe the reflection that will move triangle ABC to triangle R.

3
  a Draw a pair of axes.
     Label the x-axis from −5 to +5 and the y-axis from −5 to +5.
  b Reflect the points A(2, 1), B(5, 0), C(−3, 3), D(3, −2) in the x-axis.
  c What do you notice about the values of the coordinates of the reflected points?
  d What would the coordinates of the reflected point be if the point (a, b) were reflected in the x-axis?

4
  a Draw a pair of axes.
     Label the x-axis from −5 to +5 and the y-axis from −5 to +5.
  b Reflect the points A(2, 1), B(0, 5), C(3, −2), D(−4, −3) in the y-axis.
  c What do you notice about the values of the coordinates of the reflected points?
  d What would the coordinates of the reflected point be if the point (a, b) were reflected in the y-axis?

5 By using the middle square as a starting square called ABCD, describe how to keep reflecting the square to obtain the final shape in the diagram.

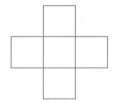

6 Triangle A is drawn on a grid.

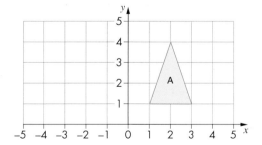

Triangle A is reflected to form a new triangle B.
The coordinates of triangle B are (−4, 4), (−3, 1) and (−5, 1).
Work out the equation of the mirror line.

Chapter 30: Transformations

Chapter 30 . Topic 3

30.3 Reflections: 2   E

You have been learning about reflections in horizontal or vertical lines.

You can reflect a shape in any line.

Example 3

Draw the reflection of triangle t in the line with equation $y = x$.

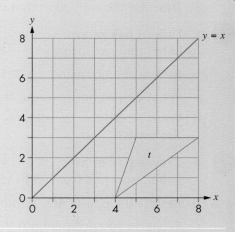

To find the image of each vertex of the triangle, draw lines perpendicular to the mirror.

Each vertex and its image are the same distance from the mirror but on opposite sides. Use the grid to help you find the new vertices.

Join the new vertices to draw the reflection M(t) of the triangle t.

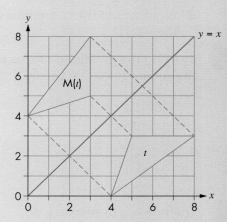

EXERCISE 30C

1 A designer used these instructions to create a design.

- Start with any rectangle ABCD.
- Reflect the rectangle ABCD in the line AC.
- Reflect the rectangle ABCD in the line BD.

Draw a rectangle and use the above to create a design.

30.3

2 Draw each of these triangles on squared paper, leaving plenty of space on the opposite side of the given mirror line. Then draw the reflection of each triangle.

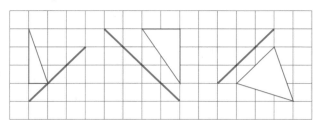

> **Advice and Tips**
>
> Turn the page around so that the mirror lines are vertical or horizontal.

3 a Draw a pair of axes and the lines $y = x$ and $y = -x$, as shown.

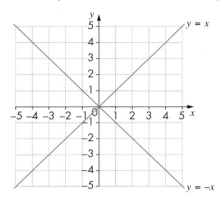

b Draw the triangle with coordinates A(2, 1), B(5, 1), C(5, 3).
c Draw the reflection of triangle ABC in the x-axis and label the image P.
d Draw the reflection of triangle P in the line $y = -x$ and label the image Q.
e Draw the reflection of triangle Q in the y-axis and label the image R.
f Draw the reflection of triangle R in the line $y = x$ and label the image S.
g Draw the reflection of triangle S in the x-axis and label the image T.
h Draw the reflection of triangle T in the line $y = -x$ and label the image U.
i Draw the reflection of triangle U in the y-axis and label the image W.
j What single reflection will move triangle W to triangle ABC?

4 Copy the diagram.

Reflect the triangle in these lines.

a $y = x$ **b** $x = 1$
c $y = -x$ **d** $y = -1$

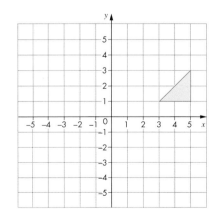

Chapter 30: Transformations **549**

Chapter 30 . Topic 4

5 **a** Draw a pair of axes.
   Label the x-axis from −5 to +5 and the y-axis from −5 to +5.
   b Draw the line $y = x$.
   c Reflect the points A(2, 1), B(5, 0), C(−3, 2), D(−2, −4) in the line $y = x$.
   d What do you notice about the values of the coordinates of the reflected points?
   e What would the coordinates of the reflected point be if the point (a, b) were reflected in the line $y = x$?

6 **a** Draw a pair of axes.
   Label the x-axis from −5 to +5 and the y-axis from −5 to +5.
   b Draw the line $y = -x$.
   c Reflect the points A(2, 1), B(0, 5), C(3, −2), D(−4, −3) in the line $y = -x$.
   d What do you notice about the values of the coordinates of the reflected points?
   e What would the coordinates of the reflected point be if the point (a, b) were reflected in the line $y = -x$?

30.4 Rotations: 1

A **rotation** transforms a shape to a new position by turning it about a fixed point called the **centre of rotation**.

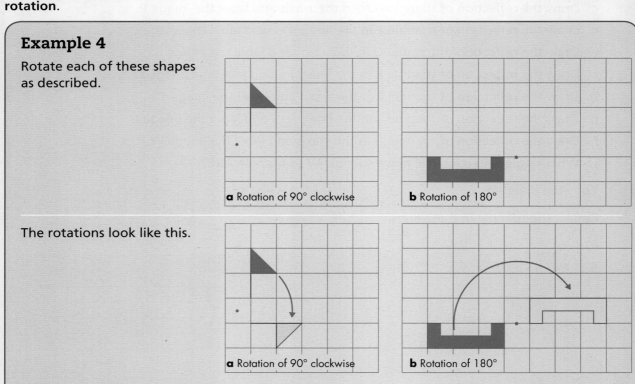

Example 4
Rotate each of these shapes as described.
a Rotation of 90° clockwise
b Rotation of 180°

The rotations look like this.
a Rotation of 90° clockwise
b Rotation of 180°

30.4

Note:
- The direction of turn or the **angle of rotation** is expressed as **clockwise** or **anticlockwise**.
- The position of the centre of rotation is always specified.
- The rotations 180° clockwise and 180° anticlockwise are the same.

The rotations that appear most frequently are 90° and 180°.

Advice and Tips

Use tracing paper to check rotations.

EXERCISE 30D

1 On squared paper, draw each of these shapes and its centre of rotation, leaving plenty of space all round the shape.

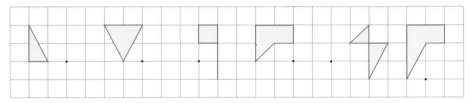

   a Rotate each shape about its centre of rotation:
   - **i** first by 90° clockwise (call the image A)
   - **ii** then by 90° anticlockwise (call the image B).

   b Describe, in each case, the rotation that would take:
   - **i** A back to its original position
   - **ii** A to B.

2 A graphics designer came up with this routine for creating a design.
   - Start with a triangle ABC.
   - Reflect the triangle in the line AB.
   - Rotate the whole shape about point C clockwise 90°, then a further clockwise 90°, then a further clockwise 90°.

   From any triangle of your choice, create a design using the above routine.

3 By using the middle square as a starting square, called ABCD, describe how to keep rotating the square to obtain the final shape in the diagram.

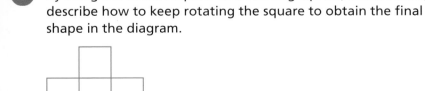

Chapter 30: Transformations 551

4 Copy the diagram. Rotate the given triangle by:

  a 90° clockwise about (0, 0)
  b 180° about (0, 0)
  c 90° anticlockwise about (1, 4)
  d 180° about (1, 3)
  e 90° clockwise about (2, 2).

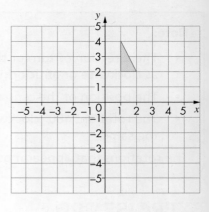

5 Give the centre and the angle for the rotations that will take:

  a A onto B
  b A onto C
  c A onto D
  d A onto E

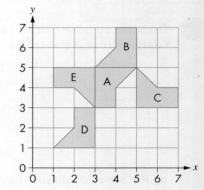

6

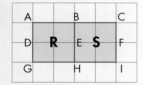

  a A 180° rotation will take R onto S. Where is the centre of rotation?
  b A 90° clockwise rotation will take R onto S. Where is the centre?

7 **a** Draw a pair of axes where both the x-values and y-values are from −5 to 5.
  b Draw the triangle ABC, where A = (1, 2), B = (2, 4) and C = (4, 1).
  c **i** Rotate triangle ABC 90° clockwise about the origin (0, 0) and label the image A′, B′, C′, where A′ is the image of A, etc.
    ii Write down the coordinates of A′, B′, C′.
    iii What connection is there between A, B, C and A′, B′, C′?
    iv Will this connection always be so for a 90° clockwise rotation about the origin?

8 Repeat question **7**, but rotate triangle ABC through 180°.

9 Show that a reflection in the x-axis followed by a reflection in the y-axis is equivalent to a rotation of 180° about the origin.

30.5 Rotations: 2

EXERCISE 30E

1 Draw a set of x and y axes and label them both from 0 to 12.

Draw the triangle with vertices at (5,5), (7,5) and (7,8). Label it T.

  a Rotate T 180° about (4,5). Label the new triangle A.
  b Rotate T 90° clockwise about (7,4). Label the new triangle B.
  c Rotate T 90° anticlockwise about (7,9). Label the new triangle C.
  d What rotation will take triangle B onto triangle C?

2 **a** A 180° rotation will take square A onto square B. Where is the centre of the rotation?

  b A 90° *clockwise* rotation will take A onto B. Where is the centre of the rotation?

  c A 90° *anticlockwise* rotation will take A onto B. Where is the centre of the rotation?

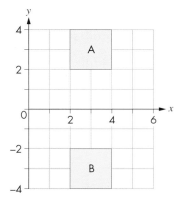

3 Give the centre and angle for the rotation of:

  a A onto B
  b B onto C
  c C onto D
  d D onto A.

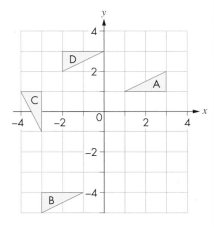

4 Show that a reflection in the line $y = x$ followed by a reflection in the line $y = -x$ is equivalent to a rotation of 180° about the origin.

Chapter 30 . Topic 6

5 a Draw a regular hexagon ABCDEF with centre O. The letters should go round the hexagon clockwise.

  b Using O as the centre of rotation, describe a transformation that will result in the following movements.

    i Triangle AOB to triangle BOC

    ii Triangle AOB to triangle COD

    iii Triangle AOB to triangle DOE

    iv Triangle AOB to triangle EOF

  c Describe the transformations that will move the rhombus ABCO to these positions.

    i Rhombus BCDO

    ii Rhombus DEFO

6 Triangle A, as shown on the grid, is rotated to form a new triangle B.

The coordinates of the vertices of B are (0, −2), (−3, −2) and (−3, −4).

Describe fully the rotation that maps triangle A onto triangle B.

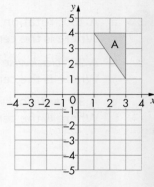

Find, if possible:

  a The equation of a mirror line that will reflect A onto B

  b The centre of a clockwise rotation of 90° that maps A onto B

  c The centre of an anticlockwise rotation of 90° that maps A onto B

  d The vector of a translation that maps A onto B

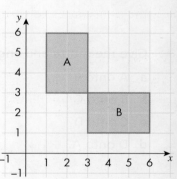

30.6 Enlargements: 1

An **enlargement** changes the size of a shape to give a similar image. It always has a **centre of enlargement** and a **scale factor**. Every length of the enlarged shape will be:

original length × scale factor

The distance of each image point on the enlargement from the centre of enlargement will be:

distance of original point from centre of enlargement × scale factor

30.6

Example 5

Enlarge triangle ABC by scale factor 3 about the centre of enlargement X.

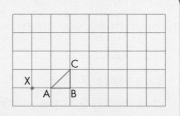

This is the completed enlargement.

Note:

- Each length on the enlargement A'B'C' is three times the corresponding length on the original shape.

    This means that the corresponding sides are in the same ratio:

    AB : A'B' = AC : A'C' = BC : B'C' = 1 : 3

- The distance of any point on the enlargement from the centre of enlargement is three times the distance from the corresponding point on the original shape to the centre of enlargement.

There are two distinct ways to enlarge a shape: the **ray method** and the **coordinate method** (counting squares).

Ray method

This is the *only* way to construct an enlargement when the diagram is not on a grid.

Example 6

Enlarge triangle ABC by scale factor 3 about the centre of enlargement X.

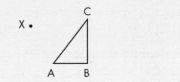

Notice that the rays have been drawn from the centre of enlargement to each vertex and beyond.

The distance from X to each vertex on triangle ABC is measured and multiplied by 3 to give the distance from X to each vertex A', B' and C' for the enlarged triangle A'B'C'.

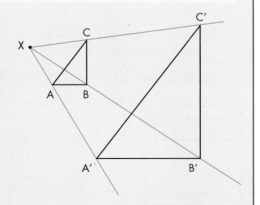

Once each image vertex has been found, the whole enlarged shape can then be drawn.

Check the measurements and see for yourself how the calculations have been done.

Notice again that the length of each side on the enlarged triangle is three times the length of the corresponding side on the original triangle.

Chapter 30: Transformations   555

Coordinate method

In this method, you use the coordinates of the vertices to 'count squares'.

Example 7

Enlarge the triangle ABC by scale factor 3 from the centre of enlargement (1, 2).

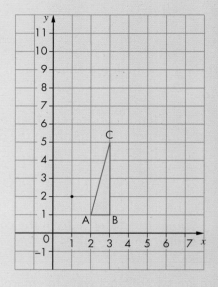

To find the coordinates of each image vertex, first work out the horizontal and vertical distances from each original vertex to the centre of enlargement.

Then multiply each of these distances by 3 to find the position of each image vertex.

For example, to find the coordinates of C' work out the distance from the centre of enlargement (1, 2) to the point C(3, 5).

horizontal distance = 2
vertical distance = 3

Make these 3 times longer to give:

new horizontal distance = 6
new vertical distance = 9

So the coordinates of C' are: (1 + 6, 2 + 9) = (7, 11)

Notice again that the length of each side is three times as long in the enlargement.

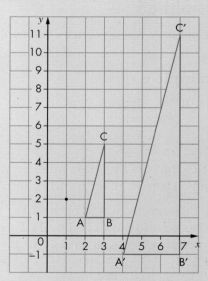

30.6

Negative enlargement

A **negative enlargement** produces an image shape on the opposite side of the centre of enlargement to the original shape.

> **Example 8**
>
> Enlarge triangle ABC by scale factor −2, with the centre of enlargement at (1, 0).
>
>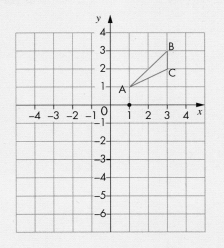
>
> You can enlarge triangle ABC to give triangle A'B'C' by either the ray method or the coordinate method. You calculate the new lengths on the opposite side of the centre of enlargement to the original shape.
>
> Notice how a negative scale factor also inverts the original shape.
>
>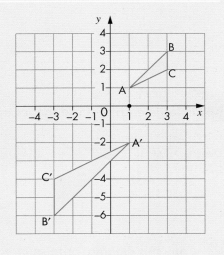

EXERCISE 30F

1 Copy each of these figures with its centre of enlargement. Then use the ray method to enlarge it by the given scale factor.

a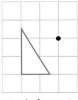
Scale factor 2

b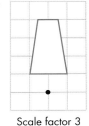
Scale factor 3

c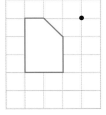
Scale factor 2

d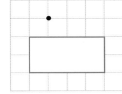
Scale factor 3

2 Copy each of these diagrams onto squared paper and enlarge it by scale factor 2, using the origin as the centre of enlargement.

a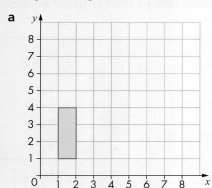

b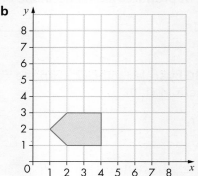

c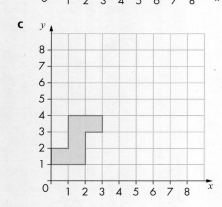

Advice and Tips

Even if you are using a counting square method, you can always check by using the ray method.

3 Enlarge each of these shapes by a scale factor of $\frac{1}{2}$ about its closest centre of enlargement.

4 Copy this diagram onto squared paper.

a Enlarge the rectangle A by scale factor $\frac{1}{3}$ about the origin. Label the image B.

b Write down the ratio of the lengths of the sides of rectangle A to the lengths of the sides of rectangle B.

c Work out the ratio of the perimeter of rectangle A to the perimeter of rectangle B.

d Work out the ratio of the area of rectangle A to the area of rectangle B.

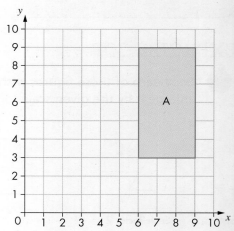

558 30.6 Enlargements: 1

30.7 Enlargements: 2

Finding the centre of enlargement

You can find the centre of an enlargement by drawing rays.

Example 9

E(*a*) is an enlargement of shape *a*.

Find the centre and the scale factor of the enlargement.

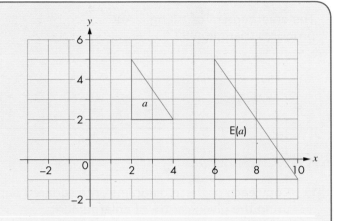

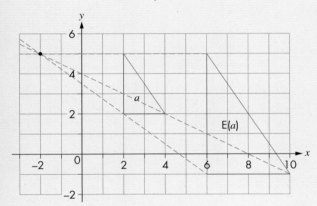

Draw rays through corresponding points on the object and image. They should all meet at one point. This is the centre of enlargement.

In this case the centre is (–2, 5).

To find the scale factor, find the ratio of corresponding sides.

Scale factor = $\dfrac{\text{height of } E(a)}{\text{height of } a}$

$= \dfrac{6}{3} = 2$

Fractional enlargement

Strange but true ... you can have an enlargement in mathematics that is actually smaller than the original shape!

Example 10

Enlarge triangle ABC by a scale factor of $\frac{1}{2}$ about the centre of enlargement O.

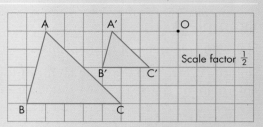

The enlargement of triangle ABC produces triangle A'B'C'.

EXERCISE 30G

1 Copy each of these diagrams onto squared paper and enlarge it by scale factor 2, using the given centre of enlargement.

a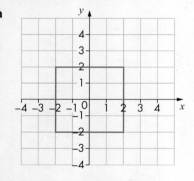

Centre of enlargement (−1, 1)

b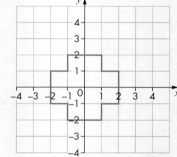

Centre of enlargement (−2, −3)

2 Copy this diagram onto squared paper.

Enlarge the triangle by scale factor −2 about the origin.

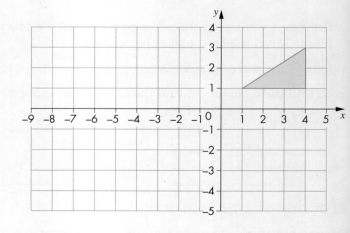

3 Copy this diagram onto squared paper.

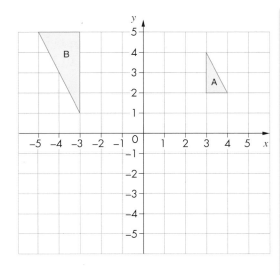

a Enlarge A by a scale factor of 3 about a centre (4, 5).

b Enlarge B by a scale factor $\frac{1}{2}$ about a centre (−1, −3).

c Enlarge B by scale factor $-\frac{1}{2}$ about a centre (−3, −1).

d What is the centre of enlargement and scale factor that maps B onto A?

e What is the centre of enlargement and scale factor that maps A onto B?

f What is the centre of enlargement and scale factor that maps the answer to part **b** to the answer to part **c**?

g What is the centre of enlargement and scale factor that maps the answer to part **c** to the answer to part **b**?

h What is the connection between the scale factors and the centres of enlargement in parts **d** and **e**, and in parts **f** and **g**?

4 Triangle A has vertices with coordinates (2, 1), (4, 1) and (4, 4).

Triangle B has vertices with coordinates (−5, 1), (−5, 7) and (−1, 7).

Describe fully the single transformation that maps triangle A onto triangle B.

5 A diagram of a shape is enlarged with a scale factor of 0.2

a The original shape was 48 cm long. Find the length of the enlargement.

b Find the ratio of the area of the original shape to the area of the enlargement.

30.8 Combined transformations

Sometimes you will need to use more than one **transformation** to produce the required image from the given object. In this exercise, you will revise the transformations you have met so far.

Remember, to describe:

- a **translation** fully, you need to use a vector
- a **reflection** fully, you need to use a mirror line
- a **rotation** fully, you need a centre of rotation, an angle of rotation and the direction of turn
- an **enlargement** fully, you need a centre of enlargement and a scale factor.

EXERCISE 30H

1 The point P(3, 4) is reflected in the *x*-axis, then rotated by 90° clockwise about the origin. What are the coordinates of the image of P?

2 A point Q(5, 2) is rotated by 180° about the origin, then reflected in the *x*-axis.

   a What are the coordinates of the image of Q?

   b What single transformation would have taken point Q directly to the image point?

3 Describe fully the transformations that will map the shaded triangle onto each of the triangles A–F.

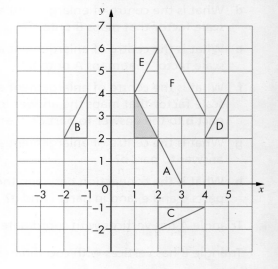

4 Describe fully the transformations that will result in the movement of:

   a T_1 to T_2

   b T_1 to T_6

   c T_2 to T_3

   d T_6 to T_2

   e T_6 to T_5

   f T_5 to T_4.

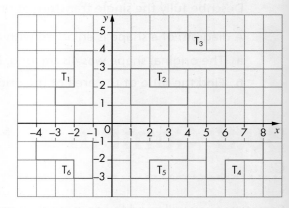

5 a Plot a triangle *t* with vertices (1, 1), (2, 1), (1, 3).

   b Reflect triangle *t* in the *y*-axis and label the image M(*t*).

   c Rotate triangle M(*t*) 90° anticlockwise about the origin and label the image RM(*t*).

   d Describe fully the transformation that will move triangle RM(*t*) back to triangle *t*.

6 Describe fully at least three different transformations that could move the square labelled S to the square labelled T.

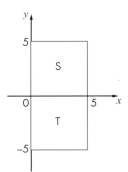

7 Copy the diagram onto squared paper.

a Triangle A is translated by the vector $\begin{pmatrix} -1.5 \\ -3 \end{pmatrix}$ to give triangle B.

Triangle B is then enlarged by a scale factor −2 about the origin to give triangle C.

Draw triangles B and C on the diagram.

b Describe fully the single transformation that maps triangle C onto triangle A.

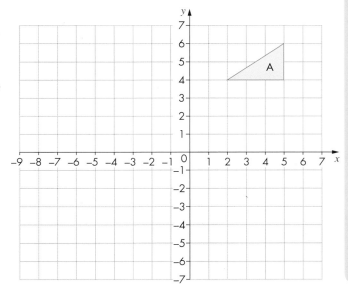

Check your progress

Core
- I can reflect a simple plane figure in a horizontal or vertical line
- I can rotate a simple plane figure through a multiple of 90° about the origin or a vertex or the midpoint of an edge
- I can translate simple plane figures
- I can enlarge simple plain figures with a positive scale factor, including fractions
- I can recognise and describe reflections, rotations, translations and enlargements

Extended
- I can reflect a simple plane figure
- I can rotate a simple plane figure
- I can enlarge simple plain figures with a negative scale factor

Examination questions: Geometry

Past paper questions reproduced by permission of Cambridge Assessment International Education.
Other exam-style questions have been written by the authors.

PAPER 1

1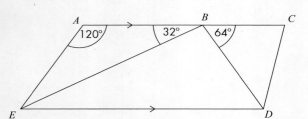

The diagram shows quadrilateral ACDE.
AC is parallel to ED and B is a point on AC.
Angle EAB = 120°, angle ABE = 32° and angle CBD = 64°.

a Work out angle EBD. [1]
b Work out angle AEB. [1]
c Complete this statement.
   Angle BED = angle ABE because they are .... angles. [1]

Cambridge International IGCSE Mathematics 0580 *Paper 11 Q13 Oct/Nov 2015*

2 Work out the size of one interior angle of a regular 15-sided polygon. [3]

Cambridge International IGCSE Mathematics 0580 *Paper 11 Q14 Oct/Nov 2015*

3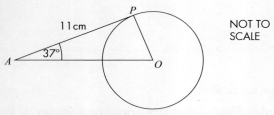

In the diagram, AP is a tangent to the circle at P.
O is the centre of the circle, angle PAO = 37° and AP = 11 cm.

a Write down the size of angle OPA. [1]
b Work out the radius of the circle. [2]

Cambridge International IGCSE Mathematics 0580 *Paper 11 Q16 Oct/Nov 2015*

Examination questions: Geometry

4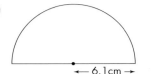

NOT TO SCALE

A protractor is a semi-circle of radius 6.1 cm.

Calculate the **perimeter** of the protractor. [3]

Cambridge International IGCSE Mathematics 0580 *Paper 11 Q18 Oct/Nov 2015*

5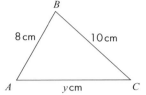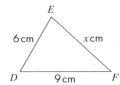

NOT TO SCALE

Triangle ABC is similar to triangle DEF.

Calculate the value of

a x, [2]

b y. [2]

Cambridge International IGCSE Mathematics 0580 *Paper 11 Q22 May/June 2015*

6 Six donkeys are **each** given two 5 ml spoons of medicine three times each day.

Calculate the number of whole days a 2 litre bottle of medicine will last. [3]

Cambridge International IGCSE Mathematics 0580 *Paper 11 Q14 May/June 2015*

7 A cuboid has volume 288 cm^3.

a The cuboid has length 12 cm and width 5 cm.

Calculate the height of the cuboid. [2]

b 1 cm^3 of the cuboid has a mass of 4 g.

Work out the mass of the cuboid. [1]

Cambridge International IGCSE Mathematics 0580 *Paper 11 Q15 May/June 2015*

8 Write each of the following as a single vector.

a $\begin{pmatrix} 6 \\ 1 \end{pmatrix} + \begin{pmatrix} -4 \\ 2 \end{pmatrix}$ [1]

b $4 \begin{pmatrix} 2 \\ -3 \end{pmatrix}$ [1]

Cambridge International IGCSE Mathematics 0580 *Paper 11 Q5 May/June 2013*

9 A cylinder has radius 3.6 cm and height 16 cm.

Calculate the volume of the cylinder. [2]

Cambridge International IGCSE Mathematics 0580 *Paper 11 Q9 Oct/Nov 2014*

Examination questions: Geometry

10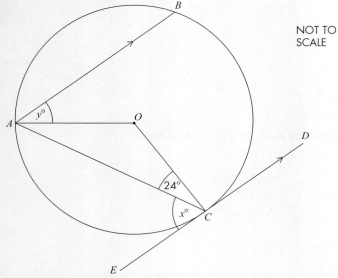

The diagram shows a circle with centre O.
ED is a tangent to the circle at C.
AB is parallel to ED and angle $ACO = 24°$.
Find the value of
- **a** x, [1]
- **b** y. [2]

Cambridge International IGCSE Mathematics 0580 *Paper 11 Q16 Oct/Nov 2014*

11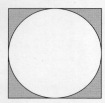

The diagram shows a circle inside a square.
The circumference of the circle touches all four sides of the square.
- **a** Calculate the area of the circle when the side of the square is 15 cm. [2]
- **b** Draw all the lines of symmetry on the diagram. [2]

Cambridge International IGCSE Mathematics 0580 *Paper 11 Q21 May/June 2014*

12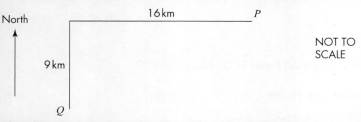

The diagram shows the route of a ship that leaves a port, P.
It travels due west for 16 km and then changes course to due south for 9 km.
- **a** Calculate the straight line distance PQ. [2]
- **b** Use trigonometry to calculate the bearing of P from Q. [2]

Cambridge International IGCSE Mathematics 0580 *Paper 11 Q22 Oct/Nov 2015*

Examination questions: Geometry

PAPER 3

1 Irina has some solid building blocks.

   a Write down the mathematical name of this solid. [1]

   b Irina describes the shape of a different block.

      She says:

      It has 12 edges and 8 vertices. All the faces are the same shape.

      Write down the mathematical name of this solid. [1]

   c The diagram shows the end face of another block.

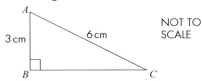

     i Show that $BC = 5.2$ cm, correct to 1 decimal place. [3]

     ii Find the area of triangle ABC. [2]

     iii This block is a triangular prism with length 8 cm.

        Calculate the volume of the block. [1]

   d The diagram shows another building block.

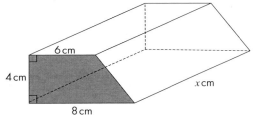

     i Calculate the area of the end face of this block. [2]

     ii The volume of this block is 336 cm³.

        Find the value of x. [1]

Cambridge International IGCSE Mathematics 0580 *Paper 31 Q6 Oct/Nov 2015*

2 a

     i Write down the order of rotational symmetry of this shape. [1]

     ii Draw the lines of symmetry on the shape. [2]

Examination questions: Geometry

b

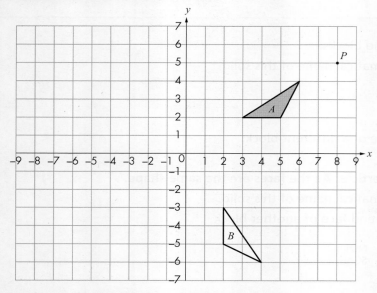

i On the grid, reflect triangle *A* in the **line** $x = -1$. [2]

ii On the grid, enlarge triangle *A* with centre *P* and scale factor 3. [2]

iii Describe fully the **single** transformation that maps triangle *A* onto triangle *B*. [3]

Cambridge International IGCSE Mathematics 0580 *Paper 31 Q8 Oct/Nov 2015*

3 a

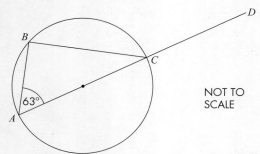

A, *B* and *C* lie on a circle with diameter *AC*.
AC is extended to *D* and angle *BAC* = 63°.
Work out angle *BCD*.
Give reasons to explain your answer. [4]

b

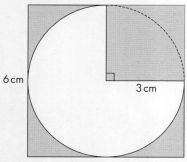

The diagram shows a circle with radius 3 cm inside a square of side 6 cm.
Calculate the shaded area. [5]

c

FGH is a right-angled triangle.

Calculate

i   *GH*, [3]

ii the perimeter of the triangle, [1]

iii the area of the triangle. [2]

Cambridge International IGCSE Mathematics 0580 *Paper 31 Q8 May/June 2015*

4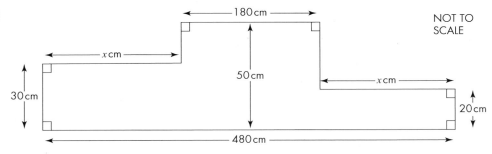

The diagram shows the cross section of a medal presentation platform.

a Show that *x* = 150. [2]
b Work out the perimeter of the cross section. [2]
c i   Calculate the area of the cross section. [2]
   ii The platform is a prism, 170 cm deep.
       Find the volume of the platform. [1]
   iii The prism is completely filled with a light material.
       1 **cubic metre** of this material has mass 16 kg.
       Calculate the mass of the material used. [2]

Cambridge International IGCSE Mathematics 0580 *Paper 31 Q4 Oct/Nov 2014*

Examination questions: Geometry

5

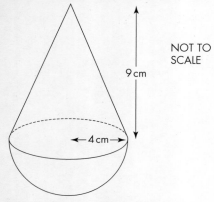

The diagram shows a toy.

The shape of the toy is a cone, with radius 4 cm and height 9 cm, on top of a hemisphere with radius 4 cm.

Calculate the volume of the toy.

Give your answer correct to the nearest cubic centimetre. [4]

[The volume, V, of a cone with radius r and height h is $V = \frac{1}{3}\pi r^2 h$.]

[The volume, V, of a sphere with radius r is $\frac{4}{3}\pi r^3$.]

Cambridge International IGCSE Mathematics 0580 *Paper 21 Q21 May/June 2015*

6

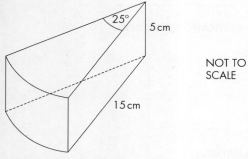

The diagram shows a wooden prism of height 5 cm.

The cross section of the prism is a sector of a circle with sector angle 25°.

The radius of the sector is 15 cm.

Calculate the total surface area of the prism. [5]

Cambridge International IGCSE Mathematics 0580 *Paper 21 Q19 Oct/Nov 2015*

PAPER 2

1 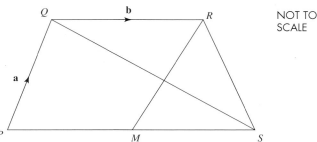 NOT TO SCALE

$PQRS$ is a quadrilateral and M is the midpoint of PS.
$\overrightarrow{PQ} = \mathbf{a}$, $\overrightarrow{QR} = \mathbf{b}$ and $\overrightarrow{SQ} = \mathbf{a} - 2\mathbf{b}$.

- **a** Show that $\overrightarrow{PS} = 2\mathbf{b}$. [1]
- **b** Write down the mathematical name for the quadrilateral $PQRM$, giving reasons for your answer. [2]

Cambridge International IGCSE Mathematics 0580 *Paper 21 Q14 May/June 2015*

2 ZEBRA

Write down the letters in the word above that have
- **a** exactly one line of symmetry, [1]
- **b** rotational symmetry of order 2. [1]

Cambridge International IGCSE Mathematics 0580 *Paper 21 Q3 Oct/Nov 2014*

3 A triangle has sides of length 2 cm, 8 cm and 9 cm.

Calculate the value of the largest angle in this triangle. [4]

Cambridge International IGCSE Mathematics 0580 *Paper 21 Q11 May/June 2014*

4 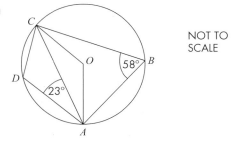 NOT TO SCALE

A, B, C and D lie on a circle centre O.

Angle $ABC = 58°$ and angle $CAD = 23°$.

Calculate
- **a** angle OCA, [2]
- **b** angle DCA. [2]

Cambridge International IGCSE Mathematics 0580 *Paper 21 Q13 May/June 2014*

Examination questions: Geometry

5

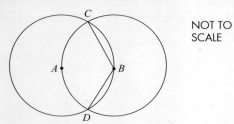

NOT TO SCALE

Two circles, centres A and B, are each of radius 8 cm and intersect at C and D.
Each circle passes through the centre of the other circle.

a Explain why angle CBD is 120°. [1]

b For the circle, centre B, find the area of the sector BCD. [2]

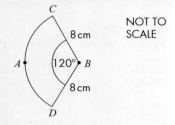

NOT TO SCALE

c i Find the area of the shaded segment CAD. [3]

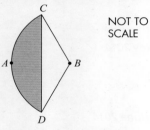

NOT TO SCALE

   ii Find the area of overlap of the two circles. [1]

Cambridge International IGCSE Mathematics 0580 *Paper 21 Q19 Oct/Nov 2014*

Examination questions: Geometry

PAPER 4

1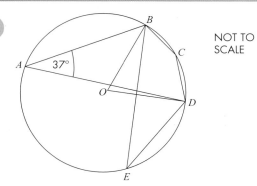

NOT TO SCALE

A, B, C, D and E are points on the circle, centre O.
Angle $BAD = 37°$.
Complete the following statements.
- **a** Angle $BED = $ .... because .... [2]
- **b** Angle $BOD = $ .... because .... [2]
- **c** Angle $BCD = $ .... because .... [2]

Cambridge International IGCSE Mathematics 0580 *Paper 41 Q5 Oct/Nov 2015*

2 **a** Andrei stands on level horizontal ground, 294 m from the foot of a vertical tower which is 55 m high.
   - **i** Calculate the angle of elevation of the top of the tower. [2]
   - **ii** Andrei walks a distance x metres directly towards the tower.
     The angle of elevation of the top of the tower is now 24.8°.
     Calculate the value of x. [4]

b The diagram shows a pyramid with a horizontal rectangular base.

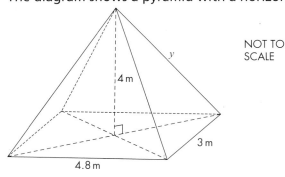

NOT TO SCALE

The rectangular base has length 4.8 m and width 3 m and the height of the pyramid is 4 m.
Calculate
- **i** y, the length of a sloping edge of the pyramid, [4]
- **ii** the angle between a sloping edge and the rectangular base of the pyramid. [2]

Cambridge International IGCSE Mathematics 0580 *Paper 41 Q5 May/June 2015*

Examination questions: Geometry

3 a

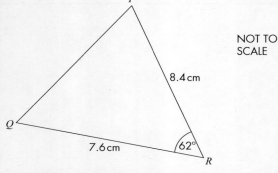

In the triangle PQR, $QR = 7.6$ cm and $PR = 8.4$ cm.
Angle $QRP = 62°$.
Calculate

i PQ, [4]

ii the area of triangle PQR. [2]

b

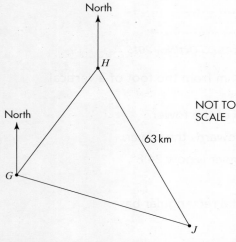

The diagram shows the positions of three small islands G, H and J.
The bearing of H from G is 045°.
The bearing of J from G is 126°.
The bearing of J from H is 164°.
The distance HJ is 63 km.
Calculate the distance GJ. [5]

Cambridge International IGCSE Mathematics 0580 Paper 41 Q7 May/June 2015

4 a $\overrightarrow{PQ} = \begin{pmatrix} -3 \\ 4 \end{pmatrix}$

i P is the point $(-2, 3)$.

Work out the co-ordinates of Q. [1]

ii Work out $|\overrightarrow{PQ}|$, the magnitude of $\overrightarrow{PQ}$. [2]

b

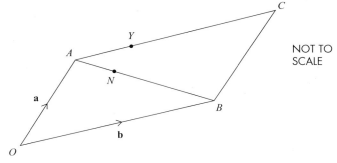

NOT TO SCALE

$OACB$ is a parallelogram.
$\overrightarrow{OA}$ = **a** and $\overrightarrow{OB}$ = **b**.
$AN:NB$ = 2:3 and $AY = \dfrac{2}{5} AC$.

  i Write each of the following in terms of **a** and/or **b**.
     Give your answers in their simplest form.

     a $\overrightarrow{ON}$ [2]

     b $\overrightarrow{NY}$ [2]

  ii Write down two conclusions you can make about the line segments NY and BC. [2]

Cambridge International IGCSE Mathematics 0580 *Paper 41 Q11 May/June 2014*

Chapter 31
Statistical representation

Topics	Level	Key words
1 Frequency tables	CORE	tally chart, frequency, frequency table, classes, class interval, grouped frequency table
2 Pictograms	CORE	pictogram, symbol, key
3 Bar charts	CORE	bar chart, axis
4 Pie charts	CORE	pie chart, angle, sector
5 Scatter diagrams	CORE	scatter diagram, variables, correlation, positive correlation, negative correlation, zero correlation, line of best fit
6 Histograms	CORE	histogram
7 Histograms with bars of unequal width	EXTENDED	class frequency, frequency density

In this chapter you will learn how to:

CORE	EXTENDED
• Collect, classify and tabulate statistical data. (C9.1 and E9.1) • Read, interpret and draw simple inferences from tables and statistical diagrams. Compare sets of data using tables, graphs and statistical measures. Appreciate restrictions on drawing conclusions from given data. (C9.2 and E9.2) • Construct and interpret bar charts, pie charts, pictograms, simple frequency distributions, histograms with equal intervals and scatter diagrams. (C9.3 and E9.3) • Understand what is meant by positive, negative and zero correlation with reference to a scatter diagram. (C9.7 and E9.7) • Draw, interpret and use lines of best fit by eye. (C9.8 and E9.8)	• Construct and read histograms with equal and unequal intervals (areas proportional to frequencies and vertical axis labelled 'frequency density'). (E9.3)

Why this chapter matters

Statistical graphs such as bar charts and line graphs are used in many areas of life from science to politics. They help us to analyse and interpret information.

One of the best ways to analyse information is to present it in a visual form. Some of the earliest types of statistical diagram were line graphs, bar charts and pie charts. They all show information in different ways.

Think about the owner of a bookshop.

He might use a graph like the one below to show how his sales go up and down over the year. Graphs like these are particularly good at showing trends in figures over time (see Chapter 14).

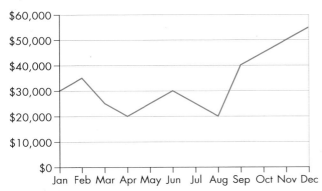

He might use a bar chart like the one on the right to show how many books they sell in different categories. Bar charts are very good at showing actual numbers.

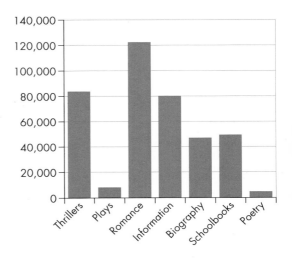

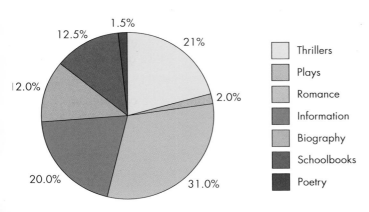

And he can get an idea of the percentage of different types of book he sells out of total sales by using a pie chart like the one on the left. Pie charts are good for analysing a whole (100%) by its parts.

This chapter introduces you to some of the most common forms of statistical representation. They fall into two groups – graphical diagrams such as bar charts and pie charts and quantitative diagrams such as frequency tables.

Chapter 31: Statistical representation

Chapter 31. Topic 1

31.1 Frequency tables

Statistics is concerned with the collection and organisation of data, the representation of data on diagrams and the interpretation of data.

When you are collecting data for simple surveys, it is usual to use a **tally chart**. For example, data collection sheets are used to gather information about how people travel to work, how students spend their free time and the amount of time people spend watching TV.

It is easy to record the data by using tally marks, as shown in Example 1. Counting up the tally marks in each row of the chart gives the **frequency** of each category. By listing the frequencies in a column on the right-hand side of the chart, you can make a **frequency table** (see Example 1). Frequency tables are an important part of making statistical calculations.

Example 1

Sandra wanted to find out about the ways in which students travelled to school. She carried out a survey. Her frequency table looked like this.

Method of travel	Tally	Frequency
Walk	⊬⊬ ⊬⊬ ⊬⊬ ⊬⊬ ⊬⊬ |||	28
Car	⊬⊬ ⊬⊬ ||	12
Bus	⊬⊬ ⊬⊬ ⊬⊬ ⊬⊬ |||	23
Bicycle	⊬⊬	5
Taxi	||	2

What does it tell you?

By adding together all the frequencies, you can see that 70 students took part in the survey. The frequencies also show you that more students travelled to school on foot than by any other method of transport.

Grouped data

Many surveys produce a lot of data that covers a wide range of values. In these cases, it is sensible to put the data into groups before attempting to compile a frequency table. These groups of data are called **classes** or **class intervals**.

Once the data has been grouped into classes, a **grouped frequency table** can be completed. The method is shown in Example 2.

Example 2

These marks are for 36 students in a Year 10 mathematics examination.

31	49	52	79	40	29	66	71	73	19	51	47
81	67	40	52	20	84	65	73	60	54	60	59
25	89	21	91	84	77	18	37	55	41	72	38

a Construct a frequency table, using classes of 1–20, 21–40 and so on.

b What was the most frequent interval of marks?

---

a Draw the grid of the table shown below and put in the headings.

Next, list the classes, in order, in the column headed 'Marks'.

Using tally marks, indicate each student's score against the class to which it belongs. For example, 81, 84, 89 and 91 belong to the class 81–100, giving five tally marks, as shown below.

Finally, count the tally marks for each class and enter the result in the column headed 'Frequency'. The table is now complete.

Marks	Tally	Frequency
1–20	|||	3
21–40	⦀⦀⦀ |||	8
41–60	⦀⦀⦀ ⦀⦀⦀ |	11
61–80	⦀⦀⦀ ||||	9
81–100	⦀⦀⦀	5

b From the grouped frequency table, you can see that the highest number of students obtained a mark in the 41–60 interval.

EXERCISE 31A

1 Kurt kept a record of the number of goals scored by his local team in the last 20 matches. These are his results.

0 1 0 3 2 1 0 2 1 1

a Draw a frequency table for his data.

b Which was the most frequent score?

c How many goals were scored in total for the 20 matches?

2 Monique was doing a geography project on the weather. As part of her work, she kept a record of the daily midday temperatures in June.

Daily temperatures for June (°C)

15 18 19 21 23 22
20 23 22 24 24 25
26 26 20 19 19 20
18 18 19 17 16 15
16 16 17 18 20 22

a Copy and complete the grouped frequency table for her data.

b In which interval do the most temperatures lie?

c Describe what the weather was probably like throughout the month.

Temperature (°C)	Tally	Frequency
14–16		
17–19		
20–22		
23–25		
26–28		

3 In a game, Mitesh used a six-sided dice. He decided to keep a record of his scores to see whether the dice was fair. These are his scores.

2 4 2 6 1 5 4 3 3 2 3 6 2 1 3
5 4 3 4 2 1 6 5 1 6 4 1 2 3 4

a Draw a frequency table for his data.

b How many throws did Mitesh have during the game?

c Do you think the dice was a fair one? Explain why.

4 The data shows the heights, in centimetres, of a sample of 32 students.

172 158 160 175 180 167 159 180
167 166 178 184 179 156 165 166
184 175 170 165 164 172 154 186
167 172 170 181 157 165 152 164

a Draw a grouped frequency table for the data, using class intervals 151–155, 156–160, …

b In which interval do the most heights lie?

c Does this agree with a survey of the students in your class?

5 A student used a stopwatch to time how long it took her rabbit to find food left in its hutch.

This is her record, in seconds.

7	30	14	27	8	31	8	28	10	41	51	37	15	21	37	16	38
23	20	9	11	55	9	33	8	35	45	35	25	25	49	23	43	55
45	8	13	9	39	12	57	16	37	26	32	19	48	29	37		

Find the best way to put this data into a frequency chart to illustrate the length of time it took the rabbit to find the food.

6 A student was doing a survey to find the ages of people at a football competition.

He said that he would make a frequency table with the regions 15–20, 20–25, 25–30.

Explain what difficulty he could have with these class divisions.

31.2 Pictograms

Data collected from a survey can be presented in pictorial or diagrammatic form to help people to understand it more quickly. You see plenty of examples of this in newspapers and magazines and on TV, where every type of visual aid is used to communicate statistical information.

Pictograms

A **pictogram** is a frequency table in which frequency is represented by a repeated **symbol**. The symbol itself usually represents a number of items, as Example 4 on the next page shows. However, sometimes it is more sensible to let a symbol represent just a single unit, as in Example 3 below. The **key** tells you how many items are represented by a symbol.

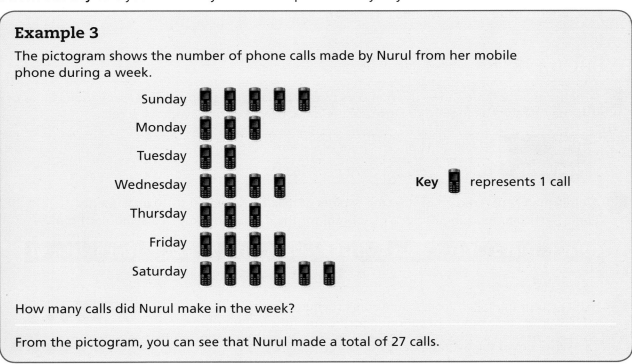

Example 3

The pictogram shows the number of phone calls made by Nurul from her mobile phone during a week.

How many calls did Nurul make in the week?

From the pictogram, you can see that Nurul made a total of 27 calls.

Although pictograms can have great visual impact (particularly as used in advertising) and are easy to understand, they have a serious drawback. Apart from a half, fractions of a symbol cannot usually be drawn accurately and so frequencies are often represented only approximately by symbols.

Example 4 on the next page highlights this difficulty.

Example 4

The pictogram shows the numbers of students who were late for school during a week.

Monday	👤👤👤👤
Tuesday	👤👤
Wednesday	👤👤👤
Thursday	👤👤👤
Friday	👤👤👤👤👤

Key 👤 represents 5 students

How many students were late on:

a Monday b Thursday?

Precisely how many students were late on Monday and Thursday respectively?

If you assume that each 'limb' of the symbol represents one student and its 'body' also represents one student, then the answers are:

a 19 students were late on Monday. b 13 on Thursday.

EXERCISE 31B

1. The frequency table shows the numbers of cars parked in a shop's car park at various times of the day. Draw a pictogram to illustrate the data. Use a key of 1 symbol = 5 cars.

Time	9 am	11 am	1 am	3 am	5 am
Frequency	40	50	70	65	45

2. A milkman kept a record of how many pints of milk he delivered to 10 apartments on a particular morning. Draw a pictogram for the data. Use a key of 1 symbol = 1 pint.

Flat 1	Flat 2	Flat 3	Flat 4	Flat 5	Flat 6	Flat 7	Flat 8	Flat 9	Flat 10
2	3	1	2	4	3	2	1	5	1

3. The pictogram, taken from a Suntours brochure, shows the average daily hours of sunshine for five months in Tenerife.

   a Write down the average daily hours of sunshine for each month.

   b Give a reason why pictograms are useful in holiday brochures.

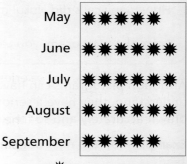

Key ✹ represents 2 hours

582   31.2 Pictograms

Chapter 31 . Topic 3

4 The pictogram shows the amounts of money collected by six students after they had completed a sponsored walk for charity.

Anthony	$ $ $ $ $
Ben	$ $ $ $ $ $
Emma	$ $ $ $ $
Leanne	$ $ $ $
Reena	$ $ $ $ $
Simon	$ $ $ $ $ $ $

Key $ represents $5

   a Who raised the most money?

   b How much money was raised altogether by the six students?

   c Robert also took part in the walk and raised $32. Why would it be difficult to include him on the pictogram?

5 A newspaper showed this pictogram about a family and the numbers of emails each family member received during one Sunday.

		Frequency
Dad	✉ ✉ ✉	
Mum	✉ ▷	
Teenage son	✉ ✉ ✉ ▷	
Teenage daughter		23
Young son		9

Key ✉ represents 4 emails

   a How many emails did:

      i Dad receive

     ii Mum receive

    iii the teenage son receive?

   b Copy and complete the pictogram.

   c How many emails were received altogether?

31.3 Bar charts

A **bar chart** consists of a series of bars or blocks of the *same* width, drawn either vertically or horizontally from an **axis**.

The heights or lengths of the bars always represent *frequencies*.

Sometimes, the bars are separated by narrow gaps of equal width, which makes the chart easier to read.

Example 5

The grouped frequency table shows the marks of 24 students in a test.

Draw a bar chart for the data.

Marks	1–10	11–20	21–30	31–40	41–50
Frequency	2	3	5	8	6

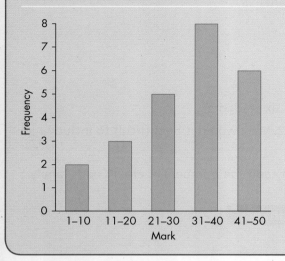

Note:
- Both axes are labelled.
- The class intervals are written under the middle of each bar.
- The bars are separated by equal spaces.

If you use a **dual bar chart**, it is easy to compare two sets of related data, as in Example 6.

Example 6

This dual bar chart shows the average daily maximum temperatures for England and Turkey over a five-month period.

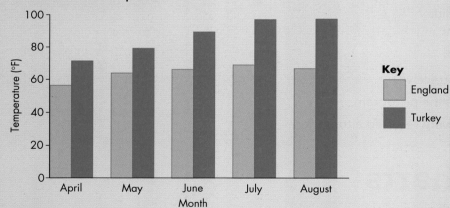

In which month was the *difference* between temperatures in England and Turkey the greatest?

The largest difference can be seen in August.

Note: You must always include a key to identify the two different sets of data.

EXERCISE 31C

1 For her survey on fitness, Samina asked a sample of people, as they left a sports centre, which activity they had taken part in. She then drew a bar chart to show her data.

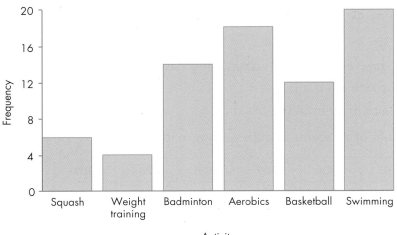

a Which was the most popular activity?

b How many people took part in Samina's survey?

2 The frequency table below shows the levels achieved by 100 students in their practice IGCSE examinations end of year tests.

Grade	F	E	D	C	B	A
Frequency	12	22	24	25	15	2

a Draw a suitable bar chart to illustrate the data.

b What fraction of the students achieved a grade C or grade B?

c Give one advantage of drawing a bar chart rather than a pictogram for this data.

3 This table shows the number of points Amir and Hasrul were each awarded in eight rounds of a general knowledge quiz.

Round	1	2	3	4	5	6	7	8
Amir	7	8	7	6	8	6	9	4
Hasrul	6	7	6	9	6	8	5	6

a Draw a dual bar chart to illustrate the data.

b Comment on how well each of them did in the quiz.

Chapter 31: Statistical representation 585

4 Mira did a survey about the time it took students in her class to get to school on a particular morning. She wrote down their times, correct to the nearest minute.

```
15  23  36  45   8  20  34  15  27  49
10  60   5  48  30  18  21   2  12  56
49  33  17  44  50  35  46  24  11  34
```

a Draw a grouped frequency table for Mira's data, using class intervals 1–10, 11–20, …
b Draw a bar chart to illustrate the data.
c What conclusions can Mira draw from the bar chart?

5 This table shows the number of accidents at a dangerous road junction over a six-year period.

Year	2005	2006	2007	2008	2009	2010
No. of accidents	6	8	7	9	6	4

a Draw a pictogram for the data.
b Draw a bar chart for the data.
c Which diagram would you use if you were going to argue that traffic lights should be installed at the junction? Explain why.

6 The diagram below shows the minimum and maximum temperatures for one day in August in five cities.

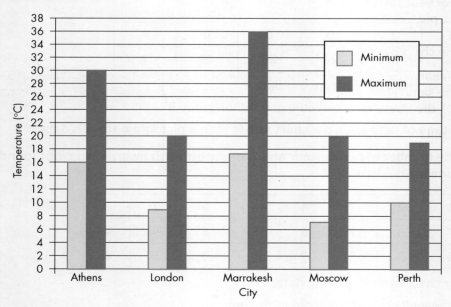

Lee says that the minimum temperature is always about half the maximum temperature for most cities.

Is Lee correct?

Give reasons to justify your answer.

31.4 Pie charts

Pictograms, bar charts and line graphs are easy to draw but they can be difficult to interpret when there is a big difference between the frequencies or there are only a few categories. In these cases, it is often more convenient to illustrate the data on a **pie chart**.

In a pie chart, the whole of the data is represented by a circle (the 'pie') and each category of it is represented by a **sector** of the circle (a 'slice of the pie'). The **angle** of each sector is proportional to the frequency of the category it represents.

So, a pie chart cannot show individual frequencies, as a bar chart can, for example. It can only show proportions.

Sometimes the pie chart will be marked off in equal sections rather than angles. In these cases, the numbers are always easy to work with.

Example 7

20 people were surveyed about their preferred drink.
Their replies are shown in the table.

Drink	Tea	Coffee	Milk	Cola
Frequency	6	7	4	3

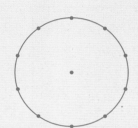

Show the results on the pie chart given.

You can see that the pie chart has 10 equally-spaced divisions.

As there are 20 people, each division is worth two people. So the sector for tea will take up 3 of these divisions. In the same way, coffee will take up $3\frac{1}{2}$ divisions, milk will take up 2 divisions and cola will take up $1\frac{1}{2}$ divisions.

The finished pie chart will look like this.

Note:

- You should always label the sectors of the pie chart (use shading and a separate key if there is not enough space to write on the pie chart).
- Give your chart a title.

Example 8

In a survey about holidays, 120 people were asked to state which type of transport they used on their last holiday. This table shows the results of the survey. Draw a pie chart to illustrate the data.

Type of transport	Train	Bus	Car	Ship	Plane
Frequency	24	12	59	11	14

You need to find the angle for the fraction of 360° that represents each type of transport. This is usually done in a table, as shown below.

Type of transport	Frequency	Calculation	Angle
Train	24	$\frac{24}{120} \times 360° = 72°$	72°
Bus	12	$\frac{12}{120} \times 360° = 36°$	36°
Car	59	$\frac{59}{120} \times 360° = 177°$	177°
Ship	11	$\frac{11}{120} \times 360° = 33°$	33°
Plane	14	$\frac{14}{120} \times 360° = 42°$	42°
Totals	120		360°

Draw the pie chart, using the calculated angle for each sector.

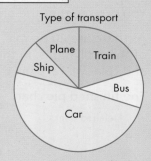

Note:

- Use the frequency total (120 in this case) to calculate each fraction.
- Check that the sum of all the angles is 360°.
- Label each sector.
- The angles or frequencies do not have to be shown on the pie chart.

EXERCISE 31D

1 Copy the diagram on the right and draw a pie chart to show each set of data.

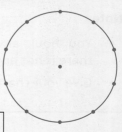

a The favourite pets of 10 children

Pet	Bird	Cat	Rabbit
Frequency	4	5	1

b The makes of cars of 20 teachers

Make of car	Ford	Toyota	BMW	Nissan	Peugeot
Frequency	4	5	2	3	6

c The newspaper read by 40 office workers

Newspaper	The Post	Today	The Mail	The Times
Frequency	14	8	6	12

2 Draw a pie chart to represent each set of data.

a The numbers of children in 40 families

No. of children	0	1	2	3	4
Frequency	4	10	14	9	3

Advice and Tips

Remember to complete a table as shown in the examples. Check that all angles add up to 360°.

b How 90 students get to school

Journey to school	Walk	Car	Bus	Cycle
Frequency	42	13	25	10

3 Mariam asked 24 of her friends which sport they preferred to play. Her data is shown in this frequency table.

Sport	Rugby	Football	Tennis	Baseball	Basketball
Frequency	4	11	3	1	5

Illustrate her data in a pie chart.

4 Ameer wrote down the number of lessons he had per week in each subject on his school timetable.

Mathematics 5   English 5
Science 8   History 6
Geography 6   Arts 4
Sport 2

a How many lessons did Ameer have on his timetable?
b Draw a pie chart to show the data.
c Draw a bar chart to show the data.
d Which diagram better illustrates the data? Give a reason for your answer.

5 A market researcher asked 720 people which new brand of tinned beans they preferred. The results are given in the table.

A	248
B	264
C	152
D	56

a Draw a pie chart to illustrate the data.
b Why do you think pie charts are used to show this sort of information?

6 This pie chart shows the proportions of the different shoe sizes worn by 144 students in one year group in a school.

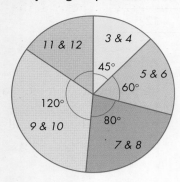

a What is the angle of the sector representing shoe sizes 11 and 12?

b How many students had a shoe size of 11 or 12?

7 The table below shows the numbers of candidates, at each grade, taking music examinations in Strings and Brass.

	Grades					Total number of candidates
	3	4	5	6	7	
Strings	300	980	1050	600	70	3000
Brass	250	360	300	120	70	1100

a Draw a pie chart to represent each of the two examinations.

b Compare the pie charts to decide which group of candidates, Strings or Brass, did better overall. Give reasons to justify your answer.

8 In a survey, a rail company asked passengers whether their service had improved.

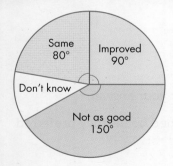

What is the probability that a person picked at random from this survey answered "Don't know"?

Chapter 31 . Topic 5

9 Two classes of 13-year-old students from different schools were give the same test.

The pie charts show the results of the test. Each pie chart was drawn by a teacher at the students' school.

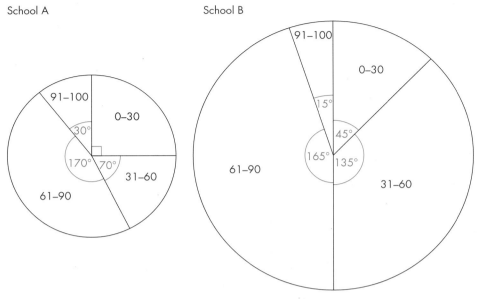

a Write a short report comparing the test results.

b State a reason why comparing pie charts like this can be unsatisfactory.

31.5 Scatter diagrams

A **scatter diagram** (also called a scattergraph or scattergram) is a method of comparing two **variables** by plotting their corresponding values on a graph. These values are usually taken from a table.

The variables are treated just like a set of (x, y) coordinates. This is shown in the scatter diagram on the right, in which the marks scored in a science test are plotted against the marks scored in a mathematics test.

This graph shows **positive correlation**. This means that students who get high marks in mathematics tests also tend to get high marks in science tests.

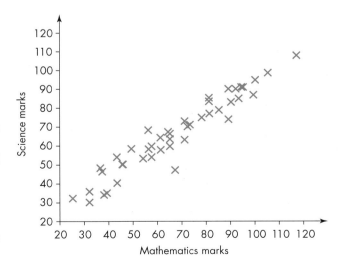

Chapter 31: Statistical representation   591

Correlation

There are different types of **correlation**. Here are three statements that may or may not be true.

- The taller people are, the wider their arm span is.
- The older a car is, the lower its value will be.
- The distance you live from your place of work will affect how much you earn.

These relationships could be tested by collecting data and plotting the points on a scatter diagram. For example, the first statement may give a scatter diagram like the one on the left below.

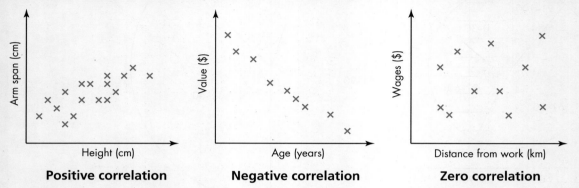

Positive correlation   **Negative correlation**   **Zero correlation**

This first diagram has **positive correlation** because, as one quantity increases, so does the other. From such a scatter diagram, you could say that the taller someone is, the wider their arm span.

Testing the second statement may give a scatter diagram like the middle one above. This has **negative correlation** because, as one quantity increases, the other quantity decreases. From such a scatter diagram, you could say that, as a car gets older, its value decreases.

Testing the third statement may give a scatter diagram like the one on the right, above. This scatter diagram has **zero correlation**. There is no relationship between the distance a person lives from their work and how much they earn.

Example 9

The graphs below show the relationship between the temperature and the amount of ice-cream sold, and the relationship between the age of people and the amount of ice-cream they eat.

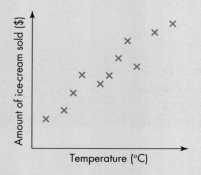

a Comment on the correlation of each graph.

b What does each graph tell you?

---

a The first graph has positive correlation and shows that, as the temperature increases, the amount of ice-cream sold increases.

b The second graph has negative correlation and shows that, as people get older, they eat less ice-cream.

31.5

Line of best fit

A **line of best fit** is a straight line that goes between all the points on a scatter diagram, passing as close as possible to all of them. You should try to have the same number of points on both sides of the line. Because you are drawing this line by eye, generous allowances are made around the correct answer. The line of best fit for the scatter diagram at the start of this section is shown below, left.

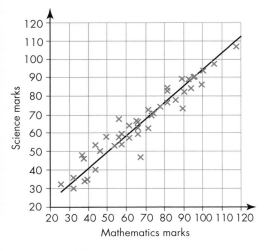
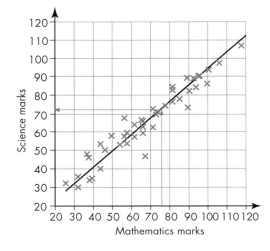

The line of best fit can be used to answer questions such as: 'A girl took the mathematics test and scored 75 marks but was ill for the science test. How many marks was she likely to have scored?'

You can find the answer by drawing a line up from 75 on the mathematics axis to the line of best fit and then drawing a line across to the science axis, as shown in the graph to the right of the graph showing the line of best fit. This gives 73, which is the mark she is likely to have scored in the science test.

Restrictions on the use of scatter diagrams

Scatter diagrams are very useful for estimating one value when given another. However, you need to be aware that there are limitations.

For example, here is a scatter diagram of the Olympic Men's 100 m winning time.

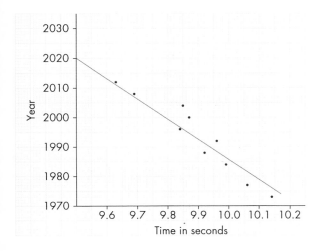

Chapter 31: Statistical representation   593

If you use the line of best fit, then the graph predicts that the time in the 2020 race should be about 9.5 seconds. However, this cannot continue forever, because it would imply that at some point the race could be won in 9 seconds, then 8 seconds, and eventually in no time at all!

Scatter diagrams and lines of best fit are very useful for predicting data within the range of values given, but are not so reliable when predicting outside of those ranges.

EXERCISE 31E

1 Describe the correlation of each of these two graphs.

a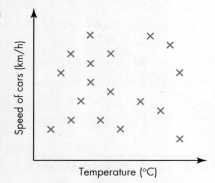

b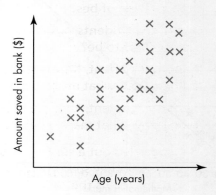

2 Explain what each graph in question **1** tells you.

3 The table below shows the results of a science experiment in which a ball is rolled along a desk top. The speed of the ball is measured at various points.

Distance from start (cm)	10	20	30	40	50	60	70	80
Speed (cm/s)	18	16	13	10	7	5	3	0

a Plot the data on a scatter diagram.
b Draw the line of best fit.
c If the ball's speed had been measured at 5 cm from the start, what is it likely to have been?
d Estimate how far the ball was from the start when its speed was 12 cm/s.

4 The heights, in centimetres, of 20 mothers and their 15-year-old daughters were measured. These are the results.

Mother	153	162	147	183	174	169	152	164	186	178
Daughter	145	155	142	167	167	151	145	152	163	168
Mother	175	173	158	168	181	173	166	162	180	156
Daughter	172	167	160	154	170	164	156	150	160	152

a Plot these results on a scatter diagram. Take the horizontal axis (the *x*-axis) for the mothers' heights from 140 to 200. Take the vertical axis (the *y*-axis) for the daughters' heights from 140 to 200.
b Is it true that the tall mothers have tall daughters?

5 The table shows the marks for ten students in their mathematics and geography examinations.

Student	Anna	Becky	Cath	Dema	Emma	Fatima	Greta	Hannah	Imogen	Sitara
Maths	145	155	142	167	167	151	145	152	163	168
Geog	175	173	158	168	181	173	166	162	180	156

a Plot the data on a scatter diagram. Take the x-axis for the mathematics scores and the y-axis for the geography scores.

b Draw the line of best fit.

c One of the students was ill when she took the geography examination. Which student was it most likely to be?

d If another student, Kate, was absent for the geography examination but scored 75 in mathematics, what mark would you expect her to have scored in geography?

e If another student, Lina, was absent for the mathematics examination but scored 65 in geography, what mark would you expect her to have scored in mathematics?

6 A teacher carried out a survey of 20 students from his class and asked them to say how many hours per week they spent playing sport and how many hours per week they spent watching TV. This table shows the results of the survey.

Student	1	2	3	4	5	6	7	8	9	10
Hours playing sport	12	3	5	15	11	0	9	7	6	12
Hours watching TV	18	26	24	16	19	27	12	13	17	14

Student	11	12	13	14	15	16	17	18	19	20
Hours playing sport	12	10	7	6	7	3	1	2	0	12
Hours watching TV	22	16	18	22	12	28	18	20	25	13

a Plot these results on a scatter diagram. Take the x-axis as the number of hours playing sport and the y-axis as the number of hours watching TV.

b If you knew that another student from the form watched 8 hours of TV a week, would you be able to predict how long they spent playing sport? Explain why.

7 The table shows the times taken and distances travelled by a taxi driver in 10 journeys on one day.

Distance (km)	1.6	8.3	5.2	6.6	4.8	7.2	3.9	5.8	8.8	5.4
Time (minutes)	3	17	11	13	9	15	8	11	16	10

a Draw a scatter diagram of this information, with time on the horizontal axis.

b Draw a line of best fit on your diagram.

c If a taxi journey takes 5 minutes, how far, in kilometres, would you expect the journey to have been?

d How much time would you expect a journey of 4 km to take?

Chapter 31. Topic 6

8 Omar records the time taken, in hours, and the average speed, in kilometres per hour (km/h), for several different journeys.

Time (h)	0.5	0.8	1.1	1.3	1.6	1.75	2	2.4	2.6
Speed (km/h)	42	38	27	30	22	23	21	9	8

Estimate the average speed for a journey of 90 minutes.

9 Describe what you would expect the scatter graph to look like if someone said that it showed negative correlation.

31.6 Histograms

You should already be familiar with bar charts like the one on the right, in which the vertical axis represents frequency, and each bar has a label to show what it represents. (Sometimes it is more convenient to have the axes the other way round.)

A **histogram** looks similar to a bar chart, but there are **three** fundamental differences.

- There are no gaps between the bars.
- The horizontal axis has a continuous scale.
- The area of each bar represents the class or group frequency of the bar.

This table shows times it takes people to walk to work.

Times, t minutes	$0 < t \leq 4$	$4 < t \leq 8$	$8 < t \leq 12$	$12 < t \leq 16$
Frequency	8	12	10	7

This **histogram** has been drawn from the table. The columns are *not* labelled $0 \leq t < 4$ and so on, as they would be on a bar chart. Instead there is a scale on the horizontal (time) axis. The first column is drawn between 0 and 4, the second between 4 and 8 and so on.

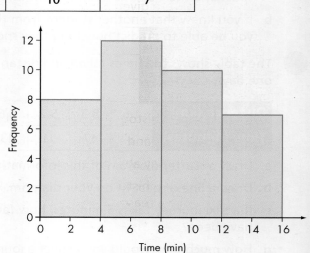

EXERCISE 31F

1 The table shows the range of heights of the girls in one year group in a school.

Height, h (cm)	120 < h ≤ 130	130 < h ≤ 140	140 < h ≤ 150	150 < h ≤ 160	160 < h ≤ 170
Frequency	8	12	10	7	−2

Draw a histogram for this data.

2 A doctor was concerned at the length of time her patients had to wait to see her when they came to the morning clinic. Her survey gave these results.

Time, m (minutes)	0 < m ≤ 10	10 < m ≤ 20	20 < m ≤ 30	30 < m ≤ 40	40 < m ≤ 50	50 < m ≤ 60
Monday	5	8	17	9	7	4
Tuesday	9	8	16	3	2	1
Wednesday	7	6	18	2	1	1

a Draw a separate histogram for each day.

b On which day did patients tend to have to wait longer?

3 These are the prices of twenty second-hand cars, in dollars.

2590, 2650, 2650, 2790, 2850, 2925, 3090, 3125, 3125, 3420,
3595, 3740, 3750, 3920, 3945, 4095, 4150, 4200, 4750, 4785

a Copy and complete the frequency table to show the prices.

Prices (dollars)	2500 < p ≤ 3000	3000 < p ≤ 3500	3500 < p ≤ 4000	4000 < p ≤ 4500	4500 < p ≤ 5000
Frequency					

b Illustrate the data on a histogram.

4 Boys and girls were given a simple task to complete. The times taken are shown below.

Times, t (minutes)	0 < t ≤ 4	4 < t ≤ 8	8 < t ≤ 12	12 < t ≤ 16	16 < t ≤ 20
Boys	3	7	21	26	15
Girls	4	8	17	23	20

a Draw separate histograms to show the boys' times and the girls' times.

b How many boys and how many girls completed the task?

c What can you say about the longest time taken?

d Look at the two histograms and say whether you think boys or girls were better at the task. Give a reason for your answer.

5. These are the results of a survey of the masses of 50 girls and 50 boys of the same age.

Mass, k (kg)	15 ≤ k < 20	20 ≤ k < 25	25 ≤ k < 30	30 ≤ k < 35	35 ≤ k < 40	40 ≤ k < 45
Girls	4	6	12	14	6	8
Boys	1	4	16	10	15	4

a Show this data on two separate histograms.

b Children of this age who weigh less than 25 kg are underweight. Shade the corresponding part of each histogram.

c Were more boys or girls underweight?

6. This histogram shows the times a group of employees took to travel to work one day.

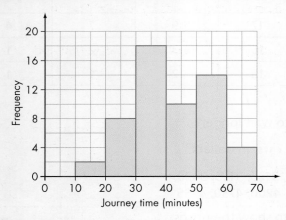

a How many took more than 50 minutes?

b How many took between 20 and 50 minutes?

c What can you say about the shortest and longest journey times?

The company introduced flexible working so that employees could start work at different times. Here are the results of a survey of journey times after the change.

Journey time, t (minutes)	0 ≤ t < 10	10 ≤ t < 20	20 ≤ t < 30	30 ≤ t < 40	40 ≤ t < 50	50 ≤ t < 50	60 ≤ t < 70
Frequency	2	4	14	12	14	5	5

d Show the new journey times on a histogram.

e Compare the two histograms. Have journey times been shortened? Justify your answer.

31.7 Histograms with bars of unequal width E

Sometimes the data in a frequency distribution are grouped into classes with intervals that are different. In this case, the resulting histogram has bars of unequal widths.

The key fact to remember is that the area of a bar in a histogram represents the **class frequency** of the bar. So, in the case of an unequal-width histogram, you find the height to draw each bar by dividing its class frequency by its **class interval** width (bar width), which is the difference between the lower and upper bounds for each interval.

Conversely, given a histogram, you can find any of its class frequencies by multiplying the height of the corresponding bar by its width.

It is for this reason that the scale on the vertical axes of these histograms is always labelled 'frequency density', where:

$$\text{frequency density} = \frac{\text{frequency of class interval}}{\text{width of class interval}}$$

Example 10

The heights of a group of girls were measured. The results were classified as shown in the table.

Height, h (cm)	$151 \leq h < 153$	$153 \leq h < 154$	$154 \leq h < 155$	$155 \leq h < 159$	$159 \leq h < 160$
Frequency	64	43	47	96	12

Draw a histogram to show the data.

It is convenient to write the table vertically and add two columns for class width and frequency density.

Calculate the class width by subtracting the lower class boundary from the upper class boundary. Calculate the frequency density by dividing the frequency by the class width.

Height, h (cm)	Frequency	Class width	Frequency density
$151 \leq h < 153$	64	2	32
$153 \leq h < 154$	43	1	43
$154 \leq h < 155$	47	1	47
$155 \leq h < 159$	96	4	24
$159 \leq h < 160$	12	1	12

The histogram can now be drawn.

The horizontal scale should be marked off as normal, from a value below the lowest value in the table to a value above the largest value in the table. In this case, mark the scale from 150 cm to 160 cm.

The vertical scale is always frequency density and is marked up to at least the largest frequency density in the table. In this case, 50 is a sensible value.

Each bar is drawn between the lower class interval and the upper class interval horizontally, and up to the frequency density vertically.

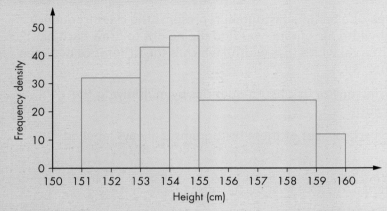

Now check that the area of each column is equal to the frequency.

151–153 is 2 × 32 = 64

153–154 is 1 × 43 = 43

and so on.

If the bars are of equal width, the frequency density and the frequency will be proportional. In that case you can use frequency on the vertical axis, as in section 31.6.

EXERCISE 31G

1 Draw histograms for these grouped frequency distributions.

a

Temperature, t (°C)	$8 \leqslant t < 10$	$10 \leqslant t < 12$	$12 \leqslant t < 15$	$15 \leqslant t < 17$	$17 \leqslant t < 20$	$20 \leqslant t < 24$
Frequency	5	13	18	4	3	6

b

Wage, w ($1000)	$6 \leqslant w < 10$	$10 \leqslant w < 12$	$12 \leqslant w < 16$	$16 \leqslant w < 24$
Frequency	16	54	60	24

c

Age, a (year)	$11 \leqslant a < 14$	$14 \leqslant a < 16$	$16 \leqslant a < 17$	$17 \leqslant w < 20$
Frequency	51	36	12	20

31.7 Histograms with bars of unequal width

d

Pressure, p (mm)	745 ⩽ p < 755	755 ⩽ p < 760	760 ⩽ p < 765	765 ⩽ p < 775
Frequency	4	6	14	10

e

Time, t (min)	0 ⩽ t < 8	8 ⩽ t < 12	12 ⩽ t < 16	16 ⩽ t < 20
Frequency	72	84	54	36

2 This information was gathered about the weekly pocket money given to 14-year-olds.

Pocket money, p ($)	0 ⩽ p < 2	2 ⩽ p < 4	4 ⩽ p < 5	5 ⩽ p < 8	8 ⩽ p < 10
Girls	8	15	22	12	4
Boys	6	11	25	15	6

Represent the information about the boys and girls on separate histograms.

3 The sales of the *Star* newspaper over 70 years are recorded in this table.

Years	1940–60	1961–80	1981–90	1991–2000	2001–05	2006–2010
Copies	62 000	68 000	71 000	75 000	63 000	52 000

Illustrate this information on a histogram.

Take the class boundaries as 1940, 1960, 1980, 1990, 2000, 2005, 2010.

4 The Madrid trains were always late, so one month a survey was undertaken to find how many trains were late, and by how many minutes.

The results are illustrated by this histogram.

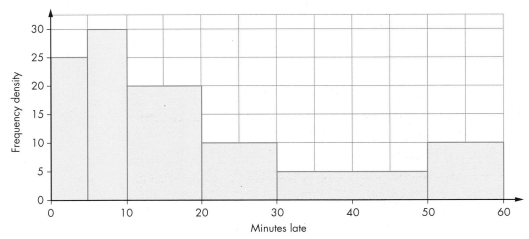

a How many trains were in the survey?

b How many trains were delayed for longer than 15 minutes?

5 For each of the frequency distributions illustrated in the histograms draw up the grouped frequency table.

a
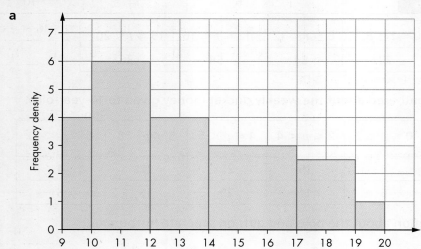

b
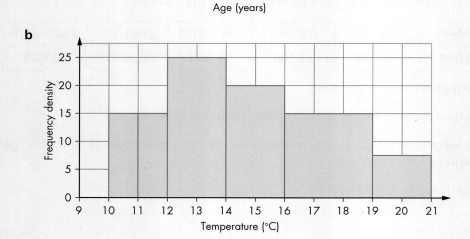

c
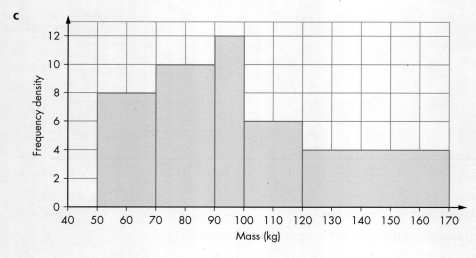

6 All the patients in a hospital were asked how long it was since they last saw a doctor. The results are shown in the table.

Hours, h	$0 \leq h < 2$	$2 \leq h < 4$	$4 \leq h < 6$	$6 \leq h < 10$	$10 \leq h < 16$	$16 \leq h < 24$
Frequency	8	12	20	30	20	10

a Draw a histogram to illustrate the data.
b Estimate how many people waited more than 8 hours.

Advice and Tips

Find the area to the right of $h = 8$.

7 One summer, Albert monitored the mass of the tomatoes grown on each of his plants. His results are summarised in this table.

Mass, m (kg)	$6 \leq m < 10$	$10 \leq m < 12$	$12 \leq m < 16$	$16 \leq m < 20$	$20 \leq m < 25$
Frequency	8	15	28	16	10

a Draw a histogram for this distribution.
b Estimate how many plants produced more than 15 kg.

8 A survey was carried out to find the speeds of cars passing a particular point on a road. The histogram illustrates the results of the survey.

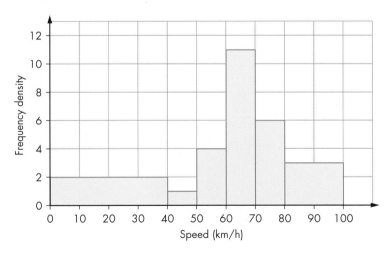

a Copy and complete this table.

Speed, v (km/h)	$0 < v \leq 40$	$40 < v \leq 50$	$50 < v \leq 60$	$60 < v \leq 70$	$70 < v \leq 80$	$80 < v \leq 100$
Frequency		10	40	110		

b Find the number of cars included in the survey.

9 The histogram shows the test scores for 320 students in a school.

   a How many students scored more than 120?

   b The pass mark was 90. What percentage of students failed the test?

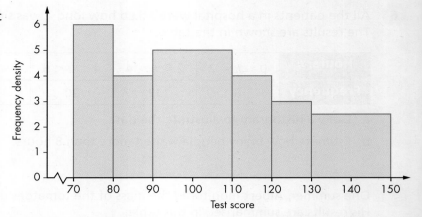

10 Adrienne and Bernice collected the same data about journey times but grouped it differently.

Here are Adrienne's figures.

Journey time (t minutes)	0 ≤ t < 5	5 ≤ t < 10	10 ≤ t < 15	15 ≤ t < 20	20 ≤ t < 25	25 ≤ t < 30
Frequency	10	15	30	12	6	3

Here are Bernice's figures.

Journey time (t minutes)	0 ≤ t < 10	10 ≤ t < 15	15 ≤ t < 30
Frequency	25	30	21

   a Draw a histogram for each set of figures. Use frequency density on the vertical axis each time.

   b Describe any similarities or differences between the histograms.

Check your progress

Core

- I can collect, classify and tabulate data
- I can read and interpret tables and statistical diagrams
- I can compare sets of data using tables and graphs
- I appreciate restrictions on drawing conclusions from data
- I can construct and interpret bar charts, pie charts, pictograms and simple frequency distributions
- I can construct and interpret histograms with equal intervals
- I can construct and interpret scatter diagrams and understand what is meant by positive, negative and zero correlation
- I can draw and interpret lines of best fit

Extended

- I can construct and interpret histograms with unequal intervals

Why this chapter matters

The idea of 'average' is important in statistics. But there are several ways of working out an average, which have been developed over a long period of time.

Mean in Ancient India

There is a story about Rtuparna who was born in India around 5000 BCE. He wanted to estimate the amount of fruit on a tree:

- He counted how much fruit there was on one branch, then estimated the number of branches on the tree.
- He multiplied the estimated number of branches by the counted fruit on one branch.

He was amazed that the total was very close to the actual counted number of fruit when it was picked.

Rtuparna was one of the first to use arithmetic **mean**.

The branch he chose was an average one representing all the branches. So the number of fruit on that branch would have been in the middle of the smallest and largest number of fruit on other branches on the tree.

Mode in Ancient Greece

This story comes from a war in Ancient Greece (431–404 BCE). It is about a battle between the Spartans and the Athenians.

The Athenians had to get over the Spartan Wall so they needed to work out its height. They started by counting the layers of bricks. This was done by hundreds of soldiers at the same time because many of them would get it wrong – but the majority would get it about right.

This is seen as an early use of the **mode**: the number of layers that occurred the most in the counting was taken as the one most likely to be correct.

They then had to guess the height of one brick and so calculate the total height of the wall. They could then make ladders long enough to reach the top of the wall.

The other average that you will use is the **median** (which finds the middle value), and there is no record of any use of this being used until the early 17th century.

These ancient examples demonstrate that you will not always work out the average in the same way – you must choose a method that is appropriate to the situation.

Chapter 32 . Topic 1

32.1 The mode

Average is a term often used when describing or comparing sets of data, for example, average rainfall over a year or the average mark in a test for a group of students.

In each of the above examples, you are representing the whole set of many values (rainfall on every day of the year or the marks of all the students) by just a single, 'typical' value, which is called the average.

The idea of an average is extremely useful, because it enables you to compare one set of data with another set by comparing just two values – their averages.

There are several ways of expressing an average, but the most commonly used averages are the **mode**, the **median** and the **mean**.

The **mode** is the value that occurs the most in a set of data. That is, it is the value with the highest **frequency**.

The mode is a useful average because it is very easy to find and it can be applied to non-numerical data (qualitative data). For example, you could find the modal style of skirts sold in a particular month.

> **Example 1**
>
> Suhail scored these numbers of goals in 12 football matches:
>
> 1 2 1 0 1 0 0 1 2 1 0 2
>
> What is the mode of his scores?
>
> The number that occurs most often in this list is 1. So, the mode is 1.
>
> You can also say that the modal score or **modal value** is 1.

EXERCISE 32A

1. Find the mode for each set of data.

   a 3, 4, 7, 3, 2, 4, 5, 3, 4, 6, 8, 4, 2, 7

   b 47, 49, 45, 50, 47, 48, 51, 48, 51, 48, 52, 48

   c −1, 1, 0, −1, 2, −2, −2, −1, 0, 1, −1, 1, 0, −1, 2, −1, 2

   d $\frac{1}{2}, \frac{1}{4}, 1, \frac{1}{2}, \frac{3}{4}, \frac{1}{4}, 0, 1, \frac{3}{4}, \frac{1}{4}, 1, \frac{1}{4}, \frac{3}{4}, \frac{1}{4}, \frac{1}{2}$

   e 100, 10, 1000, 10, 100, 1000, 10, 1000, 100, 1000, 100, 10

   f 1.23, 3.21, 2.31, 3.21, 1.23, 3.12, 2.31, 1.32, 3.21, 2.31, 3.21

   Advice and Tips

   It helps to put the data in order or group all the same things together.

2. Find the mode for each set of data.

   a red, green, red, amber, green, red, amber, green, red, amber

   b rain, sun, cloud, sun, rain, fog, snow, rain, fog, sun, snow, sun

   c α, γ, α, β, γ, α, α, γ, β, α, β, γ, β, β, α, β, γ, β

   d

3   Halima did a survey to find the shoe sizes of students in her class. The bar chart illustrates her data.

   a   How many students are there in Halima's class?
   b   What is the modal shoe size?
   c   Can you tell which are the boys' shoes sizes and which are the girls' shoe sizes?
   d   Halima then decided to draw a bar chart to show the shoe sizes of the boys and the girls separately. Do you think that the mode for the boys and the mode for the girls will be the same as the mode for the whole class? Explain your answer.

4   The frequency table shows the marks that a class obtained in a spelling test.

Mark	3	4	5	6	7	8	9	10
Frequency	1	2	6	5	5	4	3	4

   a   Write down the mode for their marks.
   b   Do you think this is a typical mark for the class? Explain your answer.

5   Explain why the mode is often referred to as the 'shopkeeper's average'.

6   This table shows the colours of eyes of the students in a class.

	Blue	Brown	Green
Boys	4	8	1
Girls	8	5	2

   a   How many students are there in the class?
   b   What is the modal eye colour for:
      i   boys   ii   girls   iii   the whole class?
   c   After two students join the class the modal eye colour for the whole class is blue. Which of these statements is true?
      • Both students had green eyes.
      • Both students had brown eyes.
      • Both students had blue eyes.
      • You cannot tell what their eye colours were.

7   Here is a large set of raw data.

   5 6 8 2 4 8 9 8 1 3 4 2 7 2 4 6 7 5 3 8

   9 1 3 1 5 6 2 5 7 9 4 1 4 3 3 5 6 8 6 9

   8 4 8 9 3 4 6 7 7 4 5 4 2 3 4 6 7 6 5 5

   a   What problems may occur if you attempted to find the mode by counting individual numbers?
   b   Explain a method that would make finding the mode more efficient and accurate.
   c   Use your method to find the mode of the data.

Chapter 32: Statistical measures   609

32.2 The median

The **median** is the **middle value** of a list of values when they are put in *order of size*, from lowest to highest.

The advantage of using the median as an average is that half the data values are below the median value and half are above it. Therefore, the average is only slightly affected by the presence of any particularly high or low values that are not typical of the data as a whole.

Example 2

Find the median for this list of numbers.

2, 3, 5, 6, 1, 2, 3, 4, 5, 4, 6

Putting the list in numerical order gives:

1, 2, 2, 3, 3, **4**, 4, 5, 5, 6, 6

There are 11 numbers in the list, so the middle of the list is the 6th number. Therefore, the median is 4.

Example 3

Find the median of the data shown in the frequency table.

Value	2	3	4	5	6	7
Frequency	2	4	6	7	8	3

First, add up the frequencies to find out how many pieces of data there are.

The total is 30 so the median value will be between the 15th and 16th values.

Now, add up the frequencies to give a running total, to find out where the 15th and 16th values are.

Value	2	3	4	5	6	7
Frequency	2	4	6	7	8	3
Running total	2	6	12	19	27	30

There are 12 data-values up to the value 4 and 19 up to the value 5.

Both the 15th and 16th values are 5, so the median is 5.

To find the median in a list of n values, written in order, use the rule:

$$\text{median} = \frac{n+1}{2}\text{th value}$$

EXERCISE 32B

1 Find the median for each set of data.

  a 7, 6, 2, 3, 1, 9, 5, 4, 8
  b 26, 34, 45, 28, 27, 38, 40, 24, 27, 33, 32, 41, 38
  c 4, 12, 7, 6, 10, 5, 11, 8, 14, 3, 2, 9
  d 12, 16, 12, 32, 28, 24, 20, 28, 24, 32, 36, 16
  e 10, 6, 0, 5, 7, 13, 11, 14, 6, 13, 15, 1, 4, 15
  f −1, −8, 5, −3, 0, 1, −2, 4, 0, 2, −4, −3, 2
  g 5.5, 5.05, 5.15, 5.2, 5.3, 5.35, 5.08, 5.9, 5.25

Advice and Tips

Remember to put the data in order before finding the median.

Advice and Tips

If there is an even number of pieces of data, the median will be halfway between the two middle values.

2 A group of 15 students had lunch in the school's cafeteria. Given below are the amounts that they spent.

$2.30, $2.20, $2, $2.50, $2.20, $3.50, $2.20, $2.25,

$2.20, $2.30, $2.40, $2.20, $2.30, $2, $2.35

  a Find the mode for the data.
  b Find the median for the data.
  c Which is the better average to use? Explain your answer.

3 a Find the median of 7, 4, 3, 8, 2, 6, 5, 2, 9, 8, 3.

  b Without putting them in numerical order, write down the median for each of these sets.
   i   17, 14, 13, 18, 12, 16, 15, 12, 19, 18, 13
   ii 217, 214, 213, 218, 212, 216, 215, 212, 219, 218, 213
   iii 12, 9, 8, 13, 7, 11, 10, 7, 14, 13, 8
   iv 14, 8, 6, 16, 4, 12, 10, 4, 18, 16, 6

Advice and Tips

Look for a connection between the original data and the new data. For example, in **i**, the numbers are each 10 more than those in part **a**.

4 Given below are the age, height and mass of each of the seven players in a netball team.

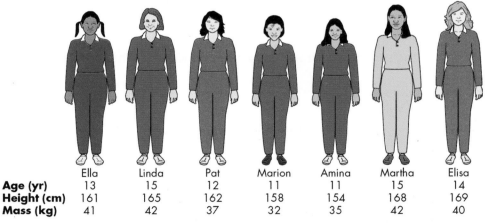

	Ella	Linda	Pat	Marion	Amina	Martha	Elisa
Age (yr)	13	15	12	11	11	15	14
Height (cm)	161	165	162	158	154	168	169
Mass (kg)	41	42	37	32	35	42	40

  a Find the median age of the team. Which player has the median age?
  b Find the median height of the team. Which player has the median height?
  c Find the median mass of the team. Which player has the median mass?
  d Who would you choose as the average player in the team? Give a reason for your answer.

Chapter 32: Statistical measures

Chapter 32 . Topic 3

5 The table shows the numbers of sandwiches sold in a shop over 25 days.

Sandwiches sold	10	11	12	13	14	15	16
Frequency	2	3	6	4	3	4	3

   a What is the modal number of sandwiches sold?

   b What is the median number of sandwiches sold?

6 **a** Write down a list of nine numbers that has a median of 12.

   b Write down a list of 10 numbers that has a median of 12.

   c Write down a list of nine numbers that has a median of 12 and a mode of 8.

   d Write down a list of 10 numbers that has a median of 12 and a mode of 8.

7 A list contains seven even numbers. The largest number is 24. The smallest number is half the largest. The mode is 14 and the median is 16. Two of the numbers add up to 42. What are the seven numbers?

8 Look at this list of numbers.

    4, 4, 5, 8, 10, 11, 12, 15, 15, 16, 20

   a Add four numbers to make the median 12.

   b Add six numbers to make the median 12.

   c What is the least number of numbers to add that will make the median 4?

9 Explain why the median is not a good average to use in this set of payments.

   Here are five payments.

     $3, $5, $8, $100, $3000

32.3 The mean

The **mean** of a set of data is the sum of all the values in the set divided by the total number of values in the set. That is:

$$\text{mean} = \frac{\text{sum of all values}}{\text{total number of values}}$$

This is what most people mean when they use the term **average**.

Another name for this average is the arithmetic **mean**.

The advantage of using the mean as an average is that it takes into account all the values in the set of data.

32.3

Example 4

The ages of 11 players in a football squad are:

21, 23, 20, 27, 25, 24, 25, 30, 21, 22, 28

What is the mean age of the squad?

Sum of all the ages = 266

Total number in squad = 11

Therefore, mean age = $\frac{266}{11}$ = 24.1818… = 24.2 (1 decimal place)

EXERCISE 32C

1 Find the mean for each set of data.

  a 7, 8, 3, 6, 7, 3, 8, 5, 4, 9
  b 47, 3, 23, 19, 30, 22
  c 42, 53, 47, 41, 37, 55, 40, 39, 44, 52
  d 1.53, 1.51, 1.64, 1.55, 1.48, 1.62, 1.58, 1.65
  e 1, 2, 0, 2, 5, 3, 1, 0, 1, 2, 3, 4

2 Calculate the mean for each set of data, giving your answer correct to 1 decimal place.

  a 34, 56, 89, 34, 37, 56, 72, 60, 35, 66, 67
  b 235, 256, 345, 267, 398, 456, 376, 307, 282
  c 50, 70, 60, 50, 40, 80, 70, 60, 80, 40, 50, 40, 70
  d 43.2, 56.5, 40.5, 37.9, 44.8, 49.7, 38.1, 41.6, 51.4
  e 2, 3, 1, 0, 2, 5, 4, 3, 2, 0, 1, 3, 4, 5, 0, 3, 1, 2

3 The table shows the marks that 10 students obtained in mathematics, English and science in their examinations.

Student	Ahmed	Badru	Camille	Dayar	Evrim	Fatima	George	Helga	Imran	Josie
Maths	45	56	47	77	82	39	78	32	92	62
English	54	55	59	69	66	49	60	56	88	44
Science	62	58	48	41	80	56	72	40	81	52

  a Work out the mean mark for mathematics.
  b Work out the mean mark for English.
  c Work out the mean mark for science.
  d Which student obtained marks closest to the mean in all three subjects?
  e How many students were above the average mark in all three subjects?

Chapter 32: Statistical measures

4 Suni kept a record of the amount of time she spent on her homework over 10 days:

$\frac{1}{2}$ h, 20 min, 35 min, $\frac{1}{4}$ h, 1 h, $\frac{1}{2}$ h, $1\frac{1}{2}$ h, 40 min, $\frac{3}{4}$ h, 55 min

Calculate the mean time, in minutes, that Suni spent on her homework.

> **Advice and Tips**
> Convert all times to minutes, for example, $\frac{1}{4}$ h = 15 minutes.

5 These are the weekly wages of 10 people working in an office.

$350 $200 $180 $200 $350 $200 $240 $480 $300 $280

a Find the modal wage.
b Find the median wage.
c Calculate the mean wage.
d Which of the three averages best represents the office staff's wages? Give a reason for your answer.

> **Advice and Tips**
> Remember that the mean can be distorted by extreme values.

6 The ages of five people in a group of walkers are 38, 28, 30, 42 and 37.

a Calculate the mean age of the group.
b Steve, who is 41, joins the group. Calculate the new mean age of the group.

7 a Calculate the mean of 3, 7, 5, 8, 4, 6, 7, 8, 9 and 3.
b Calculate the mean of 13, 17, 15, 18, 14, 16, 17, 18, 19 and 13. What do you notice?
c Write down, without calculating, the mean for each of these sets of data.
  i   53, 57, 55, 58, 54, 56, 57, 58, 59, 53
  ii 103, 107, 105, 108, 104, 106, 107, 108, 109, 103
  iii 4, 8, 6, 9, 5, 7, 8, 9, 10, 4

> **Advice and Tips**
> Look for a connection between the original data and the new data. For example in **i** the numbers are 50 more.

8 Two families were in a competition.

Speed family		Roberts family	
Brian	aged 59	Frank	aged 64
Kath	aged 54	Marylin	aged 62
James	aged 34	David	aged 34
Helen	aged 34	James	aged 32
John	aged 30	Tom	aged 30
Joseph	aged 24	Helen	aged 30
Joy	aged 19	Evie	aged 16

Each family had to choose four members with a mean age of between 35 and 36.

Choose two teams, one from each family, that have this mean age between 35 and 36.

9 Asif had an average batting score of 35 runs. He had scored 315 runs in nine games of cricket.

What is the least number of runs he needs to score in the next match if he is to get a higher average score?

10 The mean age of a group of eight walkers is 42. Joanne joins the group and the mean age changes to 40.

How old is Joanne?

32.4 The range

The **range** for a set of data is the highest value of the set minus the lowest value.

The range is *not* an average. It shows the **spread** of the data. You will use it when you are comparing two or more sets of similar data. You can also use it to comment on the **consistency** of two or more sets of data.

> **Example 5**
> Rachel's marks in 10 mental arithmetic tests were 4, 4, 7, 6, 6, 5, 7, 6, 9 and 6.
>
> Adil's marks in the same tests were 6, 7, 6, 8, 5, 6, 5, 6, 5 and 6.
>
> Compare their marks.
>
> ---
>
> Rachel's mean mark is 60 ÷ 10 = 6 and her range is 9 – 4 = 5.
>
> Adil's mean mark is 60 ÷ 10 = 6 and his range is 8 – 5 = 3.
>
> Although the means are the same, Adil has a smaller range.
>
> This shows that Adil's results are more consistent.

EXERCISE 32D

1 Find the range for each set of data.

a 3, 8, 7, 4, 5, 9, 10, 6, 7, 4

b 62, 59, 81, 56, 70, 66, 82, 78, 62, 75

c 1, 0, 4, 5, 3, 2, 5, 4, 2, 1, 0, 1, 4, 4

d 3.5, 4.2, 5.5, 3.7, 3.2, 4.8, 5.6, 3.9, 5.5, 3.8

e 2, –1, 0, 3, –1, –2, 1, –4, 2, 3, 0, 2, –2, 0, –3

2 The table shows the maximum and minimum temperatures at midday for five cities in England during a week in August.

	Birmingham	Leeds	London	Newcastle	Sheffield
Maximum temperature (°C)	28	25	26	27	24
Minimum temperature (°C)	23	22	24	20	21

  a Write down the range of the temperatures for each city.
  b What do the ranges tell you about the weather for England during the week?

3 Over a three-week period, a school sweet shop took these amounts.

	Monday	Tuesday	Wednesday	Thursday	Friday
Week 1	$32	$29	$36	$30	$28
Week 2	$34	$33	$25	$28	$20
Week 3	$35	$34	$31	$33	$32

  a Calculate the mean amount taken each week.
  b Find the range for each week.
  c What can you say about the total amounts taken for each of the three weeks?

4 In a golf tournament, the club coach had to choose either Maria or Fay to play in the first round. In the previous eight rounds, their scores were as follows.

Maria's scores: 75, 92, 80, 73, 72, 88, 86, 90

Fay's scores: 80, 87, 85, 76, 85, 79, 84, 88

  a Calculate the mean score for each golfer.
  b Find the range for each golfer.
  c Which golfer would you choose to play in the tournament? Explain why.

> **Advice and Tips**
>
> The best person to choose may not be the one with the biggest mean but could be the most consistent player.

5 Dan has a choice of two buses to get to school: Number 50 or Number 63. Over a month, he kept a record of the numbers of minutes each bus was late when it set off from his home bus stop.

No. 50: 4, 2, 0, 6, 4, 8, 8, 6, 3, 9

No. 63: 3, 4, 0, 10, 3, 5, 13, 1, 0, 1

  a For each bus, calculate the mean number of minutes late.
  b Find the range for each bus.
  c Which bus would you advise Dan to catch? Give a reason for your answer.

32.4 The range

6 The table gives the ages and heights of 10 children.

Name	Age (years)	Height (cm)
Evrim	9	121
Isaac	4	73
Lilla	8	93
Lewis	10	118
Evie	3	66
Badru	6	82
Oliver	4	78
Halima	2	69
Isambard	9	87
Chloe	7	82

a Chloe is having a party. She wants to invite as many children as possible but does not want the range of ages to be more than 5.

Who will she invite?

b This is a sign at a theme park.

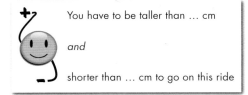

You have to be taller than ... cm

and

shorter than ... cm to go on this ride

Isaac is the shortest person who can go on the ride and Isambard is the tallest.

What are the smallest and largest missing values on the sign?

7 a The age range of a school quiz team is 20 years and the mean age is 34.

Who would you expect to be in this team?

Explain your answer.

b Another team has an average age of $15\frac{1}{2}$ and a range of 1.

Who would you expect to be in this team?

Explain your answer.

32.5 Which average to use

An average must be truly **representative** of a set of data. So, when you have to find an average, it is crucial to choose the **appropriate** type of average for this particular set of data.

If you use the wrong average, your results will be distorted and give misleading information.

This table, which compares the advantages and disadvantages of each type of average, will help you to make the correct decision.

	Mode	Median	Mean
Advantages	Very easy to find Not affected by **extreme values** Can be used for non-numerical data	Easy to find for ungrouped data Not affected by extreme values	Easy to find Uses all the values The total for a given number of values can be calculated from it
Disadvantages	Does not use all the values May not exist	Does not use all the values Often not understood	Extreme values can distort it Has to be calculated
Use for	Non-numerical data Finding the most likely value	Data with extreme values	Data with values that are spread in a balanced way

EXERCISE 32E

1 These are the ages of the members of a hockey team.

   29 26 21 24 26 28 35 23 29 28 29

   a Give:

   i the modal age

   ii the median age

   iii the mean age.

   b What is the range of the ages?

2 a For each set of data, find the mode, the median and the mean.

   i 6, 10, 3, 4, 3, 6, 2, 9, 3, 4

   ii 6, 8, 6, 10, 6, 9, 6, 10, 6, 8

   iii 7, 4, 5, 3, 28, 8, 2, 4, 10, 9

   b For each set of data, decide which average is the best one to use and give a reason.

3 These are the numbers of copies of *The Evening Star* sold on 12 consecutive evenings by a shop during a promotion exercise organised by the newspaper's publisher.

   65 73 75 86 90 112 92 87 77 73 68 62

a Find the mode, the median and the mean for the sales.
  b The shopkeeper had to report the average sale to the publisher after the promotion. Which of the three averages would you advise the shopkeeper to use? Explain why.

4 The mean age of a group of 10 young people is 15.
   a What do all their ages add up to?
   b What will be their mean age in five years' time?

5 Decide which average you would use for each statistic. Give a reason for your answer.
   a The average mark in an examination
   b The average pocket money for a group of 16-year-old students
   c The average shoe size for all the girls in one year at school
   d The average height for all the artistes on tour with a circus
   e The average hair colour for students in your school
   f The average mass of all newborn babies in a hospital's maternity ward.

6 A pack of matches consisted of 12 boxes. The contents of each box were counted as:

   34   31   29
   35   33   30
   31   28   29
   35   32   31

   On the box it stated 'Average contents 32 matches'. Is this correct?

7 Mr Brennan told each student their test mark and only gave the test statistics to the whole class. He gave the class the modal mark, the median mark and the mean mark.
   a Which average would tell a student whether they were in the top half or the bottom half of the class?
   b Which average tells the students nothing really?
   c Which average allows a student to gauge how well they have done compared with everyone else?

8 Three players were hoping to be chosen for the basketball team.

   The table shows their scores in the last few games they played.

Tom	16, 10, 12, 10, 13, 8, 10
David	16, 8, 15, 25, 8
Mohaned	15, 2, 15, 3, 5

   The teacher said they would be chosen by their best average score.

   Which average would each boy want to be chosen by?

Chapter 32 . Topic 6

9 **a** Find five numbers that have *both* the properties below:
- a range of 5
- a mean of 5.

b Find five numbers that have *all* the properties below:
- a range of 5
- a median of 5
- a mean of 5.

10 What is the average pay at a factory with 10 employees?

The boss said: '$43 295'

A worker said: '$18 210'

They were both correct.

Explain how this can be.

11 A list of nine numbers has a mean of 7.6. What number must be added to the list to give a new mean of 8?

12 A dance group of 17 people had a mean mass of 54.5 kg. To enter a competition there needed to be 18 people with an average mass of 54.4 kg or less. What is the maximum mass that the eighteenth person must have?

32.6 Stem-and-leaf diagrams

Here are the ages of 20 people

23, 13, 34, 44, 26, 12, 41, 31, 20, 18, 19, 31, 48, 32, 45, 14, 12, 27, 31, 19

Here are their ages in order.

12, 12, 13, 14, 18, 19, 19, 20, 23, 26, 27, 31, 31, 31, 32, 34, 41, 44, 45, 48

This is easier to read and analyse.

You can put the ages in a **stem-and-leaf** diagram. The number of values will be the 'stem' and the unit values will be the 'leaves'.

Key : 1 | 2 represents 12

1	2	2	3	4	8	9	9
2	0	3	6	7		
3	1	1	1	2	4	
4	1	4	5	8		

This is called a stem-and-leaf diagram. It gives a better idea of how the data is distributed.

A stem-and-leaf diagram should always have a key.

32.6

Example 6

Put the following data into a stem-and-leaf diagram.

45, 62, 58, 58, 61, 49, 61, 47, 52, 58, 48, 56, 65, 46, 54

a What is the modal value?
b What is the median value?
c What is the range of the values?

First, decide on the stem and leaf.

In this case, the tens digit will be the stem and the units digit will be the leaf.

Key: 4 | 5 represents 45

4	5	6	7	8	9	
5	2	4	6	8	8	8
6	1	1	2	5		

a The modal value is the most common, which is 58.
b There are 15 values, so the median will be the value that is, (15 + 1) ÷ 2, or the 8th value. Counting from either the top or the bottom, the median is 56.
c The range is the difference between the largest and the smallest value, which is 65 − 45 = 20

EXERCISE 32F

1 This stem-and-leaf diagram shows the marks some students scored in a test.

5	3	5	5	9	9					
6	0	0	4	5	5	6	8	8		
7	1	1	1	1	1	3	4	6	6	8	9	9
8	0	0	0	2	3	3	3	5	7		
9	0	1	4	4	5	6					

Key: 5 | 3 represents 53 marks

a How many students took the test?
b Find the median mark.
c Find the range.
d Find the modal mark.
e Explain why the median is more useful than the mode.

2 This stem-and-leaf diagram shows the times of athletes running 200 metres.

```
23. | 8 9 9
24. | 1 3 3 5 7 8
25. | 0 0 1 1 3 5 8 8 9
26. | 1 4 6 6 7 8 9 9
27. | 0 0 0 2 2 2 2 2 2 3 5 8 8 8
28. | 0 4 5 5 5 6 7
```

Key: 23. | 8 represents 23.8 seconds

a How many runners took less than 26 seconds?
b Find the median time.
c Find the range of the times.

3 This stem-and-leaf diagram shows the times 50 people take to travel to work.

```
1 | 5 5 7 8 8 9 9
2 | 0 0 0 2 3 5 5 5 5 6 9
3 | 0 0 3 3 5 5 5 5 6 6 6 7 8
4 | 0 0 0 2 5
5 | 3 5 5 5 5 6 6 9
6 | 5 5
7 | 0 6 6
8 | 5
```

Key: 5 | 3 represents 53 minutes

a How many people took more than an hour?
b Find the median time.
c Find the range of the times.

4 The heights of 15 tulips are measured.

43 cm, 39 cm, 41 cm, 29 cm, 36 cm,

34 cm, 43 cm, 48 cm, 38 cm, 35 cm,

41 cm, 38 cm, 43 cm, 28 cm, 48 cm

a Show the results in a stem-and-leaf diagram, using this key.

Key: 4 | 3 represents 43 cm

b What is the model height?
c What is the median height?
d What is the range of the heights?

5 A student records the number of text messages she receives each day for two weeks.

12, 18, 21, 9, 17, 23, 8, 2, 20, 13, 17, 22, 9, 9

  a Show the results in a stem-and-leaf diagram, using this key.

   Key: 1 | 2 represents 12 messages

  b What was the modal number of text messages received in a day?

  c What was the median number of text messages received in a day?

6 Zachia wanted to know how many people attended a daily youth club each day over a month. She recorded the data.

13, 19, 20, 9, 18, 24, 7, 8, 19, 14, 18, 23, 9, 10, 15, 31, 28, 26, 12, 24

  a Show these results in a stem-and-leaf diagram.

  b What is the median number of people at the youth club?

  c What is the range of the numbers of people who attended the youth club?

7 This stem-and-leaf diagram shows the ages of the men and women working for a company.

```
                    Men                         Women
                         9  9 | 1 | 8  9  9  9
              8  5  4  2  2  1  0 | 2 | 0  0  0  1  1  3  3  5  8  9  9
           9  9  7  6  6  4  3  1 | 3 | 0  2  2  3  3  5  6  6  7
  8  7  6  6  6  5  3  3  3  2 | 4 | 0  3  3  4  5  6  6
     7  7  6  5  4  1  0  0  0 | 5 | 1  2  8
                    4  3  1  1  1 | 6 |
```

Key: 9 | 1 | 8 represents a man age 19 and a woman age 18

Copy and complete this table.

	Men	Women
Number of people	41	
Range of ages		
Median age		

8 This chart shows the heights of some girls and boys

```
           Heights of girls                Heights of boys
                       8  7  4 | 12 | 8  9  9  9  9
                 7  7  5  5  0 | 13 | 0  4  5  5  6  7  7  7  7  8
        9  8  8  8  6  4  3  2 | 14 | 1  1  1  3  5  5  5  5  6  6  9
                 7  5  3  1  0 | 15 | 0  3  3  4  5  5  7  8  8
                       4  2  2  1 | 16 | 1  2  2  2  4  6  8  8  8  9
                                  | 17 | 0  0  1  3
```

Key: 4 | 12 | 8 represents a girl 124 cm tall and a boy 128 cm tall

a How many children are less than 130 cm tall?

b Copy and complete this table.

	Girls	Boys
Number of children		49
Median height		
Range of heights		

c Complete these sentences

  i The girls' median height is … cm more/less than the boys' median height.

  ii The girls' range is … cm more/less than the boys' range.

  Circle the correct word in more/less each time.

32.7 Using frequency tables

When you have gathered a lot of information, it is often convenient to put it together in a **frequency table**. From this table you can then find the values of the three averages and the range.

Example 7

A survey was done on the number of people in each car leaving a shopping centre. The results are summarised in the table.

Number of people in each car	1	2	3	4	5	6
Frequency	45	198	121	76	52	13

For the number of people in a car, calculate:

a the mode

b the median

c the mean.

---

a The modal number of people in a car is easy to spot. It is the number with the largest frequency, which is 198. Hence, the modal number of people in a car is 2.

b You can find the median number of people in a car by working out where the middle of the set of numbers is located. First, add up frequencies to get the total number of cars surveyed, which comes to 505. Next, calculate the middle position.

$$(505 + 1) \div 2 = 253$$

Now add the frequencies across the table to find which group contains the 253rd item. The 243rd item is the end of the group with 2 in a car. Therefore, the 253rd item must be in the group with 3 in a car. Hence, the median number of people in a car is 3.

c To calculate the mean number of people in a car, multiply the number of people in the car by the frequency. This is best done in an extra column. Add these products to find the total number of people and divide by the total frequency (the number of cars surveyed).

Number in car	Frequency	Number in these cars
1	45	1 × 45 = 45
2	198	2 × 198 = 396
3	121	3 × 121 = 363
4	76	4 × 76 = 304
5	52	5 × 52 = 260
6	13	6 × 13 = 78
Totals	505	1446

Hence, the mean number of people in a car is 1446 ÷ 505 = 2.9 (to 1 decimal place).

When you have gathered a lot of information, it is often convenient to put it together in a **frequency table**. From this table you can then find the values of the three averages and the range.

EXERCISE 32G

1. Find **i** the mode **ii** the median **iii** the mean from each frequency table below.

   a A survey of the shoe sizes of all the boys in one year of a school gave these results.

Shoe size	4	5	6	7	8	9	10
Number of students	12	30	34	35	23	8	3

   b A survey of the number of eggs laid by hens over a period of one week gave these results.

Number of eggs	0	1	2	3	4	5	6
Frequency	6	8	15	35	48	37	12

   c This is a record of the number of babies born each week over one year in a small maternity unit.

Number of babies	0	1	2	3	4	5	6	7	8	9	10	11	12	13	14
Frequency	1	1	1	2	2	2	3	5	9	8	6	4	5	2	1

   d A school did a survey on how many times in a week students arrived late at school. These are the findings.

Number of times late	0	1	2	3	4	5
Frequency	481	34	23	15	3	4

2 A survey of the number of children in each family of a school's intake gave these results.

Number of children	1	2	3	4	5
Frequency	214	328	97	26	3

a Assuming each child at the school is shown in the data, how many children are at the school?
b Calculate the mean number of children in a family.
c How many families have this mean number of children?
d How many families would consider themselves average from this survey?

3 A dentist kept records of how many teeth he extracted from his patients.

In 1989 he extracted 598 teeth from 271 patients.

In 1999 he extracted 332 teeth from 196 patients.

In 2009 he extracted 374 teeth from 288 patients.

a Calculate the average number of teeth taken from each patient in each year.
b Explain why you think the average number of teeth extracted falls each year.

4 One hundred cases of apples delivered to a supermarket were inspected and the numbers of bad apples were recorded.

Bad apples	0	1	2	3	4	5	6	7	8	9
Frequency	52	29	9	3	2	1	3	0	0	1

Give:

a the modal number of bad apples per case
b the mean number of bad apples per case.

5 Two dice are thrown together 60 times. The sums of the scores are shown below.

Score	2	3	4	5	6	7	8	9	10	11	12
Frequency	1	2	6	9	12	15	6	5	2	1	1

Find:

a the modal score
b the median score
c the mean score.

32.7 Using frequency tables

6 During a one-month period, the number of days off taken by 100 workers in a factory were noted as follows.

Number of days off	0	1	2	3	4
Number of workers	35	42	16	4	3

Calculate:

a the modal number of days off
b the median number of days off
c the mean number of days off.

7 Two friends often played golf together. They recorded the numbers of shots they made to get their balls into each hole over the last five games to compare who was more consistent and who was the better player. Their results were summarised in the table.

Number of shots	1	2	3	4	5	6	7	8	9
Roger	0	0	0	14	37	27	12	0	0
Brian	5	12	15	18	14	8	8	8	2

a What is the modal score for each player?
b What is the range of scores for each player?
c What is the median score for each player?
d What is the mean score for each player?
e Which player is the more consistent and why?
f Who would you say is the better player and why?

8 A tea stain on a newspaper removed four numbers from this frequency table of goals scored in 40 football matches one weekend.

Goals	0	1	2			5
Frequency	4	6	9			3

The mean number of goals scored is 2.4.

What could the missing four numbers be?

9 Manju made day trips to Mumbai frequently during a year.

The table shows how many days in a week she travelled.

Days	0	1	2	3	4	5
Frequency	17	2	4	13	15	1

Explain how you would find the median number of days Manju travelled in a week to Mumbai.

Chapter 32 . Topic 8
32.8 Grouped data

Sometimes the information you are given is grouped in some way (called **grouped data**), as in Example 8, which shows the range of weekly pocket money given to Year 12 students in a particular class.

Normally, grouped tables use **continuous data**, which is data that can have any value within a range of values, for example, height, mass, time, area and capacity. In these situations, the **mean** can only be **estimated** as you do not have all the information.

Discrete data is data that consists of separate numbers, for example, goals scored, marks in a test, number of children and shoe sizes.

In both cases, when using a grouped table to estimate the mean, first find the midpoint of the interval by adding the two end-values and then dividing by two.

Example 8

Pocket money, p ($)	$0 < p \leq 1$	$1 < p \leq 2$	$2 < p \leq 3$	$3 < p \leq 4$	$4 < p \leq 5$
No. of students	2	5	5	9	15

a Write down the modal class.

b Calculate an estimate of the mean weekly pocket money.

a The modal class is easy to identify, since it is simply the one with the largest frequency. Here the modal class is $4 to $5.

b To estimate the mean, assume that each person in each class has the 'midpoint' amount, then build up the following table.

To find the midpoint value, add the two end-values together and then divide by two.

Pocket money, p ($$)	Frequency (f)	Midpoint (m)	$f \times m$
$0 < p \leq 1$	2	0.50	1.00
$1 < p \leq 2$	5	1.50	7.50
$2 < p \leq 3$	5	2.50	12.50
$3 < p \leq 4$	9	3.50	31.50
$4 < p \leq 5$	15	4.50	67.50
Totals	36		120

The estimated mean will be $120 ÷ 36 = $3.33 (rounded to the nearest cent).

If you had written 0.01–1.00, 1.01–2.00 and so on for the groups, then the midpoints would have been 0.505, 1.505 and so on. This would not have had a significant effect on the final answer as it is only an estimate.

Note that you *cannot* find the **median** or the range from a grouped table as you do not know the actual values.

EXERCISE 32H

1 For each table of values given below, find:

  i the modal group

  ii an estimate for the mean.

Advice and Tips

When you copy the tables, draw them vertically as in Example 7.

a

x	$0 < x \leq 10$	$10 < x \leq 20$	$20 < x \leq 30$	$30 < x \leq 40$	$40 < x \leq 50$
Frequency	4	6	11	17	9

b

y	$0 < y \leq 100$	$100 < y \leq 200$	$200 < y \leq 300$	$300 < y \leq 400$	$400 < y \leq 500$	$500 < x \leq 600$
Frequency	95	56	32	21	9	3

c

z	$0 < z \leq 5$	$5 < z \leq 10$	$10 < z \leq 15$	$15 < z \leq 20$
Frequency	16	27	19	13

d

Weeks	1–3	4–6	7–9	10–12	13–15
Frequency	5	8	14	10	7

2 Jason brought 100 pebbles back from the beach and found their masses, recording each mass to the nearest gram. His results are summarised in the table.

Mass, m (g)	$40 < m \leq 60$	$60 < m \leq 80$	$80 < m \leq 100$
Frequency	5	9	22

Mass, m (g)	$100 < m \leq 120$	$120 < m \leq 140$	$140 < m \leq 160$
Frequency	27	26	11

Find:

a the modal class of the pebbles

b an estimate of the total mass of all the pebbles

c an estimate of the mean mass of the pebbles.

3 A gardener measured the heights of all his daffodils to the nearest centimetre and summarised his results as follows.

Height (cm)	10–14	15–18	19–22	23–26	27–40
Frequency	21	57	65	52	12

a How many daffodils did the gardener have?

b What is the modal class of the daffodils?

c What is the estimated mean height of the daffodils?

Chapter 32: Statistical measures

4 A survey was created to see how quickly an emergency service got to cars that had broken down. The table summarises the results.

Time (min)	1–15	16–30	31–45	46–60	61–75	76–90	91–105
Frequency	2	23	48	31	27	18	11

  a How many calls were used in the survey?
  b Estimate the mean time taken per call.
  c Which average would the emergency service use for the average call-out time?
  d What percentage of calls do the emergency service get to within the hour?

5 One hundred light bulbs were tested by their manufacturer to see whether the average life-span of the manufacturer's bulbs was over 200 hours. The table summarises the results.

Life span, h (hours)	$150 < h \leqslant 175$	$175 < h \leqslant 200$	$200 < h \leqslant 225$	$225 < h \leqslant 250$	$250 < h \leqslant 275$
Frequency	24	45	18	10	3

  a What is the modal length of time a bulb lasts?
  b What percentage of bulbs last longer than 200 hours?
  c Estimate the mean life-span of the light bulbs.
  d Do you think the test shows that the average life-span is over 200 hours? Explain your answer fully.

6 Three shops each claimed to have the lowest average price increase over the year. The table summarises their price increases.

Price increase (p)	1–5	6–10	11–15	16–20	21–25	26–30	31–35
Soundbuy	4	10	14	23	19	8	2
Springfields	5	11	12	19	25	9	6
Setco	3	8	15	31	21	7	3

Using their average price increases, make a comparison of the supermarkets and say which one has the lowest price increases over the year. Remember to justify your answers.

7 The table shows the distances run, over a month, by an athlete who is training for a half-marathon.

Distance, d (km)	$0 < d \leqslant 5$	$5 < d \leqslant 10$	$10 < d \leqslant 15$	$15 < d \leqslant 20$	$20 < d \leqslant 25$
Frequency	3	8	13	5	2

A half-marathon is 21 kilometres. It is recommended that an athlete's daily average distance should be at least a third of the distance of the race for which they are training. Is this athlete doing enough training?

8 The table shows the points scored in a general-knowledge competition by all the players.

Points	0–9	10–19	20–29	30–39	40–49
Frequency	8	5	10	5	2

Balvir noticed that two frequencies were the wrong way round and that this made a difference of 1.7 to the arithmetic mean.

Which two frequencies were the wrong way round?

9 The profit made each week by a charity shop is shown in the table below.

Profit	$0–$500	$501–$1000	$1001–$1500	$1501–$2000
Frequency	15	26	8	3

Estimate the mean profit made each week.

10 The table shows the number of members of 100 football clubs.

Members	20–29	30–39	40–49	50–59	60–69
Frequency	16	34	27	18	5

Roger claims that the median number of members is 39.5.

Is he correct? Explain your answer.

32.9 Cumulative frequency diagrams

The **inter-quartile range** is a measure of the **dispersion** of a set of data. The advantage of the interquartile range is that it eliminates extreme values, and bases the measure of spread on the middle 50% of the data. This section will show you how to find the interquartile range and the median of a set of data by drawing a **cumulative frequency diagram**.

Look at the marks of 50 students in a mathematics test, which have been put into a grouped table. Note that it includes a column for the **cumulative frequency**, which you can find by adding each frequency to the sum of all preceding frequencies.

Mark	No. of students	Cumulative frequency
21 to 30	1	1
31 to 40	3	4
41 to 50	6	10
51 to 60	10	20

Mark	No. of students	Cumulative frequency
61 to 70	13	33
71 to 80	6	39
81 to 90	4	43
91 to 100	3	46
101 to 110	2	48
111 to 120	2	50

This data can then be used to plot a graph of the top value of each group against its cumulative frequency. The points to be plotted are (30, 1), (40, 4), (50, 10), (60, 20), etc., which will give the graph shown below. Note that the cumulative frequency is *always* the vertical axis.

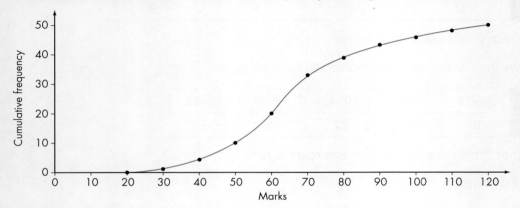

Also note that the scales on both axes are labelled at each graduation mark, in the usual way. *Do not* label the scales as shown below. It is *wrong*.

| 21–30 | 31–40 | 41–50 |

The plotted points can be joined by a freehand curve, to give a cumulative frequency diagram.

The median

The **median** is the middle item of data, once all the items have been put in order of size, from lowest to highest. So, if you have n items of data plotted as a cumulative frequency diagram, you can find the median from the middle value of the cumulative frequency, that is the $\frac{1}{2}n$th value.

But remember, if you want to find the median from a simple list of discrete data, you **must** use the $\frac{1}{2}(n + 1)$th value. The reason for the difference is that the cumulative frequency diagram treats the data as continuous, even for data such as examination marks, which are discrete. You can use the $\frac{1}{2}n$th value when working with cumulative frequency diagrams because you are only looking for an *estimate* of the median.

There are 50 values in the table on this and the previous page. To find the median:

- The middle value will be the 25th value.
- Draw a horizontal line from the 25th value to meet the graph.
- Now look down to the horizontal axis.

This will give an estimate of the median. In this example, the median is about 64 marks.

The inter-quartile range

By dividing the cumulative frequency into four parts, you can obtain **quartiles** and the inter-quartile range.

The **lower quartile** is the value one-quarter of the way up the cumulative frequency axis and is given by the $\frac{1}{4}n$th value.

The **upper quartile** is the value three-quarters of the way up the cumulative frequency axis and is given by the $\frac{3}{4}n$th value.

The inter-quartile range is the difference between the lower and upper quartiles.

These are illustrated on the graph below.

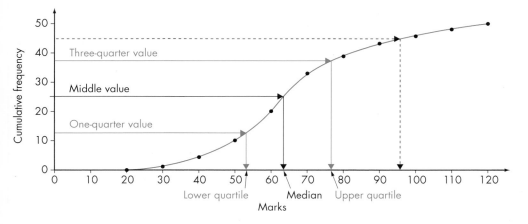

The quarter and three-quarter values out of 50 values are the 12.5th value and the 37.5th value. Draw lines across to the cumulative frequency curve from these values and down to the horizontal axis. These give the lower and upper quartiles. In this example, the lower quartile is 54 marks, the upper quartile is 77 marks and the inter-quartile range is 77 − 54 = 23 marks.

As well as the median and the quartiles we can find percentiles. For example to find the 90th percentile:

90% of 50 is 45.

Draw a line across from a cumulative frequency of 45 and then down to the horizontal axis.

This shows that a mark of 95 is the 90th percentile.

So 90% of the students scored ⩽ 95 marks.

Other percentiles:

- The median is the 50th percentile.
- The upper quartile is the 75th percentile.
- The lower quartile is the 25th percentile.

Example 9

This table shows the marks of 100 students in a mathematics test.

a Draw a cumulative frequency curve.

b Use your graph to find the median and the inter-quartile range.

c Students who score less than 44 have to have extra teaching. How many students will have to have extra teaching?

Mark, x	No. of students	Cumulative frequency
$21 \leq x \leq 30$	3	3
$31 \leq x \leq 40$	9	12
$41 \leq x \leq 50$	12	24
$51 \leq x \leq 60$	15	39
$61 \leq x \leq 70$	22	61
$71 \leq x \leq 80$	16	77
$81 \leq x \leq 90$	10	87
$91 \leq x \leq 100$	8	95
$101 \leq x \leq 110$	3	98
$111 \leq x \leq 120$	2	100

You will meet several ways of giving groups (for example, 21–30, 20, $x \leq 30$, 21, x, 30) but the important thing to remember is to plot the top point of each group against the corresponding cumulative frequency.

a and **b** Draw the graph and add the lines for the median (50th value), lower and upper quartiles (25th and 75th values).

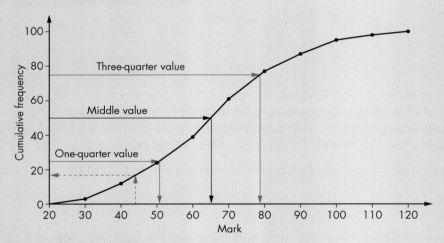

The required answers are read from the graph.

Median = 65 marks

Lower quartile = 51 marks

Upper quartile = 79 marks

Interquartile range = 79 − 51 = 28 marks

c At 44 on the mark axis, draw a perpendicular line to intersect the graph, and at the point of intersection draw a horizontal line across to the cumulative frequency axis, as shown. Number of students needing extra teaching is 18.

Note: An alternative way in which the table in Example 9 could have been set out is shown below. This arrangement has the advantage that the points to be plotted are taken straight from the last two columns. Decide which method you prefer.

32.9

Mark, x	No. of students	Less than	Cumulative frequency
$21 \leqslant x \leqslant 30$	3	30	3
$31 \leqslant x \leqslant 40$	9	40	12
$41 \leqslant x \leqslant 50$	12	50	24
$51 \leqslant x \leqslant 60$	15	60	39
$61 \leqslant x \leqslant 70$	22	70	61
$71 \leqslant x \leqslant 80$	16	80	77
$81 \leqslant x \leqslant 90$	10	90	87
$91 \leqslant x \leqslant 100$	8	100	95
$101 \leqslant x \leqslant 110$	3	110	98
$111 \leqslant x \leqslant 120$	2	120	100

EXERCISE 32I

1 A class of 30 students was asked to guess when one minute had passed. The table shows the results.

   a Copy the table and complete a cumulative frequency column.

   b Draw a cumulative frequency diagram.

   c Use your diagram to estimate the median time and the inter-quartile range.

Time, x (seconds)	No. of students
$20 < x \leqslant 30$	1
$30 < x \leqslant 40$	3
$40 < x \leqslant 50$	6
$50 < x \leqslant 60$	12
$60 < x \leqslant 70$	3
$70 < x \leqslant 80$	3
$80 < x \leqslant 90$	2

2 A group of 50 pensioners was given the task in question **1**. The results are shown in the table.

   a Copy the table and complete a cumulative frequency column.

   b Draw a cumulative frequency diagram.

   c Use your diagram to estimate the median time and the inter-quartile range.

   d Which group, the students or the pensioners, was better at estimating time? Give a reason for your answer.

Time, x (seconds)	No. of pensioners
$10 < x \leqslant 20$	1
$20 < x \leqslant 30$	2
$30 < x \leqslant 40$	2
$40 < x \leqslant 50$	9
$50 < x \leqslant 60$	17
$60 < x \leqslant 70$	13
$70 < x \leqslant 80$	3
$80 < x \leqslant 90$	2
$90 < x \leqslant 100$	1

EXTENDED

3 The sizes of 360 senior schools are recorded in the table.

  a Copy the table and complete a cumulative frequency column.
  b Draw a cumulative frequency diagram.
  c Use your diagram to estimate the median size of the schools and the inter-quartile range.
  d Schools with fewer than 350 students are threatened with closure. About how many schools are threatened with closure?
  e Use your graph to estimate the 90th percentile.
  f Use your graph to estimate the 40th percentile.

No. of students	No. of schools
100–199	12
200–299	18
300–399	33
400–499	50
500–599	63
600–699	74
700–799	64
800–899	35
900–999	11

4 The temperature at a seaside town was recorded for 50 days. It was recorded to the nearest degree. The table shows the results.

  a Copy the table and complete a cumulative frequency column.
  b Draw a cumulative frequency diagram. Note that as the temperature is to the nearest degree the top values of the groups are 7.5 °C, 10.5 °C, 13.5 °C, 16.5 °C, etc.
  c Use your diagram to estimate the median temperature and the inter-quartile range.
  d Use your diagram to estimate the 10th percentile.

Temperature (°C)	No. of days
5–7	2
8–10	3
11–13	5
14–16	6
17–19	6
20–22	9
23–25	8
26–28	6
29–31	5

5 A game consists of throwing three darts and recording the total score. The results of the first 80 people to throw are recorded in the table.

  a Draw a cumulative frequency diagram to show the data.
  b Use your diagram to estimate the median score and the inter-quartile range.
  c People who score over 90 get a prize. About what percentage of the people get a prize?

Total score, x	No. of players
$1 \leq x \leq 20$	9
$21 \leq x \leq 40$	13
$41 \leq x \leq 60$	23
$61 \leq x \leq 80$	15
$81 \leq x \leq 100$	11
$101 \leq x \leq 120$	7
$121 \leq x \leq 140$	2

6 One hundred children were asked to say how much pocket money they got in a week. The results are in the table.

  a Copy the table and complete a cumulative frequency column.
  b Draw a cumulative frequency diagram.
  c Use your diagram to estimate the median amount of pocket money and the inter-quartile range.
  d Estimate the 10th percentile and the 90th percentile.

Amount of pocket money (cents)	No. of children
51–100	6
101–150	10
151–200	20
201–250	28
251–300	18
301–350	11
351–400	5
401–450	2

7 Johan set his class an end-of-course test with two papers, A and B. He produced the cumulative frequency graphs, as shown.

  a What is the median score for each paper?
  b What is the inter-quartile range for each paper?
  c Which is the harder paper? Explain how you know.

Johan wanted 80% of the students to pass each paper and 20% of the students to get a top grade in each paper.

  d What marks for each paper give:
    i a pass
    ii the top grade?

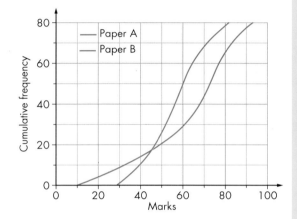

8 The lengths of time taken by 60 helpline telephone calls were recorded.

A cumulative frequency diagram of this data is shown here.

  a Estimate the percentage of calls lasting more than 10 minutes.
  b What is the 15th percentile for call lengths?

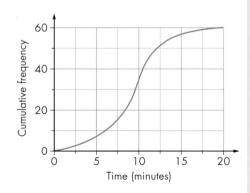

9 Byron was given a cumulative frequency diagram showing the marks obtained by students in a mental maths test.

He was told the top 10% were given the top grade.

How would he find the marks needed to gain this top award?

32.10 Box-and-whisker plots

Another way of displaying data for comparison is by means of a **box-and-whisker plot**. This requires five pieces of data. These are the **lowest value**, the **lower quartile** (Q_1), the **median** (Q_2), the **upper quartile** (Q_3) and the **highest value**. They are drawn in the following way.

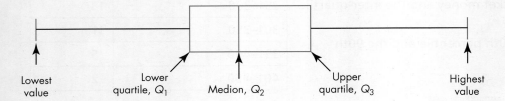

These data values are always placed against a scale so that they are accurately plotted.

The following diagrams show how the cumulative frequency curve, the frequency curve and the box-and-whisker plot are connected for three common types of distribution.

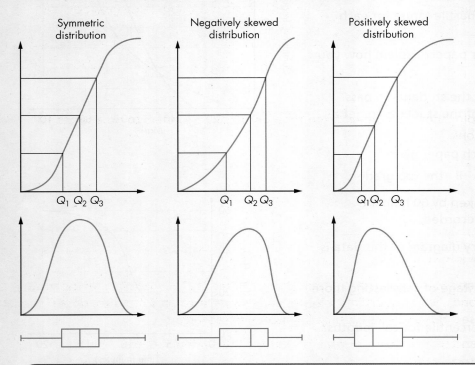

Example 10

The box-and-whisker plot for the girls' marks in last year's examination is shown here.

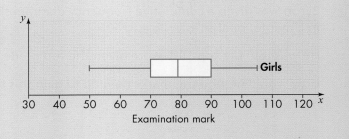

The boys' results for the same examination are: lowest mark 39, lower quartile 65, median 78, upper quartile 87, highest mark 112.

a On the same grid, draw the box-and-whisker plot for the boys' marks.

b Comment on the differences between the two distributions of marks.

a The data for boys and girls is plotted on the grid below.

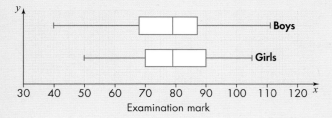

b The girls and boys have the same median mark but both the lower and upper quartiles for the girls are higher than those for the boys, and the girls' range is slightly smaller than the boys'.

This suggests that the girls did better than the boys overall, even though a boy got the highest mark.

EXERCISE 32J

1 The box-and-whisker plot shows the times taken for a group of pensioners to do a set of 10 long-division calculations.

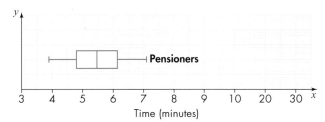

The same set of calculations was given to some students in Year 11. Their results are: shortest time 3 minutes 20 seconds, lower quartile 6 minutes 10 seconds, median 7 minutes, upper quartile 7 minutes 50 seconds and longest time 9 minutes 40 seconds.

a Copy the diagram and draw a box-and-whisker plot for the students' times.

b Comment on the differences between the two distributions.

2 The box-and-whisker plot shows the sizes of secondary schools in Dorset.

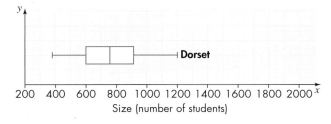

The data for schools in Rotherham is: smallest 280 students, lower quartile 1100 students, median 1400 students, upper quartile 1600 students, largest 1820 students.

a Copy the diagram and draw a box-and-whisker plot for the sizes of schools in Rotherham.

b Comment on the differences between the two distributions.

3 The box-and-whisker plots for the noon temperature at two resorts, recorded over a year, are shown on the grid below.

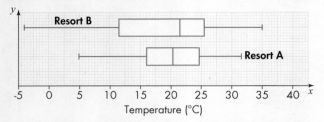

a Comment on the differences in the two distributions.

b Mary wants to go on holiday in July. Which resort would you recommend? Why?

4 The following table shows some data on the annual salaries for 100 men and 100 women.

	Lowest Salary	Lower quartile	Median salary	Upper quartile	Highest salary
Men	$6500	$16 000	$20 000	$22 000	$44 500
Women	$7000	$14 000	$16 000	$21 500	$33 500

a Draw box-and-whisker plots to compare both sets of data.

b Comment on the differences between the distributions.

5 The table shows the monthly salaries of 100 families.

Monthly salary ($)	No. of families
1451–1500	8
1501–1550	14
1551–1600	25
1601–1650	35
1651–1700	14
1701–1750	4

a Draw a cumulative frequency diagram to show the data.

b Estimate the median monthly salary and the interquartile range.

c The lowest monthly salary was $1480 and the highest was $1740.

   i Draw a box-and-whisker plot to show the distribution of salaries.

   ii Is the distribution symmetric, negatively skewed or positively skewed?

6 A health practice had two doctors, Dr Excel and Dr Collins.

The following box-and-whisker plots were created to illustrate the waiting times for their patients during October.

640   32.10 Box-and-whisker plots

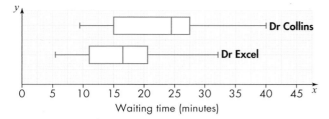

a For Dr Collins, what is:
   i the median waiting time
   ii the interquartile range for his waiting time
   iii the longest time a patient had to wait on October?
b For Dr Excel, what is:
   i the shortest waiting time for any patient in October
   ii the median waiting time
   iii the interquartile range for his waiting time?
c Anwar was deciding which doctor to try to see. Which one would you advise he sees? Why?

7 The box-and-whisker plot for a school's end-of-year mathematics tests are shown below.

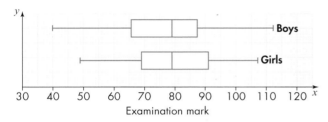

What is the difference between the means of the boys' and the girls' test results?

Check your progress

Core

- I can calculate the mean, median, mode and range for individual and discrete data
- I can distinguish between the uses for the mean, median, mode and range
- I can construct and interpret stem-and-leaf tables
- I can calculate an estimate of the mean for grouped and continuous data
- I can identify the modal class from a grouped frequency distribution
- I can construct and use cumulative frequency diagrams
- I can estimate and interpret the median, percentiles, quartiles and inter-quartile range

Extended

- I can construct and interpret box-and-whisker plots

Chapter 33
Probability

Topics	Level	Key words
1 The probability scale	CORE	chance, outcome, event, probability, impossible, certain
2 Calculating probabilities	CORE	outcome, equally likely, probability fraction, random
3 Probability that an event will not happen	CORE	outcome
4 Probability in practice	CORE	experimental probability, trials, relative frequency
5 Using Venn diagrams	CORE	
6 Possibility diagrams	CORE	possibility diagram
7 Tree diagrams	CORE	tree diagram
8 Conditional probability	EXTENDED	

In this chapter you will learn how to:

CORE	EXTENDED
• Calculate the probability of a single event as either a fraction, decimal or percentage. (C8.1 and E8.1) • Understand and use the probability scale from 0 to 1. (C8.2 and E8.2) • Understand that: *the probability of an event occurring = 1 − the probability of the event not occurring*. (C8.3 and E8.3) • Understand relative frequency as an estimate of probability. Expected frequency of occurrences. (C8.4 and E8.4) • Calculate the probability of simple combined events, using possibility diagrams and tree diagrams and Venn diagrams. (C8.5 and E8.5)	• Calculate conditional probability using Venn diagrams, tree diagrams and tables. (E8.6)

Why this chapter matters

Chance is a part of everyday life. Judgements are frequently made based on the probability of something happening.

For example:

- there is an 80% chance that my team will win the game tomorrow
- there is a 40% chance of rain tomorrow
- she has a 50–50 chance of having a baby girl
- there is a 10% chance of the bus being on time tonight.

In everyday life we talk about the probability of something happening. Two people might give different probabilities to the same events because of their different views. For example, some people might not agree that there is an 80% chance of your team winning the game. They might say that there is only a 70% chance of them winning tomorrow. A lot depends on what people believe or have experienced.

When people first started to predict the weather scientifically over 150 years ago, they used probabilities to do it. For example, meteorologists looked for three important indicators of rain:

- the number of nimbus clouds in the sky
- falling pressure on a barometer
- the direction of the wind and whether it was blowing from a part of the country with high rainfall.

If all three of these things occurred together rain would almost certainly follow soon.

Now, in the 21st century, probability theory is used to control the flow of traffic through road systems (below left) or the running of telephone exchanges (below right), and to look at patterns of the spread of infections.

Chapter 33: Probability

Chapter 33. Topic 1
33.1 The probability scale

Almost daily, you hear somebody talking about the probability of whether something will happen. They usually use words such as '**chance**', '**likelihood**' or '**risk**' rather than 'probability'. For example:

'What is the likelihood of rain tomorrow?'

'What chance does she have of winning the 100 metre sprint?'

'Is there a risk that his company will go bankrupt?'

You can give a value to the chance of any of these **outcomes** or **events** happening – and millions of others, as well. This value is called the **probability**.

It is true that some things are certain to happen and that some things cannot happen; that is, the chance of something happening can be anywhere between **impossible** and **certain**. This situation is represented on a sliding scale called the **probability scale**, as shown here.

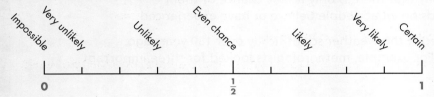

Note: All probabilities lie somewhere in the range of **0** to **1**.

An outcome or an event that cannot happen (is impossible) has a probability of 0. For example, the probability that donkeys will fly is 0.

An outcome or an event that is certain to happen has a probability of 1. For example, the probability that the sun will rise tomorrow is 1.

Example 1

Put arrows on the probability scale to show the probability of each of the outcomes of these events.

a   You will get a head when throwing a coin.

b   You will get a six when throwing a dice.

c   You will have maths homework this week.

---

a   This outcome is an even chance. (Commonly described as a fifty-fifty chance.)

b   This outcome is fairly unlikely.

c   This outcome is likely.

The arrows show the approximate probabilities on the probability scale.

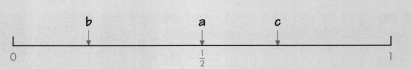

EXERCISE 33A

1 State whether each of these events is impossible, very unlikely, unlikely, even chance, likely, very likely or certain.

  a Someone in your class is left-handed.
  b You will live to be 100.
  c You get a score of seven when you throw a dice.
  d You will watch some TV this evening.
  e A new-born baby will be a girl.

2 Draw a probability scale and put an arrow to show the approximate probability of each of these events happening.

  a The next car you see will have been made in Japan.
  b A person in your class will have been born in the 20th century.
  c It will rain tomorrow.
  d In the next Olympic Games, someone will run the 1500 m race in 3 minutes.
  e During this week, you will have noodles with a meal.

3 a Draw a probability scale and mark an arrow to show the approximate probability of each of these events.

  A The next person to come into the room will be male.
  B The person sitting next to you in mathematics is over 16 years old.
  C Someone in the class will have a mobile phone.

  b What number on your scale corresponds to each arrow?

4 a Give two events of your own for which you think the probability of an outcome is:

  A impossible
  B very unlikely
  C evens
  D likely
  E certain.

  b Draw a probability scale numbered from 0 to 1 and put an arrow for each of your events.
  c What number on your scale corresponds to each arrow?

5 'The train was late yesterday so it is very likely that it will be late today.'

Is this true?

Chapter 33 . Topic 2
33.2 Calculating probabilities

In Exercise 33A, you may have had difficulty in knowing exactly where to put some of the arrows on the probability scale. It would have been easier for you if each result of the event could have been given a value, from 0 to 1, to represent the probability for that result.

For some events, this can be done by first finding all the possible results, or **outcomes**, for a particular event. For example, when you throw a coin there are two **equally likely** outcomes: it lands heads up or tails up. (The 'head' of a coin is the side which usually shows a head, the 'tail' is the side which shows the value of the coin.)

If you want to calculate the probability of getting a head, there is only one outcome that is possible. So, you can say that there is a 1 in 2, or 1 out of 2, chance of getting a head. This is usually given as a **probability fraction**, namely $\frac{1}{2}$. So, you would write the event as:

$$P(\text{head}) = \frac{1}{2}$$

Probabilities can also be written as decimals or percentages, so that:

$$P(\text{head}) = \frac{1}{2} \text{ or } 0.5 \text{ or } 50\%$$

The probability of an outcome is defined as:

$$P(\text{outcome}) = \frac{\text{number of ways the outcome can happen}}{\text{total number of possible outcomes}}$$

This definition always leads to a fraction, which should be cancelled to its simplest form.

Another probability term you will meet is at **random**. This means that the outcome cannot be predicted or affected by anyone.

Example 2

The spinner shown here is spun and the score on the side on which it lands is recorded.

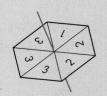

What is the probability that the score is:

a 2

b odd

c less than 5?

---

a There are two 2s out of six sides, so $P(2) = \frac{2}{6} = \frac{1}{3}$.

b There are four odd numbers, so $P(\text{odd}) = \frac{4}{6} = \frac{2}{3}$.

c All of the numbers are less than 5, so this is a certain event.

  $P(\text{less than 5}) = 1$

33.2

Example 3

Bernice is always early, just on time or late for work.

The probability that she is early is 0.1, the probability she is just on time is 0.5.

What is the probability that she is late?

As all the possibilities are covered – that is 'early', 'on time' and 'late' – the total probability is 1. So:

P(early) + P(on time) = 0.1 + 0.5 = 0.6

So, the probability of Bernice being late is 1 − 0.6 = 0.4.

EXERCISE 33B

1 There are ten balls in a bag. One is red, two are blue, three are yellow and four are green. A ball is taken out without looking.

What is the probability that it is:

a red

b green

c green or yellow

d red or green

e white?

Advice and Tips

If an event is impossible, just write the probability as 0, not as a fraction such as $\frac{0}{6}$. If it is certain, write the probability as 1, not as a fraction such as $\frac{6}{6}$.

2 An 8-sided spinner has the numbers 1, 2, 3, 4, 5, 6, 7 and 8 on it. It is spun once.

What is the probability that the score is:

a 3

b more than 3

c an even number?

Advice and Tips

Remember to cancel the fractions if possible.

3 A bag contains only blue balls. If I take one out at random, what is the probability of each of these outcomes?

a I get a black ball.

b I get a blue ball.

4 Ten number cards with the numbers 1 to 10 inclusive are placed in a hat. Amir takes a number card out of the bag without looking. What is the probability that he draws:

a the number 7

b an even number

c a number greater than 6

d a number less than 3

e a number between 3 and 8?

Chapter 33: Probability 647

5 A pencil case contains six red pens and five blue pens. Paulo takes out a pen without looking at it. What is the probability that he takes out:

  a a red pen
  b a blue pen
  c a pen that is not blue?

6 A bag contains 50 balls. 10 are green, 15 are red and the rest are white. Galenia takes a ball from the bag at random. What is the probability that she takes:

  a a green ball
  b a white ball
  c a ball that is not white
  d a ball that is green or white?

7 There are 500 students in a school and 20 students in Ali's class. One person is chosen at random to welcome a special visitor.

What is the probability the person is in Ali's class?

8 Anton, Bianca, Charlie, Debbie and Elisabeth are in the same class. Their teacher wants two students to do a special job.

  a Write down all the possible combinations of two people, for example, Anton and Bianca, Anton and Charlie. (There are 10 combinations altogether.)
  b How many pairs give two boys?
  c What is the probability of choosing two boys?
  d How many pairs give a boy and a girl?
  e What is the probability of choosing a boy and a girl?
  f What is the probability of choosing two girls?

> **Advice and Tips**
>
> Try to be systematic when writing out all the pairs.

9 A bag contains 25 coloured balls. 12 are red, 7 are blue and the rest are green. Ravi takes a ball at random from the bag.

  a Find:
    i P(he takes a red)
    ii P(he takes a blue)
    iii P(he takes a green).
  b Add together the three probabilities. What do you notice?
  c Explain your answer to part **b**.

10 The weather tomorrow will be sunny, cloudy or raining.

If P(sunny) = 40%, P(cloudy) = 25%, what is P(raining)?

11 At morning break, Priya has a choice of coffee, tea or hot chocolate.

If P(she chooses coffee) = 0.3 and P(she chooses hot chocolate) = 0.2, what is P(she chooses tea)?

Chapter 33 . Topic 3

12 The following information is known about the classes at Bradway School.

Year	Y1		Y2		Y3		Y4		Y5		Y6	
Class	P	Q	R	S	T	U	W	X	Y	Z	K	L
Girls	7	8	8	10	10	10	9	11	8	12	14	15
Boys	9	10	9	10	12	13	11	12	10	8	16	17

A class representative is chosen at random from each class.

Which class has the best chance of choosing a boy as the representative?

13 The teacher chooses, at random, a student to ring the school bell.

Tom says: 'It's even chances that the teacher chooses a boy or a girl.'

Explain why Tom might not be correct.

33.3 Probability that an event will not happen

In some questions in Exercise 33B, you were asked for the probability of something not happening. For example, in question 5 you were asked for the probability of picking a pen that is not blue. You could answer this because you knew how many pens there were in the case. However, sometimes you do not have this type of information.

The probability of throwing a six on a fair, six-sided dice is $P(6) = \frac{1}{6}$.

There are five **outcomes** that are not sixes: 1, 2, 3, 4, 5.

So, the probability of *not* throwing a six on a dice is:

$P(\text{not a 6}) = \frac{5}{6}$

Notice that:

$P(6) = \frac{1}{6}$ and $P(\text{not a 6}) = \frac{5}{6}$

So,

$P(6) + P(\text{not a 6}) = 1$

If you know that $P(6) = \frac{1}{6}$, then $P(\text{not a 6})$ is:

$1 - \frac{1}{6} = \frac{5}{6}$

So, if you know P(outcome happening), then:

P(outcome not happening) = 1 − P(outcome happening)

Chapter 33: Probability

Example 4

A box of coloured pencils has 20 different pencils. There are four red pencils, five blue, one green, three yellow, two brown, one black, and four other colours.

A pencil is chosen at random. What is the probability that it is not red?

There are 4 red pencils out of 20.

The probability that a red is chosen is $= \frac{4}{20} = \frac{1}{5}$.

The probability that a red is **not** chosen is $1 - \frac{1}{5} = \frac{4}{5}$.

EXERCISE 33C

1. **a** The probability that a football team will win their next match is $\frac{1}{4}$. What is the probability that the team will not win?

   b The probability that snow will fall during the winter holidays is 0.45. What is the probability that it will not snow?

   c The probability that Paddy wins a game of chess is 0.7 and the probability that he draws the game is 0.1. What is the probability that he loses the game?

2. Look at Example 4.

   What is the probability that the pencil is:

   a not blue

   b not yellow

   c not black?

3. These letter cards are put into a bag.

   M A T H E M
   A T I C A L

   a Lee takes a letter card at random.
      i What is the probability he takes a letter A?
      ii What is the probability he does not take a letter A?

   b Ziad picks an M and keeps it. Tasnim now takes a letter from those remaining.
      i What is the probability she takes a letter A?
      ii What is the probability she does not take a letter A?

4. Hamzah is told: 'The chance of your winning this game is 0.3.'

   Hamzah says: 'So I have a chance of 0.7 of losing.'

   Explain why Hamzah might be wrong.

33.4 Probability in practice

Suppose you toss a coin many times and count how many times it lands with the 'heads' side showing.

The value of 'number of heads ÷ number of tosses' is called an **experimental probability**. As the number of **trials**, or experiments, increases, the value of the experimental probability gets closer to the theoretical probability, which in this case is $\frac{1}{2}$.

Experimental probability is also known as the **relative frequency** of an event. The relative frequency of an event is an estimate for the theoretical probability. It is given by:

$$\text{relative frequency of an outcome or event} = \frac{\text{frequency of the outcome or event}}{\text{total number of trials}}$$

Example 5

The frequency table shows the speeds of 160 vehicles that pass a radar speed check on a fast road.

Speed (km/h)	20–29	30–39	40–49	50–59	60–69	70+
Frequency	14	23	28	35	52	8

a What is the experimental probability that a car is travelling faster than 70 km/h?

b If 500 vehicles pass the speed check, estimate how many will be travelling faster than 70 km/h.

a The experimental probability is the relative frequency, which is $\frac{8}{160} = \frac{1}{20}$.

b The number of vehicles travelling faster than 70 km/h will be $\frac{1}{20}$ of 500.

That is:

$500 ÷ 20 = 25$

EXERCISE 33D

1. Naseer throws a fair, six-sided dice and records the number of sixes that he gets after various numbers of throws. The table shows his results.

Number of throws	10	50	100	200	500	1000	2000
Number of sixes	2	4	10	21	74	163	329

   a Calculate the experimental probability of scoring a 6 at each stage that Naseer recorded his results.

   b How many ways can a dice land?

   c How many of these ways give a 6?

   d What is the theoretical probability of throwing a 6 with a dice?

   e If Naseer threw the dice a total of 6000 times, how many sixes would you expect him to get?

2 Marie made a five-sided spinner, like the one shown in the diagram. She used it to play a board game with her friend Sarah.

The girls thought that the spinner was not very fair as it seemed to land on some numbers more than others. They threw the spinner 200 times and recorded the results. The results are shown in the table.

Side spinner lands on	1	2	3	4	5
Number of times	19	27	32	53	69

a Work out the experimental probability of each number.

b How many times would you expect each number to occur if the spinner is fair?

c Do you think that the spinner is fair? Give a reason for your answer.

3 A bottle contains 20 balls. The balls are either black or white.

Kenny conducts an experiment to see how many black balls there are in the bottle. He tips one ball at a time into a clear sealed tube at the end of the bottle. He records the number of black balls and tips them back into the bottle.

The results are shown in the table.

Number of samples	Number of black balls	Experimental probability
10	2	
100	25	
200	76	
500	210	
1000	385	
5000	1987	

a Copy the table and complete it by calculating the experimental probability of getting a black ball at each stage.

b Using this information, how many black balls do you think there are in the bottle?

4 Use a set of number cards from 1 to 10 and work with a partner. Take turns to choose a card and keep a record each time of what card you get, before returning it to the pack. Shuffle the cards each time and repeat the experiment 60 times. Put your results in a copy of this table.

Score	1	2	3	4	5	6	7	8	9	10
Total										

a How many times would you expect to get each number?

b Do you think you and your partner conducted this experiment fairly?

c Explain your answer to part **b**.

5 A four-sided dice has faces numbered 1, 2, 3 and 4. The score is the face on which it lands. Five students throw the dice to see if it is biased. They each throw it a different number of times. Their results are shown in the table.

Student	Total number of throws	Score			
		1	2	3	4
Ali	20	7	6	3	4
Balvir	50	19	16	8	7
Caryl	250	102	76	42	30
Deema	80	25	25	12	18
Emma	150	61	46	26	17

a Which student will have the best set of results for finding the probability of each score? Why?

b Add up all the score columns and work out the relative frequency of each score. Give your answers to 2 decimal places.

c Is the dice biased? Explain your answer.

6 Andrew made an eight-sided spinner.

He tested it to see if it was fair.

He spun the spinner and recorded the results.

Unfortunately his little sister spilt something over his results table, so he could not see the middle part.

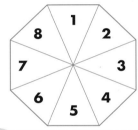

Number spinner lands on	1	2	3			6	7	8
Frequency	18	19	22			19	20	22

Assuming the spinner was a fair one, try to complete the missing parts of the table for Andrew.

7 At a computer factory, tests were carried out to see how many faulty computer chips were produced in one week.

	Monday	Tuesday	Wednesday	Thursday	Friday
Sample	850	630	1055	896	450
Number faulty	10	7	12	11	4

On which day was it most likely that the highest number of faulty computer chips were produced?

8 Steve tossed a coin 1000 times to see how many heads he got.

He said: 'If this is a fair coin, I should get 500 heads.'

Explain why he is wrong.

33.5 Using Venn diagrams

Sometimes you want to find the probability of two events happening at the same time. Venn diagrams are one way of answering these questions.

Example 6

There is a set of 15 cards. Each card has an integer from 1 to 15.

You take a card at random. Find the probability that it is

a A multiple of 3 and a factor of 12

b A multiple of 3 and not a factor of 12

c Neither a multiple of 3 nor a factor of 12.

Show the sets M = {multiples of 3} and F = {factors of 12} in a Venn diagram.

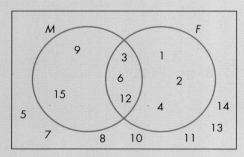

a The numbers 3, 6 and 12 are in both M and F so $n(M \cap F) = 3$. The probability is $\frac{3}{15} = \frac{1}{5}$.

b The numbers 9 and 15 are in M but not in F. The probability is $\frac{2}{15}$.

c There are 7 numbers outside both sets. The probability is $\frac{7}{15}$.

EXERCISE 33E

1. ξ = {integers from 1 to 21} E = {even numbers} T = {multiples of 3}

    a Show these sets on this Venn diagram.

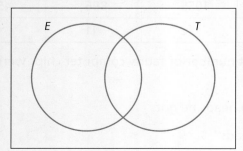

b A set of cards have the integers from 1 to 21.

A card is chosen at random.

Find the probability that it is

  i an even number

  ii a multiple of 3

  iii both an even number and a multiple of 3.

2 There are 100 people in a survey.

This Venn diagram shows how many went last year to the doctor or the dentist or both.

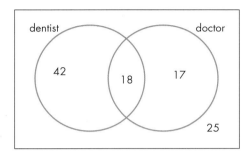

For example, 42 people went to the dentist but not the doctor.

One person is chosen at random. Find the probability that the person went to

a the dentist **b** the doctor

c both the dentist and the doctor **d** neither.

3 60 people are asked if they play football (F) or basketball (B)

Here are the results

	Football and basketball	Only football	Only basketball	Neither
Number of people	12	30	8	10

a Show the numbers in this Venn diagram.

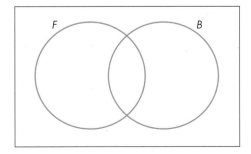

b A person is chosen at random.

Find the probability the person plays

  i football and basketball **ii** football **iii** basketball

4 ξ = {integers from 1 to 100} T = {multiples of 10} F = {multiples of 15}

a Put the elements of T and F in this Venn diagram

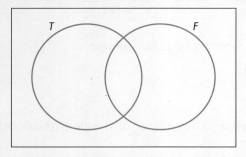

b A number between 1 and 100 is chosen at random.
   Find the probability that the number is
   i   a multiple of 15
   ii a multiple of both 10 and 15
   iii not a multiple of 10 or 15.

5

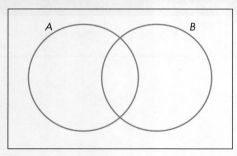

ξ = {points marked with a cross}

A = {points on the line $y = x$} B = {point on the line $y = 4 - x$}

a Put the coordinates of the points in this Venn diagram.

b One of the points is chosen at random.
   Find the probability that it is
   i   on line A
   ii on line B
   iii on both lines

6 ξ = {integers from 1 to 100}

T = {multiples of 10}   F = {multiples of 4}

a How many elements are in set T?

b On this Venn diagram show the number of elements in each region.

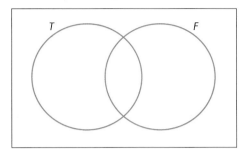

c A computer generates an integer between 1 and 100 at random.
Find the probability that it is
  i a multiple of 10
  ii a multiple of 4
  iii a multiple of both 10 and 4
  iv not a multiple of 10 or 4

33.6 Possibility diagrams

Venn diagrams are not the only way to show two events happening at the same time.

Suppose that you throw two dice. One is red and one is blue.

You can show all the possible outcomes in a **possibility diagram**.

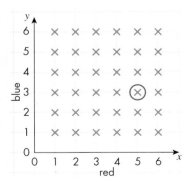

There are 36 different outcomes.

The cross with a circle round it is 5 on the red dice and 3 0n the blue dice.

Example 7

Two dice are thrown. Find the probability of getting

a two 3s **b** exactly one 3 **c** at least one 3 **d** no 3s

Here is a possibility diagram.

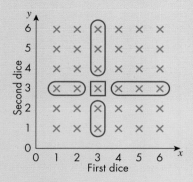

a There are 36 outcomes. There is only one way to get two 3s. It is shown with a square box.

The probability is $\frac{1}{36}$

b There are 10 ways to get one 3. They are in the four loops.

The probability is $\frac{10}{36} = \frac{5}{18}$

c At least one 3 means getting one or two. There are $10 + 1 = 11$ ways.

The probability $= \frac{11}{36}$

d No 3s means any of the 25 crosses that are not in a loop.

Probability $= \frac{25}{36}$

Example 8

Two dice are thrown and the numbers are added. Find the probability of a total of

a 8 **b** 8 or more **c** less than 8

Here is a possibility diagram. The totals have been written for each outcome.

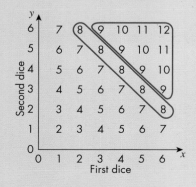

a The diagonal loop shows 5 ways to get a total of 9.

Probability = $\frac{5}{36}$

b The triangular loop shows 10 ways to get a total more than 8.

To get a total of 8 or more there are 5 + 10 = 15 ways.

Probability = $\frac{15}{36} = \frac{5}{12}$

c The crosses that are not in a loop are the ways to get a total of less than 8.

There are 21 of these.

Probability = $\frac{21}{36} = \frac{7}{12}$

EXERCISE 33F

1 Two dice are thrown. Use a possibility diagram like the one in Example 7 to answer these questions.

  a What is the most likely score?

  b Which two scores are least likely?

  c Write down the probabilities of throwing all the scores from 2 to 12.

  d What is the probability of a score that is:

   i bigger than 10

   ii between 3 and 7

   iii even

   iv a square number

   v a prime number

   vi a triangular number?

2 Two dice are thrown. Use a possibility diagram like the one in Example 8 to answer these questions.

  a the score is an even 'double'

  b at least one of the dice shows 2

  c the score on one dice is twice the score on the other dice

  d at least one of the dice shows a multiple of 3?

3 Two dice are thrown. Use a possibility diagram to find these probabilities.

  a both dice show a 6

  b at least one of the dice will show a 6

  c exactly one dice shows a 6

4 The diagram shows the scores for the event 'the difference between the scores when two fair, six-sided dice are thrown'. Copy and complete the diagram.

For the event described above, what is the probability of a difference of:

a 1
b 0
c 4
d 6
e an odd number?

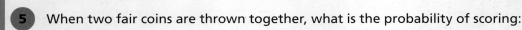

5 When two fair coins are thrown together, what is the probability of scoring:

a two heads
b a head and a tail
c at least one tail
d no tails?

Use a diagram of the outcomes when two coins are thrown together.

6 Two five-sided spinners are spun together and the total score of the faces that they land on is worked out. Copy and complete the possibility diagram shown.

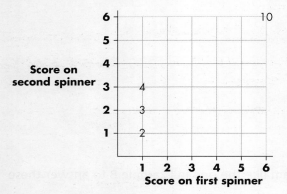
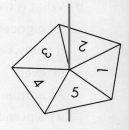

a What is the most likely score?
b When two five-sided spinners are spun together, what is the probability that:

  i the total score is 5
  ii the total score is an even number
  iii the score is a 'double'
  iv the total score is less than 7?

7 Two eight-sided spinners showing the numbers 1 to 8 were thrown at the same time.

a Draw a possibility diagram to show the product of the two scores.
b What is the probability that the product of the two spinners is an even square number?

8 Isaac throws two dice. He multiplies the numbers together. Find the probability that the product is between 19 and 35.

33.6 Possibility diagrams

33.7 Tree diagrams

Here is the question in Example 7 again.

Two dice are thrown.

Find the probability of getting

a two 3s   **b** exactly one 3   **c** at least one 3   **d** no 3s

You answered this question with a possibility diagram.

You can also use a **tree diagram**.

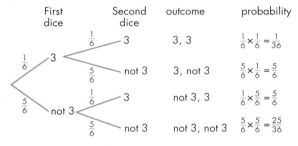

The probability that a dice shows 3 is $\frac{1}{6}$.

The probability that a dice does not show 3 (a 'not 3') is $\frac{5}{6}$.

Write these probabilities on each branch.

The branches show four outcomes, with each dice either '3' or 'not 3'.

To find the probability of each outcome, *multiply* the probabilities on the branches.

a The probability of 3, 3 is $\frac{1}{6} \times \frac{1}{6} = \frac{1}{36}$.   This is the answer to part **a**.

b The probability of 3, not 3 is $\frac{1}{6} \times \frac{5}{6} = \frac{5}{36}$.   This is 3 on the first but not the second.

The probability of not 3, 3 is $\frac{5}{6} \times \frac{1}{6} = \frac{5}{36}$.   This is 3 on the second but not the first.

Add these two: $\frac{5}{36} + \frac{5}{36} = \frac{10}{36} = \frac{5}{18}$.   This is the answer to part **b**.

c For at least one 3, add the three probabilities:

$\frac{1}{36} + \frac{5}{36} + \frac{5}{36} = \frac{11}{36}$ This is the answer to part **c**.

d The probability of not 3, not 3 is $\frac{5}{6} \times \frac{5}{6} = \frac{25}{36}$.   This is the answer to part **d**.

EXERCISE 33G

1 A coin is tossed twice. Copy and complete this tree diagram to show all the possible outcomes.

First event   Second event   Outcome   Probability

H — $\frac{1}{2}$ → H   (H, H) $\frac{1}{2} \times \frac{1}{2} = \frac{1}{4}$

$\frac{1}{2}$ H
    T

$\frac{1}{2}$ T
    H
    T

Use your tree diagram to work out the probability of getting:

    a two heads   **b** a head and a tail   **c** at least one tail.

2 There are two black balls and one white one in a bag.

Luis takes one at random.

He replaces it and then takes another one at random.

    a Copy and complete this tree diagram.

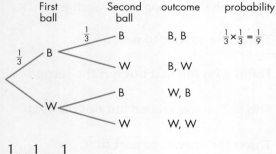

    b Find the probability that he takes

       i two white balls   **ii** one ball of each colour   **iii** at least one black ball

3 On my way to work, I drive through two sets of road works with traffic lights that only show green or red. I know that the probability of the first set being green is $\frac{1}{3}$ and the probability of the second set being green is $\frac{1}{2}$.

    a What is the probability that the first set of lights will be red?

    b What is the probability that the second set of lights will be red?

    c Copy and complete this tree diagram to show the possible outcomes when passing through both sets of lights.

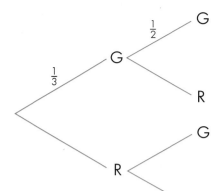

First event   Second event   Outcome   Probability

(G, G) $\frac{1}{3} \times \frac{1}{2} = \frac{1}{6}$

d Using the tree diagram, what is the probability that:
  i I do not get held up at either set of lights
  ii I get held up at exactly one set of lights
  iii I get held up at least once.

e Over a school term I make 90 journeys to work. On how many days can I expect to get two green lights?

4 There are two bags of sweets.

The first bag has 5 sweets and 4 of them are red.

The second bag has 5 sweets and 2 of them are red.

Delia takes one sweet from each bag.

a Copy and complete this tree diagram.

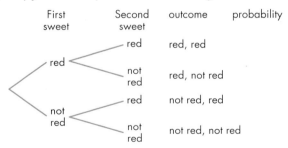

b Find the probability that she takes
  i two red sweets
  ii at least one red sweet
  iii no red sweets

5 An English exam has two parts, an oral test and a written test.

The probability that Chen passes the oral test is 0.9

The probability that Chen passes the written test is 0.6

Chapter 33: Probability   **663**

a Copy and complete this tree diagram

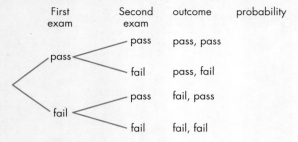

b Find the probability that Chen
  i   passes both parts   ii passes only one of the two.

6 A football team is playing two matches.

The probability that the team will win the first match is 0.6

The probability that the team will win the second match is 0.7

a Copy and complete this tree diagram to show probabilities.

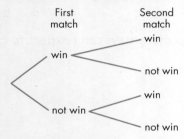

b Find the probability that the team will win at least one of the two matches.

7 Ahmed is playing a computer game.

The probability that he wins each time is 0.3.

He plays 2 games. Find the probability that:

a Ahmed wins both games   b the computer wins both games
c Ahmed and the computer both win one game.

8 The probability of rain on Monday is 0.25.

The probability of rain on Tuesday is 0.6.

Find the probability of:

a rain on Monday but not on Tuesday
b no rain on either day
c rain on at least one day.

9 Abebe is running in an 800 m race and a 1500 m race.

The probability that he wins the 800 m is 0.65

The probability that he wins the 1500 m is 0.4

Find the probability that he wins one race and loses the other.

664   33.7 Tree diagrams

10 A fruit bowl contains six oranges and eight lemons. Kevin takes two pieces of fruit at random.

  a If the first piece is an orange, what is the probability that the second is:
   i an orange
   ii a lemon?
  b What is the probability that:
   i both are oranges
   ii both are lemons?

11 A bag contains three black balls and seven red balls. A ball is taken out and not replaced. This is repeated twice. What is the probability that:

  a all three are black
  b exactly two are black
  c exactly one is black
  d none are black?

12 On my way to work, I pass two sets of traffic lights. The probability that the first is green is $\frac{1}{3}$. If the first is green, the probability that the second is green is $\frac{1}{3}$. If the first is red, the probability that the second is green is $\frac{2}{3}$. What is the probability that:

  a both are green
  b none are green
  c exactly one is green
  d at least one is green?

13 An engineering test is in two parts, a written test and a practical test. 90% of those who take the written test pass. When a person passes the written test, the probability that he or she will also pass the practical test is 60%. When a person fails the written test, the probability that he or she will pass the practical test is 20%.

  a What is the probability that someone passes both tests?
  b What is the probability that someone passes one test?
  c What is the probability that someone fails both tests?
  d What is the combined probability of the answers to parts **a**, **b** and **c**?

33.8 Conditional probability

Here is the Venn diagram from Example 6.

There is a set of 15 cards. Each card has an integer from 1 to 15.

You take a card at random.

M = {multiples of 3} and F = {factors of 12}

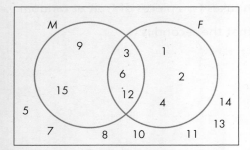

There are 5 numbers in set M.

The probability of picking a multiple of 3 is $\frac{5}{15}$.

There are 6 numbers in set F.

The probability of picking a factor of 12 is $\frac{6}{15} = \frac{2}{5}$.

Suppose you know the card is a factor of 12.

Now the probability that is a multiple of three is different.

You want the probability that it is in M, given that it is in F.

This is $\frac{3}{6} = \frac{1}{2}$.

This is called a **conditional probability**.

Suppose you know the card is a multiple of 3.

What is the conditional probability that the number is a factor of 12 given that it is a multiple of 3?

You want the probability that it is in F, given that it is in M.

This is $\frac{3}{5}$.

You sometimes need conditional probabilities on a tree diagram.

Example 9

There are 3 red pens and 2 blue pens in a box.

Two pens are taken out at random.

Find the probability that they are

　　a　both red　　**b**　both blue　　**c**　one of each colour.

Here is the tree diagram.

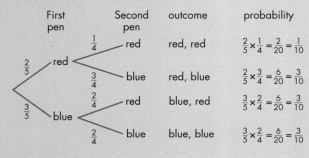

The probabilities on the branches are not the same each time.

When the first pen is taken, the probabilities are red = $\frac{2}{5}$ and blue = $\frac{3}{5}$.

These are on the first pair of branches.

If the first pen is red, there are 1 red and 3 blue pens left.

The conditional probabilities are red = $\frac{1}{4}$ and blue = $\frac{3}{4}$.

These go on the top pair of branches for the second pen.

If the first pen is blue, there are 2 red and 2 blue pens left.

The conditional probabilities are red = $\frac{2}{4}$ (or $\frac{1}{2}$) and blue = $\frac{2}{4}$ (or $\frac{1}{2}$).

These go on the bottom pair of branches for the second pen.

 a The probability they are both red is $\frac{2}{5} \times \frac{1}{4} = \frac{2}{20} = \frac{1}{10}$.

 b The probability they are both blue is $\frac{3}{5} \times \frac{2}{4} = \frac{6}{20} = \frac{3}{10}$.

 c One of each colour is either the second or the third outcome.

    Add the two probabilities.

    $\left(\frac{2}{5} \times \frac{3}{4}\right) + \left(\frac{3}{5} \times \frac{2}{4}\right) = \frac{6}{20} + \frac{6}{20} = \frac{12}{20} = \frac{3}{5}$

EXERCISE 33H

1 100 people are asked if they have a brother or a sister.

The numbers are in this Venn diagram.

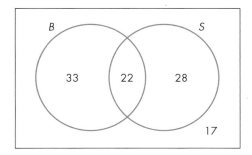

B = {people with a brother} S = {people with a sister}

One person is chosen at random.

 a Find the probability that the person has a brother.
 b The person has a brother. Find the probability that the person has a sister.
 c Another person is chosen at random. This person has a sister. Find the probability that this person has a brother.

2 ξ = {integers from 1 to 25}

S = {factors of 16}

T = {factors of 24}

a Put the elements of ξ in this Venn diagram.

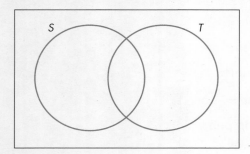

An integer between 1 and 25 is chosen at random.

b Find the probability that it is a factor of both 16 and 24.

c If the number is a factor if 24, find the probability that it is a factor of 16.

d If the number is a factor if 16, find the probability that it is a factor of 24.

e If the number is a not a factor if 16, find the probability that it is a not a factor of 24.

3 ξ = {integers from 1 to 30}

E = {even numbers}

T = {multiples of 3}

F = {multiples of 5}

a Put the elements of ξ in this Venn diagram.

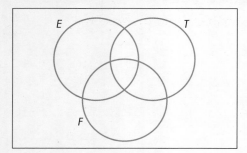

An integer between 1 and 30 is chosen at random.

b Find the probability that it is a multiple of 3.

c Find the probability that it is a multiple of 3 and 5.

d If the number is a multiple of 5, find the probability that it is also a multiple of 3.

e If the number is a multiple of 5, find the probability that it is an even multiple of 3.

f If the number is a multiple of 3, find the probability that it is also a multiple of 5.

4 120 students take exams in English (*E*), maths (*M*) and science (*S*).

This Venn diagram shows how many students pass each exam.

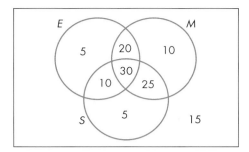

A student is chosen at random.

a Find the probability that the student passed maths.

b Find the probability that the student passed all three subjects.

A student is chosen at random. This student passed maths.

c Find the probability that the student passed science.

d Find the probability that the student passed English.

A student is chosen at random. This student failed maths.

e Find the probability that the student also failed science.

5 Two dice are thrown. The numbers are added together.

a Show the possible totals on a possibility diagram.

b Find the probability of a total of 6.

c If the total is 6, find the probability that one of the numbers is 2.

d Find the probability that at least one of the numbers is 2.

e If at least one of the numbers is 2, find the probability that the total is 6.

6

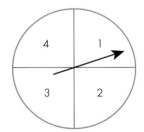

A spinner can show the numbers 1, 2, 3 and 4 with equal probabilities.

The spinner is spun twice and the numbers are added.

a Find the probability that the total is less than 5.

b If the total is 5, find the probability that the numbers on each spin are the same.

c If both numbers are the same, find the probability that the total is less than 5.

7 Driving to work, Karim passes two sets of traffic lights.

The probability that the first set is red is 0.3.

If the first set is red, the probability that the second set is red is 0.8.

If the first set is green, the probability that the second set is red is 0.4.

   a Put probabilities on the branches of this tree diagram.

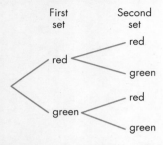

   b Find the probability that both sets of lights are red.
   c Find the probability that both sets of lights are green.
   d Find the probability that one set is red and the other is green.

8 Laura has 4 $5 notes and 6 $10 notes in her purse.

She takes out two notes at random.

   a Find the probability that the first note is $5.
   b If the first note is $5, what is the probability that the second note is also $5?
   c Put probabilities on the branches of this tree diagram.

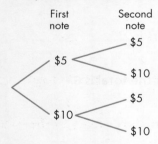

   d Find the probability that the total value of the two notes is:
      i $10   **ii** $15   **iii** $20

9 An engineering exam has two parts, a written test and a practical test.

The probability that a student passes the written test is 0.9.

If a student passes the written test, the probability of passing the practical test is 0.8.

If a student fails the written test, the probability of failing the practical test is 0.4.

   a Put probabilities on the branches of this tree diagram.

670 33.8 Conditional probability

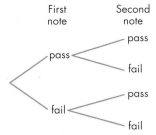

b Find the probability that a student passes both parts.

c Find the probability that a student passes one part and fails the other.

10 A box has 3 red counters and 5 white counters.

Sara takes out counters at random, one at a time.

a Find the probability that the first two counters are red.

b Find the probability that the first two counters are white.

c Sara takes out 3 counters all together. Find the probability that all are the same colour.

11 Aaron and Barak play a game.

There are 5 white balls and one red ball in a bag.

Aaron takes out 3 balls, one at a time, without looking.

If one of the balls is red, Aaron wins the game. If they are all white Barak wins.

What is the probability that Aaron wins? Give a reason for your answer.

Check your progress

Core

- I can calculate the probability of a single event as a fraction, a decimal or a percentage
- I can understand and use the probability scale from 0 to 1
- I understand that the probability of an event not occurring = 1 – the probability of the event occurring
- I understand relative frequency as an estimate of probability
- I can calculate the probability of simple combined events using possibility diagrams, tree diagrams or Venn diagrams with two sets
- I can calculate the probability of simple combined events using Venn diagrams with three sets

Extended

- I can calculate conditional probability using tables, tree diagrams or Venn diagrams

Examination questions: Statistics and probability

Past paper questions reproduced by permission of Cambridge Assessment International Education. Other exam-style questions have been written by the authors.

PAPER 1

1 Jim scores the following marks in 8 tests.

7 8 8 y 6 9 10 5

His mean mark is 7.5.

Calculate the value of y. [2]

Cambridge International IGCSE Mathematics 0580 Paper 11 Q9 Oct/Nov 2015

2 Chico has a bag of sweets.

He takes a sweet from the bag at random.

The table shows the probabilities of taking each flavour of sweet.

Flavour	Lemon	Lime	Strawberry	Blackcurrant	Orange
Probability	0.15	0.22		0.18	0.24

a Complete the table. [2]

b Find the probability that the sweet is lemon or lime. [1]

Cambridge International IGCSE Mathematics 0580 Paper 11 Q15 Oct/Nov 2015

3 The scatter diagram shows the number of sun hats and ice creams sold by a shop each day for two weeks.

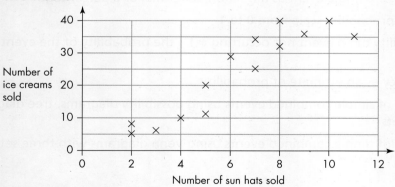

a Write down the type of correlation shown by the diagram. [1]

b Describe the relationship between the number of sun hats sold and the number of ice creams sold. [1]

Cambridge International IGCSE Mathematics 0580 Paper 11 Q9 May/June 2015

Examination questions: Statistics and probability

4 These are the heights, correct to the nearest centimetre, of 12 children.
132 114 151 130 132 145 163 142 153 170 132 125
Find the median height. [2]

Cambridge International IGCSE Mathematics 0580 *Paper 11 Q5 Oct/Nov 2014*

5 Cheryl recorded the midday temperatures in Seoul for one week in January.

Day	Mon	Tue	Wed	Thu	Fri	Sat	Sun
Temperature (°C)	−4	−5	−3	−11	−8	−3	−1

a Write down the mode. [1]
b On how many days was the temperature lower than the mode? [1]

Cambridge International IGCSE Mathematics 0580 *Paper 11 Q10 Oct/Nov 2014*

6 The table shows the average monthly temperature (°C) for Fairbanks, Alaska.

Month	Jan	Feb	Mar	Apr	May	Jun	Jul	Aug	Sep	Oct	Nov	Dec
Temperature (°C)	−23.4	−19.8	−11.7	−0.8	9.2	15.4	16.9	13.8	7.5	−5.8	−21.4	−21.8

a Find
  i the difference between the highest and the lowest temperatures,
  ii the median. [1] [2]
b A month is chosen at random from the table.
  Find the probability that its average temperature is below zero. [1]

Cambridge International IGCSE Mathematics 0580 *Paper 11 Q19 May/June 2014*

7

One of the 6 letters is taken at random.
a Write down the probability that the letter is S. [1]
b The letter is replaced and again a letter is taken at random.
  This is repeated 600 times.
  How many times would you expect the letter to be S? [1]

Cambridge International IGCSE Mathematics 0580 *Paper 11 Q14 Oct/Nov 2013*

8 The probability of Sachin's team winning any match is 0.45.
a Write down the probability of Sachin's team not winning any match. [1]
b In a season there are 40 matches.
  How many matches should Sachin's team expect to win in a season? [2]

Cambridge International IGCSE Mathematics 0580 *Paper 11 Q12 May/June 2013*

9 Marco throws a six-sided dice 27 times.
The bar chart shows his results.

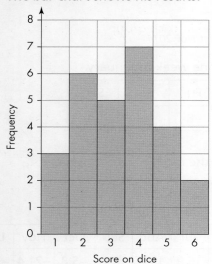

a Write down the mode. [1]
b Work out the probability that Marco throws a number less than 5. [2]
c Calculate the mean. [3]

Cambridge International IGCSE Mathematics 0580 *Paper 11 Q20 May/June 2013*

10 8 15 7 8 7 15 4 13 4 3 10 2 9 4 5
a Write down the mode. [1]
b Work out the median. [2]

Cambridge International IGCSE Mathematics 0580 *Paper 11 Q16 Oct/Nov 2013*

Examination questions: Statistics and probability

PAPER 3

1 a 120 children take part in an athletics competition.
  i Complete the table to show the number of children in each group.

	Girls	Boys	Total
Age 15			65
Age 16	44		
Total	70		120

  [2]
  ii One child is selected at random.
  Find the probability that it is a girl aged 16.
  Give your answer as a fraction in its lowest terms. [2]
  iii Write down the ratio number of girls aged 15 : number of boys aged 15.
  Give your answer in its simplest form. [2]

b Here are the distances, in metres, recorded in the boys' shot putt.
  9.23 6.21 9.86 8.64 7.15 7.72 9.01 7.34 6.53 6.89
  i Find the median. [2]
  ii Find the range. [1]
  iii Another boy was a late entry to the competition.
  After his attempt, the range increased by 20 cm.
  Work out the two possible distances of his attempt. [2]

Cambridge International IGCSE Mathematics 0580 *Paper 31 Q1 Oct/Nov 2015*

2 All the children in a school are asked to choose their favourite colour.
The pie chart shows the results.

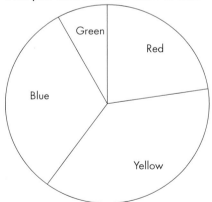

a Write down the least favourite colour chosen. [1]
b 27 children choose yellow as their favourite colour.
  Work out the total number of children in the school. [3]
c Work out the percentage of the children in the school who choose red. [2]

Cambridge International IGCSE Mathematics 0580 *Paper 31 Q5 May/June 2015*

Examination questions: Statistics and probability   675

Examination questions: Statistics and probability

3 12 athletes took part in the 100 metres race.

11 of these athletes also took part in the long jump.

The times and distances, each measured correct to 3 significant figures, for these athletes are shown in the table.

Athlete	A	B	C	D	E	F	G	H	I	J	K	L
100 m time (seconds)	12.1	10.3	12.8	10.7	12.6	11.2	12.0	12.4	10.6	12.7	11.8	11.1
Long jump (metres)	X	7.60	5.15	7.25	6.72	6.30	5.60	6.20	6.90	5.70	6.85	6.70

a The scatter diagram shows the times and distances for athletes B to H.

  i Plot the times and distances for athletes I, J, K and L. [2]

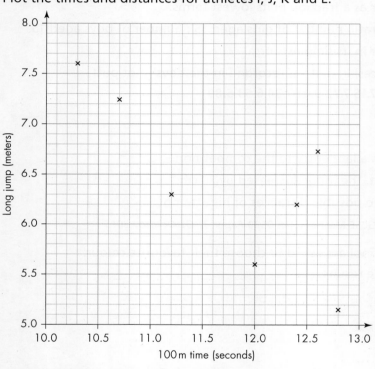

  ii On the scatter diagram, draw a line of best fit. [1]

  iii Athlete A did not take part in the long jump.

  Use your line of best fit to estimate a long jump distance for athlete A. [1]

  iv What type of correlation is shown on the scatter diagram? [1]

  v Describe in words the relationship between the time for 100 metres and the distance in the long jump. [1]

b Use the table of times and distances to work out

  i the mean of the 100 metres times, [2]

  ii the percentage of athletes who ran 100 metres in less than 11.5 seconds, [2]

  iii the range of the distances jumped by the 11 athletes, B to L. [1]

Cambridge International IGCSE Mathematics 0580 *Paper 31 Q3 Oct/Nov 2014*

Examination questions: Statistics and probability

4 a Amir asked 15 friends how many hours they spent playing sport last weekend.
His results are shown in the table below.

Number of hours	0	1	2	3	4	5
Frequency	6	2	3	1	2	1

   i Write down the mode. [1]
   ii Find the median. [1]
   iii Calculate the mean. [3]
   iv On the grid, draw a bar chart to show the information given in the table. [4]

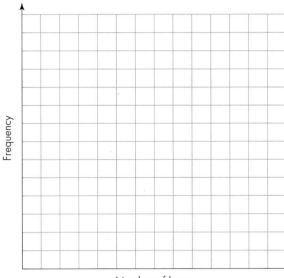

b Amir also asked these 15 friends which was their favourite sport.
His results are shown in the table below.

Football	4
Cricket	5
Basketball	2
Badminton	4

Amir picks one of these friends at random.
Write down the probability that his friend's favourite sport is
   i cricket, [1]
   ii **not** football, [1]
   iii basketball or badminton. [1]

Cambridge International IGCSE Mathematics 0580 *Paper 31 Q3 Oct/Nov 2012*

Examination questions: Statistics and probability

PAPER 2

1 Paul and Sammy take part in a race.

The probability that Paul wins the race is $\frac{9}{35}$.

The probability that Sammy wins the race is 26%.

Who is more likely to win the race?

Give a reason for your answer. [2]

Cambridge International IGCSE Mathematics 0580 Paper 21 Q5 May/June 2015

2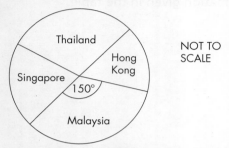

A travel brochure has 72 holidays in four different countries.

The pie chart shows this information

a There are 24 holidays in Thailand.
Show that the sector angle for Thailand is 120°. [2]

b The sector angle for Malaysia is 150°.
The sector angle for Singapore is twice the sector angle for Hong Kong.
Calculate the number of holidays in Hong Kong. [3]

Cambridge International IGCSE Mathematics 0580 Paper 21 Q17 May/June 2014

3 Leon scores the following marks in 5 tests.

8 4 8 y 9

His mean mark is 7.2.

Calculate the value of y. [2]

Cambridge International IGCSE Mathematics 0580 Paper 21 Q6 May/June 2012

Examination questions: Statistics and probability

4 If it rains today the probability that it will rain tomorrow is 0.4.
If it does not rain today the probability that it will rain tomorrow is 0.2.
On Sunday it rained.

a Complete the tree diagram for Monday and Tuesday. [2]

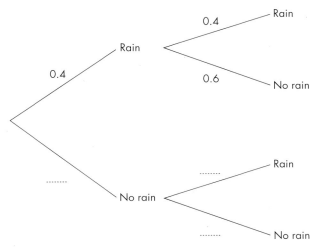

b Find the probability that it rains on at least one of the two days shown in the tree diagram. [3]

Cambridge International IGCSE Mathematics 0580 *Paper 21 Q18 Oct/Nov 2014*

5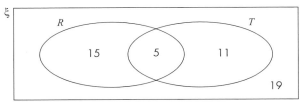

The Venn diagram shows the number of red cars and the number of two-door cars in a car park.
There is a total of 50 cars in the car park.
R = {red cars} and T = {two-door cars}.

a A car is chosen at random.
Write down the probability that
   i it is red and it is a two-door car, [1]
   ii it is not red and it is a two-door car. [1]

b A two-door car is chosen at random.
Write down the probability that it is not red. [1]

c Two cars are chosen at random.
Find the probability that they are both red. [2]

Cambridge International IGCSE Mathematics 0580 *Paper 21 Q22 Oct/Nov 2013*

Examination questions: Statistics and probability

6 The table shows the probability that a person has blue, brown or green eyes.

Eye colour	Blue	Brown	Green
Probability	0.4	0.5	0.1

Use the table to work out the probability that two people, chosen at random,

a have blue eyes, [2]

b have different coloured eyes. [4]

Cambridge International IGCSE Mathematics 0580 *Paper 21 Q20 Oct/Nov 2015*

7 In this question, give all your answers as fractions.
A box contains 3 red pencils, 2 blue pencils and 4 green pencils.
Raj chooses 2 pencils at random, without replacement.
Calculate the probability that

a they are both red, [2]

b they are both the same colour, [3]

c exactly one of the two pencils is green. [3]

Cambridge International IGCSE Mathematics 0580 *Paper 21 Q21 May/June 2012*

Examination questions: Statistics and probability

PAPER 4

1

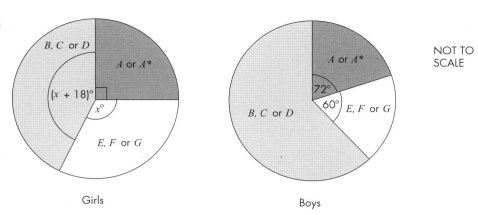

The pie charts show information on the grades achieved in mathematics by the girls and boys at a school.

a For the **Girls'** pie chart, calculate
  i x, [2]
  ii the angle for grades B, C or D. [1]

b Calculate the percentage of the **Boys** who achieved grades E, F or G. [2]

c There were 140 girls and 180 boys.
  i Calculate the percentage of students (girls and boys) who achieved grades A or $A*$. [3]
  ii How many more boys than girls achieved grades B, C or D? [2]

d The table shows information about the times, t minutes, taken by 80 of the girls to complete their mathematics examination.

Time (t minutes)	$40 < t \leq 60$	$60 < t \leq 80$	$80 < t \leq 120$	$120 < t \leq 150$
Frequency	5	14	29	32

  i Calculate an estimate of the mean time taken by these 80 girls to complete the examination. [4]

  ii On a histogram, the height of the column for the interval $60 < t \leq 80$ is 2.8 cm.
  Calculate the heights of the other three columns.
  Do not draw the histogram. [4]

  iii $40 < t \leq 60$ column height = ... cm
  $80 < t \leq 120$ column height = ... cm
  $120 < t \leq 150$ column height = ... cm [4]

Cambridge International IGCSE Mathematics 0580 *Paper 41 Q1 Oct/Nov 2012*

2 120 students take a mathematics examination.

a The time taken, m minutes, for each student to answer question 1 is shown in this table.

Time (m minutes)	$0 < m \leq 1$	$1 < m \leq 2$	$2 < m \leq 3$	$3 < m \leq 4$	$4 < m \leq 5$	$5 < m \leq 6$
Frequency	72	21	9	11	5	2

Calculate an estimate of the mean time taken. [4]

Examination questions: Statistics and probability

b i Using the table in part (a), complete this cumulative frequency table. [2]

Time (*m* minutes)	$m \leq 1$	$m \leq 2$	$m \leq 3$	$m \leq 4$	$m \leq 5$	$m \leq 6$
Cumulative frequency	72					120

ii Draw a cumulative frequency diagram to show the time taken. [3]

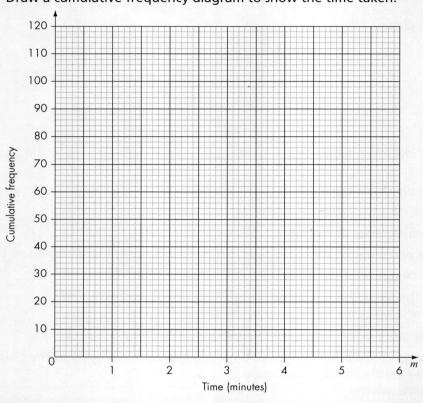

iii Use your cumulative frequency diagram to find

  a the median, [1]

  b the inter-quartile range, [2]

  c the 35th percentile. [2]

c A new frequency table is made from the table shown in **part (a)**.

Time (*m* minutes)	$0 < m \leq 1$	$1 < m \leq 3$	$3 < m \leq 6$
Frequency	72		

  i Complete the table above. [2]

  ii A histogram was drawn and the height of the first block representing the time $0 < m \leq 1$ was 3.6 cm.

  Calculate the heights of the other two blocks. [3]

Cambridge International IGCSE Mathematics 0580 *Paper 41 Q6 Oct/Nov 2015*

3 30 students were asked if they had a bicycle (*B*), a mobile phone (*M*) and a computer (*C*). The results are shown in the Venn diagram.

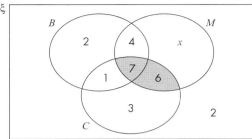

- **a** Work out the value of *x*. [1]
- **b** Use set notation to describe the shaded region in the Venn diagram. [1]
- **c** Find n($C \cap (M \cup B)'$). [1]
- **d** A student is chosen at random.
  - **i** Write down the probability that the student is a member of the set *M*'. [1]
  - **ii** Write down the probability that the student has a bicycle. [1]
- **e** Two students are chosen at random from the students who have computers.
  Find the probability that each of these students has a mobile phone but no bicycle. [3]

Cambridge International IGCSE Mathematics 0580 Paper 41 Q4 M/J 2015

4 a A square spinner is biased.
The probabilities of obtaining the scores 1, 2, 3 and 4 when it is spun are given in the table.

Score	1	2	3	4
Probability	0.1	0.2	0.4	0.3

  - **i** Work out the probability that on one spin the score is 2 or 3. [2]
  - **ii** In 5000 spins, how many times would you expect to score 4 with this spinner? [1]
  - **iii** Work out the probability of scoring 1 on the first spin and 4 on the second spin. [2]

b In a bag there are 7 red discs and 5 blue discs.
From the bag a disc is chosen at random and not replaced.
A second disc is then chosen at random.
Work out the probability that at least one of the discs is red.
Give your answer as a fraction. [3]

Cambridge International IGCSE Mathematics 0580 Paper 41 Q6 May/June 2014

Examination questions: Statistics and probability

5 Lauris records the mass and grade of 300 eggs. The table shows the results.

Mass (x grams)	$30 < x \leq 40$	$40 < x \leq 50$	$50 < x \leq 60$	$60 < x \leq 70$	$70 < x \leq 80$	$80 < x \leq 90$
Frequency	15	48	72	81	54	30
Grade	small	medium	large		very large	

a Find the probability that an egg chosen at random is graded very large. [1]

b The cumulative frequency diagram shows the results from the table.

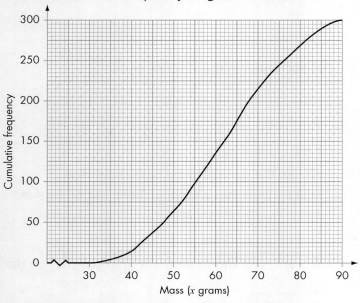

Use the cumulative frequency diagram to find

  i the median, [1]
  ii the lower quartile, [1]
  iii the inter-quartile range, [1]
  iv the number of eggs with a mass greater than 65 grams. [2]

Cambridge International IGCSE Mathematics 0580 *Paper 21 Q18 Oct/Nov 2012*

Examination questions: Mixed type

Past paper questions reproduced by permission of Cambridge Assessment International Education. Other exam-style questions have been written by the authors.

PAPER 3

1 a Luis buys a season ticket to watch his local football team.
The season ticket costs $595.
  i Luis buys the season ticket online and gets a 5% discount on the $595.
  Work out how much Luis pays for the season ticket online. [2]
  ii A ticket to watch one match costs $3 8.
  Luis watches 16 matches.
  How much did Luis save by buying a season ticket online instead of 16 tickets at $38 each? [2]
b The football stadium has 26 272 seats.
The number of people who attend one match is 23 854,
Calculate the percentage of the 26 272 seats that are **empty**. [2]
c The total number of people attending matches at the stadium last season was 506 762.
Write 506 762 in standard form, correct to 3 significant figures. [2]

Cambridge International IGCSE Mathematics 0580 Paper 31 Q3 Oct/Nov 2015

2 Three friends are going on holiday.
They travel by plane.
a Ahmed's suitcase has mass m kilograms.
  i The mass of Sonia's suitcase is 5 kg more than the mass of Ahmed's suitcase.
  Write down an expression, in terms of m, for the mass of Sonia's suitcase. [1]
  ii The mass of Hala's suitcase is twice the mass of Ahmed's suitcase.
  Write down an expression, in terms of m, for the mass of Hala's suitcase. [1]
  iii The total mass of the three suitcases is 47 kg.
  Write down an equation in terms of m. [1]
  iv Solve your equation and find the mass of each suitcase. [3]
  Ahmed's suitcase .... kg
  Sonia's suitcase .... kg
  Hala's suitcase .... kg

Cambridge International IGCSE Mathematics 0580 Paper 31 Q4 Oct/Nov 2015

Examination questions: Mixed type

3 Denzil grows tomatoes. He selects a random sample of 25 tomatoes.
The mass of each tomato, to the nearest 5 grams, is shown below.

55	65	50	75	65
80	70	70	55	60
70	60	65	50	75
65	70	75	80	70
55	65	70	80	55

a i Complete the frequency table.
You may use the tally column to help you. [2]

Mass (grams)	Tally	Frequency
50		
55		
60		
65		
70		
75		
80		

  ii Write down the mode. [1]
  iii Find the range. [1]
  iv Show that the mean mass is 66 g. [2]

b Denzil picks 800 tomatoes.
4% of the 800 tomatoes are damaged.
How many of these tomatoes are not damaged? [2]

e Denzil sells 750 of his tomatoes.
  i The mean mass of a tomato is 66 g.
  Calculate the mass of the 750 tomatoes in kilograms. [3]
  ii Denzil sells his tomatoes at $1.40 per kilogram.
  Calculate the total amount he receives from selling all the 750 tomatoes. [1]
  iii The cost of growing these tomatoes was $33.
  Calculate his percentage profit. [3]

Cambridge International IGCSE Mathematics 0580 *Paper 31 Q4 May/June 2014*

Examination questions: Mixed type

4 Sonia works in a toy shop.

   a **i** One week she works for 30 hours and is paid $180.

         Calculate the amount she is paid per hour. [1]

     ii The next week Sonia works for 38 hours and is paid $220.

         Find the difference in her pay per hour for these two weeks. [2]

   b The shop sells bags of 40 marbles.

      One bag has marbles in the ratio red : blue : green = 1 : 3 : 4.

     i Calculate the number of marbles of each colour. [2]

     ii A second bag of 40 marbles contains 11 red marbles, 9 blue marbles and 20 green marbles.

        All the marbles from the two bags are mixed together.

        Write down the ratio of marbles red : blue : green.

        Give your answer in its simplest form. [2]

Cambridge International IGCSE Mathematics 0580 *Paper 31 Q3 May/June 2015*

PAPER 4

1 a Luc is painting the doors in his house.

He uses $\frac{3}{4}$ of a tin of paint for each door.

Work out the least number of tins of paint Luc needs to paint 7 doors. [3]

b Jan buys tins of paint for $17.16 each.

He sells the paint at a profit of 25%.

For how much does Jan sell each tin of paint? [2]

c The cost of $17.16 for each tin of paint is 4% more than the cost in the previous year.

Work out the cost of each tin of paint in the previous year. [3]

d In America a tin of paint costs $17.16.

In Italy the same tin of paint costs €13.32.

The exchange rate is $1 = €0.72.

Calculate, in dollars, the difference in the cost of the tin of paint. [2]

Cambridge International IGCSE Mathematics 0580 Paper 41 Q1 Oct/Nov 2015

2 a A company makes compost by mixing loam, sand and coir in the following ratio.

loam : sand : coir = 7 : 2 : 3

  i How much loam is there in a 72 litre bag of the compost? [2]

  ii In a small bag of the compost there are 13.5 litres of coir.

  How much compost is in a small bag? [2]

  iii The price of a large bag of compost is $8.40.

  This is an increase of 12% on the price last year.

  Calculate the price last year. [3]

b Teresa builds a raised garden bed in the shape of a hexagonal prism.

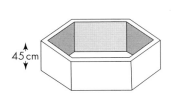 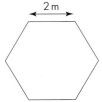

NOT TO SCALE

The garden bed has a height of 45 cm.

The cross section of the inside of the garden bed is a regular hexagon of side 2 m.

  i Show that the area of the cross section of the inside of the garden bed is 10.4 m², correct to 3 significant figures. [3]

  ii Calculate the volume of soil needed to fill the garden bed. [2]

  iii Teresa wants to fill the garden bed with organic top soil.

Examination questions: Mixed type

She sees this advertisement in the local garden centre.

ORGANIC TOP SOIL	Number of tonnes purchased		
	1 to 5	6 to 10	Over 10
Cost per tonne	$47.00	$45.50	$44.00

Organic top soil is sold in one tonne bags.

$1 m^3$ of organic top soil has a mass of 1250 kg.

Calculate the cost of the organic top soil needed to fill the garden bed completely. [1 tonne = 1000 kg] [4]

Cambridge International IGCSE Mathematics 0580 *Paper 41 Q1 Oct/Nov 2014*

3 a

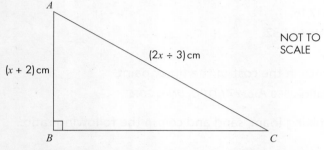

In triangle ABC, $AB = (x + 2)$ cm and $AC = (2x + 3)$ cm.

$\sin ACB = \dfrac{9}{16}$

Find the length of BC. [6]

b A bag contains 7 white beads and 5 red beads.

  i The mass of a red bead is 2.5 grams more than the mass of a white bead.

  The total mass of all the 12 beads is 114.5 grams.

  Find the mass of a white bead and the mass of a red bead. [5]

  ii Two beads are taken out of the bag at random, without replacement.

  Find the probability that

  a they are both white, [2]

  b one is white and one is red. [3]

Cambridge International IGCSE Mathematics 0580 *Paper 41 Q7 May/June 2013*

4 a The running costs for a papermill are $75 246.
This amount is divided in the ratio labour costs : materials = 5 : 1.
Calculate the labour costs. [2]

b In 2012 the company made a profit of $135 890.
In 2013 the profit was $150 675.
Calculate the percentage increase in the profit from 2012 to 2013. [3]

c The profit of $135 890 in 2012 was an increase of 7% on the profit in 2011.
Calculate the profit in 2011. [3]

d

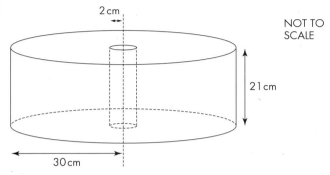

Paper is sold in cylindrical rolls.

There is a wooden cylinder of radius 2 cm and height 21 cm in the centre of each roll.
The outer radius of a roll of paper is 30 cm.

   i Calculate the volume of paper in a roll. [3]

   ii The paper is cut into sheets which measure 21 cm by 29.7 cm.

   The thickness of each sheet is 0.125 mm.

   a Change 0.125 millimetres into centimetres. [1]

   b Work out how many whole sheets of paper can be cut from a roll. [4]

Cambridge International IGCSE Mathematics 0580 *Paper 41 Q3 May/June 2014*

Glossary

π pi the value of the circumference of any circle divided by its diameter. Approximately 3.142

12-hour clock used to give a time of day using times up to 12 and am for morning times and pm for afternoon times

24-hour-clock used to give a time of day using times up to 24 and does not need am or pm.

Absolute value the positive value of the difference between a number and zero

Acceleration the rate at which the velocity of a moving object changes

Acute angles are angles which are smaller than 90°

Adjacent side the side adjacent (next) to the known or required angle in a right-angled triangle

Adjacent

Allied angles are made when a line crosses a pair of parallel lines. Allied angles are sometimes called interior angles

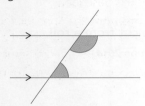

Alternate angles are made when a line crosses a pair of parallel lines. The alternate angles are on alternate sides of the line

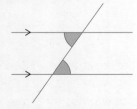

Amplitude the maximum displacement of a cyclical function from its central position

Angle an angle measures the amount of turn or the change in direction between two lines

Angle bisector a line which divides an angle into two equal parts

Angle of depression looking down this is the angle between the horizontal and the line of sight

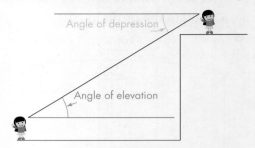

Angle of elevation looking up this is the angle between the horizontal and the line of sight

Angle of rotation the angle through which a shape is turned when it is rotated about a point

Angles at a point the sum of the angles at a point is 360° For example: $a + b + c + d + e = 360°$

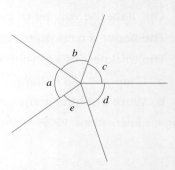

Angles on straight lines the angles at a point on a straight line add up to 180°, for example:

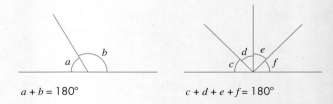

$a + b = 180°$ $c + d + e + f = 180°$

Annual rate the amount something changes in a year, often referring to an interest rate which is given as a percentage

Anticlockwise the opposite direction to which the hands of a clock turn

Glossary

Appropriate sensible for the context being considered

Approximation a value that is close but not exactly Appropriate sensible for the context being considered equal to another value which can be used to give an idea of the size of the value. For example, a journey taking 58 minutes may be described taking about an hour. The sign ≈ indicates 'is approximately equal to'

(Arbitrary) constant of integration the constant introduced due to integrating a function can take any value so is an arbitrary constant

Arc part of the circumference of a circle

Area the amount of space in a 2-D shape

Area scale factor the factor by which an area is multiplied

Area sine rule a formula used to find the area of a triangle

Arithmetic progression a sequence of numbers are in arithmetic progression if they difference between consecutive numbers are constant

Average a value which is chosen to represent a set of data. Mode, mean, and median are common examples of an average

Average speed the ratio of distance ÷ time for a journey

Axis (plural **axes**) a fixed reference line for the measurement of coordinates

Axis of symmetry a line through a shape so that one side is a reflection of the other.

Bar chart a type of frequency diagram drawn using bars or rectangles of equal widths to display discrete data

Base
1) base number is the number which is being raised to a power
2) the base of a 2-D shape is the horizontal line drawn at the bottom of the shape
3) the base of a 3-D object is the flat part of the object upon which it stands

Bearing an angle measured clockwise from North to describe a direction

Bisect cut in half

Boundary
1) edge of an area
2) class boundary is the largest or smallest value in a class

Box plot a graph that shows the distribution of data along a number line

Brackets used to group terms together in algebra

Cancel dividing both the numerator and denominator in a fraction by the same number

Capacity the amount a container holds when it is full

Cartesian coordinates are an ordered pair (x, y) of numbers used to specify the position of a point where the x-value, or x-coordinate, gives the distance parallel to the x-axis and the y-value, or y-coordinate, gives the distance parallel to the y-axis

Cartesian plane the plane in which the x-axis and y-axis lie

Centilitre one hundredth of a litre

Centimetre one hundredth of a metre

Centre
1) of a circle is where the compass point is placed when drawing a circle using a pair of compasses
2) of a transformation is a fixed point from which a transformation is described

Centre of enlargement the lines joining corresponding points in an object and its image all meet at the centre of enlargement

Centre of rotation the fixed point about which an object is rotated

Certain having a probability of 1

Chance likelihood

Chord a line across a circle

Circumference the distance around the outside (perimeter) of a circle

Class frequency the number of values in a class

Class interval the width of a group in a grouped frequency distribution

Classes are groups

Clockwise the hands of a clock turn in this direction

693

Glossary

Coefficient a constant term which is multiplied by a variable. For example, $2x$, or $5x^2$, the '2' and the '5' are coefficients of x and x^2 respectively

Collinear lying on the same straight line

Column vector an ordered set of 2 or 3 numbers used to give the position of a point or to describe the change in position of a point

Common factors factors which are common to more than one term or number

Compasses drawing instruments used to draw arcs and circles

Complement the complement of set A is everything outside of set A

Completing the square a way of simplifying or solving a quadratic equation by adding an expression to both sides to make one part of the equation a perfect square

Composite function a function that is made from two or more separate functions

Compound interest the overall interest earned on investment when the total interest earned in each period is added back to the original capital

Cone a 3-D shape with a circular base and a curved sloping face

Congruent two shapes are congruent if they have exactly the same shape and size

Consecutive next to each other

Consistency how varied a set of values are

Constant a value that does not change

Constant of proportionality the constant value of the ratio between two proportional quantities

Constant speed a particle has constant speed if its speed does not change

Construct use only pencil, straight edge and compasses

Continuous data data that can have any value in a range

Conversion graph a graph used to convert from one unit to another

Coordinate method the way in which an object or an image can be found in 'enlargements' by counting squares

Coordinate grid a grid used to show the positions of points

Coordinates used to specify the location of a point. *See* Cartesian coordinates

Correlation a connection between two sets of data

Corresponding angles angles which are in the same position and are equal

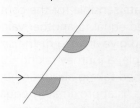

Corresponding sides sides which are in the same position

Cosine the ratio of the adjacent side to the hypotenuse

Cosine rule a rule connecting sides and an angle of any triangle usually used when the triangle is not right-angled. $a^2 = b^2 + c^2 - 2bc \cos A$

Cross-section a face formed by cutting through a 3-D object

Cube a 3-D solid consisting of six square faces

Cube number the number you get when you multiply a number by itself and then again. For example 8 is a cube number as $2 \times 2 \times 2 = 8$. 8 is called the cube of 2 and can be written as 2^3, 2 cubed

Cube root the opposite of cubing a number, so the cube root of 8 is 2

Cubic function is a function in which the highest power of x is x^3

Cubic sequence a sequence in which the values are obtained using n^3 in some way.

Cuboid a 3-D solid consisting of 6 rectangular faces

Cumulative frequency obtained by adding frequencies together to accumulate them

Curved surface a part of a cylinder

Cyclic quadrilateral a quadrilateral whose vertices lie on a circle

Cylinder a prism whose constant cross-section is a circle.

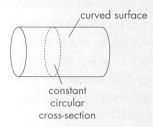

Glossary

Dashed line a line which has regular gaps, sometimes called broken

Decagon a polygon with 10 sides

Decay a reduction which follows a predictable pattern

Deceleration the rate at which speed or velocity decreases with time

Decimal a number written using only digits and a decimal point, for example 12.34

Decimal equivalent a number written as a decimal which has the same value as a number written in a different form

Decimal place the position of a digit after the decimal point in a number

Decrease become smaller or reduce

Denominator the number on the 'bottom' of a fraction

Derivative the result of differentiation, the gradient function

Diameter the distance from one point on a circle to another passing through the centre of the circle

Difference
1) the result when one number is subtracted from another
2) the gap between consecutive numbers in a sequence

Difference of two squares a results used to factorise algebraic expressions
$x^2 - y^2 = (x + y)(x - y)$

Differentiate find the gradient function

Differentiation the process used to find the gradient function

Digit a single number, for example in the number 234 the digit in the 'tens column' is '3'

Direct proportion two quantities are in direct proportion if one increases as the other increases

Direct variation same as direct proportion

Directed number a positive or negative number or zero

Direction in vectors the direction is indicated by an arrow on the vector

Discrete data data that takes only particular values in a range

Dispersion spread or variation

Displacement when an object moves from a position A to a position B the displacement is the magnitude of the vector AB

Distance a measured length between two points

Distance travelled the measured length between two points of a journey

Distance–time graph a graph plotting the distance travelled against the time taken

Dividend a number being divided by another. For example 12 ÷ 4 = 3, here 12 is the dividend

Divisor a number being divided into another. For example 12 ÷ 4 = 3, here 4 is the divisor.

Edge the line where two faces of a solid meet

Element a member of a set

Eliminate remove. For example when two simultaneous equations are solved you eliminate one of the letters, that is you combine the two equations by removing one of the letters

Empty set a set with no elements. Also called the null set

Enlargement a transformation in which the shape of an object remains the same but the size usually changes

Equally likely have the same probability

Equals has the same value as

Equation a statement which involves two expressions which have the same value so have an '=' sign between them

Equation of line the relationship between the x-coordinate and the y-coordinate for each of a set of points

Equidistant at the same distance

Equilateral triangle an equilateral triangle is a triangle with all its sides equal. All three angles in an equilateral triangle are 60°

Equivalent fractions two or more fractions which have the same value as each other. For example $\frac{2}{4}$ and $\frac{3}{6}$ are equivalent fractions, both are equivalent to $\frac{1}{2}$

Estimate find an approximate value for

Estimated a value which has been found approximately

Glossary

Event a set of outcomes in probability

Exact form written without using decimals or any approximations

Exchange rate the equivalence between two different currencies

Expand multiply all the terms inside the brackets by those outside the brackets. (opposite of factorise)

Expansion when brackets are expanded the result is called the expansion of the brackets

Experimental probability the ratio of the number of times the event occurs to the total number of trials

Exponential having a constant base raised to a variable power

Exponential decay a reduction which follows a pattern predictable using an exponential function

Exponential form written in the form of a^x

Exponential functions functions which involve a constant base raised to a variable power

Exponential growth an increase which follows a pattern predictable using an exponential function

Exponential sequence a set of numbers which follow an exponential pattern

Expression a series of terms connected by plus and minus signs

External angles angles turned through when going round the perimeter of a polygon

Extreme values values which stand out as being particularly large or small relative to the other values in a set.

Face one flat surface of a solid which is enclosed by edges

Factor a whole number which divides exactly into another whole number

Factor pair a pair of numbers which multiply to give another number. For example, 3 and 4 are a factor pair of 12.

Factorise take all common factors outside brackets. (opposite of expand)

Formula (plural **formulae** or **formulas**) a rule expressed in words or letters

Fraction a number which is written using two parts called the numerator and denominator

Frequency the number of times a value occurs in a set of data

Frequency density the ratio of the frequency to the class-width in a frequency distribution. Frequency density is used to draw histograms

Frequency table a table showing frequencies

Function a rule which takes one number and changes it into another.

Gradient a measure of how steep a line is

Gram (g) a basic unit of mass in the metric system

Greater than 'greater than' means the same as 'more than'

Grouped data data which has been sorted into groups or classes

Grouped frequency table a table which shows the number of values in each of a set of groups or classess.

Heptagon a polygon with 7 sides

Hexagon a six-sided polygon

Highest common factor (HCF) the largest factor which is common to two or more other numbers

Histogram a chart drawn using rectangles that uses the area to represent frequencies

Hypotenuse the longest side of a right-angled triangle. It is always opposite the right angle.

Image the new shape after a transformation

Impossible an event with probability 0

Improper fraction a fraction in which the numerator is greater than the denominator

Included in inequalities lines on the boundary of a region which are included in that region are drawn using a solid line

Included angle the angle between two adjacent sides

Increase go up in value

Index (plural **indices**)
1) the power of a number
2) a quantity which allows comparisons to be made over time, for example a cost of living index allows costs over time to be compared

Glossary

Inequality a statement about the relative size of two values or expressions using the symbols < (less than), ≤ (less than or equal to), > (greater than), ≥ (greater than or equal to)

Infinite going on for ever

Integer a whole number

Intercept the place a line crosses an axis, for example the y-intercept is the place a line crosses the y-axis

Interior angles
1) Interior angles are the angles inside a polygon. The sum, S, of the interior angles of a polygon with n sides is given by the formula $S = 180(n − 2)°$
2) Allied angles are sometimes called interior angles

Inter-quartile range the distance between the lower and upper quartiles

Intersection
1) a set of elements that belong to both of two other sets
2) the point where two lines cross

Inverse something that has an opposite or reverse effect

Inverse functions functions which have the reverse effect to each other

Inverse operations operations that have the reverse or opposite effect to each other. For example, addition and subtraction are inverse operations, multiplication and division are inverse operations

Inverse proportion the relationship between two variables where one decreases as the other increases

Irrational number a number that cannot be written as a fraction

Irregular polygon any polygon which is not regular

Isosceles triangle a triangle with two equal sides. An isosceles triangle has two equal angles (at the foot of the equal sides).

Key an explanation of what a diagram shows. For example, in a pictogram the key will show a symbol and how many items it represents and it states what the items are

Kilogram (kg) a measure of mass in the metric system. 1 kg = 1000 g

Kilometre (km) a measure of distance in the metric system. 1 km = 1000 m

Kite a kite is a quadrilateral with two pairs of equal adjacent sides.

Length how long an object is

Less than the sign < is used to mean 'less than'. For example, 2 < 3 means 2 is smaller than 3, or 2 is less than 3

Like terms terms containing the same variable raised to the same power. These terms can then be added or subtracted to be combined

Limits of accuracy the upper and lower bounds when approximating

Line a one-dimensional object extending infinitely in both directions

Line bisector a line which cuts another line in two equal parts at right angles

Line of best fit a single straight line which passes through a set of points and is as close as possible to as many of them as possible

Line of symmetry a line drawn so that one side of the line is a reflection of the other side

Line segment the part of a line that joins two points

Linear of a straight line

Linear equations equations that do not contain any powers or roots such as $\sqrt{x}$, or y^2

Linear programming problem solving using graphs of straight lines

Linear scale factor a multiplier which is linear

Linear sequence a sequence of numbers so that the difference between consecutive numbers is constant

Litre a basic unit of capacity in the SI system

Locus (plural **loci**) the path of a moving point

Loss the difference between the amount taken when it is sold and the amount paid for it initially when the amount taken is smaller than the amount paid initially

Lower bound the smallest possible value of a rounded quantity

Lower quartile the value which 25% of the data are below or equal to

Lowest common multiple (LCM) the smallest number which is a multiple of two or more other numbers. For example the LCM of 6 and 10 is 30

Lowest terms a fraction which has been cancelled as much as possible so that it is not possible to cancel it further is said to be in its lowest terms.

Glossary

Magnitude the size or length of a vector

Map scale the scale on a map indicates how many centimetres on the ground are represented by one centimetre on the map

Mapping the process which changes one number to another

Mapping diagram a diagram showing two sets of numbers and how each number in one set is mapped to each number in the other

Mass the amount of matter in an object

Maximum the largest possible value

Mean a measure of average, found by adding all the values and dividing by how many there are

Median a measure of average, found by listing all the values in order and taking the value in the middle

Metric system the system of weights and measures most commonly in use

Middle value the value in the middle

Midpoint the point which is exactly half way between two others

Millilitre (ml) a unit used for measuring capacity. 1000 millilitres = 1 litre

Millimetre (mm) a unit used for measuring distance. 1000 millimetres = 1 metre

Minimum the smallest possible value

Mirror line a line of symmetry

Mixed number a number containing a whole number part and a fraction

Modal class the class which has the highest frequency

Modal value the value with the highest frequency

Mode the value with the highest frequency

More than the sign > is used to mean 'more than'. For example, 5 > 4 means 5 is larger than 4, or 5 is more than 4

Multiple a number which is obtained by multiplying two other numbers

Multiplier a number which is used to multiply another number is called a multiplier

Mutually exclusive two events are mutually exclusive if one prevents the other from happening. For example, when you follow a maze and get to a junction where there are exactly two possible choices: you can turn left or right, these are mutually exclusive as you cannot go both left and right at the same time.

Natural number the counting numbers 1, 2, 3, ...

Negative less than 0

Negative correlation a correlation where one variable decreases as the other increases

Negative enlargement an enlargement involving a negative scale factor. A negative enlargement produces an image shape on the opposite side of the centre of enlargement to the original shape

Negative index an index which is below 0

Net a 2-D pattern that can be cut out and folded into a 3-D shape

Nonagon a polygon with 9 sides

Non-linear not following the pattern of a straight line

nth term counting from the first term in a sequence it is the term in position n. It is used to give a general formula that can be used to find every term in the sequence

Null set a set with no elements

Number line a line labelled with numbers. Sometimes used to help working with negative numbers

Numerator the number which is on the 'top' of a fraction.

Object the original shape before a transformation

Obtuse angles greater than 90° but less than 180°

Obtuse-angled triangle a triangle with one angle greater than 90°

Octagon a regular eight-sided polygon

Operation +, −, ×, ÷ are all operations

Opposite angles are made by two straight lines crossing each other. The diagram shows a and c are opposite angles, and b and d are opposite angles. Opposite angles are equal so $a = c$ and $b = d$. Opposite angles are sometimes called vertically opposite angles

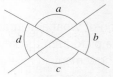

Opposite side the side opposite the known or required angle in a right-angled triangle

Order when numbers are organised into a sequence they are said to be in order

Glossary

Order of rotational symmetry the number of times a shape will fit onto itself exactly in one full rotation when rotated about a point

Origin (0, 0) the point where the x-axis and y-axis cross

Outcome result.

Parabola the shape of the graph of a quadratic function

Parallel lines which have exactly the same direction, are always the same distance apart and never meet

Parallelogram a quadrilateral in which opposite sides are parallel

Patterns predictable arrangements or sequences of numbers

Pentagon a five-sided polygon

Percentage out of each hundred

Percentage change the difference between the current and the original value as a fraction of the original value multiplied by 100

Percentage decrease a negative percentage change where the current value is lower than the original value

Percentage increase a positive percentage change where the current value is higher than the original value

Percentage loss the difference between the current and the original value as a fraction of the original value multiplied by 100 when the values are money and the current value is lower than the initial value

Percentage profit the difference between the current and the original value as a fraction of the original value multiplied by 100 when the values are money and the current value is higher than the initial value

Perimeter the distance around the outside of a 2-D shape

Perpendicular two lines are perpendicular if they are at right angles to each other

Perpendicular bisector a line that cuts a line in half at right angles

Perpendicular height the height of a 2-D shape measured from the base of the shape and at right angles to it. The perpendicular height of a shape does not have to be inside the shape, as seen in this parallelogram, the perpendicular height is labelled h in the diagram

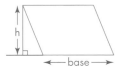

Pictogram a chart which uses identical pictures to show frequencies

Pie chart a circular chart showing how a whole set of data is divided into parts. The angles, or areas, are used to represent frequencies

Plane area the area of a flat shape

Plane of symmetry a plane in a solid which divides the solid into two parts each of which is a reflection of the other

Point of contact the position where a line touches a curve

Point of inflexion a point on a curve where the gradient changes from decreasing to increasing or from increasing to decreasing

Polygon a 2-dimensional shape made using only straight lines

Position vector a vector which gives the position of a point

Positive greater than 0

Positive correlation a correlation where one variable increases as the other increases

Possibility diagram a diagram which shows all the possible outcomes of an event

Power an index

Power 0 any value or expression raised to the power 0 has a value of 1. For example $3^0 = 1$, and $(3x)^0 = 1$

Power 1 raising any value or expression to the power 1 does not change its value. For example $3^1 = 3$, and $(3x)^1 = 3x$

Prime factorisation writing a number as a product of all its factors which are prime

Prime number a number which has exactly 2 factors

Principal the amount of money invested

Prism a 3-D solid which has the same cross-section throughout its shape (a uniform cross-section). Triangular prisms (with a triangular cross-section) and cylinders (with a circular cross-section) are examples of prisms

Glossary

Probability how likely an individual outcome of an event is to occur. Probability is measured on a scale from 0 to 1

Probability fraction a probability which is written as a fraction

Product the product of two or more numbers is obtained by multiplying the numbers together

Product of prime factors see prime factorisation

Profit when you sell something for more than you paid for it the difference is the profit

Proof a set of statements which together form a step by step mathematical argument

Proper fraction a fraction in which the numerator is less than the denominator

Proper subset a set of elements which is contained within another set, but not the same as that set. For example, Set $A = \{1, 2, 3, 4, 5\}$ then set $D = \{1, 2, 3\}$ is a proper subset of A

Protractor used to measure angles

Prove make a proof

Pyramid a solid shape with triangular faces. The base of a pyramid does not have to be triangular

Pythagoras' theorem a relationship between the sides of a right angled triangle. $a^2 + b^2 = c^2$ where c is the hypotenuse of the triangle.

Quadrant the axes divide a page into four quadrants

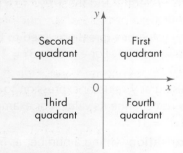

Quadratic equation an equation in which the highest power of x is x^2

Quadratic expression an expression in which the highest power of x is x^2

Quadratic formula the formula used to solve a quadratic equation $\frac{-b \pm \sqrt{b^2 - 4ac}}{2a}$, where a, b, and c are the coefficients of x^2, x, and the constant term, respectively

Quadratic graph a graph of a quadratic equation

Quadratic sequence a sequence of numbers form a quadratic sequence if they differences between consecutive numbers are in an arithmetic progression. The formula for the nth term of a quadratic sequence is a formula in which the highest power of n is n^2

Quadrilateral a polygon with four sides

Qualitative descriptive rather than numerical

Quantity an amount of something

Quartile a value which is one quarter way through a set of data when it is listed in order. Lower quartile is the value which is one quarter way from the lowest value, upper quartile is the value which is one quarter way from the highest value.

Radius (plural **radii**) the distance from one point on a circle to the centre

Random a random number is a number which is not predictable

Range
1) the distance between the largest and smallest value in a set of data
2) the set of y values for a function

Rate the rate of interest when borrowing or saving money

Rate of increase how quickly a variable or value increases relative to another variable

Ratio a way to compare two quantities with each other

Rational number a number which can be written as a fraction $\left(\frac{a}{b}\right)$ where a and b are integers

Ray methods the way in which an object or an image can be found in 'enlargements' by drawing lines or 'rays' from the centre of enlargement

Real number any rational or irrational number

Rearrange rewrite in a different order

Reciprocal the reciprocal of a number is 1 divided by the number. For example, the reciprocal of 3 is $\frac{1}{3}$, and the reciprocal of $\frac{3}{5}$ is $\frac{5}{3}$

Rectangle a quadrilateral with two pairs of parallel sides and four right angles

Recurring decimal a decimal with an infinite number of decimal places the last digit or group of digits repeats

Glossary

Reflection a transformation which results in a copy of the object looking as though it is the object viewed in a mirror

Reflex angles are greater than 180°

Regular polygon A polygon is regular if all its interior angles are equal and all of its sides are the same length

Relative frequency the ratio of the number of the successful outcomes to the total number of trials, which can then be used as an estimate of probability

$$\text{relative frequency} = \frac{\text{number of successful outcomes}}{\text{number of trials}}$$

Representative in place of, for example when a single value is used to represent a set of values it is used in place of all the values

Required to be found. The required angle is an angle whose size is to be found

Rhombus a parallelogram with all its sides equal

Right angle an angle of 90°

Right-angled triangle a triangle which has an angle of 90°

Rotation turning an object or shape about a point by a given angle in either a clockwise or anti-clockwise direction

Rotational symmetry a shape has rotational symmetry if when it is rotated about a point it fits exactly onto itself at least once before being turned 360°

Round
1) circular in shape
2) replace a number with one which is approximately equal

Rounded down replace a number with one which is approximately equal but smaller than the original number

Rounded up replace a number with one which is approximately equal but smaller than the original number

Rule a general statement

Ruler used to draw straight lines and measure lengths.

Sample a selection from a larger population

Scalar a single number

Scale the ratio between the lengths on a scale drawing and the actual length represented

Scale drawing an accurate drawing in which the lengths are in proportion to the original

Scale factor a number which tells you how many times larger an image is of an object

Scatter diagram points plotted on Cartesian axes which are used to see any correlation or to make predictions

Sector part of a circle enclosed between part of the circumference and two radii

Segment part of a circle enclosed between part of the circumference and a chord

Semi-circle half a circle

Sequence a set of numbers which follow a rule

Set a collection of items

Set square used to find right angles

Significant figures the digits of a number

Similar two shapes are similar if they have the same shape but not the same size

Simple interest - a charge for borrowing or lending money. $I = \dfrac{PRT}{100}$

Simplest form the way in which a fraction or ratio is written so that the smallest possible whole numbers are used

Simplify
1) in an expression or equation this means to collect any like terms so there are as few terms as possible
2) in a fraction or ratio this means write a fraction so that the smallest possible whole numbers are used

Simultaneous linear equations two equations with two unknowns

Sine the ratio of the opposite side to the hypotenuse

Sine rule a rule connecting the sides and angles of a triangle which is not right-angled

$$\frac{a}{\sin A} = \frac{b}{\sin B} = \frac{c}{\sin C} \text{ or } \frac{\sin A}{a} = \frac{\sin B}{b} = \frac{\sin C}{c}$$

Single fraction a fraction which consists of exactly one numerator and one denominator only

Single unit value in unitary method it is the cost of one unit or the time it takes for one item

Slant height the sloping distance from the circular base of a cone to the apex at the top

Solid line a line which has no breaks or gaps

Solid shapes 3-D objects

Glossary

Soluble can be solved or worked out

Solution answer to a problem

Solve
1) a triangle work out any missing sides or angles
2) an equation work out the value of the letter

Speed the rate of change of distance with time

Speed–time graph a graph which shows how speed varies with time

Sphere a solid shape which appears round no matter what position it is viewed from, for example a football

Spread the distance(s) between the values

Square
1) a square is a regular four-sided polygon
2) to square a number is to multiply it by itself, for example the square of 5 is 25

Square number the result of squaring an integer

Square root the opposite of squaring a number. For example the square root of 36 is 6

Standard form a way to write very large and very small numbers using a number between 1 and 10 multiplied by a power of 10

Stem-and-leaf diagram shows data arranged by place value, for the purpose of comparing frequencies

Straight-line graph the graph of a linear function

Subject the subject of a formula is the letter in front of the equals sign. For example, in $t = 2s + 3$, t is the subject of the formula

Subset if all the elements of a set C are in a set A then C is a subset of A

Substitute replace a letter with either an expression or a value

Subtended made. The angle subtended at the circumference of a circle is the angle made there

Supplementary supplementary angles add up to 180°

Surface area the total area of the faces of a solid

Symbol each of $+$, $-$, $\times$, $\div$ is a symbol.

Tally chart a chart used to collect data with a tally used to record each value. Tallies are marked as |s and grouped in 5s as a 5-bar gate. For example this tally represents 7 values being recorded Note the 5-bar gate is 4 vertical lines with one diagonal line

Tangent
1) a straight line that touches a circle at one point only
2) the ratio of the opposite side to the adjacent side in a right-angled triangle

Term
1) an expression forming part of a larger equation or part of an equation
2) a single item in a sequence

Term to term rule a rule which links one term in a sequence to the next term

Terminating decimal a decimal that has a finite number of digits

Three-figure bearing a bearing which has 3 digits exactly

Time in speed distance and time, time is how long a journey takes, measured in hours minutes or seconds

Timetable a table giving arrival and/or departure times

Tonne a metric unit for measuring mass. 1 tonne = 1000 kg

Top heavy a fraction in which the numerator is greater than the denominator. *See* improper fraction

Transformation a change in the position or size of a shape. Reflections, rotations, translations or enlargements are all examples of transformations

Translation a transformation which moves a shape from one position to another without changing the orientation or size of the shape

Transversal a line crossing a pair of parallel lines

Trapezium a quadrilateral with two parallel sides

Tree diagram a diagram with branches which shows all the possible outcomes of an event and their probabilities

Trial an experiment which is repeated a number of times

Turning point a point on a graph where the gradient changes from being either positive to negative or from negative to positive.

Union the union of sets A and B is the set of all elements which are in either set A or in set B or in both

Unitary method a way to solve proportion problems by finding the value of a single item

Glossary

Universal set the set containing all the things being considered

Upper bound the largest possible value of a rounded quantity

Upper quartile the value below which three-quarters of the data lie.

Variable a quantity that can take different values

Variation
1) *see* proportion
2) how the values in a data set are spread out

Vector a way of writing the position of a point or describing a movement from one position to another

Venn diagram a way to show the elements of different sets

Vertex a corner of a shape, where two edges meet

Vertical height distance of one point above another point or line

Vertically opposite angles vertically opposite angles are the same as opposite angles

Volume the amount of space a solid shape occupies

Volume scale factor a multiplier for enlargements using volumes.

Weak correlation when there is little connection between two sets of data

Width distance from one side to the other.

Zero correlation when there is no connection between two sets of data.

Answers to Chapter 1

1.1 Square numbers and cube numbers

Exercise 1A

1 36, 49, 64, 81, 100, 121, 144, 169, 196, 225, 256, 289, 324, 361, 400

2 4, 9, 16, 25, 36, 49

3 **a** 50, 65, 82 **b** 98, 128, 162
c 51, 66, 83 **d** 48, 63, 80

4 **a** 25, 169, 625, 1681
b Answers in each row are the same

5 **a** 13 **b** 5 **c** 9 **d** 11 **e** 20

6 36 and 49

Exercise 1B

1 **a** 12, 24, 36 **b** 20, 40, 60 **c** 15, 30, 45
d 18, 36, 54 **e** 35, 70, 105

2

	Square number	Factor of 70
Even number	16	10
Multiple of 7	49	35

3 4761 (69^2)

4 24 seconds

5 30 seconds

6 **a** 12 **b** 9 **c** 6 **d** 13
e 15 **f** 16 **g** 10 **h** 17

Exercise 1C

1 1, 8, 27, 64, 125, 216, 343, 512, 729, 1000, 1331, 1728

2 9, 36, 100: square numbers
1 + 8 + 27 + 64 + 125 = 225
1 + 8 + 27 + 64 + 125 + 216 = 441
1 + 8 + 27 + 64 + 125 + 216 + 343 = 784

3 **a** 126 217 344
b 124 215 342
c 250 432 686
d 216 125 64

4 **a** 153, 370, 371
b Each answer is the sum of the cubes of its digits

5 1729

6 692 = 4761 and 693 = 328509 The answers use all the digits from 0 to 9 exactly once.

1.2 Multiples of whole numbers

Exercise 1D

1 **a** 3, 6, 9, 12, 15 **b** 7, 14, 21, 28, 35
c 9, 18, 27, 36, 45 **d** 11, 22, 33, 44, 55
e 16, 32, 48, 64, 80

2 **a** 72, 132, 216 **b** 161, 91 **c** 72, 102, 132, 78, 216

3 **a** 98 **b** 99 **c** 96 **d** 95 **e** 98 **f** 96

4 4 or 5 (as 2, 10 and 20 are not realistic answers)

5 **a** 18 **b** 28 **c** 15

6 5 numbers: 18, 36, 54, 72, 90

1.3 Factors of whole numbers

Exercise 1E

1 **a** 1, 2, 5, 10 **b** 1, 2, 4, 7, 14, 28
c 1, 2, 3, 6, 9, 18 **d** 1, 17
e 1, 5, 25 **f** 1, 2, 4, 5, 8, 10, 20, 40
g 1, 2, 3, 5, 6, 10, 15, 30 **h** 1, 3, 5, 9, 15, 45
i 1, 2, 3, 4, 6, 8, 12, 24 **j** 1, 2, 4, 8, 16

2 **a** 55 **b** 67 **c** 29
d 39 **e** 65 **f** 80

3 **a** 2 **b** 2 **c** 3 **d** 5 **e** 3
f 3 **g** 7 **h** 5 **i** 10 **j** 11

4 5

Exercise 1F

1 **a** 168 **b** 105 **c** 84 **d** 84
e 96 **f** 54 **g** 75 **h** 144

2 **a** 8 **b** 7 **c** 4 **d** 14
e 4 **f** 9 **g** 5 **h** 4
i 3 **j** 16 **k** 5 **l** 18

3 18 and 24

1.4 Prime numbers

Exercise 1G

1 23 and 29

2 97

3 All these numbers are not prime.

4 3, 5, 7

5 Only if all 31 bars are in a single row, as 31 is a prime number and its only factors are 1 and 31.

1.5 Prime factorisation

Exercise 1H

1 **a** 36 **b** 105 **c** 250 **d** 816
e 714 **f** 1715 **g** 1089 **h** 1352

2 **a** $2 \times 3^2 \times 5$ **b** $2^3 \times 19$ **c** 2^6
d $2 \times 3 \times 5 \times 11$ **e** 17^2 **f** $2^5 \times 5^2$
g It is a prime number and cannot be factorised.
h $7 \times 11 \times 13$

3 $77 = 7 \times 11$; $129 = 3 \times 43$; $221 = 13 \times 17$

4 **a** $900 = 2^2 \times 3^2 \times 5^2$ **b** $1800 = 2^3 \times 3^2 \times 5^2$
c $1350 = 2 \times 3^3 \times 5^2$

5 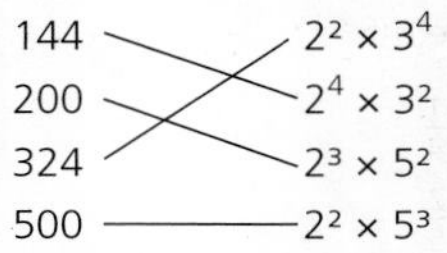

Answers to Chapter 2

6 **a** $2 \times 3 \times 5 \times 7 = 210$
   b The answer to **a** $\times 11 = 2 \times 3 \times 5 \times 7 \times 11 = 2310$

7 **a** $2^2 \times 3^2 \times 17$ **b** $2 \times 3^2 \times 17$ **c** $2^3 \times 3 \times 17$

8 71, 73 and 79 because they are prime numbers

9 $456\,533 = 7^3 \times 11^3$

1.6 More about HCF and LCM

Exercise 1I

1 **a** $2 \times 3^2 = 18$ **b** $2^3 \times 3^4 = 648$
2 **a** 35   **b** 735
3 **a** $2^4 \times 3 \times 5$ **b** $2 \times 3^2 \times 7$ **c** 6   **d** 5040
4 **a** $2^3 \times 3^2$ and $2^2 \times 3^3$ **b** 36   **c** 216
5 **a** 16   **b** 576
6 **a** 33   **b** 2772
7 **a** $2^2 \times 3 = 12$ **b** $2^3 \times 3^2 \times 5 = 360$
8 **a** 5   **b** 1575
9 **a** 15   **b** 23 625
10 **a** 72   **b** $2^5 \times 3^4$
11 **a** 1   **b** 12 600
12 **a** $72 \times 162 = 11664$ **b** $18 \times 648 = 11664$
   c You could do a similar calculation for the numbers in questions 2 to 6. You should find that the two products are equal each time.
   d There is not a similar result in this case.

1.7 Real numbers

Exercise 1J

1 **a** yes   **b** no   **c** yes   **d** yes   **e** no
   f yes   **g** no   **h** yes   **i** yes

2 **a** rational   **b** rational   **c** irrational
   d rational   **e** irrational   **f** rational
   g rational   **h** irrational   **i** irrational

3 **a** $\frac{1}{300}$ **b** $\frac{10}{3}$ or $3\frac{1}{3}$ **c** $\frac{4}{17}$ **d** 0.4

4 $\frac{1}{2}$ and 0.5

5 $\frac{5}{6}$

6 **a** 2.5 and 3.5 is one possible answer
   b 0.4×2.5 is one possible answer
   c Not possible

7 $\sqrt{2} \times \sqrt{8}$ is a possible answer

8 π and $4 - \pi$ is a possible answer

9 There are many possible answers. You could just give the same answer as question 5.

10 **a** $\frac{1}{28}$ **b** $\frac{1}{2.8} = \frac{10}{28} = \frac{5}{14}$ **c** $\frac{8}{1} = 8$ **d** $5\frac{3}{4} = \frac{23}{4}$. The reciprocal is $\frac{4}{23}$

Notice that in part **b** you could use a calculator to get $1 \div 2.8 = 0.3571$ to 4 d.p.

This is only an approximate answer. For an exact answer you must use fractions.

Answers to Chapter 2

2.1 Equivalent fractions

Exercise 2A

1 **a** $\frac{8}{20}$ **b** $\frac{3}{12}$ **c** $\frac{15}{40}$
   d $\times 6, \frac{12}{18}$ **e** $\times 3, \frac{9}{12}$ **f** $\times 5, \frac{25}{40}$

2 **a** $\frac{2}{3}$ **b** $\frac{4}{5}$ **c** $\frac{5}{7}$ **d** $\div 6, \frac{2}{3}$
   e $25 \div 5, \frac{3}{5}$ **f** $30 \div 3, \frac{7}{10}$

3 **a** $\frac{2}{3}$ **b** $\frac{1}{3}$ **c** $\frac{2}{3}$ **d** $\frac{3}{4}$ **e** $\frac{1}{3}$
   f $\frac{1}{2}$ **g** $\frac{7}{8}$ **h** $\frac{4}{5}$ **i** $\frac{1}{2}$ **j** $\frac{1}{4}$

4 **a** $\frac{1}{2}, \frac{2}{3}, \frac{5}{6}$ **b** $\frac{1}{2}, \frac{5}{8}, \frac{3}{4}$ **c** $\frac{2}{5}, \frac{1}{2}, \frac{7}{10}$
   d $\frac{7}{12}, \frac{2}{3}, \frac{3}{4}$ **e** $\frac{1}{6}, \frac{1}{4}, \frac{1}{3}$ **f** $\frac{3}{4}, \frac{4}{5}, \frac{9}{10}$

5 **a** $\frac{1}{3} + \frac{1}{4} = \frac{4}{12} + \frac{3}{12} = \frac{7}{12}$
   Explanations may involve ruling out other combinations
   b $\frac{1}{2}$ as the smallest denominator is the biggest unit fraction
   Diagrams may be used but must be based on equal sized area.

6 **a** $2\frac{1}{3}$ **b** $2\frac{2}{3}$ **c** $2\frac{1}{4}$
   d $1\frac{3}{7}$ **e** $2\frac{2}{5}$ **f** $1\frac{2}{5}$

7 **a** $\frac{10}{3}$ **b** $\frac{35}{6}$ **c** $\frac{9}{5}$ **d** $\frac{37}{7}$ **e** $\frac{41}{10}$ **f** $\frac{17}{3}$
   g $\frac{5}{2}$ **h** $\frac{13}{4}$ **i** $\frac{43}{6}$ **j** $\frac{29}{8}$ **k** $\frac{19}{3}$ **l** $\frac{89}{9}$

8 Students check their own answers.

9 $\frac{27}{4} = 6\frac{3}{4}, \frac{31}{5} = 6\frac{1}{5}, \frac{13}{2} = 6\frac{1}{2}$, so $\frac{27}{4}$ is the biggest since $\frac{1}{5}$ is less than $\frac{1}{2}$ and $\frac{3}{4}$ is greater than $\frac{1}{2}$

10 Any mixed number which is between 7.7272 and 7.9. For example $7\frac{4}{5}$

2.2 Fractions and decimals

Exercise 2B

1 **a** $\frac{7}{10}$ **b** $\frac{2}{5}$ **c** $\frac{1}{2}$ **d** $\frac{3}{100}$ **e** $\frac{3}{50}$
   f $\frac{13}{100}$ **g** $\frac{1}{4}$ **h** $\frac{19}{50}$ **i** $\frac{11}{20}$ **j** $\frac{16}{25}$

2 **a** 0.5   **b** 0.75   **c** 0.6   **d** 0.9   **e** 0.333
   f 0.625   **g** 0.667   **h** 0.35   **i** 0.636   **j** 0.444

3 **a** $0.3, \frac{1}{2}, 0.6$ **b** $0.3, \frac{2}{5}, 0.8$ **c** $0.15, \frac{1}{4}, 0.35$
   d $\frac{7}{10}, 0.71, 0.72$ **e** $0.7, \frac{3}{4}, 0.8$ **f** $\frac{1}{20}, 0.08, 0.1$
   g $0.4, \frac{1}{2}, 0.55$ **h** $1.2, 1.23, 1\frac{1}{4}$

4 Store A – $\frac{1}{3}$ (0.33) is greater than $\frac{1}{4}$ (0.25)

705

Answers to Chapter 2

5 a $\frac{12}{30} = \frac{2}{5}$ **b** 0.4

6 $\frac{7}{8}$ (= 0.875)

7 $\frac{2}{3}$ (= 0.67)

2.3 Recurring decimals

Exercise 2C

1 a 0.333... or $0.\dot{3}$ **b** 0.75 **c** 0.8333... or $0.8\dot{3}$
  d 0.222... or $0.\dot{2}$ **e** 0.65 **f** 0.8181... or $0.\dot{8}\dot{1}$
  g 0.1875 **h** 0.916 66... or $0.91\dot{6}$

2 a 0.4666... or $0.4\dot{6}$ **b** 0.9333... or $0.9\dot{3}$

3 a 0.1111... **b** 0.1666... **c** 0.2777... **d** 0.0555...

4 $\frac{8}{9}$

5 $\frac{8}{33}$

6 $\frac{11}{30}$

7 $\frac{1}{12}$

8 $2\frac{7}{15}$

9 $0.\dot{2}3076\dot{9}$

10 a $0.\dot{0}\dot{9}$ **b** $0.\dot{1}\dot{8}$ **c** $0.\dot{2}\dot{7}, 0.\dot{3}\dot{6}$ and $0.\dot{6}\dot{3}$

11 a $0.\dot{2}8571\dot{4}$ **b** $0.\dot{4}2857\dot{1}$ **c** $\frac{4}{7} = 0.\dot{5}7142\dot{8}$, $\frac{5}{7} = 0.\dot{7}1428\dot{5}$ and $\frac{6}{7} = 0.\dot{8}5714\dot{2}$

12 a $\frac{1}{5}, \frac{1}{8}, \frac{1}{10}$
  b $\frac{1}{N}$ is a terminating decimal if the only prime factors of N are 2 or 5. Otherwise it is a recurring decimal.

2.4 Percentages, fractions and decimals

Exercise 2D

1 a $\frac{2}{25}$ **b** $\frac{1}{2}$ **c** $\frac{1}{4}$ **d** $\frac{7}{20}$ **e** $\frac{9}{10}$ **f** $\frac{3}{4}$

2 a 0.27 **b** 0.85 **c** 0.13 **d** 0.06 **e** 0.8 **f** 0.32

3 a $\frac{3}{25}$ **b** $\frac{2}{5}$ **c** $\frac{9}{20}$ **d** $\frac{17}{25}$ **e** $\frac{1}{4}$ **f** $\frac{5}{8}$

4 a 29% **b** 55% **c** 3% **d** 16% **e** 60% **f** 125%

5 a 28% **b** 30% **c** 95% **d** 34% **e** 27.5% **f** 87.5%

6 a 0.6 **b** 0.075 **c** 0.76 **d** 0.3125 **e** 0.05 **f** 0.125

7 a 63%, 83%, 39%, 62%, 77% **b** English

8 34%, 0.34, $\frac{17}{50}$, 85%, 0.85, $\frac{17}{20}$, 7.5%, 0.075, $\frac{3}{40}$, 45%, 0.45, $\frac{9}{20}$, 30%, 0.3, $\frac{3}{10}$, 67%, 0.67, $\frac{2}{3}$, 84%, 0.84, $\frac{21}{25}$, 45%, 0.45, $\frac{9}{20}$, 37.5%, 0.375, $\frac{3}{8}$

2.5 Calculating a percentage

Exercise 2E

1 a 0.88 **b** 0.3 **c** 0.25 **d** 0.08 **e** 1.15

2 a 78% **b** 40% **c** 75% **d** 5% **e** 110%

3 a $45 **b** $6.30 **c** 128.8 kg **d** 1.125 kg
  e 1.08 h **f** 37.8 cm **g** $0.12 **h** 2.94 m
  i $7.60 **j** 33.88 min **k** 136 kg **l** $162

4 $2410

5 a 86% **b** 215

6 8520

7 287

8 990

9 Mon: 816, Tue: 833, Wed: 850, Thu: 799, Fri: 748

10 a $3.25 **b** 2.21 kg **c** $562.80
   d $6.51 **e** 42.93 m **f** $24

11 480 cm³ nitrogen, 120 cm³ oxygen

12 13

13 $270

14 More this year as it was 3% of a higher amount than last year.

2.6 Increasing or decreasing quantities by a percentage

Exercise 2F

1 a 1.1 **b** 1.03 **c** 1.2 **d** 1.07 **e** 1.12

2 a $62.40 **b** 12.96 kg **c** 472.5 g **d** 599.5 m
  e $38.08 **f** $90 **g** 391 kg **h** 824.1 cm
  i 253.5 g **j** $143.50 **k** 736 m **l** $30.24

3 $29 425

4 1 690 200

5 a Caretaker: $17 325, Driver: $18 165, Supervisor: $20 475, Manager: $26 565
  b 5% of different amounts is not a fixed amount. The more pay to start with, the more the increase (5%) will be.

6 $411.95

7 193 800

8 575 g

9 918

10 60

11 TV: $287.88, microwave: $84.60, CD player: $135.13, stereo: $34.66

12 $10

13 c Both the same as 1.05 × 1.03 = 1.03 × 1.05

14 a Shop A, as 1.04 × 1.04 = 1.0816, so an 8.16% increase.

15 $540.96

Exercise 2G

1 a 0.92 **b** 0.85 **c** 0.75 **d** 0.91 **e** 0.88

2 a $9.40 **b** 23 kg **c** 212.4 g **d** 339.5 m
  e $4.90 **f** 39.6 m **g** 731 m **h** 83.52 g
  i 360 cm **j** 117 min **k** 81.7 kg **l** $37.70

3 $5525

4 a 52.8 kg **b** 66 kg **c** 45.76 kg

5 Mr Patel $176, Mrs Patel $297.50, Sandeep $341, Priyanka $562.50

6 448

7 705

8 a 66.5 km/h **b** 73.5 km/h

Answers to Chapter 3

9 No, as the total is $101. She will save $20.20, which is less than the $25 it would cost to join the club.
10 10% off $50 is $45; 10% off $45 is $40.50; 20% off $50 is $40
11 $765
12 $1.10 \times 0.9 = 0.99$ (99%)
13 Offer A gives 360 grams for $1.40, i.e. 0.388 cents per gram. Offer B gives 300 grams for $1.12, i.e 0.373 cents per gram, so Offer B is the better offer.
Or Offer A is 360 for 1.40 = 2.6 grams per cent, offer B is 300 for 1.12 = 2.7 grams per cent, so offer B is better.

2.7 One quantity as a percentage of another

Exercise 2H
1 **a** 25% **b** 60.6% **c** 46.3% **d** 12.5%
   e 41.7% **f** 60% **g** 20.8% **h** 10%
   i 1.9% **j** 8.3% **k** 45.5% **l** 10.5%
2 32%
3 6.5%
4 33.7%
5 **a** 49.2% **b** 64.5% **c** 10.6%
6 17.9%
7 4.9%
8 90.5%
9 **a** Brit Com: 20.9%, USA: 26.5%, France: 10.3%, Other 42.3%
   b Total 100%, all imports
10 Nadia had the greater percentage increase.
   Nadia: $(20 - 14) \times 100 \div 14 = 42.9\%$.
   Imran: $(17 - 12) \times 100 \div 12 = 41.7\%$.
11 Yes, as 38 out of 46 is over 80% (82.6%)
12 Vase 20% loss, radio 25% profit, doll 175% profit, toy train 64% loss

2.8 Simple interest and compound interest

Exercise 2I
1 7420 dollars
2 3600 dollars
3 4 years
4 **a** $15 600 **b** $16 224
5 **a** $1272 **b** $1348.32 **c** $1429.22
6 **a** Amar 3200 $, Mona 3328 $ **b** Mona, 128 $
7 **a** $9528.13 **b** £1528.13
8 £3840

9 **a** Simple **b** 6.5%
10 **a** $13 800 **b** $15 870
    c Student's own explanation
11 **a** 2652.25 and 5304.50 **b** £796.37

2.9 A formula for compound interest

Exercise 2J
1 $2249.73
2 $5681.15
3 **a** $5071.50 **b** $5591.33 **c** $6164.44
4 **a** $3589.07 **b** $4458.69
5 $4272.64
6 **a** $3941.57 **b** $441.57
7 8 years
8 The interest in the second five years will be more than the interest in the first five years. The missing number is $5000 \times 1.06^{10} = 8954.24$.
9 **a** $15 000 **b** $16 288.95
10 **a** $1268.24
    b The interest over the year is $268.24.
    This is $\frac{268.24}{1000} \times 100\% = 26.824\%$ of $1000.

2.10 Reverse percentage

Exercise 2K
1 **a** 800 g **b** 250 m **c** 60 cm
   d $3075 **e** $200 **f** $400
2 80
3 T shirt: $8.40, Tights: $1.20, Shorts: $5.20, Sweater: $10.74, Trainers: $24.80, Boots: $32.40
4 $833.33
5 $300
6 240
7 537.63 dollars
8 4750 blue bottles
9 $2585
10 $1440
11 $2450
12 95 dollars
13 $140
14 $945
15 $1325
16 $1300
17 Lee has assumed that 291.2 is 100% instead of 112%. He rounded his wrong answer to the correct answer of $260.

Answers to Chapter 3

3.1 Order of operations

Exercise 3A
1 **a** 11 **b** 6 **c** 10 **d** 12 **e** 11 **f** 13
   g 11 **h** 12 **i** 12 **j** 4 **k** 13 **l** 3

2 **a** 16 **b** 2 **c** 10 **d** 10 **e** 6 **f** 18
   g 6 **h** 15 **i** 9 **j** 12 **k** 3 **l** 8
3 **a** (4 + 1) **b** No brackets needed
   c (2 + 1) **d** No brackets needed
   e (4 + 4) **f** (16 − 4)

707

Answers to Chapter 3

g No brackets needed
h No brackets needed
i (20 − 10)
j No brackets needed
k (5 + 5)
l (4 + 2)
m (15 − 5)
n (7 − 2)
o (3 + 3)
p No brackets needed
q No brackets needed
r (8 − 2)

4 No, correct answer is 5 + 42 = 47
5 a 2 × 3 + 5 = 11   b 2 × (3 + 5) = 16
  c 2 + 3 × 5 = 17   d 5 − (3 − 2) = 4
  e 5 × 3 − 2 = 13   f 5 × 3 × 2 = 30
6 4 + 5 × 3 = 19
  (4 + 5) × 3 = 27. So 4 + 5 × 3 is smaller
7 (5 − 2) × 6 = 18
8 8 ÷ (5 − 3) = 4

3.2 Choosing the correct operation

Exercise 3B

1 a 6000
  b 5 cans cost $1.95, so 6 cans cost $1.95. 32 = (5 × 6) + 2. Cost is $10.53.
2 a 288   b 16
3 a 38
  b Coach price for adults = $8, coach price for juniors = $4, money for coaches raised by tickets = $12 400, cost of coaches = $12 160, profit = $240
4 (39 × 20) + (90 × 30) = 1050 = $10.50
5 (18.81…) Kirsty can buy 18 models.
6 (7.58 …) Michaela must work for 8 weeks.
7 $8.40 per year, 70 cents per copy
8 $450
9 15
10 Gustav pays 2296.25 − 1840 = $456.25

3.3 Finding a fraction of a quantity

Exercise 3C

1 a 18   b 10   c 18   d 28
2 a $1800   b 128 g   c 160 kg
  d $116   e 65 litres f 90 min
3 a $\frac{5}{8}$ of 40 = 25   b $\frac{3}{4}$ of 280 = 210
  c $\frac{4}{5}$ of 70 = 56   d $\frac{5}{6}$ of 72 = 60
4 $6080
5 $31 500
6 52 kg
7 a 856   b 187 675
8 a $50   b $550
9 a $120   b $240
10 Lion Autos
11 Offer B

3.4 Adding and subtracting fractions

Exercise 3D

1 a $\frac{5}{7}$ b $\frac{7}{9}$ c $\frac{4}{5}$ d $\frac{6}{7}$
2 a $\frac{3}{7}$ b $\frac{1}{9}$ c $\frac{4}{11}$ d $\frac{7}{13}$
3 a $\frac{6}{8}=\frac{3}{4}$ b $\frac{4}{10}=\frac{2}{5}$ c $\frac{6}{9}=\frac{2}{3}$ d $\frac{2}{4}=\frac{1}{2}$
4 a $\frac{4}{8}=\frac{1}{2}$ b $\frac{4}{10}=\frac{2}{5}$ c $\frac{4}{6}=\frac{2}{3}$ d $\frac{8}{10}=\frac{4}{5}$
5 a $\frac{12}{10}=\frac{6}{5}=1\frac{1}{5}$ b $\frac{9}{8}=1\frac{1}{8}$ c $\frac{9}{8}=1\frac{1}{8}$
  d $\frac{13}{8}=1\frac{5}{8}$ e $\frac{11}{8}=1\frac{3}{8}$ f $\frac{7}{6}=1\frac{1}{6}$
  g $\frac{9}{6}=\frac{3}{2}=1\frac{1}{2}$ h $\frac{5}{4}=1\frac{1}{4}$
6 a $\frac{10}{8}=\frac{5}{4}=1\frac{1}{4}$ b $\frac{6}{4}=\frac{3}{2}=1\frac{1}{2}$
  c $\frac{5}{5}=1$ d $\frac{16}{10}=\frac{8}{5}=1\frac{3}{5}$
7 a $\frac{5}{8}$ b $\frac{5}{10}=\frac{1}{2}$ c $\frac{1}{4}$ d $\frac{3}{8}$
  e $\frac{1}{4}$ f $\frac{3}{8}$ g $\frac{4}{10}=\frac{2}{5}$ h $\frac{5}{16}$

Exercise 3E

1 a $\frac{8}{15}$ b $\frac{7}{12}$ c $\frac{3}{10}$ d $\frac{11}{12}$ e $\frac{7}{8}$ f $\frac{1}{2}$
  g $\frac{1}{6}$ h $\frac{1}{20}$ i $\frac{1}{10}$ j $\frac{1}{8}$ k $\frac{1}{12}$ l $\frac{1}{3}$
  m $\frac{1}{6}$ n $\frac{7}{9}$ o $\frac{5}{8}$ p $\frac{3}{8}$ q $\frac{1}{15}$ r $1\frac{13}{24}$
  s $\frac{59}{80}$ t $\frac{22}{63}$ u $\frac{37}{54}$
2 a $3\frac{5}{14}$ b $10\frac{3}{5}$ c $2\frac{1}{6}$ d $3\frac{31}{45}$
  e $4\frac{47}{60}$ f $\frac{41}{72}$ g $\frac{29}{48}$ h $1\frac{43}{48}$
  i $1\frac{109}{120}$ j $1\frac{23}{30}$ k $1\frac{31}{84}$
3 $\frac{1}{20}$
4 a $\frac{1}{6}$ b 30, must be divisible by 2 and 3

3.5 Multiplying and dividing fractions

Exercise 3F

1 a $\frac{1}{6}$ b $\frac{1}{10}$ c $\frac{3}{8}$ d $\frac{3}{14}$ e $\frac{8}{15}$
  f $\frac{1}{5}$ g $\frac{2}{7}$ h $\frac{3}{10}$ i $\frac{1}{2}$ j $\frac{2}{5}$
2 a $\frac{3}{32}$ b $\frac{3}{8}$ c $\frac{7}{20}$
  d $\frac{16}{45}$ e $\frac{3}{5}$ f $\frac{5}{8}$
3 $\frac{1}{12}$
4 $\frac{3}{8}$

Answers to Chapter 4

5 a $\frac{5}{12}$ b $2\frac{1}{12}$
 c $6\frac{1}{4}$ d $2\frac{11}{12}$
 e $3\frac{9}{10}$ f $3\frac{1}{3}$
 g $12\frac{1}{2}$ h 30

6 $\frac{2}{5}$ of $6\frac{1}{2} = 2\frac{3}{5}$

Exercise 3G
1 a $\frac{3}{4}$ b $1\frac{2}{5}$ c $1\frac{1}{15}$ d $1\frac{1}{14}$ e 4
 f 4 g 5 h $1\frac{5}{7}$ i $\frac{4}{9}$ j $1\frac{3}{5}$

2 18
3 40
4 15
5 16
6 a $2\frac{2}{15}$ b 38 c $1\frac{7}{8}$ d $\frac{9}{32}$ e $\frac{1}{16}$ f $\frac{256}{625}$
7 a $1\frac{1}{3}$ b $\frac{3}{4}$

Answers to Chapter 4

4.1 Introduction to directed numbers

Exercise 4A
1 a 0 °C b 5 °C c –2 °C d –5 °C e –1 °C
2 a 11 degrees Celsius b 9 degrees Celsius
3 8 degrees Celsius
4 38 degrees Celcius
5 a 2 degrees Celcius between Helsinki and Moscow
 b 34 degrees Celcius between Dubai and Helsinki

4.2 Everyday use of directed numbers

Exercise 4B
1 –$5
2 –200 m
3 above
4 –5 h
5 –2 °C
6 – 70 km
7 +5 minutes
8 –5 km/h
9 –2
10 a –11 °C b 6 degrees Celsius
11 1.54 am

4.3 The number line

Exercise 4C
1 a < b > c < d < e > f <
 g < h > i > j < k < l >
2 a < b < c < d > e < f <
3 a –5 –4 –3 –2 –1 0 1 2 3 4 5
 c –25 –20 –15 –10 –5 0 5 10 15 20 25
 d –10 –8 –6 –4 –2 0 2 4 6 8 10
 e –50 –40 –30 –20 –10 0 10 20 30 40 50

4 6 °C –2 °C –4 °C 2 °C
5 a 1 or 0 or –1 or –2 are the possible answers b No solution
 c Any integer larger than 2. That is 3 or 4 or 5 or …
 d Any integer smaller than –3. That is –4 or –5 or –6 or …

4.4 Adding and subtracting directed numbers

Exercise 4D
1 a –2° b –3° c –2° d –3° e –2° f –3°
 g 3 h 3 i –1 j –1 k 2 l –3
 m –4 n –6 o –6 p –1 q –5 r –4
 s 4 t –1 u –5 v –4 w –5 x –5
2 a 7 degrees Celsius b –6 °C
3 a 2 – 8
 b 2 + 5 – 8 or 2 + 4 – 7 or 8 – 4 – 5 or 8 – 2 – 7 or 5 – 4 – 2
 c 2 – 5 – 7 – 8
 d 2 + 5 – 4 – 7 – 8
4 250 metres

Exercise 4E
1 a –8 b –10 c –11 d –3 e 2 f –5
 g 1 h 4 i 7 j –8 k –5 l –11
 m 11 n 6 o 8 p 8 q –2 r –1
 s –9 t –5
2 a 10 degrees Celsius
 b 7 degrees Celsius
 c 9 degrees Celsius
3 a 2 b –3 c –5 d –7 e –10 f –20
4 a 2 b 4 c –1 d –5 e –11 f 8
5 a 13 b 2 c 5 d 4 e 11 f –2
6 a –10 b –5 c –2 d 4 e 7 f –4
7 a +6 + +5 = 11 b +6 + –9 = –3
 c +6 – –9 = 15 d +6 – +5 = 1
8 It may not come on as the thermometer inaccuracy might be between 0° and 2° or 2° and 4°
9 –1 and 6

Answers to Chapter 5

4.5 Multiplying and dividing directed numbers

Exercise 4F

1. a −15 b −14 c −24 d 6 e 14 f 2
   g −2 h −8 i −4 j 3 k −24 l −10
   m −18 n 16 o 36 p −4 q −12 r −4
   s 7 t 25 u 18
2. a −9 b 16 c −3 d −32 e 18 f 18
   g 6 h −4 i 20 j 16 k 8 l −48
   m 13 n −13 o −8 p 0 q 16 r −42
3. a −2 b 30 c 15 d −27 e −7
4. a 4 b −9 c −3 d 6 e −4
5. a −9 b 3 c 1
6. a 16 b −2 c −12
7. a 24 b 6 c −4 d −2
8. For example: 1 × (−12), −1 × 12, 2 × (−6), 6 × (−2), 3 × (−4), 4 × (−3)
9. For example: 4 ÷ (−1), 8 ÷ (−2), 12 ÷ (−3), 16 ÷ (−4), 20 ÷ (−5), 24 ÷ (−6)
10. −5 × 4, 3 × −6, −20 ÷ 2, −16 ÷ −4
11. a 4 b 25 c 12 d 1
12.

×	−2	3	−4
−5	10	−15	20
2	−4	6	−8
−6	12	−18	24

Answers to Chapter 5

5.1 Squares and square roots

Exercise 5A

1. a 49 b 100 c 1.44 d 6.25 e 256 f 400
   g 9.61 h 20.25 i 9 j 64 k 0.25 l 0.25
2. a 3 and −3 b 10 and −10 c 11 and −11
   d 1.2 and −1.2 e 20 and −20 f 3.5 and −3.5
   g 1 and −1 h 100 and −100
3. a 5 b 6 c 10 d 7 e 8
   f 1.5 g 5.5 h 1.2 i 20 j 0.5
4. a 81 b 40 c 100 d 14 e 36
   f 15 g 49 h 12 i 25 j 21
5. a 24 b 31 c 45 d 40 e 67
   f 101 g 3.6 h 6.5 i 13.9 j 22.2
6. $\sqrt{50}$, 3^2, $\sqrt{90}$, 4^2
7. a 6^2 is 36 and 7^2 is 49; 40 is between 36 and 49
   b 6.3245553……
8. 4 and 5
9. a 8 and 9 b 9 and 10
   c 12 and 13 d 15 and 16
10. $\sqrt{324} = 18$
11. 15

5.2 Cubes and cube roots

Exercise 5B

1. a 8 b 27 c 512 d 1000
   e 1.331 f 15.625 g −27 h −125
   i 8000 j 68.921 k −68.921
2. a 2 b 5 c 9 d 1 e 3
   f −3 g 10 h 1.5 i 4.5 j 0.5
3. a 5 and 6 b 6 and 7
   c 7 and 8 d −8 and −7
4. 2^3 because it equals 8, the rest equal 9
5. One possible answer is $8^2 = 4^3$
6. $\sqrt[3]{2000}$, $\sqrt{225}$, 2.5^3, 4^2
7.

Number	Square	Cube
10	100	1000
5	25	125
4	16	64
11	121	1331
9	81	729

8. 0.8^3, 0.8^2, $\sqrt{0.8}$, $\sqrt[3]{0.8}$.

5.3 More powers and roots

Exercise 5C

1. a 243 b 2401 c 1 000 000 d 256
2. a 0 b 118 c 513
3. a 2592 b 227 c or 0.3789 to 4 d.p.
4. LHS = 31; RHS = 32 − 1 = 31
5. a 5 b 11 c 3 d 20
6. a 20 000 b or 1.125
7. $2^3 \times \sqrt[3]{8} = 16$ $3^2 \times \sqrt[10]{1024} = 18$ $\sqrt[3]{64} \times \sqrt[3]{125} = 20$
   $\sqrt[4]{14641} \times \sqrt[6]{64} = 22$
8. a 32 b 12.5
9. a $8\frac{1}{3}$ b $\frac{7}{8}$
10. a 3 b 8 c 5
11. a 3 b 4 c 8
12. a 15 625 b 1 953 125
13. a $x = 4$ and $y = 2$ is one possible pair.
    b $x = 8$ and $y = 4$ is a second possible pair; $x = 12$ and $y = 6$ is a third possible pair
14. a 1024 b 1 048 576 c 32 d 4
    e 2

Answers to Chapter 6

5.4 Exponential growth and decay

Exercise 5D
1. a i 10 million ii 20 million iii 40 million
   b i 15 million ii 45 million iii 135 million
2. a 6000 b 9000 c 13 500 d 20 250
3. a 6000 b 1500 c 375
4. a 4800 b 768 c 123
5. 21.8, 23.8, 25.9
6. 18.2, 16.6, 15.1
7. a $6312 b 6 years
8. a $100 000 b $195 313
9. The correct value is 150 000 × 1.2^5 = 373 248
10. a $20 b $20 480
11. a 1185 b 351
12. 272 million
13. a $18 b $32 c $340 d $11568

Answers to Chapter 6

6.1 Inequalities

Exercise 6A
1. a > b < c < d =
   e = f > g > h <
2. $\frac{1}{3} < \frac{1}{2} < \frac{3}{5}$
3. a 4,5,6 b 1,2 c 6 d 1,2,3,4,5
   e 2,3,4 f 4,5 g 1,2,3 h 6
4. a underweight b overweight
   c normal d normal
5. 20, 22, 26, 28
6. a 49 b 45
   c 3,6,9 d 16,17,18,19,20
7. a true b false c true
   d true e false f true
8. a 6,7,8 b 26, 27, 28 c −7, −6, −5, −4
   d −2, −1, 0, 1 e there are none f 33
9. a $N \geq 8$ or $N \leq -8$ b $N \geq 4$ c $N < -4$ d no solution

6.2 Sets and Venn diagrams

Exercise 6B
1. a The elements can be listed in any order.
      i {2, 3, 12, 4, 10, 11} ii {4, 5, 11, 10}
      iii {4, 10, 11}
   b i 6 ii 7 iii 12
   c The first 12 natural numbers.
2. a i 7 ii 5 iii 2 iv 10
   b a, p, r, l or n c g or d
3.

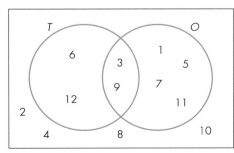

4. a 100 is in A but not in B b 6
   c multiples of 6.

5.

(Venn diagram with X and Y: X contains r; intersection contains p, a, i, s; Y contains n; outside: g, o, e)

6. a i A = {4, 22, 2, 20, 12, 10, 28, 30}
      ii B = {2, 20, 12, 6, 26, 16, 18}
      iii $A \cap B$ = {10, 12, 28, 30}
      iv $A \cap C$ = {2, 20, 12} $A \cap B \cap C$ = {12}
   b i 7 ii 10 iii 12 iv 13 c The even numbers up to 30.
7. a i X = {1, 2, 5, 10} ii Y = {1, 2, 3, 4, 6, 8, 12, 24}
   b i factors of 10 ii factors of 24
   c i {1, 2, 3, 6} ii {1, 2} iii [1, 2, 3]
   d i 8 ii 10 iii 12
8. a i {10} ii {7, 15} iii none
   b i 6 ii 9 iii 7
      iv 14 v 12
9. a i {4, 8, 12} ii {2, 4, 6, 8, 10, 12, 14}
      iii {10} iv {4, 8, 12}
   b i 1 ii 0
10. a 15 b 21 c 35 d 105
11. a 8 b 10 c {1, 2, 4} d {2, 7}
12. a i {a, b, c, d, e, f} ii {d, e, f}
       iii {b, c, d, e, f, g} iv {b, d, e}
    b 3
13. a {1, 2, 3, 4, 5, 6, 7, 8, 10, 13} b {1, 2, 4, 6, 7, 8, 10, 13}
    c {1, 4, 7, 10} d {1, 2, 4, 6, 7, 10, 13}

6.3 More about Venn diagrams

Exercise 6C
1. a {8, 11, 12, 14, 15} b {6, 9, 14, 15} c {14, 15}
   d {8, 11, 12} e {6, 9}
2. a $6 \in B'$ b $7 \notin A'$ c $8 \in (A \cap B)'$
   d $9 \in A \cup B'$ e $9 \notin A' \cup B$ f $9 \notin (A \cup B)'$

3 a i $X \cap Y'$ is 3   **ii** $(X \cap Y)'$ is 2   **iii** $X \cup Y'$ is 4
 iv $(X \cup Y)'$ is 1   **v** $X' \cap Y'$ is 1   **vi** $X' \cup Y'$ is 2
 b $(X \cap Y)' = X' \cup Y'$ because they have the same diagram. Also $(X \cup Y)' = X' \cap Y'$

4 a

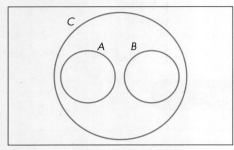

 b C

5 $n(A) + n(B)$ counts the elements in the intersection twice but $n(A \cup B)$ only counts them once. This means that $n(A \cup B) = n(A) + n(B) - n(A \cap B)$

If $n(A \cup B) = n(A) + n(B)$ there are no elements in the intersection. Hence $A \cap B = \emptyset$

6 $x = 2$ because 2 is the only even prime number.

 b

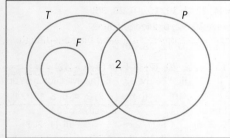

 c T = {even numbers} so T' = {odd numbers}

7

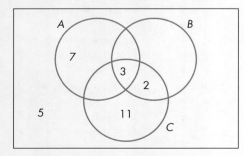

8 a $C \subseteq B$ is true   **b** $A \cap C$ is true
 c It is false; $C \cup B = B$ **d** It is false; $B \cap C = C$
 e It is false; $A' \cap C = C$ **f** It is false; $C' \cup A' = \xi$

9 a and b

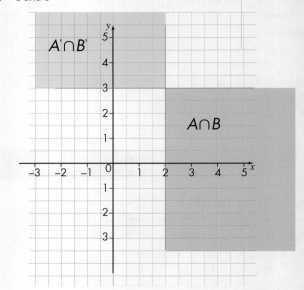

10

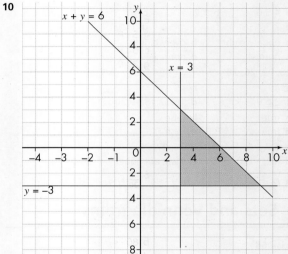

 b Any coordinates of the form $(0, c)$ where $-3 \leqslant c \leqslant 6$

11 a

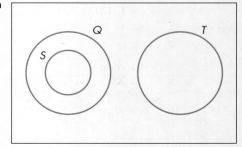

 b $R \cap T$ = {equilateral triangles}

Answers to Chapter 7

12 a

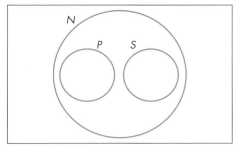

b No prime number is a square number and so $S \cap P = \emptyset$
c The smallest natural number that is not prime or square is 6.

13 A Venn diagram shows the sets

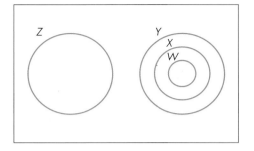

a $W \cap Y = W$ **b** $W \cup Y = Y$ **c** $X' \cap Z = Z$ **d** $Z' \cup Y = Z'$

Answers to Chapter 7

7.1 Ratio

Exercise 7A
1. **a** 1 : 3   **b** 1 : 4   **c** 2 : 3   **d** 2 : 1
   e 2 : 5   **f** 2 : 5   **g** 5 : 8   **h** 5 : 1
2. **a** 8 : 1   **b** 12 : 1   **c** 5 : 6   **d** 1 : 24
   e 48 : 1   **f** 5 : 2   **g** 3 : 8   **h** 1 : 5
3. $\frac{7}{10}$
4. $\frac{10}{25} = \frac{2}{5}$
5. **a** $\frac{2}{5}$ **b** $\frac{3}{5}$
6. **a** $\frac{7}{10}$ **b** $\frac{3}{10}$
7. **a** $\frac{1}{2}$ **b** $\frac{7}{20}$ **c** $\frac{3}{20}$
8. 3 : 1

Exercise 7B
1. **a** 160 g, 240 g   **b** 80 kg, 200 kg
   c 150, 350   **d** 950 m, 50 m
   e 175 min, 125 min   **f** $20, $30, $50
   g $36, $60, $144   **h** 50 g, 250 g, 300 g
   i $1.40, $2, $1.60   **j** 120 kg, 72 kg, 8 kg
2. **a** 175   **b** 30%
3. **a** 28   **b** 42
4. 21
5. Joshua $2500, Aicha $3500, Mariam $4000
6. **a** 1 : 400 000   **b** 1 : 125 000   **c** 1 : 250 000
   d 1 : 25 000   **e** 1 : 20 000   **f** 1 : 40 000
   g 1 : 62 500   **h** 1 : 10 000   **i** 1 : 60 000
7. **a** 1 : 1 000 000   **b** 47 km   **c** 8 mm
8. **a** 1 : 250 000   **b** 2 km   **c** 4.8 cm
9. **a** 1 : 20 000   **b** 0.54 km   **c** 40 cm
10. **a** 1 : 1.6   **b** 1 : 3.25   **c** 1 : 1.125
    d 1 : 1.44   **e** 1 : 5.4   **f** 1 : 1.5
    g 1 : 4.8   **h** 1 : 42   **i** 1 : 1.25
11. **c** 1 : 250 000   At this scale 134 km is 53.6 cm which is a sensible size. The others are too small (5.36 mm or 5.36 cm) or too large (5.36 m).

Exercise 7C
1. **a** 3 : 2   **b** 32   **c** 80
2. 1000 g
3. 10 125
4. **a** 14 min   **b** 75 min
5. **a** 11 pages   **b** 32%
6. Ren $2040, Shota $2720
7. **a** lemonade 20 litres, ginger 0.5 litres
   b This one, one-thirteenth is greater than one-fiftieth.

7.2 Increases and decreases using ratios

Exercise 7D
1. **a** 600   **b** 300   **c** 2000   **d** 350   **e** 240   **f** 220
2. **a** 20   **b** 60   **c** 8   **d** 56   **e** 16   **f** 64
3. **a** 160 cm by 120 cm   **b** 60 cm by 45 cm
4. 12.5 cm by 15 cm   **b** 3 : 2
5. **a** 9 cm by 15 cm   **b** 6.75 cm by 11.25 cm
6. **a** $7500   **b** 50%   **c** $9375   **d** 25%
   e 87.5%   **f** One way is to cancel 9375 : 5000
7. **a** One possible answer is to say $\frac{6}{5} = 1.2$ and this is the multiplier for a 20% increase
   b 11:10   **c** 9:10
8. 100
9. 40 cm^3

7.3 Speed

Exercise 7E
1. 18 km/hour
2. 440 kilometres
3. 52.5 km/hour
4. 11.50 am
5. 500 s
6. **a** 75 km/hour   **b** 6.5 hours   **c** 175 km   **d** 240 km
   e 64 km/h   **f** 325 km   **g** 4.3 hours (4 h 18 min)

Answers to Chapter 8

7	a 7.75 h	b 85.2 km/hour		
8	a 2.25 h	b 157.5 km		
9	a 1.25 h	b 1 h 15 min		
10	a 48 km/hour	b 6 h 40 min		
11	a 120 km	b 48 km/h		
12	a 30 min	b 12 km/h		
13	a 10 m/s	b 3.3 m/s	c 16.7 m/s	d 41.7 m/s
	e 20.8 m/s			
14	a 90 km/h	b 43.2 km/h	c 14.4 km/h	d 108 km/h
	e 1.8 km/h			
15	a 64.8 km/h	b 28 s	c 8.07 (37 min journey)	
16	a 6.7 m/s	b 66 km	c 5 minutes	d 133.3 metres
17	6.6 minutes			

7.4 Rates

Exercise 7F

1 a 3.5 cm   b 20 days
2 a 15 litres   b 20 seconds
3 a 10.59 g/cm³   b 50 cm³
4 1600 days (4.38 years)
5 a 0.5 mm/year   b 12.5 mm   c 40 years
6 a 4.44 cm³ to 3 d.p.   b 360 g
   c No. The density is 8 g/cm³, not 9 g/cm³
7 a 62.5 Pa or N/m²   b 250 Pa or N/m²
   c 187.5 Pa or N/m²
8 a 31.8 litres   b 5.3 litres   c 1.06   d 0.106
   e 943 km
9 a 14.5 kg   b 2900 kg   c About 8 kg/day
10 724 N

7.5 Direct proportion

Exercise 7G

1 60 g
2 $5.22
3 45
4 $6.72
5 a $312.50   b 8
6 a 56 litres   b 350 km
7 a 300 kg   b 9 weeks
8 40 s
9 a i 100 g margarine, 200 g sugar, 250 g flour, 150 g ground rice
      ii 150 g margarine, 300 g sugar, 375 g flour, 225 g ground rice
      iii 250 g margarine, 500 g sugar, 625 g flour, 375 g ground rice
   b 24
10 Pieter's shop as I can buy 24. At Paulo's shop I can only buy 20.

7.6 Inverse proportion

Exercise 7H

1 20 minutes
2 16
3 a 36   b 48
4 $10.80
5 a 16 days   b 20
6 The missing times are 5 hours, 1 hour 40 minutes, 1 hour 15 minutes
7 a 6   b 15
8 $8

Answers to Chapter 8

8.1 Rounding whole numbers

Exercise 8A

1	a 20	b 60	c 80	d 50	e 100				
	f 20	g 90	h 70	i 10	j 30				
2	a 200	b 600	c 800	d 500	e 1000				
	f 100	g 600	h 400	i 1000	j 1100				
3	a 2000	b 6000	c 8000	d 5000					
	e 10 000	f 1000	g 6000	h 3000					
	i 9000	j 2000							
4	a True	b False	c True	d True	e True	f False			

5 a Highest Germany, lowest Italy
   b 36 000, 43 000 , 25 000, 29 000
   c 25 499 and 24 500
6 a 375   b 25 (350 to 374 inclusive)
7 A number between 75 and 84 inclusive added to a number between 45 and 54 inclusive with a total not equal to 130, for example 79 + 49 = 128

8.2 Rounding decimals

Exercise 8B

1	a 4.8	b 3.8	c 2.2	d 8.3	e 3.7	
	f 46.9	g 23.9	h 9.5	i 11.1	j 33.5	
2	a 5.78	b 2.36	c 0.98	d 33.09	e 6.01	
	f 23.57	g 91.79	h 8.00	i 2.31	j 23.92	
3	a 4.6	b 0.08	c 45.716	d 94.85	e 602.1	
	f 671.76	g 7.1	h 6.904	i 13.78	j 0.1	
4	a 8	b 3	c 8	d 6	e 4	
	f 7	g 2	h 47	i 23	j 96	

5 3 + 9 + 6 + 4 = 22 dollars
6 3, 3.46, 3.5
7 4.7275 or 4.7282

Answers to Chapter 9

8.3 Rounding to significant figures

Exercise 8C

1. a 50000 b 60000 c 30000 d 90000
   e 90000 f 0.5 g 0.3 h 0.006
   i 0.05 j 0.0009 k 10 l 90
   m 90 n 200 o 1000
2. a 56000 b 27000 c 80000 d 31000
   e 14000 f 1.7 g 4.1 h 2.7
   i 8.0 j 42 k 0.80 l 0.46
   m 0.066 n 1.0 o 0.0098
3. a 60000 b 5300 c 89.7 d 110
   e 9 f 1.1 g 0.3 h 0.7
   i 0.4 j 0.8 k 0.2 l 0.7
4. a 65, 74 b 95, 149 c 950, 1499
5. Satora 750, 849, Nimral 1150, 1249, Korput 164 500, 165 499
6. One, because there could be 450 then 449.
7. Vashti has rounded to 2 significant figures or nearest 10 000.

8.4 Upper and lower bounds

Exercise 8D

1. a 6.5 and 7.5 b 115 and 125
   c 3350 and 3450 d 49.5 and 50.5
   e 5.50 and 6.50 f 16.75 and 16.85
   g 15.5 and 16.5 h 14 450 and 14 550
   i 54.5 and 55.5 j 52.5 and 57.5
2. a $5.5 \leq$ length in cm < 6.5
   b $16.5 \leq$ mass in kg < 17.5
   c $31.5 \leq$ time in minutes < 32.5
   d $237.5 \leq$ distance in km < 238.5
   e $7.25 \leq$ distance in m < 7.35
   f $25.75 \leq$ mass in kg < 25.85
   g $3.35 \leq$ time in hours < 3.45
   h $86.5 \leq$ mass in g < 87.5
   i $4.225 \leq$ distance in mm < 4.235
   j $2.185 \leq$ mass in kg < 2.195
   k $12.665 \leq$ time in minutes < 12.675
   l $24.5 \leq$ distance in metres < 25.5
   m $35 \leq$ length in cm < 45
   n $595 \leq$ mass in g < 605
   o $25 \leq$ time in minutes < 35
   p $995 \leq$ distance in metres < 1050
   q $3.95 \leq$ distance in metres < 4.05
   r $7.035 \leq$ mass in kg < 7.045
   s $11.95 \leq$ time in seconds < 12.05
   t $6.995 \leq$ distance in metres < 7.005
3. a 7.5, 8.5 b 25.5, 26.5
   c 24.5, 25.5 d 84.5, 85.5
   e 2.395, 2.405 f 0.15, 0.25
   g 0.055, 0.065 h 250 g, 350 g
   i 0.65, 0.75 j 365.5, 366.5
   k 165, 175 l 205, 215
4. C: The chain and distance are both any value between 29.5 and 30.5 metres, so there is no way of knowing if the chain is longer or shorter than the distance.
5. 2 kg 450 grams
6. a <65.5 g b 64.5 g
   c <2620 g d 2580 g

8.5 Upper and lower bounds for calculations

Exercise 8E

1. 65 kg and 75 kg
2. a 12.5 kg b 20
3. 9 kg   53.5 – 44.5
4. a 26 cm $\leq$ perimeter < 30 cm
   b 25.6 cm $\leq$ perimeter < 26.0 cm
   c 50.5 cm $\leq$ perimeter < 52.7 cm
5. a 38.25 cm^2 $\leq$ area < 52.25 cm^2
   b 37.1575 cm^2 $\leq$ area < 38.4475 cm^2
   c 135.625 cm^2 $\leq$ area < 145.225 cm^2
6. 79.75 m^2 $\leq$ area < 100.75 m^2
7. 216.125 cm^3 $\leq$ volume < 354.375 cm^3
8. 12.5 metres
9. Yes, because they could be walking at 4.5 km/h and 2.5 km/h meaning that they would cover 4.5 km + 2.5 km = 7 km in 1 hour
10. 20.9 m $\leq$ length < 22.9 m (3 sf)
11. a 14.65 s $\leq$ time < 14.75 s
    b 99.5 m $\leq$ length < 100.5 m
    c 6.86 m/s (3 sf)
12. 14 s $\leq$ time < 30 s
13. 337.75 and 334.21
14. 177.3 and 169.4

Answers to Chapter 9

9.1 Standard form

Exercise 9A

1. a 250 b 34.5 c 0.00467 d 34.6
   e 0.020789 f 5678 g 246 h 7600
   i 897 000 j 0.00865 k 60 000 000 l 0.000567
2. a 2.5×10^2 b 3.45×10^{-1} c 4.67×10^4
   d 3.4×10^9 e 2.078×10^{10} f 5.678×10^{-4}
   g 2.46×10^3 h 7.6×10^{-2} i 7.6×10^{-4}
   j 9.99×10^{-1} k 2.3456×10^2 l 9.87654×10^1
   m 6×10^{-4} n 5.67×10^{-3} o 5.60045×10^1
3. 2.7797×10^4
4. $3.211\,97 \times 10^5$, $4.491\,863 \times 10^6$
5. 1.298×10^7, 2.997×10^9, 9.3×10^4
6. 100
7. 61.8 kilometres.
8. 7.78×10^8; 5.8×10^7; 5.92×10^9

Answers to Chapter 10

9.2 Calculating with standard form

Exercise 9B
1. **a** 5.67×10^3 **b** 6×10^2 **c** 3.46×10^{-1}
   d 7×10^{-4} **e** 5.6×10^2 **f** 6×10^5
   g 7×10^3 **h** 1.6 **i** 2.3×10^7
2. **a** 1.08×10^8 **b** 4.8×10^6 **c** 1.2×10^9
   d 1.08 **e** 6.4×10^2 **f** 1.2×10^1
   g 2.88 **h** 2.5×10^7 **i** 8×10^{-6}
3. **a** 2.7×10 **b** 1.6×10^{-2} **c** 2×10^{-1}
   d 4×10^{-8} **e** 2×10^5 **f** 6×10^{-2}
4. 2×10^{13}, 1×10^{-10}, mass $= 2 \times 10^3$ g
5. **a** (2^{63}), 9.2×10^{18} grains
   b $2^{64} - 1 = 1.8 \times 10^{19}$
6. **a** 1.0×10^8 sq km
   b 31%
7. 3.80×10^7 sq km
8. 5×10^4
9. 2.3×10^5
10. 4.55×10^8 kg or 455 070 tonnes
11. **a** 100 000 000 (100 million) **b** 1.4%
12. **a** 2.048×10^6 **b** 4.816×10^6
13. 9.41×10^4
14. Any value from 1.00000001×10^8 to 1×10^9 (excluding 1×10^9), i.e. any value of the form $a \times 10^8$ where $1 < a < 10$
15. **a** India **b** India and Jamaica
    c 2.2×10^7 **d** 21 or 22 **e** 480
16. **a** Togo **b** Sri Lanka
    c Sri Lanka **d** Russian Federation
    e $\frac{1}{261}$

Answers to Chapter 10

10.1 Units of measurement

Exercise 10A
1. **a** metres **b** kilometres
   c millimetres **d** kilograms or grams
   e litres **f** tonnes
   g millilitres **h** metres
   i kilograms **j** millimetres
2. Check individual answers.
3. The 5 metre since his height is about 175 cm, the lamp post will be about 525 cm

10.2 Converting between metric units

Exercise 10B
1. **a** 1.25 m **b** 8.2 cm **c** 0.55 m **d** 4.2 kg
   e 5.75 t **f** 8.5 cl **g** 0.755 kg **h** 0.8 l
   i 2 l **j** 1.035 m³ **k** 0.53 m³ **l** 34 000 m
2. **a** 3400 mm **b** 135 mm **c** 67 cm **d** 640 m
   e 2400 ml **f** 590 cl **g** 3750 kg **h** 0.00094 l
   i 2160 cl **j** 15 200 g **k** 14 000 l **l** 0.19 ml
3. He should choose the 2000 mm × 15 mm × 20 mm
4. 1 000 000
5. 400 hours
6. 7.5×10^9

10.3 Time

Exercise 10C
1. **a** 1 hour 10 minutes; 2 hours 3 minutes; 2 hours 9 minutes; 1 hour 45 minutes
   b the 0900
2. **a** 9:45am, 10:36am, 1:33pm, 4:49pm
   b 3 hours and 48 minutes, 6 hours and 13 minutes
3. **a** 1605 **b** 0715 **c** 6 hours 45minutes
4. **a** 1050 **b** 1635 **c** 5 hours 45 minutes
5. **a** 1210 **b** 2hours 50 minutes
6. **a** 12 minutes **b** 40 minutes **c** 54 minutes
7. 1 hour 13 minutes
8. 1415
9. 0715 the next day
10. **a** 0330 **b** 0600

10.4 Currency conversions

Exercise 10D
1. 3197.41
2. 164
3. The missing values are 4.30, 7.76, 38.78, 193.88, 387.75, 775.50
4. 43.01
5. **a** 224.91 **b** 172.74
6. **a i** 349.83 **ii** 24692 **iii** 432.90
   b 54000 yen, 500 euros, 650 dollars
7. **a** 2391.38 **b** 3489.75
   c Taiwan dollar **d** 1.4593
8. **a** 74.7755 **b** 0.14747

10.5 Using a calculator efficiently

Exercise 10E
1. **a** 144 **b** 108
2. **a** 12.54 **b** 27.45
3. **a** 196.48 **b** 1.023 **c** 0.236 **d** 4.219
4. **a** 3.58 **b** 6
5. **a** 497.952 **b** 110.978625
6. **a** 3.12 **b** 0.749 **c** 90.47
   d 184.96 **e** 5.56 **f** 27.52

Answers to Chapter 11

11.1 The language of algebra

Exercise 11A

1. **a** $x + 2$ **b** $x - 6$ **c** $x + k$ **d** $x - t$
   e $x + 3$ **f** $d + m$ **g** $b - y$ **h** $p + t + w$
   i $8x$ **j** hj **k** $x \div 4$ or $\frac{x}{4}$ **l** $2 \div x$ or $\frac{2}{x}$
   m $y \div t$ or $\frac{y}{t}$ **n** wt **o** a^2 **p** g^2
2. **a** $x + 3$ years **b** $x - 4$ years
3. $F = 2C + 30$
4. Rule **c**
5. **a** $3n$ **b** $3n + 3$ **c** $n + 1$ **d** $n - 1$
6. Anil: $2n$, Reza: $n + 2$, Dale: $n - 3$, Chen: $2n + 3$
7. **a** $4 **b** $(10 - x)$ **c** $(y - x)$ **d** $2x$
8. **a** $75 **b** $15x$ **c** $4A$ **d** Ay
9. $(A - B)$ dollars
10. $A \div 5$ or $\frac{A}{5}$.
11. **a** Dad: $(72 + x)$ years, me: $(T + x)$ years **b** 31 years
12. **a** $T \div 2$ or $\frac{T}{2}$ **b** $T \div 2 + 4$ or $\frac{T}{2} + 4$ **c** $T - x$
13. **a** $8x$ **b** $12m$ **c** $18t$
14. Andrea: $3n - 3$, Barak: $3n - 1$, Ahmed: $3n - 6$ or $3(n - 2)$, Dina: 0, Emma: $3n - n = 2n$. Hana: $3n - 3m$
15. For example, $2 \times 6m$, $1 \times 12m$, $6m + 6m$, etc.

11.2 Substitution into formulae

Exercise 11B

1. **a** 8 **b** 17 **c** 32
2. **a** 3 **b** 11 **c** 43
3. **a** 9 **b** 15 **c** 29
4. **a** 9 **b** 5 **c** −1
5. **a** 13 **b** 33 **c** 78
6. **a** 10 **b** 13 **c** 58
7. **a** $4 **b** 13 km **c** Yes, the fare is $5.00
8. **a** $2 \times 8 + 6 \times 11 - 3 \times 2 = 76$
   b $5 \times 2 - 2 \times 11 + 3 \times 8 = 12$
9. Any values such that $lw = \frac{1}{2}bh$ or $bh = 2lw$
10. **a** 32 **b** 64 **c** 160
11. **a** 6.5 **b** 0.5 **c** −2.5
12. **a** 2 **b** 8 **c** −10
13. **a** 3 **b** 2.5 **c** −5
14. **a** 6 **b** 3 **c** 2
15. **a** 12 **b** 8 **c** $1\frac{1}{2}$
16. **a** $\frac{1050}{n}$ **b** $925
17. **a i** odd **ii** odd **iii** even **iv** odd
    b Any valid expression such as $xy + z$
18. **a** $20
    b i −$40 **ii** Delivery cost will be zero. **c** 40 kilometres

11.3 Rearranging formulae

Exercise 11C

1. $k = \frac{T}{3}$
2. $y = X + 1$
3. $p = 3Q$
4. $r = \frac{A - 9}{4}$
5. $n = \frac{W + 1}{3}$
6. **a** $m = p - t$
   b $t = p - m$
7. $m = gv$
8. $m = \sqrt{t}$
9. $r = \frac{C}{2\pi}$
10. $b = \frac{A}{h}$
11. $l = \frac{P - 2w}{2}$
12. $p = \sqrt{m - 2}$
13. **a** $-40 - 32 = -72$, $-72 \div 9 = -8$, $5 \times -8 = -40$
    b $68 - 32 = 36$, $36 \div 9 = 4$, $4 \times 5 = 20$
    c student's own demonstration
14. **a** $a = \frac{v - u}{t}$ **b** $t = \frac{v - u}{a}$
15. $d = \sqrt{\frac{4A}{\pi}}$
16. **a** $n = \frac{W - t}{3}$ **b** $t = W - 3n$
17. **a** $y = \frac{x + w}{5}$ **b** $w = 5y - x$
18. $p = \sqrt{\frac{k}{2}}$
19. **a** $t = u^2 - v$ **b** $u = \sqrt{v + t}$
20. **a** $m = k - n^2$ **b** $n = \sqrt{k - m}$
21. $r = \sqrt{\frac{T}{5}}$
22. **a** $w = K - 5n^2$ **b** $n = \sqrt{\frac{K - w}{5}}$

11.4 More complicated formulae

Exercise 11D

1. **a** 2.5 **b** $a = \sqrt{c^2 - b^2}$
2. **a** 60 **b** $a = \frac{2(s - ut)}{t^2}$
3. **a** $b = ac - 2$ **b** $c = \frac{b + 2}{a}$
4. $t = \frac{r}{p} + 3$
5. $e = \left(\frac{12}{d} - 1\right)^2$
6. **a** 5 **b** $u = \sqrt{v^2 - 2as}$ **c** $s = \frac{v^2 - u^2}{2a}$
7. **a** $L = \left(\frac{T}{2\pi}\right)^2 G$ **b** Student's proof

717

Answers to Chapter 12

8   **a** $R = \sqrt{\frac{D + \pi r^2}{\pi}}$ **b** $r = \sqrt{\frac{\pi R^2 - D}{\pi}}$ **c** $\pi = \frac{D}{R^2 - r^2}$

9   **a** $x = 5$ or -5 **b** $x = \sqrt{\frac{11 + 4y^2}{3}}$ **c** $y = \sqrt{\frac{3x^2 - 11}{4}}$

10   **a** $a = \left(\frac{T}{2}\right)^2 (c + 3)$ **b** $c = a\left(\frac{2}{T}\right)^2 - 3$

11    $T = \frac{b^2 + c^2 - a^2}{2bc}$

12   **a** 12   **b** $f = \frac{uv}{u+v}$ **c** $u = \frac{fv}{v-f}$ **d** $v = \frac{fu}{u-f}$

Answers to Chapter 12

12.1 Simplifying expressions

Exercise 12A

1   **a** $6t$ **b** $15y$ **c** $8w$ **d** $5b^2$ **e** $2w^2$
    f $8p^2$ **g** $6t^2$ **h** $15t^2$ **i** $2mt$ **j** $5qt$
    k $6mn$ **l** $6qt$ **m** $10hk$ **n** $21pr$

2   **a** All except $2m \times 6m$ **b** 2 and 0

3    $4x$ cm

4   **a** y^3 **b** $3m^3$ **c** $4t^3$ **d** $6n^3$ **e** t^4
    f h^5 **g** $12n^5$ **h** $6a^7$ **i** $4k^7$ **j** t^3
    k $12d^3$ **l** $15p^6$ **m** $3mp^2$ **n** $6m^2n$ **o** $8m^2p^2$

Exercise 12B

1   **a** $\$t$ **b** $\$(4t + 3)$

2   **a** $10x + 2y$ **b** $7x + y$ **c** $6x + y$

3   **a** $5a$ **b** $6c$ **c** $9e$ **d** $6f$
    e $4j$ **f** $3q$ **g** 0   **h** $-w$
    i $6x^2$ **j** $5y^2$ **k** 0

4   **a** $7x$ **b** $3t$ **c** $-5x$ **d** $-5k$
    e $2m^2$ **f** 0

5   **a** $7x + 5$ **b** $5x + 6$ **c** $5p$ **d** $5x + 6$
    e $5p + t + 5$ **f** $8w - 5k$ **g** c **h** $8k - 6y + 10$

6   **a** $2c + 3d$ **b** $5d + 2e$ **c** $f + 3g + 4h$
    d $6u - 3v$ **e** $7m - 7n$ **f** $3k + 2m + 5p$
    g $2v$ **h** $2w - 3y$ **i** $11x^2 - 5y$
    j $-y^2 - 2z$ **k** $x^2 - z^2$

7   **a** $8x + 6$ **b** $3x + 16$ **c** $2x + 2y + 8$

8   Any acceptable answers, e.g. $x + 4x + 2y + 2y$ or $6x - x + 6y - 2y$

9   **a** $2x$ and $2y$ **b** a and $7b$

10   **a** $3x - 1 - x$ **b** $10x$ **c** 25 cm

11   Maria is correct, as the two short horizontal lengths are equal to the bottom length and the two short vertical lengths are equal to the side length.

12.2 Expanding brackets

Exercise 12C

1   **a** $6 + 2m$ **b** $10 + 5l$ **c** $12 - 3y$
    d $20 + 8k$ **e** $6 - 12f$ **f** $10 - 6w$
    g $10k + 15m$ **h** $12d - 8n$ **i** $t^2 + 3t$
    j $k^2 - 3k$ **k** $4t^2 - 4t$ **l** $8k - 2k^2$
    m $8g^2 + 20g$ **n** $15h^2 - 10h$ **o** $y^3 + 5y$
    p $h^4 + 7h$ **q** $k^3 - 5k$ **r** $3t^3 + 12t$
    s $15d^3 - 3d^4$ **t** $6w^3 + 3tw$ **u** $15a^3 - 10ab$
    v $12p^4 - 15mp$ **w** $12h^3 + 8h^2g$ **x** $8m^3 + 2m^4$

2   **a** $5(t - 1)$ and $5t - 5$
    b Yes, as $5(t - 1)$ when $t = 4.50$ is $5 \times 3.50 = \$17.50$

3   He has worked out 3×5 as 8 instead of 15 and he has not multiplied the second term by 3. Answer should be $15x - 12$.

4   **a** $3(2y + 3)$ **b** $2(6z + 4)$ or $4(3z + 2)$

Exercise 12D

1   **a** $7t$ **b** $9d$ **c** $3e$ **d** $2t$
    e $5t^2$ **f** $4y^2$ **g** $5ab$ **h** $3a^2d$

2   **a** $2x$ and $11y$ **b** a and $8b$

3   **a** $2x - 3$ **b** $10x - 16$ or $2(5x - 8)$

4   **a** $22 + 5t$ **b** $21 + 19k$ **c** $22 + 2f$ **d** $14 + 3g$

5   **a** $2 + 2h$ **b** $9g + 5$ **c** $17k + 16$ **d** $6e + 20$

6   **a** $4m + 3p + 2mp$ **b** $3k + 4h + 5hk$
    c $12r + 24p + 13pr$ **d** $19km + 20k - 6m$

7   **a** $9t^2 + 13t$ **b** $13y^2 + 5y$ **c** $10e^2 - 6e$ **d** $14k^2 - 3kp$

8   **a** $17ab + 12ac + 6bc$ **b** $18wy + 6ty - 8tw$
    c $14mn - 15mp - 6np$ **d** $8r^3 - 6r^2$

9   For x-coefficients, 3 and 1 or 1 and 4; for y-coefficients, 5 and 1 or 3 and 4 or 1 and 7.

10    $5(3x + 2) - 3(2x - 1) = 9x + 13$

12.3 Factorisation

Exercise 12E

1   **a** $6(m + 2t)$ **b** $3(3t + p)$ **c** $4(2m + 3k)$
    d $4(r + 2t)$ **e** $m(n + 3)$ **f** $g(5g + 3)$
    g $2(2w - 3t)$ **h** $y(3y + 2)$ **i** $t(4t - 3)$
    j $3m(m - p)$ **k** $3p(2p + 3t)$ **l** $2p(4t + 3m)$
    m $4b(2a - c)$ **n** $5bc(b - 2)$ **o** $2b(4ac + 3de)$
    p $2(2a^2 + 3a + 4)$ **q** $3b(2a + 3c + d)$ **r** $t(5t + 4 + a)$
    s $3mt(2t - 1 + 3m)$ **t** $2ab(4b + 1 - 2a)$ **u** $5pt(2t + 3 + p)$

2   **a** Suni has taken out a common factor.
    b Because the bracket adds up to $10.
    c $30

3   **a**, **d**, **f** and **h** do not factorise.
    b $m(5 + 2p)$ **c** $t(t - 7)$ **e** $2m(2m - 3p)$
    g $a(4a - 5b)$ **i** $b(5a - 3bc)$

4   **a** Bernice
    b Ahmed has not taken out the largest possible common factor. Craig has taken m out of both terms but there isn't an m in the second term.

5   There are no common factors.

6   numerator $4x^2 - 12x$, denominator $2x - 6$

7   **a** $4(x + 1)$ **b** $2(x + 4)$ **c** $4(x + 1)$ **d** $2(3x + 2)$

Answers to Chapter 12

12.4 Multiplying two brackets: 1

Exercise 12F
1. $x^2 + 5x + 6$
2. $t^2 + 7t + 12$
3. $w^2 + 4w + 3$
4. $m^2 + 6m + 5$
5. $k^2 + 8k + 15$
6. $a^2 + 5a + 4$
7. $x^2 + 2x - 8$
8. $t^2 + 2t - 15$
9. $w^2 + 2w - 3$
10. $f^2 - f - 6$
11. $g^2 - 3g - 4$
12. $y^2 + y - 12$
13. $x^2 + x - 12$
14. $p^2 - p - 2$
15. $k^2 - 2k - 8$
16. $y^2 + 3y - 10$
17. $a^2 + 2a - 3$
18. $x^2 - 9$
19. $t^2 - 25$
20. $m^2 - 16$
21. $t^2 - 4$
22. $y^2 - 64$
23. $p^2 - 1$
24. $25 - x^2$
25. $49 - g^2$
26. $x^2 - 36$
27. $(x + 2)$ and $(x + 3)$
28. **a** B: $1 \times (x - 2)$ C: 1×2 D: $2 \times (x - 1)$
    b $(x - 2) + 2 + 2(x - 1) = 3x - 2$
    c Area A = $(x - 1)(x - 2)$ = area of square minus areas (B + C + D)
    $= x^2 - (3x - 2)$
    $= x^2 - 3x + 2$
29. **a** $x^2 - 9$
    b i 9991   **ii** 39991

12.5 Multiplying two brackets: 2

Exercise 12G
1. $6x^2 + 11x + 3$
2. $12y^2 + 17y + 6$
3. $6t^2 + 17t + 5$
4. $8t^2 + 2t - 3$
5. $10m^2 - 11m - 6$
6. $12k^2 - 11k - 15$
7. $6p^2 + 11p - 10$
8. $10w^2 + 19w + 6$
9. $6a^2 - 7a - 3$
10. $8r^2 - 10r + 3$
11. $15g^2 - 16g + 4$
12. $12d^2 + 5d - 2$
13. $8p^2 + 26p + 15$
14. $6t^2 + 7t + 2$
15. $6p^2 + 11p + 4$
16. $-10t^2 - 7t + 6$
17. $-6n^2 + n + 12$
18. $6f^2 - 5f - 6$
19. $-10q^2 + 7q + 12$
20. $-6p^2 - 7p + 3$
21. $-6t^2 + 10t + 4$
22. **a** $x^2 - 1$ **b** $4x^2 - 1$ **c** $4x^2 - 9$
    d $9x^2 - 25$
23. **a** $(3x - 2)(2x + 1) = 6x^2 - x - 2$
    $(2x - 1)(2x - 1) = 4x^2 - 4x + 1$
    $(6x - 3)(x + 1) = 6x^2 + 3x - 3$
    $(3x + 2)(2x + 1) = 6x^2 + 7x + 2$
    b Multiply the x terms to match the x^2 term and/or multiply the constant terms to get the constant term in the answer.

Exercise 12H
1. $4x^2 - 1$
2. $9t^2 - 4$
3. $25y^2 - 9$
4. $16m^2 - 9$
5. $4k^2 - 9$
6. $16h^2 - 1$
7. $4 - 9x^2$
8. $25 - 4t^2$
9. $36 - 25y^2$
10. $a^2 - b^2$
11. $9t^2 - k^2$
12. $4m^2 - 9p^2$
13. $25k^2 - g^2$
14. $a^2b^2 - c^2d^2$
15. $a^4 - b^4$
16. **a** $a^2 - b^2$
    b Dimensions: $a + b$ by $a - b$; Area: $a^2 - b^2$
    c Areas are the same, so $a^2 - b^2 = (a + b) \times (a - b)$
17. First shaded area is $(2k)^2 - 1^2 = 4k^2 - 1$
    Second shaded area is $(2k + 1)(2k - 1) = 4k^2 - 1$

Exercise 12I
1. $x^2 + 10x + 25$
2. $m^2 + 8m + 16$
3. $t^2 + 12t + 36$
4. $p^2 + 6p + 9$
5. $m^2 - 6m + 9$
6. $t^2 - 10t + 25$
7. $m^2 - 8m + 16$
8. $k^2 - 14k + 49$
9. $9x^2 + 6x + 1$
10. $16t^2 + 24t + 9$
11. $25y^2 + 20y + 4$

Answers to Chapter 12

12 $4m^2 + 12m + 9$
13 $16t^2 - 24t + 9$
14 $9x^2 - 12x + 4$
15 $25t^2 - 20t + 4$
16 $25r^2 - 60r + 36$
17 $x^2 + 2xy + y^2$
18 $m^2 - 2mn + n^2$
19 $4t^2 + 4ty + y^2$
20 $m^2 - 6mn + 9n^2$
21 $x^2 + 4x$
22 $x^2 - 10x$
23 $x^2 + 12x$
24 $x^2 - 4x$
25 **a** Marcela has just squared the first term and the second term. She hasn't written down the brackets twice.
    b Paulo has written down the brackets twice but has worked out $(3x)^2$ as $3x^2$ and not $9x^2$.
    c $9x^2 + 6x + 1$
26 Whole square is $(2x)^2 = 4x^2$.
    Three areas are $2x - 1$, $2x - 1$ and 1.
    $4x^2 - (2x - 1 + 2x - 1 + 1) = 4x^2 - (4x - 1) = 4x^2 - 4x + 1$

12.6 Expanding three brackets

Exercise 12J

1 **a** $x^2 + 2x - 3$ **b** $x^3 + 2x^2 - 3x$ **c** $x^3 + 4x^2 + x - 6$
2 **a** $x^2 - 7x + 10$ **b** $x^3 - 6x^2 + 3x - 10$
   c $2x^3 - 13x^2 + 13x + 10$
3 **a** $x^3 - 3x^2 - 13x + 15$ **b** $3x^3 + 31x^2 + 78x + 56$
   c $x^3 - 14x^2 + 53x - 40$
4 **a** $x^2 + 4x + 4$ **b** $x^3 + 6x^2 + 12x + 8$
   c $8x^3 + 12x^2 + 6x + 1$
5 **a** $x^3 + x^2 - 4x - 4$ **b** $2x^3 - 3x^2 - 11x + 6$
   c $x^3 + 4x^2 - 4x - 16$
6 **a** $x^3 + 6x^2 + 11x + 6$ **b** $x^3 - 6x^2 + 11x - 6$
7 **a** $x^3 + 4x^2 - 3x - 18$ **b** $x^3 - 6x^2 - 15x + 100$
   c $9x^3 + 78x^2 - 116x + 40$
8 **a** $(x + 1)^3 - (x - 1)^3 = x^3 + 3x^2 + 3x + 1 - (x^3 - 3x^2 + 3x - 1)$
   $= x^3 + 3x^2 + 3x + 1 - x^3 + 3x^2 - 3x + 1$
   $= 6x^2 + 2 = 2(3x^2 + 1)$
   b $4(3x^2 + 4)$
9 The volume of the cube is $(x + 1)^3$
   One of the eight pieces is a cube of side x and volume x^3
   Three of the eight pieces are cuboids, with sides x, x and 1 and each has volume x^2
   Three of the eight pieces are cuboids with sides x, 1 and 1 and each has volume x
   One of the eight pieces is a cube of side 1 and volume 1
   Add these eight volumes to get $x^3 + 3x^2 + 3x + 1$ which is $(x + 1)^3$
10 **a** $a = 6$ **b** $b = 5$ **c** $c = -8$
11 **a** $x^3 - 1$ **b** $x^3 - 8$
   c $x^3 - 27 = (x^2 + 3x + 9)(x - 3)$
12 $6x^3 + 11x^2 + 6x + 1$ cm³

12.7 Quadratic factorisation

Exercise 12K

1 $(x + 2)(x + 3)$
2 $(t + 1)(t + 4)$
3 $(m + 2)(m + 5)$
4 $(k + 4)(k + 6)$
5 $(p + 2)(p + 12)$
6 $(r + 3)(r + 6)$
7 $(w + 2)(w + 9)$
8 $(x + 3)(x + 4)$
9 $(a + 2)(a + 6)$
10 $(k + 3)(k + 7)$
11 $(f + 1)(f + 21)$
12 $(b + 8)(b + 12)$
13 $(t - 2)(t - 3)$
14 $(d - 4)(d - 1)$
15 $(g - 2)(g - 5)$
16 $(x - 3)(x - 12)$
17 $(c - 2)(c - 16)$
18 $(t - 4)(t - 9)$
19 $(y - 4)(y - 12)$
20 $(j - 6)(j - 8)$
21 $(p - 3)(p - 5)$
22 $(y + 6)(y - 1)$
23 $(t + 4)(t - 2)$
24 $(x + 5)(x - 2)$
25 $(m + 2)(m - 6)$
26 $(r + 1)(r - 7)$
27 $(n + 3)(n - 6)$
28 $(m + 4)(m - 11)$
29 $(w + 4)(w - 6)$
30 $(t + 9)(t - 10)$
31 $(h + 8)(h - 9)$
32 $(t + 7)(t - 9)$
33 $(d + 1)^2$
34 $(y + 10)^2$
35 $(t - 4)^2$
36 $(m - 9)^2$
37 $(x - 12)^2$
38 $(d + 3)(d - 4)$
39 $(t + 4)(t - 5)$
40 $(q + 7)(q - 8)$
41 $(x + 2)(x + 3)$, giving areas of $2x$ and $3x$, or $(x + 1)(x + 6)$, giving areas of x and $6x$.

Exercise 12L

1 $(x + 3)(x - 3)$
2 $(t + 5)(t - 5)$
3 $(m + 4)(m - 4)$
4 $(3 + x)(3 - x)$

Answers to Chapter 13

5 $(7 + t)(7 - t)$
6 $(k + 10)(k - 10)$
7 $(2 + y)(2 - y)$
8 $(x + 8)(x - 8)$
9 $(t + 9)(t - 9)$
10 a x^2
  b i $(x - 2)$ **ii** $(x + 2)$ **iii** $x(x - 2) = x^2 - 2x$ **iv** 4
  c $A + B - C = x^2 - 4$, which is the area of D, which is $(x + 2)(x - 2)$.
11 a $x^2 + 4x + 4 - (x^2 + 2x + 1) = 2x + 3$
   b $(a + b)(a - b)$
   c $(x + 2 + x + 1)(x + 2 - x - 1) = (2x + 3)(1) = 2x + 3$
   d The answers are the same.
   e $(x + 1 + x - 1)(x + 1 - x + 1) = (2x)(2) = 4x$
12 $(x + y)(x - y)$
13 $(x + 2y)(x - 2y)$
14 $(x + 3y)(x - 3y)$
15 $(3x + 1)(3x - 1)$
16 $(4x + 3)(4x - 3)$
17 $(5x + 8)(5x - 8)$
18 $(2x + 3y)(2x - 3y)$
19 $(3t + 2w)(3t - 2w)$
20 $(4y + 5x)(4y - 5x)$

Exercise 12M
1 $(2x + 1)(x + 2)$
2 $(7x + 1)(x + 1)$
3 $(4x + 7)(x - 1)$
4 $(3t + 2)(8t + 1)$
5 $(3t + 1)(5t - 1)$
6 $(4x - 1)^2$
7 $3(y + 7)(2y - 3)$
8 $4(y + 6)(y - 4)$
9 $(2x + 3)(4x - 1)$
10 $(2t + 1)(3t + 5)$
11 $(x - 6)(3x + 2)$
12 $(x - 5)(7x - 2)$
13 $4x + 1$ and $3x + 2$

14 a All the terms in the quadratic have a common factor of 6.
   b $6(x + 2)(x + 3)$. This has the highest common factor taken out.
   c For example, 'A rectangle could be split in many different ways.'

12.8 Algebraic fractions

Exercise 12N
1 a $\frac{5x}{6}$ **b** $\frac{19x}{20}$ **c** $\frac{23x}{20}$ **d** $\frac{3x + 2y}{6}$
   e $\frac{x^2y + 8}{4x}$ **f** $\frac{5x + 7}{6}$ **g** $\frac{7x + 3}{4}$ **h** $\frac{13x + 5}{15}$
   i $\frac{3x - 7}{4}$ **j** $\frac{5x - 10}{4}$

2 a $\frac{x}{6}$ **b** $\frac{11x}{20}$ **c** $\frac{7x}{20}$ **d** $\frac{3x - 2y}{6}$
   e $\frac{xy^2 - 8}{4y}$ **f** $\frac{x - 1}{6}$ **g** $\frac{x + 1}{4}$ **h** $\frac{-7x - 5}{15}$
   i $\frac{x - 1}{4}$ **j** $\frac{2 - 3x}{4}$

3 a $\frac{x^2}{6}$ **b** $\frac{3xy}{14}$ **c** $\frac{8}{3}$ **d** $\frac{2xy}{3}$
   e $\frac{x^2 - 2x}{10}$ **f** $\frac{1}{6}$ **g** $\frac{6x^2 + 5x + 1}{8}$ **h** $\frac{2x^2 + x}{15}$
   i $\frac{2x - 4}{x - 3}$ **j** $\frac{1}{2x}$

4 a x **b** $\frac{x}{2}$ **c** $\frac{3x^2}{16}$ **d** 3
   e $\frac{17x + 1}{10}$ **f** $\frac{13x + 9}{10}$ **g** $\frac{3x^2 - 5x - 2}{10}$ **h** $\frac{x + 3}{2}$
   i $\frac{2x^2 - 6y^2}{9}$

5 a $\frac{7x + 9}{(x + 1)(x + 2)}$ **b** $\frac{11x - 10}{(x - 2)(x + 1)}$ **c** $\frac{2 - 13x}{(4x + 1)(x + 2)}$
   d $\frac{8 - 10x}{(2x - 1)(x + 1)}$ **e** $\frac{x + 1}{(2x - 1)(3x - 1)}$

6 First, he did not factorise and just cancelled the x^2s. Then he cancelled 2 and 6 with the wrong signs. Then he said two minuses make a plus when adding, which is not true.

7 $\frac{2x^2 + x - 3}{4x^2 - 9}$

8 a $\frac{9x + 13}{(x + 1)(x + 2)}$ **b** $\frac{14x + 19}{(4x - 1)(x + 1)}$ **c** $\frac{2x^2 + x - 13}{2(x + 1)}$ **d** $\frac{x + 1}{(2x - 1)(3x - 1)}$

9 a $\frac{x - 1}{2x + 1}$ **b** $\frac{2x + 1}{x + 3}$ **c** $\frac{2x - 1}{3x - 2}$ **d** $\frac{x + 1}{x - 1}$ **e** $\frac{2x + 5}{4x - 1}$

Answers to Chapter 13

13.1 Solving linear equations

Exercise 13A
1 a 56   **b** 2   **c** 6   **d** 3   **e** 4
   f $2\frac{1}{2}$ **g** $3\frac{1}{2}$ **h** $2\frac{1}{2}$ **i** 4   **j** 21
   k 72   **l** 56   **m** 0   **n** −7   **o** −18
   p 36   **q** 36   **r** 60   **s** 7   **t** 11
   u 2   **v** 7   **w** 2.8   **x** 1   **y** 11.5
   z 0.2

2 a −4   **b** 15

3 a Elif
   b Second line: Mustafa subtracts 1 instead of adding 1; fourth line: Mustafa subtracts 2 instead of dividing by 2.

Exercise 13B
1 a 3   **b** 7   **c** 5   **d** 3   **e** 4   **f** 6
   g 8   **h** 1   **i** $1\frac{1}{2}$ **j** $2\frac{1}{2}$ **k** $\frac{1}{2}$ **l** $1\frac{1}{5}$
   m 2   **n** −2   **o** −1   **p** −2   **q** −2   **r** −1

2 Any values that work, e.g. $a = 2$, $b = 3$ and $c = 30$.

3 55

Answers to Chapter 13

Exercise 13C

1. **a** $x = 2$ **b** $y = 1$ **c** $a = 7$ **d** $t = 4$
   e $p = 2$ **f** $k = -1$ **g** $m = 3$ **h** $s = -2$
2. $3x - 2 = 2x + 5$, $x = 7$
3. **a** $d = 6$ **b** $x = 11$ **c** $y = 1$ **d** $h = 4$
   e $b = 9$ **f** $c = 6$
4. $6x + 3 = 6x + 10$; $6x - 6x = 10 - 3$; $0 = 7$, which is obviously false. Both sides have $6x$, which cancels out.
5. Check student's example.

13.2 Setting up equations

Exercise 13D

1. 90 cents or 0.90 dollars
2. **a** 1.5 **b** 2
3. **a** 1.5 cm **b** 6.75 cm^2
4. 17
5. 8
6. **a** $8c - 10 = 56$ **b** $8.25
7. **a** B: 450 cars, C: 450 cars, D: 300 cars
   b 800 **c** 750
8. 360 dollars
9. 3 years
10. 9 years
11. 3 cm
12. 5
13. **a** $4x + 40 = 180$ **b** $x = 35°$
14. **a** $\frac{x + 10}{5} = 9.50$ **b** $37.50
15. No, as $x + x + 2 + x + 4 + x + 6 = 360$ gives $x = 87°$ so the consecutive numbers (87, 89, 91, 93) are not even but odd
16. $4x + 18 = 3x + 1 + 50$, $x = 33$. Large bottle 1.5 litres, small bottle 1 litre

13.3 Solving quadratic equations by factorisation

Exercise 13E

1. $-2, -5$
2. $-3, -1$
3. $-6, -4$
4. $-3, 2$
5. $-1, 3$
6. $-4, 5$
7. $1, -2$
8. $2, -5$
9. $7, -4$
10. $3, 2$
11. $1, 5$
12. $4, 3$
13. $-4, -1$
14. $-9, -2$
15. $2, 4$
16. $3, 5$
17. $-2, 5$
18. $-3, 5$
19. $-6, 2$
20. $-6, 3$
21. $-1, 2$
22. -2
23. -5
24. 4
25. $-2, -6$
26. 7
27. **a** $x(x - 3) = 550$, $x^2 - 3x - 550 = 0$
    b $(x - 25)(x + 22) = 0$, $x = 25$ years
28. $x(x + 40) = 48000$, $x^2 + 40x - 48000 = 0$, $(x + 240)(x - 200) = 0$. Fence is $2 \times 200 + 2 \times 240 = 880$ m.
29. $-6, -4$
30. $2, 16$
31. $-6, 4$
32. $-9, 6$
33. $-10, 3$
34. $-4, 11$
35. $-8, 9$
36. $8, 9$
37. 1
38. Mario was correct. Sylvan did not make it into a standard quadratic and only factorised the x terms. She also incorrectly solved the equation $x - 3 = 4$.

Exercise 13F

1. **a** $\frac{1}{3}, -3$ **b** $1\frac{1}{3}, -\frac{1}{2}$ **c** $-\frac{1}{5}, 2$
   d $-2\frac{1}{2}, 3\frac{1}{2}$ **e** $-\frac{1}{6}, -\frac{1}{3}$ **f** $\frac{2}{3}, 4$
   g $\frac{1}{2}, -3$ **h** $\frac{5}{2}, -\frac{7}{6}$ **i** $-1\frac{2}{3}, 1\frac{2}{5}$
   j $1\frac{3}{4}, 1\frac{2}{7}$ **k** $\frac{2}{3}, \frac{1}{8}$ **l** $\pm\frac{1}{4}$
   m $-2\frac{1}{4}, 0$ **n** $\pm 1\frac{2}{5}$ **o** $-\frac{1}{3}, 3$
2. **a** $-6, 7$ **b** $-\frac{5}{2}, \frac{3}{2}$ **c** $-6, 7$
   d $-1, \frac{11}{13}$ **e** $-2, 3$ **f** $-\frac{2}{5}, \frac{1}{2}$
   g $\frac{1}{2}, -\frac{1}{3}$ **h** $-2, \frac{1}{5}$ **i** 4
   j $-2, \frac{1}{8}$ **k** $-\frac{1}{3}, 0$ **l** $-5, 5$
   m $-\frac{5}{3}$ **n** $-\frac{7}{2}, \frac{7}{2}$ **o** $-\frac{5}{2}, 3$
3. **a** Both have only one solution: $x = 1$.
   b B is a linear equation, but A and C are quadratic equations.
4. **a** $(5x - 1)^2 = (2x + 3)^2 + (x + 1)^2$, when expanded and collected into the general quadratics, gives the required equation.
   b $(10x + 3)(2x - 3)$, $x = 1.5$; area $= 7.5$ cm^2.

13.4 Solving quadratic equations by the quadratic formula

Exercise 13G

1. $1.77, -2.27$
2. $-0.23, -1.43$

Answers to Chapter 13

3 3.70, −2.70
4 0.29, −0.69
5 −0.19, −1.53
6 −1.23, −2.43
7 −0.41, −1.84
8 −1.39, −2.27
9 1.37, −4.37
10 2.18, 0.15
11 −0.39, −5.11
12 0.44, −1.69
13 1.64, 0.61
14 0.36, −0.79
15 1.89, 0.11
16 13
17 $x^2 - 3x - 7 = 0$
18 Hasan gets $x = \frac{4 \pm \sqrt{0}}{8}$ and Miriam gets $(2x - 1)^2 = 0$; each method only gives one solution, $x = \frac{1}{2}$

13.5 Solving quadratic equations by completing the square

Exercise 13H

1 a $(x + 2)^2 - 4$ b $(x + 7)^2 - 49$
   c $(x - 3)^2 - 9$ d $(x + 3)^2 - 9$
   e $(x - 1.5)^2 - 2.25$ f $(x - 4.5)^2 - 20.25$
   g $(x + 6.5)^2 - 42.25$ h $(x + 5)^2 - 25$
   i $(x + 4)^2 - 16$ j $(x - 1)^2 - 1$
   k $(x + 1)^2 - 1$

2 a $(x + 2)^2 - 5$ b $(x + 7)^2 - 54$
   c $(x - 3)^2 - 6$ d $(x + 3)^2 - 2$
   e $(x - 1.5)^2 - 3.25$ f $(x + 3)^2 - 6$
   g $(x - 4.5)^2 - 10.25$ h $(x + 6.5)^2 - 7.25$
   i $(x + 4)^2 - 22$ j $(x + 1)^2 - 2$
   k $(x - 1)^2 - 8$ l $(x + 1)^2 - 10$

3 a $-2 \pm \sqrt{5}$ b $-7 \pm 3\sqrt{6}$ c $3 \pm \sqrt{6}$
   d $-3 \pm \sqrt{2}$ e $1.5 \pm \sqrt{3.25}$ f $-3 \pm \sqrt{6}$
   g $4.5 \pm \sqrt{10.25}$ h $-6.5 \pm \sqrt{7.25}$ i $-4 \pm \sqrt{22}$
   j $-1 \pm \sqrt{2}$ k $1 \pm 2\sqrt{2}$ l $-1 \pm \sqrt{10}$

4 a 1.45, −3.45   b 5.32, −1.32   c −4.16, 2.16

5 a $x = 1.5 \pm \sqrt{3.75}$ b $x = 1 \pm \sqrt{0.75}$
   c $x = -1.25 \pm \sqrt{6.5625}$ d $x = 7.5 \pm \sqrt{40.25}$

6 $p = -14$, $q = -3$

7 a 3rd, 1st, 4th and 2nd – in that order

13.6 Simultaneous equations

Exercise 13I

1 a $x = 5$, $y = 10$ b $x = 18$, $y = 6$ c $x = 12$, $y = 48$
2 a $x = 6$, $y = 18$ b $x = 12.5$, $y = 2.5$ c $x = 0.5$, $y = 4.5$
3 a $x = 13$, $y = 7$ b $x = 9$, $y = 14$ c $x = 10$, $y = -4$
4 a $x = 0.5$, $y = 4$ b $x = 5.5$, $y = 14.5$ c $x = 2$, $y = 8$
5 Carmen 32, Anish 8

6 11.5 and 25.5
7 8 and −3
8 a $x + y = 75$ b $y = 2x$ c $x = 25$, $y = 50$
9 a $x + y = 300$ b $x = y + 60$ or $y = x - 60$
   c $x = 180$ and $y = 120$
10 a $x = y - 26$ or $y = x + 26$ or $y - x = 26$ b $x + y = 50$
   c Ahmed is 12 and his mother is 38.
11 a $x = y - 0.4$ or $y = x + 0.4$ b $x + y = 8.6$ c 4.5 m

Exercise 13J

1 a $x = 4$, $y = 1$ b $x = 1$, $y = 4$
   c $x = 3$, $y = 1$ d $x = 5$, $y = -2$
   e $x = 7$, $y = 1$ f $x = 5$, $y = \frac{1}{2}$
   g $x = 4\frac{1}{2}$, $y = 1\frac{1}{2}$ h $x = -2$, $y = 4$
   i $x = 2\frac{1}{2}$, $y = -1\frac{1}{2}$ j $x = 2\frac{1}{4}$, $y = 6\frac{1}{2}$
   k $x = 4$, $y = 3$ l $x = 5$, $y = 3$

2 a 3 is the first term. The next term is $3 \times a + b$, which equals 14.
   b $14a + b = 47$
   c $a = 3$, $b = 5$
   d 146, 443

Exercise 13K

1 a $x = 2$, $y = -3$ b $x = 7$, $y = 3$
   c $x = 4$, $y = 1$ d $x = 2$, $y = 5$
   e $x = 4$, $y = -3$ f $x = 1$, $y = 7$
   g $x = 2\frac{1}{2}$, $y = 1\frac{1}{2}$ h $x = -1$, $y = 2\frac{1}{2}$
   i $x = 6$, $y = 3$ j $x = \frac{1}{2}$, $y = -\frac{3}{4}$
   k $x = -1$, $y = 5$ l $x = 1\frac{1}{2}$, $y = \frac{3}{4}$

2 a They are the same equation. Divide the first by 2 and it is the second, so they have an infinite number of solutions.
   b Double the second equation to get $6x + 2y = 14$ and subtract to get $9 = 14$. The left-hand sides are the same if the second is doubled so they cannot have different values.

Exercise 13L

1 a $x = 5$, $y = 1$ b $x = 3$, $y = 8$
   c $x = 9$, $y = 1$ d $x = 7$, $y = 3$
   e $x = 4$, $y = 2$ f $x = 6$, $y = 5$
   g $x = 3$, $y = -2$ h $x = 2$, $y = \frac{1}{2}$
   i $x = -2$, $y = -3$ j $x = -1$, $y = 2\frac{1}{2}$
   k $x = 2\frac{1}{2}$, $y = -\frac{1}{2}$ l $x = -1\frac{1}{2}$, $y = 4\frac{1}{2}$
   m $x = -\frac{1}{2}$, $y = -6\frac{1}{2}$ n $x = 3\frac{1}{2}$, $y = 1\frac{1}{2}$
   o $x = -2\frac{1}{2}$, $y = -3\frac{1}{2}$

2 (1, −2) is the solution to equations A and C; (−1, 3) is the solution to equations A and D; (2, 1) is the solution to B and C; (3, −3) is the solution to B and D.

3 Intersection points are (0, 6), (1, 3) and (2, 4). Area is 2 cm²

4 Intersection points are (0, 3), (6, 0) and (4, −1). Area is 6 cm²

Answers to Chapter 14

13.7 Linear and non-linear simultaneous equations

Exercise 13M

1. **a** (5, −1)   **b** (4, 1)   **c** (8, −1)
2. **a** (1, 2) and (−2, −1)   **b** (−4, 1) and (−2, 2)
3. **a** (3, 4) and (4, 3)   **b** (0, 3) and (−3, 0)   **c** (3, 2) and (−2, 3)
4. **a** (2, 5) and (−2, −3)   **b** (−1, −2) and (4, 3)   **c** (3, 3) and (1, −1)
5. **a** (−3, −3), (1, 1)   **b** (3, −2), (−2, 3)   **c** (−2, −1), (1, 2)
   d (2, −1), (3, 1)   **e** (−2, 1), (3, 6)   **f** (1, −4), (4, 2)
   g (4, 5), (−5, −4)
6. **a** $x + y = 12$; $x^2 + y^2 = 90$ **b** Either 391290 or 931290
7. 12 years old
8. **a** $x^2 + y^2 = 85$ and $(x + y)^2 = 121$ **b** 2 and 9

13.8 Solving inequalities

Exercise 13N

1. **a** $x < 3$
   b $t > −2$
   c $p \geq −10$
   d $x < 5$
   e $y \leq 3$
   f $t > 3$
   g $x < 6$
   h $y \leq −15$
   i $t \geq 18$
   j $x < 7$
   k $x \leq 3$
   l $t \geq 5.25$
2. **a** 8   **b** 6   **c** 16
   d 3   **e** 7
3. **a** 11   **b** 16   **c** 16
4. $2x + 3 < 20$, $x < 8.50$, so the most each could cost is $8.49
5. **a** Because $3 + 4 = 7$, which is less than the third side of length 8
   b $x + x + 2 > 10$, $2x + 2 > 10$, $2x > 8$, $x > 4$, so smallest value of x is 5
6. **a** $x = 6$ and $x < 3$ scores −1 (nothing in common), $x < 3$ and $x > 0$ scores 1 (1 in common for example), $x > 0$ and $x = 2$ scores 1 (2 in common), $x = 2$ and $x \geq 4$ scores −1 (nothing in common), so we get −1 + 1 + 1 − 1 = 0
   b $x > 0$ and $x = 6$ scores +1 (6 in common), $x = 6$ and $x \geq 4$ scores +1 (6 in common), $x \geq 4$ and $x = 2$ scores −1 (nothing in common), $x = 2$ and $x \leq 3$ scores +1 (2 in common), +1 + 1 − 1 + 1 = 2
   c Any acceptable combination, e.g. $x = 2$, $x \leq 3$, $x > 0$, $x \geq 4$, $x = 6$
7. **a** $x \geq −6$ **b** $t \leq \frac{8}{3}$
   c $y \leq 4$ **d** $x \geq −2$
   e $w \leq 5.5$ **f** $x \leq \frac{14}{5}$
8. **a** $x \leq 2$ **b** $x > 38$
   c $x < 6\frac{1}{2}$ **d** $x \geq 7$
   e $t > 15$ **f** $y \leq \frac{7}{5}$
9. **a** 4   **b** 99   **c** 11   **d** 11   **e** 6
10. **a** 0, 10 − 10   **b** $x < 16$
11. **a** $x < 9$ **b** $x \geq 11$ **c** $x \geq 3$
12. **a** $x \geq 7.5$ **b** $x \leq −2$ **c** $x < 6$
    d $x > 1.5$ **e** $x \geq −5$ **f** $x < 0.5$

Answers to Chapter 14

14.1 Conversion graphs

Exercise 14A

1. **a i** $8\frac{1}{4}$ kg   **ii** $2\frac{1}{4}$ kg   **iii** 9 lb   **iv** 22 lb
   b 2.2 lb
   c Read off the value for 12 lb (5.4 kg) and multiply this by 4 (21.6 kg)
2. **a i** 10 cm   **ii** 23 cm   **iii** 2 in   **iv** $8\frac{3}{4}$ in
   b $2\frac{1}{2}$ cm
   c Read off the value for 9 in (23 cm) and multiply this by 2 (46 cm)
3. **a i** $320   **ii** $100   **iii** £45   **iv** £78
   b $3.20
   c It would become more steep.
4. **a i** $120   **ii** $82
   b i 32   **ii** 48
5. **a i** $100   **ii** $325
   b i 500   **ii** 250
6. **a i** $70   **ii** $29
   b i $85   **ii** $38
7. **a i** 95 °F   **ii** 68 °F   **iii** 10 °C   **iv** 32 °C
   b 32°F
8. **a** Check student's graph   **b** $50
9. **a** Student's own graph   **b** about 48 kilometres
   c about 16 miles
10. **a** Student's own graph   **b** about 9 centimetres
    c about 4 hours
11. **a** Student's own graph   **b** about 23 minutes

14.2 Travel graphs

Exercise 14B

1. **a i** 2 h   **ii** 3 h   **iii** 5 h
   b i 40 km/h   **ii** 120 km/h   **iii** 40 km/h
   c 5.30 am
2. **a i** 125 km   **ii** about 25 km/h
   b i Between 2 pm and 3 pm   **ii** About 12 km/h

3 a 30 km **b** 40 km **c** 100 km/h
4 a i 263 m/min (3 sf)
  ii 15.8 km/h (3 sf)
  b 500 m/min
  c Yuto by 1 minute
5 a Patrick ran quickly at first, then had a slow middle section but he won the race with a final sprint. Araf ran steadily all the way and came second. Sean set off the slowest, speeded up towards the end but still came third.
  b i 1.67 m/s **ii** 6 km/h
6 a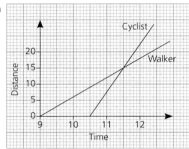
  b At 1130
7 a i Because it stopped several times
  ii Ravinder
  b Ravinder at 1558, Sue at 1620, Michael at 1635
  c i 24 km/h
  ii 20.6 km/h
  iii 5
8 a 50 metres **b** student's graph **c** 1 metre/second
9 a student's graph **b** 80 km/hour
10 a 1300 **b** 15 km **c** student's graph
  d For the three stages, 5 km/hour, 4 km/hour and 2 km/hour. For the whole trip 3.75 km/hour

14.3 Speed–time graphs

Exercise 14C
1 a 20 m/s **b** 0.5 m/s^2
  c 1 m/s^2 **d** 600 metres
  e 10 m/s
2 a 0.6 m/s^2 **b** 750 m
3 a 0.2 m/s^2 **b** 0.1 m/s^2
  c 75 metres **d** 2.5 m/s
4 a 1 m/s^2 **b** $\frac{2}{3}$ m/s^2
  c 6 kilometres (or 6000 metres)
  d 30 m/s
5 a student's graph **b** 0.8 m/s^2 **c** 80 metres
6 a 26 m/s **b** student's graph **c** 144 metres
7 a $1\frac{1}{3}$ m/s^2
  b They are together. They have both travelled 450 metres
8 a 2 m/s^2
  b i after 20 seconds
  ii 100 metres
  c 1150 metres
9 a 15 seconds **b** $1\frac{1}{3}$ m/s^2
10 a 6 m/s **b** student's own graph
  c 15 metres

14.4 Curved graphs

Exercise 14D
1 a and b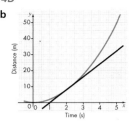
  c 8 m/s **d** 16 m/s
2 a
  b 10 m/s **c** 30 m/s **d** 0 m/s
  e 20 m/s downwards
3 a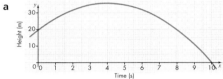
  b 4 m/s **c** 6 m/s downwards **d** after 4 s
  e 12 m/s
4 a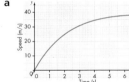
  b about 12 m/s^2 **c** about 7.4 m/s^2 **d** about 2.7 m/s^2
5 a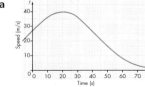
  b about 0.73 m/s^2
  c after 20 seconds
  d about 0.65
6 a
  b about 0.72 m/s^2 **c** about 0.36 m/s^2
  d about 0.72 m/s^2 **e** after about 23 s and 57 s

Answers to Chapter 15

15.1 Drawing straight-line graphs

Exercise 15A

1. Extreme points are (0, 4), (5, 19)
2. Extreme points are (0, −5), (5, 5)
3. Extreme points are (0, −3), (10, 2)
4. Extreme points are (−3, −4), (3, 14)
5. Extreme points are (−6, 2), (6, 6)
6. **a** Extreme points are (0, −2), (5, 13) and (0, 1), (5, 11)
   b (3, 7)
7. **a** Extreme points are (0, −5), (5, 15) and (0, 3), (5, 13)
   b (4, 11)
8. **a** Extreme points are (0, −1), (12, 3) and (0, −2), (12, 4)
   b (6, 1)
9. **a** Extreme points are (0, 1), (4, 13) and (0, −2), (4, 10)
   b Do not cross because they are parallel
10. **a** Values of y: 5, 4, 3, 2, 1, 0. Extreme points are (0, 5), (5, 0)
    b Extreme points are (0, 7), (7, 0)
11. **a** yes **b** no **c** yes **d** no **e** yes **f** no
12. **a** 6 **b** 3.5 **c** 2
13. **a** 20 **b** −10

15.2 The equation $y = mx + c$

Exercise 15B

1. **a** 3 **b** 2 **c** $\frac{1}{2}$
   d 3 **e** $\frac{1}{3}$
2. **a** $y = 2x − 2$ **b** $y = x + 1$ **c** $y = 2x − 3$
   d $2y = x + 6$ **e** $y = x$ **f** $y = 2x$
3. **a** $y = 2x + 1$, $y = −2x + 1$ **b** $5y = 2x − 5$, $5y = −2x − 5$
   c $y = x + 1$, $y = −x + 1$
4. **a** $y = −2x + 1$ **b** $2y = −x$ **c** $y = −x + 1$
   d $5y = −2x − 5$ **e** $y = −\frac{3}{2}x − 3$ or $2y = −3x − 6$
5. **a** 3 **b** (0,3)
6. **a** 4 **b** 4
7. The first and last are parallel because they both have a gradient of 4. The middle one has a gradient of 3.
8. $y = 0.6x$
9. **a**, **b** and **d**

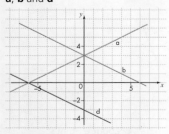

c $y = −0.5x + 3$ **e** $y = −0.5x − 3$

15.3 More about straight-line graphs

Exercise 15C

1. **a** $y = −2.5x + 5$ **b** −2.5 and (0, 5)
   c
   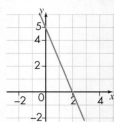

2. **a** $y = −\frac{1}{3}x + 5$ **b** $−\frac{1}{3}$ and (0, 5)
   c
   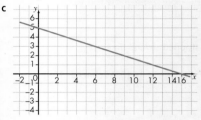

3. **a** $y = −0.5x − 3$ **b** −0.5 and (0, −3)
   c
   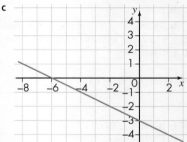

4. **a** $y = −x − 20$ **b** $y = 3x + 15$ **c** $y = −0.7x + 3$
   d $y = 0.2x − 8$ **e** $y = 1.5x − 1$ **f** $y = −0.5x + 6$
5. **a** −1 and (0, −20) **b** 3 and (0, 15) **c** −0.7 and (0, 3)
   d 0.2 and (0, −8) **e** 1.5 and (0, −1) **f** −0.5 and (0, 6)
6. **a** −1 **b** $\frac{1}{3}$ **c** 6
   d −0.5 **e** 2 **f** 5.6
7. **a** C **b** A **c** D **d** B
8. line d, $3x − 2y = 12$, all the rest are the same line.
9. **a** (9, 0) and (0, 15)
   b (20, 0) and (0, 10)
   c (0, −10) and (−5, 0)

15.4 Solving equations graphically

Exercise 15D

1. **a** 1.8 **b** −0.4 **c** 2.7
2. **a** −1.6 **b** 3.8 or 3.9 **c** −3.8

3	**a** 2.3	**b** 1.6	**c** 0.9			
4	**a** 23	**b** −29	**c** 71			
5	**a** 6.9	**b** 2.9	**c** 4.2 or 4.3			
6	**a** −0.8	**b** 2.6	**c** 0.3			

15.5 Parallel lines

Exercise 15E

1 **a** $2 \times 3 + 6 = 12$ **b** 2 **c** student's graph
  d $y = 2x$ **e** $y = 2x + 3$
2 **a** (0, −1) and (4, 0) **b** $y = \frac{1}{4}x + 3$
3 **a** −2 **b** (4,0) **c** student's graph
  d $y = −2x$ **e** $y = −2x + 14$
4 **a** 1 **b** $y = 5x − 9$
5 $y = \frac{2}{3}x − 3$
6 **a** $y = 2x − 4$ **b** $y = 2x + 8$
7 **a** $4 + 2 \times 1 = 6$ **b** $y = −2x + 6$
  c −2 **d** $y = −2x$

15.6 Points and lines

Exercise 15F

1 **a** 3 **b** $\frac{1}{2}$ **c** 4 **d** −1 **e** $-\frac{1}{2}$ **f** $\frac{2}{3}$

Answers to Chapter 16

2 **a** $y = 2x − 3$ **b** $y = \frac{1}{2}x + 4$ **c** $y = 4x − 2$ **d** $y = −3x + 8$
3 **a** (5, 3) **b** (4, 5) **c** (3, 2)
  d (3, 3) **e** (1, 3.5) **f** (−0.5, 0)
4 **a** student's graph **b** $y = 0.5x + 6.5$ **c** (−1,3)
  d $y = −x + 8$
5 **a** 5 **b** 13 **c** 10 **d** 17
6 Show that the distance from each point to (2,1) is 5.
7 $AB = \sqrt{32}$, $AC = \sqrt{80}$, $BC = \sqrt{80}$, so two of the sides are the same length.

15.7 Perpendicular lines

Exercise 15G

1 **a** −2 **b** $-\frac{1}{4}$ **c** $\frac{3}{2}$
2 The gradients are 5 and $-\frac{1}{5}$; $5 \times -\frac{1}{5} = -1$
3 **a** $-\frac{1}{3}$ **b** −2 **c** 2.5 **d** $-\frac{2}{15}$
4 **a** 1 **b** 2 **c** $\frac{4}{3}$ **d** −6
5 **a** $y = -\frac{1}{5}x$ **b** $y = -\frac{1}{5}x + 10$ **c** $y = -\frac{1}{5}x + 2$
6 $y = \frac{3}{5}x + 2$
7 $y = -\frac{1}{4}x + \frac{1}{2}$
8 $y = −5x + 11$
9 $6x + 3y = 7$ is the odd one out
10 $y = 2x + 6$

Answers to Chapter 16

16.1 Quadratic graphs

Exercise 16A

1 **a** x: −3, −2, −1, 0, 1, 2, 3
    y: 11, 6, 3, 2, 3, 6, 11
  b student's graph
2 **a** x: −3, −2, −1, 0, 1, 2, 3, 4, 5W
    x^2: 9, 4, 1, 0, 1, 4, 9, 16, 25
    $−3x$: 9, 6, 3, 0, −3, −6, −9, −12, −15
    y: 18, 10, 4, 0, −2, −2, 0, 4, 10
  b 1.8 **c** (1.5, −2.25)
  d $x = 1.5$ **e** $x = 4.2$ or −1.2
3 **a** y: 7, 0, −5, −8, −9, −8, −5, 0, 7 **b** $x = 4$ or −2
  c The graph should give a value of about −8.75
  d The graph should give values of about 4.5 and −2.5
4 **a** y: 10, 4, 0, −2, −2, 0, 4, 10
  b (2.5, −2.25)
  c $x = 2.5$
  d The graph should give a value of about 6.75
  e $x = 1$ or 4
  f The graph should give values of about 0.2 and 4.8
5 **a** x: −4, −3, −2, −1, 0, 1, 2
    y: 7, 2, −1, −2, −1, 2, 7
  b 1.6, 0.2
  c 0.5
  d $x = −2.7$ or 0.7
6 **a** x: −4, −3, −2, −1, 0, 1, 2, 3, 4
    y: −4, 3, 8, 11, 12, 11, 8, 3, −4
  b 9.75
  c ±3.5
  d 2.2 and −2.2
7 **a**

x	−5	−4	−3	−2	−1	0	1	2
x^2	25	16	9	4	1	0	1	4
$+4x$	−20	−16	−12	−8	−4	0	4	8
y	5	0	−3	−4	−3	0	5	12

  b $x = −4$ and 0
  c −3.8
  d −4, 0
8 **a** x: −1, 0, 1, 2, 3, 4, 5, 6, 7
    y: 10, 3, −2, −5, −6, −5, −2, 3, 10
  b $x = 0.6$ or 5.5
  c −5.8
  d −0.3, 6.5
9 **a** y values: −6, 0, 4, 6, 6, 4, 0, −6
  b student's graph
  c (2.5, 6.25)
  d $x = 2.5$
  e $x = 4.6$ and 0.4

727

Answers to Chapter 16

16.2 Turning points on a quadratic graph

Exercise 16B

1. **a** $x^2 - 2x - 8 = (x - 1)^2 - 9$ **b** (1, −9) **c** $x = 4$ or -2
   d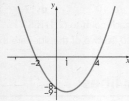

2. **a** $x^2 + 10x + 21 = (x + 5)^2 - 4$ **b** (−5, −4) **c** $x = -3$ or -7
   d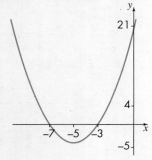

3. **a** $x^2 - 7x + 10 = (x - 3.5)^2 - 2.25$ **b** (3.5, −2.25)
   c $x = 2$ or 5 **d**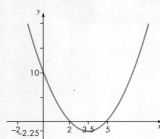

4. **a** (0, 12) **b** (3, 3)
   c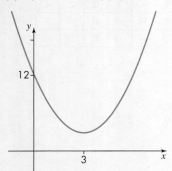
   d The graph does not cross the x-axis so there is no value of x for which $x^2 - 6x + 12 = 0$

5. **a** (−10, 60) **b**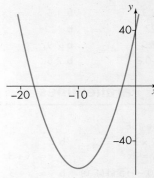

6. −56.5
7. $b = -10$ and $c = 14$

16.3 Reciprocal graphs

Exercise 16C

1. **a** y values: 10, 5, 4, 2.5, 2, 1.33, 1, 0.67, 0.5
   b 0.8 **c** 0.3 **d** −1.6
2. **a** y values: 25, 12.5, 10, 5, 2.5, 1, 0.5, 0.33, 0.25
   b student's graph **c** −0.5 and −9.5
3. student's own graph
4. **a** y values 20, 10, 5, 4, 2.5, 2, 1
   b student's graph **c** student's graph **d** $x = 6.5$ or -1.5

16.4 More graphs

Exercise 16D

1. student's own graph.
2. **a** y values: −7.81, −4, −1.69, −0.5, −0.06, 0, 0.06, 0.5, 1.69, 4, 7.81
   b 2.3
3. **a** y values: −12.63, −5, −0.38, 2, 2.89, 3, 3.13, 4, 6.38, 11, 18.63
   b −1.4
4. **a** y values: 1, 4.63, 6, 5.88, 5, 4.13, 4, 5.38, 9
   b $x = -1.8$ **c** $x = 1.8$ **d** (0.8, 3.9) and (−0.8, 6.1)
5. **a**

x	−3	−2	−1	0	1	2	3	4
y	−18	0	4	0	−6	−8	0	24

   b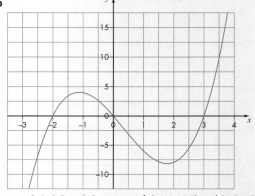
   c $x = -2.4$, 0.8 or 2.6 **d** (−1.1, 4.1) and (1.8, −8.2)

Answers to Chapter 16

6 a *y* values: 20, 5, 2.22, 1.25, 0.8 **b** student's graph
  c The asymptotes are the *x*-axis and the *y*-axis
     (or $y = 0$ and $x = 0$)

7 a *y* values: 4.25, 1.5, −0.22, −1.46, −2.44
  b

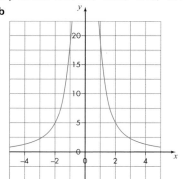

  c About 5.85

8 a

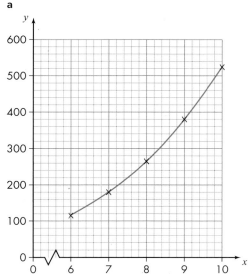

  b 8.3 cm

9 a $x^2 y = 1000 \Rightarrow y = \frac{1000}{x^2}$
  b

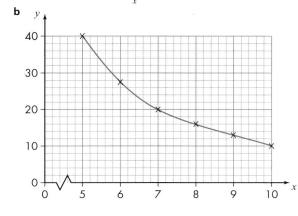

10 a

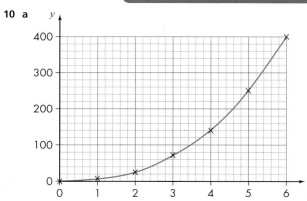

  b 4.6 cm, 4.6 cm and 9.6 cm

11 a

x	−5	−4	−3	−2	−1	0	1	2	3	4	5
y	−5.8	−5	−4.33	−4	−5	-	5	4	4.33	5	5.8

  b and c

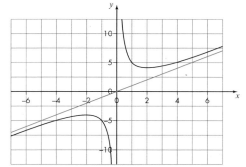

  d If *x* is large, the curve and the straight line are very close together.
  e The *y*-axis (or $x = 0$)

16.5 Exponential graphs

Exercise 16E

1 a The missing values are 48, 24, 12 and 6
  b

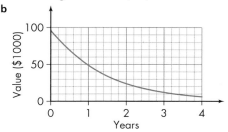

2 a 25 000 × 0.8 = 20 000
  b The missing values are 16 000, 12 800, 10 240, 8192

Answers to Chapter 16

c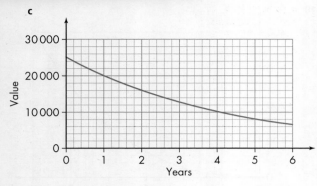

d 3.1 years

3 a The missing numbers, rounded to one decimal place, are 79.4, 91.3 and 104.9

b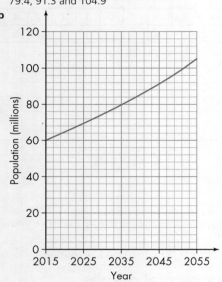

c 2051 or 2052

4 a the missing values, to the nearest whole number, are 650, 845, 1099, 1428

b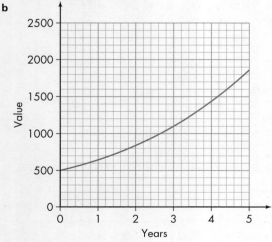

c about 2.6 or 2.7 years

5 a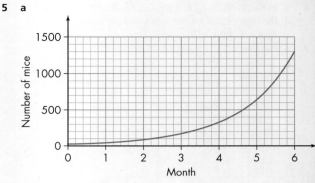

b about 3.3 months

6 a 20% b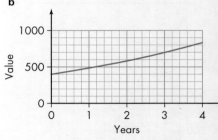

c About $630

7 a The missing values are 4500, 3645 and 3280.5

b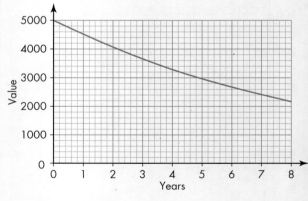

c About 6.6 years

8 It is not exponential growth because the gradient is decreasing. The gradient increases as time passes in exponential growth.

9 a 200 b 50% c about 2.7 hours

10 a The initial value is $1000. After two years it is $500. After four years it is $250. After six years it is $125. The value halves every two years.

b 29 or 30%

16.6 Estimating gradients

Exercise 16F

Gradients found in this exercise may vary from the answers given due to variations in drawings of the tangents

1 0.67

Answers to Chapter 17

2 A: 0.5 B: –2
3 **a** student's drawing
  b student's drawing
  c about 1.8
  d (1, 1.5)
4 **a** y values: 0, 0.01, 0.1, 0.34, 0.8, 1.56, 2.7
  b student's drawing
  c student's drawing
  d about 1.2

5 **a** y values: 0.25, 0.5, 1, 2, 4
  b student's drawing
  c student's drawing
  d about 0.7
6 **a** y values: 2.5, 1.67, 1.25, 1, 0.83
  b student's drawing
  c about –0.3

Answers to Chapter 17

17.1 Patterns in number sequences

Exercise 17A

1 **a** 9, 11, 13: add 2
  b 10, 12, 14: add 2
  c 80, 160, 320: double
  d 81, 243, 729: multiply by 3
  e 28, 34, 40: add 6
  f 23, 28, 33: add 5
  g 20 000, 200 000, 2 000 000: multiply by 10
  h 19, 22, 25: add 3
  i 114, 105, 96: subtract 9
  j 405, 1215, 3645: multiply by 3
  k 25, 12.5, 6.25; halve
  l 625, 3125, 15 625: multiply by 5
2 **a** 16, 22   **b** 26, 37
  c 31, 43   **d** 46, 64
  e 121, 169   **f** 782, 3907
  g 22 223, 222 223   **h** 11, 13
  i 33, 65   **j** 78, 108
3 **a** 48, 96, 192   **b** 33, 39, 45
  c 4, 2, 1   **d** 38, 35, 32
  e 37, 50, 65   **f** 26, 33, 41
  g 14, 16, 17   **h** 25, 22, 19
  i 28, 36, 45   **j** 5, 6, 7
  k 0.16, 0.032, 0.0064
  l 0.0625, 0.031 25, 0.015 625
4 **a** 21, 34: add previous 2 terms
  b 49, 64: next square number
  c 47, 76: add previous 2 terms
  d 216, 343: cube numbers
5 15, 21, 28, 36
6 61, 91, 127
7 29 and 41
8 No, they both increase by the same number (3).
9 10, 45 and 80

17.2 The nth term of a sequence

Exercise 17B

1 **a** 3, 5, 7, 9, 11   **b** 1, 4, 7, 10, 13
  c 7, 12, 17, 22, 27   **d** 1, 4, 9, 16, 25
  e 4, 7, 12, 19, 28   **f** 18, 16, 14, 12, 10

2 **a** 4, 5, 6, 7, 8   **b** 2, 5, 8, 11, 14
  c 3, 8, 13, 18, 23   **d** 0, 3, 8, 15, 24
  e 9, 13, 17, 21, 25   **f** 42, 39, 36, 33, 30
3 **a** 94, 88, 82, 76   **b** the 17th term, –2
4 **a** 1, 4, 9, 16   **b** 3, 6, 11, 18   **c** 2, 8, 18, 32
  d 3, 15, 35, 63   **e** 199, 196, 191, 184
  f 0.25, 1, 2.25, 4
5 **a** 1, 8, 27, 64   **b** 2, 9, 28, 65   **c** –1, 6, 25, 62
  d 2, 23, 80, 191   **e** 0.5, 4, 13.5, 32   **f** 108, 101, 82, 45
6 **a** $305   **b** $600   **c** 3
  d 5 (the amount is $250)
7 $4n - 2 = 3n + 7$ rearranges as $4n - 3n = 7 + 2$, $n = 9$

Exercise 17C

1 **a** 13, 15, $2n + 1$ **b** 25, 29, $4n + 1$
  c 33, 38, $5n + 3$ **d** 32, 38, $6n - 4$
  e 20, 23, $3n + 2$ **f** 37, 44, $7n - 5$
  g 17, 15; $29 - 2n$ **h** 22, 18; $46 - 4n$
  i 17, 20, $3n - 1$ **j** 42, 52, $10n - 8$
  k 24, 28, $4n + 4$ **l** 29, 34, $5n - 1$
2 **a** $3n + 1$, 151   **b** $2n + 5$, 105
  c $5n - 2$, 248   **d** $4n - 3$, 197
  e $8n - 6$, 394   **f** $n + 4$, 54
  g $5n + 1$, 251   **h** $8n - 5$, 395
  i $3n - 2$, 148   **j** $3n + 18$, 168
  k $7n + 5$, 355   **l** $8n - 7$, 393
3 **a i** $4n + 1$ **ii** 401
  b i $2n + 1$ **ii** 201
  c i $3n + 1$ **ii** 301
  d i $2n + 6$ **ii** 206
  e i $4n + 5$ **ii** 405
  f i $5n + 1$ **ii** 501
  g i $3n - 3$ **ii** 297
  h i $6n - 4$ **ii** 596
  i i $8n - 1$ **ii** 799
  j i $2n + 23$ **ii** 223
4 **a** $8n + 2$ **b** $8n + 1$ **c** $8n$ **d** $8
5 **a** n^2 **b** $n^2 + 2$ **c** $2n^2$ **d** $n^2 - 1$
6 **a** n^3 **b** $n^3 + 10$ **c** $0.5n^3$ **d** $10n^3$
7 **a** $n + 5$ **b** $n^2 + 5$ **c** $n^3 + 5$
  d $5n + 1$ **e** $5n^2$ **f** $5n^3$

Answers to Chapter 18

17.3 General rules from patterns

Exercise 17D

1. a

   b The missing number is 13
   c $4n - 3$
   d 97
   e 50th diagram

2. a △▽△
   b The bottom line is 3, 5, 7, 9, 11
   c $2n + 1$
   d 121
   e 49th set

3. a 18
   b the bottom line is 6, 10, 14, 18
   c $4n + 2$
   d 12

4. a i 24 ii $5n - 1$ iii 224
   b 25

5. a i 20 cm ii $(3n + 2)$ cm iii 152 cm
   b 332

6. a i 10 ii $2n + 2$ iii 162
   b 79.8 km

7. a i 14 ii $3n + 2$ iii 41
   b 66

8. a i 5 ii n iii 18
   b Formula gives 3 and 6
   c 55

9. a

1	2	3
3	9	27
1	4	13
4	13	40

   b the numbers in column 4 (top to bottom) are: 4, 81, 40, 121. Student's explanation of method.

10. a Student's drawing – one complete recurring sequence should be added to each one.
    b bottom row: 5, 9, 13, 17
    c $4n + 1$
    d $3n + 1$
    e $2n + 1$
    f $9n + 3$

17.4 Further sequences

Exercise 17E

1. a 432 b 1053 c 1250 d 41.472
   e 640 000 f 15 g 1.6 h 32

2. a 6, 1296 b 16, 128 c 15, 405
   d 20, 2.5 e 54, 16

3. a 25×2^n b 1.5×2^n c 2×3^n d 240×0.5^n
   e 50×0.9^n f 64×1.25^n

4. a 13 720, 19 208 b 5000×1.4^n

5. a $n^2 + 1$ b $n^2 + 6$ c $n^2 - n$ d $3n^2$
   e $3n^2 - 2$ f $3n^2 + n$

6. a 15, 21 b $n^2 + n$ c $0.5(n^2 + n)$ or $0.5n^2 + 0.5n$
   d 210 e It is the 50th triangular number because $(50^2 + 50) \div 2 = 1275$

7. a $n^3 - 1$ b $n^3 + 50$ c $n^3 + n$ d $n^3 + 3n$ e $4n^3$ f $4n^3 - n$

8. a 20 b $(4 \times 5 \times 6) \div 6 = 20$ c 7 (using 84 oranges)
   d The layers of the tetrahedron are triangular numbers. 20 layers have $(20 \times 21 \times 22) \div 6 = 1540$ oranges.

9. a When $n = 1$ the first term is $a + b$ and this is 6.
   b $2a + 4b = 16$ c $a = 4, b = 2$; the nth term is $4n + 2n^2$

Answers to Chapter 18

18.1 Using indices

Exercise 18A

1. a 2^4 b 3^5 c 7^2 d 5^3 e 10^7
   f 6^4 g 4^1 h 1^7 i 0.5^4 j 100^3

2. a $3 \times 3 \times 3 \times 3$
   b $9 \times 9 \times 9$
   c 6×6
   d $10 \times 10 \times 10 \times 10 \times 10$
   e $2 \times 2 \times 2 \times 2 \times 2 \times 2 \times 2 \times 2$
   f 8
   g $0.1 \times 0.1 \times 0.1$
   h 2.5×2.5
   i $0.7 \times 0.7 \times 0.7$
   j 1000×1000

3. a 16 b 243
   c 49 d 125
   e 10 000 000 f 1296
   g 4 h 1
   i 0.0625 j 1 000 000

4. a 81 b 729
   c 36 d 100 000
   e 1024 f 8
   g 0.001 h 6.25
   i 0.343 j 1 000 000

732

Answers to Chapter 18

5 125 m³
6 b 10^2 c 2^3 d 5^2
7 a 1   b 4   c 1   d 1   e 1
8 Any power of 1 is equal to 1.
9 10^6
10 10^6
11 a 1   b −1   c 1   d 1   e −1
12 a 1   b −1   c −1   d 1   e 1
13 2^{24}, 4^{12}, 8^8, 16^6

18.2 Negative indices

Exercise 18B

1 a $\frac{1}{5^3}$ b $\frac{1}{6}$ c $\frac{1}{10^5}$ d $\frac{1}{3^2}$
   e $\frac{1}{8^2}$ f $\frac{1}{9}$ g $\frac{1}{w^2}$ h $\frac{1}{t}$
   i $\frac{1}{x^m}$ j $\frac{4}{m^3}$
2 a 3^{-2} b 5^{-1} c 10^{-3} d m^{-1} e t^{-n}
3 a i 2^4 ii 2^{-1} iii 2^{-4} iv -2^3
   b i 10^3 ii 10^{-1} iii 10^{-2} iv 10^6
   c i 5^3 ii 5^{-1} iii 5^{-2} iv 5^{-4}
   d i 3^2 ii 3^{-3} iii 3^{-4} iv -3^5
4 a $\frac{5}{x^3}$ b $\frac{6}{t}$ c $\frac{7}{m^2}$ d $\frac{4}{q^4}$
   e $\frac{10}{y^5}$ f $\frac{1}{2x^3}$ g $\frac{1}{2m}$ h $\frac{3}{4t^4}$
   i $\frac{4}{5y^3}$ j $\frac{7}{8x^5}$
5 a $7x^{-3}$ b $10p^{-1}$ c $5t^{-2}$ d $8m^{-5}$ e $3y^{-1}$
6 a i 25   ii $\frac{1}{125}$ iii $\frac{4}{5}$
   b i 64   ii $\frac{1}{16}$ iii $\frac{5}{256}$
   c i 8   ii $\frac{1}{32}$ iii $\frac{9}{2}$ or $4\frac{1}{2}$
   d i 1 000 000   ii $\frac{1}{1000}$ iii $\frac{1}{4}$
7 24 (32 − 8)
8 $x = 8$ and $y = 4$ (or $x = y = 1$)
9 $\frac{1}{2097152}$
10 a x^{-5}, x^0, x^5 b x^5, x^0, x^{-5} c x^5, x^{-5}, x^0
11 a $\frac{M}{3}$ b 3M   c 27M

18.3 Multiplying and dividing with indices

Exercise 18C

1 a 5^4 b 5^3 c 5^2 d 5^3 e 5^{-5}
2 a 6^3 b 6^0 c 6^6 d 6^{-7} e 6^2
3 a a^3 b a^5 c a^7 d a^4 e a^2 f a^1
4 a Any two values such that $x + y = 10$
   b Any two values such that $x - y = 10$
5 a 4^6 b 4^{15} c 4^6
   d 4^{-6} e 4^6 f 4^0
6 a $6a^5$ b $9a^2$ c $8a^6$
   d $-6a^4$ e $8a^8$ f $-10a^{-3}$
7 a $3a$ b $4a^3$ c $3a^4$
   d $6a^{-1}$ e $4a^7$ f $5a^{-4}$
8 a $8a^5b^4$ b $10a^3b$ c $30a^{-2}b^{-2}$
   d $2ab^3$ e $8a^{-5}b^7$
9 a $3a^3b^2$ b $3a^2c^4$ c $8a^2b^2c^3$
10 a Possible answer: $6x^2 \times 2y^5$ and $3xy \times 4xy^4$
   b Possible answer: $24x^2y^7 \div 2y^2$ and $12x^6y^8 \div x^4y^3$
11 12 ($a = 2, b = 1, c = 3$)
12 a A^2 b A^{-1} c $A^{\frac{1}{2}}$ or $\sqrt{A}$ d $A^{\frac{1}{3}}$ or $\sqrt[3]{A}$
13 a $x^{2n+1} = x^{2n} \times x = (x^n)^2 \times x = xy^2$ b $\frac{y^2}{x}$

18.4 Fractional indices

Exercise 18D

1 a 5   b 10   c 8   d 9   e 25
   f 3   g 4   h 10   i 5   j 8
   k 12   l 20   m 5   n 3   o 10
   p 3   q 2   r 2   s 6   t 6
   u $\frac{1}{4}$ v $\frac{1}{2}$ w $\frac{1}{3}$ x $\frac{1}{5}$ y $\frac{1}{10}$
2 a $\frac{5}{6}$ b $1\frac{2}{3}$ c $\frac{8}{9}$ d $1\frac{4}{5}$ e $\frac{5}{8}$
   f $\frac{3}{5}$ g $\frac{1}{4}$ h $2\frac{1}{2}$ i $\frac{4}{5}$ j $1\frac{1}{7}$
3 $(x^{\frac{1}{n}})^n = x^{\frac{1}{n} \times n} = x^1 = x$, but $(\sqrt[n]{x})^n = \sqrt[n]{x} \times \sqrt[n]{x} \dots n$ times $= x$, so $x^{\frac{1}{n}} = \sqrt[n]{x}$
4 $64^{-\frac{1}{2}} = \frac{1}{8}$, others are both $\frac{1}{2}$
5 Possible answer: The negative power gives the reciprocal, so $27^{-\frac{1}{3}} = \frac{1}{27^{\frac{1}{3}}}$. The power one-third means cube root, so you need the cube root of 27 which is 3, so $27^{\frac{1}{3}} = 3$ and $\frac{1}{27^{\frac{1}{3}}} = \frac{1}{3}$
6 Possible answer: $x = 1$ and $y = 1$, $x = 8$ and $y = \frac{1}{64}$.
7 a 3   b $\frac{1}{3}$ c 0   d $\frac{1}{2}$
   e $\frac{1}{2}$ f $\frac{1}{4}$ g $\frac{1}{4}$ h $\frac{1}{3}$
   i $\frac{1}{3}$ j $\frac{1}{2}$ k $\frac{1}{3}$ l $\frac{1}{7}$

Exercise 18E

1 a 16   b 25   c 216   d 81
2 a $t^{\frac{2}{3}}$ b $m^{\frac{3}{4}}$ c $k^{\frac{2}{5}}$ d $x^{\frac{3}{2}}$
3 a 4   b 9   c 64   d 3125
4 a $\frac{1}{5}$ b $\frac{1}{6}$ c $\frac{1}{2}$ d $\frac{1}{3}$
   e $\frac{1}{4}$ f $\frac{1}{2}$ g $\frac{1}{2}$ h $\frac{1}{3}$
5 a $\frac{1}{125}$ b $\frac{1}{216}$ c $\frac{1}{8}$ d $\frac{1}{27}$
   e $\frac{1}{256}$ f $\frac{1}{4}$ g $\frac{1}{4}$ h $\frac{1}{9}$
6 a $\frac{1}{100 000}$ b $\frac{1}{12}$ c $\frac{1}{25}$ d $\frac{1}{27}$
   e $\frac{1}{32}$ f $\frac{1}{32}$ g $\frac{1}{81}$ h $\frac{1}{13}$
7 $8^{-\frac{2}{3}} = \frac{1}{4}$, others are both $\frac{1}{8}$

Answers to Chapter 20

8 Possible answer: The negative power gives the reciprocal, so the power one-third means cube root, so we need the cube root of 27 which is 3 and the power 2 means square, so

$3^2 = 9$, so $27^{\frac{2}{3}} = 9$ and $\frac{1}{27^{\frac{2}{3}}} = \frac{1}{9}$

9 a $\frac{27}{8}$ b $\frac{9}{25}$ c $\frac{1024}{243}$ d $\frac{8}{343}$
   e $\frac{16}{9}$ f $\frac{8}{27}$ g $\frac{625}{256}$ h $\frac{32}{243}$

10 a $\frac{25}{9}$ b $\frac{27}{64}$ c $\frac{125}{729}$ d $\frac{243}{32}$
   e $\frac{8}{27}$ f $\frac{243}{32}$ g $\frac{9}{4}$ h $\frac{125}{343}$

   i $\frac{16}{25}$ j $\frac{512}{125}$ k $\frac{243}{32}$ l $\frac{32}{243}$

11 a x^4 b x^{-1} c $4y^2$
   d $10x^2$ e $20x^{-1}$ f $\frac{1}{3}y$

12 a x b d^{-1} c $t^{\frac{3}{2}}$
   d x^2 e $y^{\frac{1}{2}}$ f a^4

13 a $x^{\frac{1}{2}}$ b y^{-1} c $a^{\frac{5}{3}}$
   d t^{-2} e d^2 f 1

14 $y^{\frac{9}{4}}$

Answers to Chapter 19

19.1 Direct proportion

Exercise 19A

1 a 15   b 2
2 a 75   b 6
3 a 150   b 6
4 a 22.5   b 12
5 a 175 kilometres   b 8 hours
6 a 66.50 dollars   b 175 kg
7 a 44   b 84 m²
8 a 33 spaces
   b 66 spaces since new car park has 366 spaces
9 17 minutes 30 seconds

Exercise 19B

1 a 100   b 10
2 a 27   b 5
3 a 56   b 1.69
4 a 192   b 2.25
5 a 25.6   b 5
6 a 80   b 8
7 a $50   b 225
8 a 3.2°C   b 10 atm
9 a 388.8 g   b 3 mm
10 a 2 J   b 40 m/s

11 a 78 dollars   b 400 miles
12 4000 cm³
13 $250
14 a B   b A   c C
15 a B   b A

19.2 Inverse proportion

Exercise 19C

1 $Tm = 12$ a 3   b 2.5
2 $Wx = 60$ a 20   b 6
3 $Q(5 - t) = 16$ a −3.2   b 4
4 $Mt^2 = 36$ a 4   b 5
5 $W\sqrt{T} = 24$ a 4.8   b 100
6 $x^3y = 32$ a 32   b 4
7 $gp = 1800$ a $15   b 36
8 $td = 24$ a 3°C   b 12 km
9 $ds^2 = 432$ a 1.92 km   b 8 m/s
10 $p\sqrt{h} = 7.2$ a 2.4 atm   b 100 m
11 $W\sqrt{F} = 0.5$ a 5 t/h   b 0.58 t/h
12 B – This is inverse proportion, as x increases y decreases.
13

x	8	27	64
y	1	$\frac{2}{3}$	$\frac{1}{2}$

Answers to Chapter 20

20.1 Graphical inequalities

Exercise 20A

1 a & b

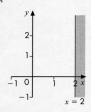

$x = 2$

2 a & b

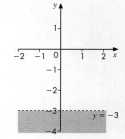

$y = -3$

Answers to Chapter 20

3 a–c

4 a–c

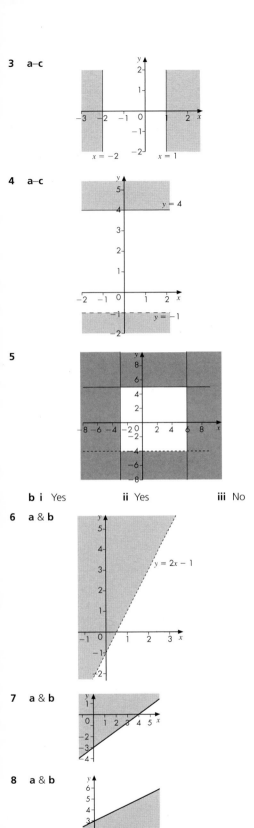

5

b i Yes   ii Yes   iii No

6 a & b

7 a & b

8 a & b

9

10 a–d

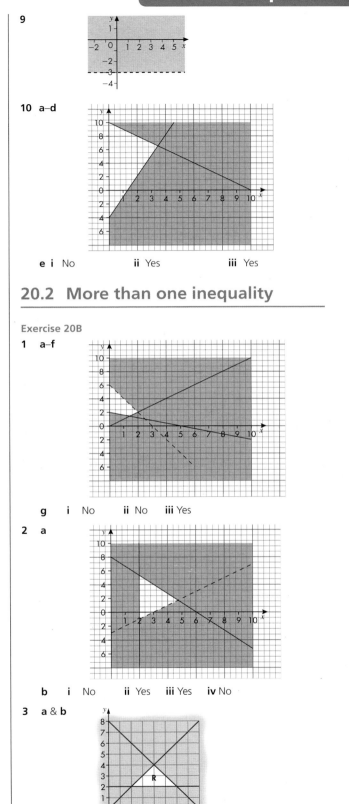

e i No   ii Yes   iii Yes

20.2 More than one inequality

Exercise 20B

1 a–f

g i No   ii No   iii Yes

2 a

b i No   ii Yes   iii Yes   iv No

3 a & b

735

Answers to Chapter 20

4 a & b
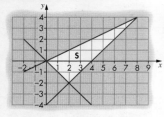

c 4   **d** −2   **e** 12

5 Test a point such as the origin (0, 0), so $0 < 0 + 2$, which is true. So the side that includes the origin is the required side.

6 a $x + y \geq 3$, $y \leq \frac{1}{2}x + 3$ and $y \geq 5x − 15$
   b 9
   c 3 at (3,0)

20.3 Linear programming

Exercise 20C

1 a He buys at least 3 cartons of milk
   b $y \geq 2$
   c $x + y \leq 8$
   d
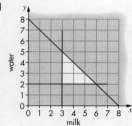

2 a $x + y \leq 10$
   b $2x + 5y \geq 30$
   c
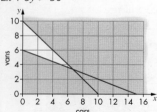

   d 6 because cars & vans (x, y) must both be integers
   e 4

3 a $x > y$ $x < 2y$ $x + y < 12$
   b
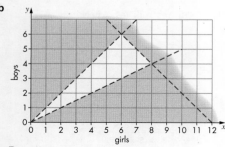

   c 7

4 a $x + y \geq 6$ $x + y \leq 12$ $y > x$ $x \geq 2$

b
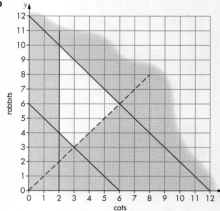

c 4   **d** 5

5 a Follows from $15x + 10y \geq 600$
   b $x + y \leq 50$
   c
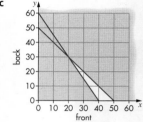

   d 30

6 a i $x + y \leq 15$ **ii** $x + y \geq 5$ **iii** $y \geq \frac{1}{2}x$
   b the number of small coaches is not more than 2 more than the number of large coaches
   c
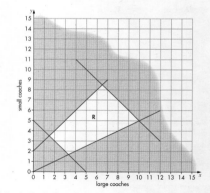

   d 3, 4, 5, 6, 7 or 8

7 a $3x + y$ **b** $3x + y \leq 18$ $x > y$ $y \geq 2$
   c
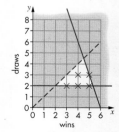

   d 3 wins and 2 draws; 4 wins and 2 draws; 5 wins and 2 draws; 4 wins and 3 draws; 5 wins and 3 draws

Answers to Chapter 21

21.1 Function notation

Exercise 21A

1. **a** 12 **b** 26 **c** 7 **d** −2 **e** 3
2. **a** 0.5 **b** 5 **c** 50.5 **d** 2.5 **e** 0.625 or $\frac{5}{8}$
3. **a** 5 **b** −3 **c** 999801 **d** 1 **e** $\frac{1}{8}$
4. **a** 4 **b** 32 **c** 1 **d** $\frac{1}{2}$ **e** $\frac{1}{8}$
5. **a** 3 **b** 2 **c** 0 **d** −1 **e** 5
6. **a** 7.5 **b** −2.5 **c** −5
7. **a** 6 **b** 97 **c** 3.25
8. **a** 6 **b** at (6, 4)
9. **a** 3 **b** $-\frac{1}{2}$ **c** $x = 2$ **d** $x = 0$

21.2 Inverse functions

Exercise 21B

1. **a** $x - 7$ **b** $\frac{x}{8}$ **c** $5x$ **d** $x + 3$
2. **a** 8 **b** 4 **c** 5 **d** −2
3. **a** $3(x + 2)$ **b** $\frac{x}{4} + 5$ **c** $5x - 4$ **d** $\frac{5x + 6}{3}$ **e** $2(\frac{x}{3} - 4)$ **f** $\sqrt[3]{\frac{x}{4}}$
4. **a** 2 **b** $\frac{1}{2}$ **c** −2.5
5. **a** $10 - x$ **b** They are identical
6. **a** $\frac{8}{x}$ **b** $\frac{20}{x + 1}$ **c** $\frac{2}{x} - 1$
7. **a** $\frac{x + 4}{2}$ **b** student's graph **c** (4,4)
8. 5
9. $\frac{x + 2}{3}$

21.3 Composite functions

Exercise 21C

1. **a** 6 and 3 **b** 7 and 3.5 **c** 10 and 5 **d** $\frac{x + 4}{2}$ **e** 1 and 5 **f** 1.5 and 5.5 **g** −5 and −1 **h** $\frac{x}{2} + 4$
2. **a** 6 and 216 **b** 10 and 1000 **c** $(2x)^3$ or $8x^3$ **d** 64 and 128 **e** $2x^3$
3. **a** 1, 9, 25 **b** 1, 3, 5 **c** $\sqrt{2x + 1}$
4. **a** 6 and 18 **b** 12 and 36 **c** $9x$
5. **a** $3(x - 6)$ **b** $3x - 6$
6. Both are $x - 3$

21.4 More about composite functions

Exercise 21D

1. **a** 3.5 **b** 1 **c** 8 **d** 5.5
2. **a** 20 **b** 9 **c** 8.75 **d** 3
3. **a** 7 **b** 8 **c** 256 **d** 21
4. **a** $6x$ **b** $6x - 5$
5. **a** $9x^2 + 24x + 16$ or $(3x + 4)^2$ **b** $6x - 5$ **c** $2x + 3$ **d** $4 - 2x$
6. **a** $x - 10$ **b** $x + 10$ **c** x
7. **a** x^4 **b** $\left(\frac{12}{x}\right)^2$ or $\frac{144}{x^2}$ **c** $\frac{12}{x^2}$ **d** x
8. **a** 80 **b** $2(2x - 1) - 1$ simplified **c** $(2x - 1)^2 + 2(2x - 1)$ simplified
9. **a** $\frac{1}{3x - 3}$ **b** $\frac{x - 1}{x - 4}$
10. **a** $6x - 14$ **b** $\frac{x + 12}{4}$
11. **a** $0.5(1 + 9) = 5$ **b** 7 **c** 8 **d** 8.5, 8.75, 8.875, 8.9375, 8.96875, 8.984375 **e** Getting closer and closer to 9, halving the difference from 9 each time.
12. Student's own description of the convergence towards 9.

Answers to Chapter 22

22.1 The gradient of a curve

Exercise 22A

1   **a** The missing numbers are 0, −1, 0
    b
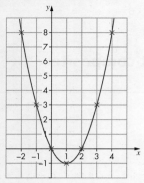
    c $2x - 2$ **d** 4
    e 6   **f** student's choice
    g $(1, -1)$ **h** student's check
2   **a** $2x - 6$ **b** −6   **c** 4   **d** $(4, 7)$
3   **a** $4x$ **b** 8   **c** −4 **d** $(3, 8)$
4   **a** $4 - 2x$ **b** 4 and −4 **c** $(1, 3)$ **d** $(1.5, 3.75)$
5   **a** $2x + 1$ **b** $2x - 7$ **c** $8x - 1$ **d** $0.6x - 1.5$
    e $-2 + 2x$ **f** $3 - 2x$ **g** 2 **h** 0
6   $2x + 2$
7   **a** $4x + 2$ **b** $2x + 7$ **c** $2x$
8   **a** $(0, -5)$ **b** 2

22.2 More complex curves

Exercise 22B

1   **a** $6x^2$ **b** 6 and 24
2   **a** $3x^2 - 12x + 8$
    b If $x = 0$ or 2 or 4, $y = 0$
    c 8; −4; 8
3   **a** $1.5x^2 - 6x + 4$
    b 4 at $(0, 0)$ and $(4, 0)$; −2 at $(2, 0)$
4   **a** $8x^3$ **b** $6x^2 + 10x$
    c $15x^2 - 2$ **d** $-1 - 2x^2$
    e $9x^2 + 5$ **f** $-3x^2$
    g $4x^3 - 1$ **h** $8x^3 + 18x^2$

5   16 at $(2, 0)$; −16 at $(−2, 0)$; 0 at $(0, 0)$
6   **a** $dy/dx = 4x^3 - 6x^2$ and if $x = 0$ then $dy/dx = 0$
    b −10   **c** 8
7   $x^2 - 5 = 4$ has two solutions, $x = 3$ or −3. Points are $(3, -2)$ and $(-3, 10)$
8   $y = 1.5x - 2$
9   **a** 12   **b** 24

22.3 Turning points

Exercise 22C

1   **a** $2x - 4$ **b** $2x - 4 = 0 \Rightarrow x = 2; (2, -1)$
    c Minimum
2   **a** $(-3, -12)$ **b** Minimum
3   **a** $5 - 2x$ **b** $(2.5, 7.25)$
    c Maximum
4   2 and −3
5   **a** $3x^2 - 6x$ **b** $x = 0$ or 2
    c $(0, 0)$ and $(2, -4)$
6   **a** If $x = -2$ or 5, $y = 0$ **b** $2x - 3$
    c $(1.5, -12.25)$; Minimum **d** $x = 1.5$
7   **a** $(-3, 81)$, $(0, 0)$ and $(3, 81)$
    b
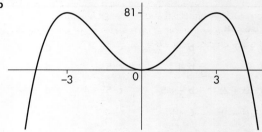
8   **a** $6x^2 - 6$ **b** $(1, 0)$ minimum, $(-1, 8)$ maximum
9   **a** The two sides add up to half the perimeter
    b $15 - 2x$
    c $(7.5, 56.25)$
    d Maximum
    e The largest possible area is 56.25 cm², when the rectangle is a square of side 7.5 cm

Answers to Chapter 23

23.1 Angle facts

Exercise 23A

1   **a** 48°   **b** 307°   **c** 108°   **d** 52°
    e 59°   **f** 81°   **g** 139°   **h** 58°

2   **a** 82°   **b** 105°   **c** 75°
3   45° + 125° = 170° and for a straight line it should be 180°.
4   **a** $x = 100°$ **b** $x = 110°$ **c** $x = 30°$
5   **a** $x = 55°$ **b** $x = 45°$ **c** $x = 12.5°$
6   **a** $x = 34°, y = 98°$ **b** $x = 70°, y = 120°$ **c** $x = 20°, y = 80°$

Answers to Chapter 23

7 6 × 60° = 360°; imagine six of the triangles meeting at a point.
8 $x = 35°$, $y = 75°$; $2x = 70°$ (opposite angles), so $x = 35°$ and $x + y = 110°$ (angles on a line), so $y = 75°$
9 a = 88   **b** = 132

23.2 Parallel lines

Exercise 23B

1 a 40°   **b** $b = c = 70°$
  c $d = 75$, $e = f = 105°$ **d** $g = 50°$, $h = i = 130°$
  e $j = k = l = 70°$ **f** $n = m = 80°$
2 a $a = 50°$, $b = 130°$ **b** $c = d = 65°$, $e = f = 115°$
  c $g = i = 65°$, $h = 115°$ **d** $j = k = 72°$, $l = 108°$
  e $m = n = o = p = 105°$ **f** $q = r = s = 125°$
3 a $a = 95°$ **b** $b = 66°$, $c = 114°$
4 a $x = 30°$, $y = 120°$ **b** $x = 25°$, $y = 105°$
  c $x = 30°$, $y = 100°$
5 a $x = 50°$, $y = 110°$ **b** $x = 25°$, $y = 55°$
  c $x = 20$, $y = 140°$
6 290°; x is double the angle allied to 35°, so is 2 × 145°
7 $a = 66$
8 Angle $PQD = 64°$ (alternate angles), so angle $DQY = 116°$ (angles on a line = 180°)
9 Use alternate angles to see b, a and c are all angles on a straight line, and so total 180°.
10 Third angle in triangle equals q (alternate angle), angle sum of triangle is 180°.
11 $A + D = 180°$ because they are allied angles. $C + D = 180°$ because they are allied angles. Hence $A = C$.
In the same way $B + C = 180° = D + C$ because they are pairs of allied angles. Hence $B = D$.

23.3 Angles in a triangle

Exercise 23C

1 a 70°   **b** 50°   **c** 80°   **d** 60°
  e 75°   **f** 109°   **g** 38°   **h** 63°
2 a No, total is 190°   **b** Yes, total is 180°   **c** No, total is 170°
  d Yes, total is 180°   **e** Yes, total is 180°   **f** No, total is 170°
3 a 60°   **b** Equilateral triangle   **c** Same length
4 a 70° each   **b** Isosceles triangle   **c** Same length
5 a 109°   **b** 130°   **c** 135°
6 Isosceles triangle; angle DFE ∠30° (opposite angles), angle DEF ∠75° (angles on a line), angle FDE ∠75° (angles in a triangle), so there are two equal angles in the triangle and hence it is an isosceles triangle
7 a is ∠80° (opposite angles), b is 65° (angles on a line), c is 35° (angles in a triangle)
8 Missing angle = y, $x + y = 180°$ and $a + b + y = 180°$ so $x = a + b$
9 $b = 240 - a$

23.4 Angles in a quadrilateral

Exercise 23D

1 a $a = 110°$, $b = 55°$ **b** $c = e = 105°$, $d = 75°$
  c $f = 135°$, $g = 25°$ **d** $h = i = 94°$
  e $j = l = 105°$, $k = 75°$ **f** $m = o = 49°$, $n = 131°$
2 a $x = 25°$, $y = 15°$ **b** $x = 7°$, $y = 31°$ **c** $x = 60°$, $y = 30°$
3 a $x = 50°$: 60°, 70°, 120°, 110° – possibly trapezium
  b $x = 60°$: 50°, 130°, 50°, 130° – parallelogram or isosceles trapezium
  c $x = 30°$: 20°, 60°, 140°, 140° – possibly kite
  d $x = 20°$: 90°, 90°, 90°, 90° – square or rectangle
4 52°
5 Both 129°
6 $y = 360° - 4x$
7 36°, 72°, 108° and 144°

23.5 Regular polygons

Exercise 23E

1 a i 45°   **ii** 8   **iii** 1080°
  b i 20°   **ii** 18   **iii** 2880°
  c i 15°   **ii** 24   **iii** 3960°
  d i 36°   **ii** 10   **iii** 1440°
2 a i 172°   **ii** 45   **iii** 7740°
  b i 174°   **ii** 60   **iii** 10 440°
  c i 156°   **ii** 15   **iii** 2340°
  d i 177°   **ii** 120   **iii** 21 240°
3 a Exterior angle is 7°, which does not divide exactly into 360°
  b Exterior angle is 19°, which does not divide exactly into 360°
  c Exterior angle is 11°, which does divide exactly into 360°
  d Exterior angle is 70°, which does not divide exactly into 360°
4 a 7° does not divide exactly into 360°
  b 26° does not divide exactly into 360°
  c 44° does not divide exactly into 360°
  d 13° does not divide exactly into 360°
5 $x = 45°$, they are the same, true for all polygons
6 a 36°   **b** 10
7 a The exterior angle is 180 − 170 = 10°; 360 ÷ 10 = 36 so a regular polygon with 36 sides is possible.
  b The exterior angle is 180 − 169 = 11°; 360 ÷ 11 is not a whole number so a regular polygon is not possible.

23.6 Irregular polygons

Exercise 23F

1 a 1440°   **b** 2340°   **c** 17 640°   **d** 7740°
2 a 9   **b** 15   **c** 102   **d** 50
3 a 130°   **b** 95°   **c** 130°
4 a 50°   **b** 40°   **c** 59°
5 Hexagon
6 a Octagon   **b** 89°
7 a i 71°   **ii** 109°   **iii** Equal
  b If S = sum of the two opposite interior angles, then $S + I = 180°$ (angles in a triangle), and we know $E + I = 180°$ (angles on a straight line), so $S + I = E + I$, therefore $S = E$
8 144°; 360 − (2 × 108)°
9 Three angles are 135° and two angles are 67.5°

23.7 Tangents and diameters

Exercise 23G

1 a 38°   **b** 110°   **c** 15°   **d** 45°
2 a $x = 12°$, $y = 156°$ **b** $x = 100°$, $y = 50°$
  c $x = 62°$, $y = 28°$ **d** $x = 30°$, $y = 60°$
3 Angle $OCD = 58°$ (triangle OCD is isosceles), angle $OCB = 90°$ (tangent/radius theorem), so angle $DCB = 32°$, hence triangle BCD is isosceles (2 equal angles)

Answers to Chapter 24

23.8 Angles in a circle

Exercise 23H

1. a 56° b 62° c 105° d 55°
   e 45° f 30° g 60° h 145°
2. a 55° b 52° c 50° d 24°
   e 39° f 80° g 34° h 30°
3. a 41° b 49° c 41°
4. a 72° b 37° c 72°
5. Angle AZY = 35° (angles in a triangle), a = 55° (angle in a semicircle = 90°)
6. a $x = y = 40°$ b $x = 131°, y = 111°$
   c $x = 134°, y = 23°$ d $x = 32°, y = 19°$
   e $x = 59°, y = 121°$ f $x = 155°, y = 12.5°$
7. 68°
8. Angle ABC = 180° – x (angles on a line), angle AOC = 360° – $2x$ (angle at centre is twice angle at circumference), reflex angle AOC = 360° – (360° – $2x$) = $2x$ (angles at a point)
9. a x
   b $2x$
   c From part **b**, angle $AOD = 2x$
   Similarly, angle $COD = 2y$
   So angle $AOC = AOD + COD = 2x + 2y = 2(x + y)$
   = 2 × angle ABC

23.9 Cyclic quadrilaterals

Exercise 23I

1. a $a = 50°, b = 95°$ b $c = 92°, x = 90°$
   c $d = 110°, e = 110°, f = 70°$ d $g = 105°, h = 99°$
   e $j = 89°, k = 89°, l = 91°$ f $m = 120°, n = 40°$
   g $p = 44°, q = 68°$ h $x = 40°, y = 34°$
2. a $x = 26°, y = 128°$ b $x = 48°, y = 78°$
   c $x = 133°, y = 47°$ d $x = 36°, y = 72°$
   e $x = 55°, y = 125°$ f $x = 35°$
   g $x = 48°, y = 45°$ h $x = 66°, y = 52°$
3. a $x = 49°, y = 49°$ b $x = 70°, y = 20°$
   c $x = 80°, y = 100°$ d $x = 100°, y = 75°$
4. a $x = 50°, y = 62°$ b $x = 92°, y = 88°$
   c $x = 93°, y = 42°$ d $x = 55°, y = 75°$
5. a $x = 95°, y = 138°$ b $x = 14°, y = 62°$
   c $x = 32°, y = 48°$ d 52°
6. a 71° b 125.5° c 54.5°
7. a $x + 2x - 30° = 180°$ (opposite angles in a cyclic quadrilateral), so $3x - 30° = 180°$
   b $x = 70°$, so $2x - 30° = 110°$ angle $DOB = 140°$ (angle at centre equals twice angle at circumference), $y = 80°$ (angles in a quadrilateral)
8. a x b $360° - 2x$
   c Angle $ADC = \frac{1}{2}$ reflex angle $AOC = 180° - x$, so angle ADC + angle $ABC = 180°$
9. Let angle $AED = x$, then angle $ABC = x$ (opposite angles are equal in a parallelogram), angle $ADC = 180° - x$ (opposite angles in a cyclic quadrilateral), so angle $ADE = x$ (angles on a line)
10. Let angle $ABC = x$ and angle $EFG = y$.
    Then angle $ADC = 180 - x°$ (opposite angles in a cyclic quadrilateral) and angle $EDG = 180 - y°$.
    But angle ADC = angle EDG (opposite angles).
    $180 - x° = 180 - y°$ and therefore $x = y$.

23.10 Alternate segment theorem

Exercise 23J

1. a $a = 65°, b = 75°, c = 40°$
   b $d = 79°, e = 58°, f = 43°$
   c $g = 41°, h = 76°, i = 76°$
   d $k = 80°, m = 52°, n = 80°$
2. a $a = 75°, b = 75°, c = 75°, d = 30°$
   b $a = 47°, b = 86°, c = 86°, d = 47°$
   c $a = 53°, b = 53°$
   d $a = 55°$
3. a 36° b 70°
4. a $x = 25°$ b $x = 46°, y = 69°, z = 65°$
   c $x = 38°, y = 70°, z = 20°$ d $x = 48°, y = 42°$
5. Angle ACB = 64° (angle in alternate segment), angle ACX = 116° (angles on a line), angle CAX = 32° (angles in a triangle), so triangle ACX is isosceles (two equal angles)
6. Angle AXY = 69° (tangents equal and so triangle AXY is isosceles), angle XZY = 69° (alternate segment), angle XYZ = 55° (angles in a triangle)
7. a $2x$ b $90° - x$ c angle OPT = 90°, so angle $APT = x$

Answers to Chapter 24

24.1 Measuring and drawing angles

Exercise 24A

1. a 40° b 125° c 340° d 225°
2. student's drawings of angles
3. AC and BE; AD and CE; AE and CF.
4. Yes, the angle is 75°.
5. Any angle between 90° and 100°.
6. a 80° b 50° c 25°

24.2 Bearings

Exercise 24B

1. a 110° b 250° c 091° d 270° e 130° f 180°
2. student's sketches

Answers to Chapter 24

3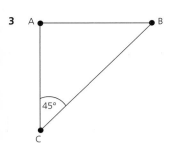

4 a 090°, 180°, 270° **b** 000°, 270°, 180°

5

Leg	Actual distance	Bearing
1	50 km	060°
2	70 km	355°
3	65 km	260°
4	46 km	204°
5	60 km	130°

6 a 045° **b** 286°
7 a 250° **b** 325° **c** 144°
8 a 900 m **b** 280°
   c angle NHS = 150° and HS = 3 cm
9 108°
10 255°

24.3 Nets

Exercise 24C

1 a b c

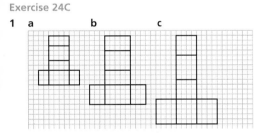

2 Yes.

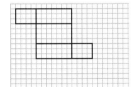

3

4 a b

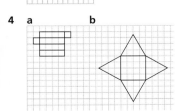

c d

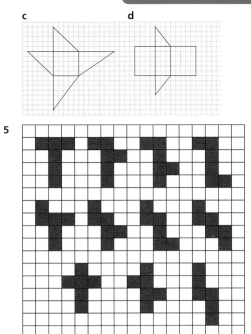

5

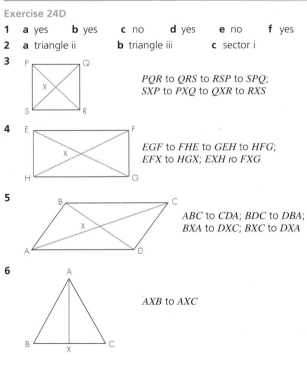

6 a and b

24.4 Congruent shapes

Exercise 24D

1 a yes **b** yes **c** no **d** yes **e** no **f** yes
2 a triangle ii **b** triangle iii **c** sector i
3

PQR to QRS to RSP to SPQ;
SXP to PXQ to QXR to RXS

4

EGF to FHE to GEH to HFG;
EFX to HGX; EXH to FXG

5

ABC to CDA; BDC to DBA;
BXA to DXC; BXC to DXA

6

AXB to AXC

24.5 Congruent triangles

Exercise 24E

1 a SAS **b** SSS **c** ASA **d** RHS **e** SSS **f** ASA
2 a SSS. A to R, B to P, C to Q
   b SAS. A to R, B to Q, C to P

741

Answers to Chapter 25

3   **a** 60°   **b** 80°   **c** 40°   **d** 5 cm
4   **a** 110°   **b** 55°   **c** 85°   **d** 110°   **e** 4 cm
5   SSS or RHS
6   SSS or SAS or RHS
7   For example, use $\triangle ADE$ and $\triangle CDG$. $AD = CD$ (sides of large square), $DE = DG$ (sides of small square), angle ADE = angle CDG (angles sum to 90° with angle ADG), so angle ADE = angle CDG (SAS), so $AE = CG$
8   AB and PQ are the corresponding sides to the 42° angle, but they are not equal in length.

24.6 Similar shapes

Exercise 24F

1   **a** 2   **b** 3
2   **a** Yes, 4
    b No, corresponding sides have different ratios.
3   **a** PQR is an enlargement of ABC
    b 1 : 3   **c** Angle R **d** BA
4   **a** Sides in same ratio
    b Angle P **c** PR
5   **a** Same angles   **b** Angle Q **c** AR
6   **a** 8 cm   **b** $x = 45$ cm, $y = 9$ cm
    c $x = 19.5$ cm, $y = 10.8$ cm   **d** 4.2 cm
7   **a** The angles are all 90 degrees. The sides of a square are all equal so the ratio between sides of two different squares will be the same, whatever two sides are chosen.
    b No. They will only be similar if they have the same ratio of length to width.
8   5.2 m

24.7 Areas of similar triangles

Exercise 24G

1   **a** 2.5   **b** 125 cm²
2   **a** All equilateral triangles are similar   **b** 3.8 cm² (to 2 sf)
3   40.32 cm²
4   75 cm²
5   **a** 144 cm²   **b** 69.4 cm²
6   **a** All angles are the same   **b** 247.7 cm²
7   **a** 2   **b** 10 cm   **c** 7.1 cm
8   354.9 cm²
9   It will double the area
10   28.3 cm²

24.8 Areas and volumes of similar shapes

Exercise 24H

1   **a** 4 : 25   **b** 8 : 125
2   **a** 16 : 49   **b** 64 : 343
3

Linear scale factor	Linear ratio	Linear fraction	Area scale factor	Volume scale factor
2	1 : 2	$\frac{2}{1}$	4	8
3	1 : 3	$\frac{3}{1}$	9	27
$\frac{1}{4}$	4 : 1	$\frac{1}{4}$	$\frac{1}{16}$	$\frac{1}{64}$
5	1 : 5	$\frac{5}{1}$	25	125
$\frac{1}{10}$	10 : 1	$\frac{1}{10}$	$\frac{1}{100}$	$\frac{1}{1000}$

4   135 cm²
5   **a** 56 cm²   **b** 126 cm²
6   **a** 48 m²   **b** 3 m²
7   **a** 2400 cm³   **b** 8100 cm³
8   4 litres
9   1.38 m³
10   $6
11   4 cm
12   8 × 0.60 = $4.80 which is greater than $4.00 so the large tub is better value
13   **a** 3 : 4   **b** 9 : 16   **c** 27 : 64
14   720 ÷ 8 = 90 cm³

Exercise 24I

1   6.2 cm, 10.1 cm
2   4.26 cm, 6.74 cm
3   9.56 cm
4   3.38 m
5   8.4 cm
6   26.5 cm
7   47.8 cm
8   **a** 4.33 cm, 7.81 cm   **b** 143 g, 839 g
9   53.8 kg
10   1.73 kg
11   8.8 cm
12   7.9 cm and 12.6 cm
13   b

Answers to Chapter 25

25.1 Constructing shapes

Exercise 25A

1   **a** $BC = 2.9$ cm, angle $B = 53°$, angle $C = 92°$
    b $EF = 7.4$ cm, $ED = 6.8$ cm, angle $E = 50°$
    c angle $G = 105°$, angle $H = 29°$, angle $I = 46°$
    d angle $J = 48°$, angle $L = 32°$, $JK = 4.3$ cm
    e angle $N = 55°$, $ON = OM = 7$ cm
    f angle $P = 51°$, angle $R = 39°$, $QP = 5.7$ cm

2   **a** Students can check one another's triangles.
    b angle $ABC = 44°$, angle $BCA = 79°$, angle $CAB = 57°$
3   5.9 cm
4   student drawing.
5   student drawing.
6   4.3 cm
7   4.3 cm

Answers to Chapter 26

8
 a Right-angled triangle constructed with sides 3, 4, 5 and 4.5, 6, 7.5, and scale marked 1 cm : 1 m
 b Right-angled triangle constructed with 12 equally spaced dots.
9 An equilateral triangle of side 4 cm.
10 Even with all three angles, you need to know at least one length.

25.2 Scale drawings

Exercise 25B

1 **a** pond: 40 m × 10 m, fruit: 50 m × 10 m, trees: 20 m × 20 m, lawn: 30 m × 20 m, vegetables: 50 m × 20 m
 b pond: 400 m², fruit: 500 m², trees: 400 m², lawn: 600 m², vegetables: 1000 m²
2 **a** 33 cm **b** 9 cm
3 **a** 30 cm × 30 cm **b** 40 cm × 10 cm **c** 20 cm × 15 cm
 d 30 cm × 20 cm **e** 30 cm × 20 cm **f** 10 cm × 5 cm
4 **a** student's scale drawing. **b** 39 plants
5 **a** 8.4 km **b** 4.6 km **c** 6.2 km
 d 6.4 km **e** 7.6 km **f** 2.4 km
6 **a** student's drawing **b** 12.9 metres
7 **a** 900 km **b** 1100 km **c** 860 km
8 c – 7 cm represents 210 m, so 1 cm represents 30 m

Answers to Chapter 26

26.1 Pythagoras' theorem

Exercise 26A

1 10.3 cm
2 5.9 cm
3 8.5 cm
4 20.6 cm
5 18.6 cm
6 17.5 cm
7 5 cm
8 13 cm
9 10 cm
10 The smaller square in the first diagram and the two smaller squares in the second have the same area.

Exercise 26B

1 **a** 15 cm **b** 14.7 cm **c** 6.3 cm **d** 18.3 cm
2 **a** 20.8 m **b** 15.5 m **c** 15.5 m **d** 12.4 cm
3 **a** 5 m **b** 6 m **c** 3 m **d** 50 cm
4 There are infinite possibilities, e.g. any multiple of 3, 4, 5 such as 6, 8, 10; 9, 12, 15; 12, 16, 20; multiples of 5, 12, 13 and of 8, 15, 17.
5 42.6 cm

26.2 Trigonometric ratios

Exercise 26C

1 **a** 0.682 **b** 0.829 **c** 0.922 **d** 1
2 **a** 0.731 **b** 0.559 **c** 0.388 **d** 0
3 **a** **i** 0.574 **ii** 0.574
 b **i** 0.208 **ii** 0.208
 c **i** 0.391 **ii** 0.391
 d They are the same.
 e **i** sin 15° is the same as cos 75°
  ii cos 82° is the same as sin 8°
  iii sin x is the same as cos (90° – x)
4 **a** 0.933 **b** 1.48 **c** 2.38 **d** Infinite
 e 1 **f** 0.364 **g** 0.404 **h** 0
5 Tan has values > 1
6 **a** 3.56 **b** 8.96 **c** 28.4 **d** 8.91
7 **a** 5.61 **b** 7.08 **c** 1.46 **d** 7.77
8 **a** $\frac{4}{5}, \frac{3}{5}, \frac{4}{3}$ **b** $\frac{5}{13}, \frac{12}{13}, \frac{5}{12}$ **c** $\frac{7}{25}, \frac{24}{25}, \frac{7}{24}$

26.3 Calculating angles

Exercise 26D

1 **a** 30° **b** 51.7° **c** 39.8°
 d 61.3° **e** 87.4° **f** 45.0°
2 **a** 60° **b** 50.2° **c** 2.6°
 d 45° **e** 78.5° **f** 45.6°
3 **a** 31.0° **b** 20.8° **c** 41.8°
 d 46.4° **e** 69.5° **f** 77.1°
4 Error message, largest value 1, smallest value –1
5 **a** **i** 17.5° **ii** 72.5° **iii** 90°
 b Yes

26.4 Using sine, cosine and tangent functions

Exercise 26E

1 **a** 17.5° **b** 22.0° **c** 32.2°
2 **a** 5.29 cm **b** 5.75 cm **c** 13.2 cm
3 **a** 4.57 cm **b** 6.86 cm **c** 100 cm
4 **a** 5.12 cm **b** 9.77 cm **c** 11.7 cm **d** 15.5 cm

Exercise 26F

1 **a** 51.3° **b** 75.5° **c** 51.3°
2 **a** 5.35 cm **b** 14.8 cm **c** 12.0 cm **d** 8.62 cm
3 **a** 5.59 cm **b** 46.6° **c** 9.91 cm **d** 40.1°

Exercise 26G

1 **a** 33.7° **b** 36.9° **c** 52.1°
2 **a** 9.02 cm **b** 7.51 cm **c** 7.14 cm **d** 8.90 cm
3 **a** 13.7 cm **b** 48.4° **c** 7.03 cm **d** 41.2°

26.5 Which ratio to use

Exercise 26H

1 **a** 12.6 **b** 59.6 **c** 74.7 **d** 16.0
 e 67.9 **f** 20.1
2 **a** 44.4° **b** 39.8° **c** 44.4° **d** 49.5°
 e 58.7° **f** 38.7°
3 **a** 67.4° **b** 11.3 **c** 134 **d** 28.1°
 e 39.7 **f** 263 **g** 50.2° **h** 51.3°
 i 138 **j** 22.8

Answers to Chapter 26

4 a Sides of right-hand triangle are sine θ and cosine θ
   b Pythagoras' theorem
   c Students should check the formulae

26.6 Application of trigonometric ratios

Exercise 26I
1 14.0 cm
2 a 24.5 cm   b 20.6 cm   c 19.4 cm
3 1.46 km
4 3.33
5 10.1 km
6 22°
7 429 m
8 a 156 m
   b No. the new angle of depression is $\tan^{-1}\left(\frac{200}{312}\right) = 33°$ and half of 52° is 26°
9 a 222 m   b 42°
10 a 21.5 m   b 17.8 m
11 13.4 m
12 19°
13 The angle is 16° so Cara is not quite correct.

26.7 Problems in three dimensions

Exercise 26J
1 25.1°
2 a 25 cm   b 58.6°   c 20.5 cm
3 a 3.46 m   b 75.5°   c 73.2°
4 a 24.0°   b 48.0°   c 13.5 cm   d 16.6°
5 It is 44.6°; use triangle XDM where M is the midpoint of BD; triangle DXB is isosceles, as X is over the point where the diagonals of the base cross; the length of DB is $\sqrt{656}$. and the cosine of the required angle is $0.5\sqrt{656} \div 18$

26.8 Sine and cosine of obtuse angles

Exercise 26K
1 a The bottom row of the table is 0.174, 0.5, 0.766, 0.996, 1, 0.996, 0.766, 0.5, 0.174.
   b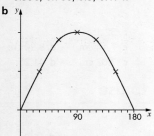
   c It has reflection symmetry. The line of symmetry is $x = 90$.
   d You could choose 10° and 170°, 30° and 150°, 50° and 130° or 85° and 95°
2 30° and 150°.
3 46° and 134°.
4 122.9°

5 a The bottom row of the table is 0.966, 0.819, 0.5, 0.174, 0, −0.174, −0.5, −0.819, −0.966.
   b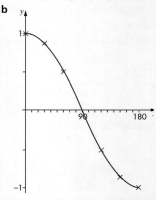
   c It has rotational symmetry of order 2 about the point (90, 0)
6 a 31.8°   b 148.2°   c 120°
   d 90°   e 82.8°   f 97.2°
7 a 53°   b 104°   c 49°, 131°   d 90°
   e 90°   f 72°, 108° g no solution h 45°

26.9 The sine rule and the cosine rule

Exercise 26L
1 a 3.64 m   b 8.05 cm   c 19.4 cm
2 a 46.6°   b 68°   c 36.2°
3 a i 30°   ii 40°
   b 19.4 m
4 36.5 m
5 22.2 m
6 3.47 m
7 64.6 km
8 134°

Exercise 26M
1 a 7.71 m   b 29.1 cm   c 27.4 cm
2 a 76.2°   b 125.1°   c 90°
   d Right-angled triangle
3 5.16 cm
4 65.5 cm
5 a 10.7 cm   b 41.7°   c 38.3°   d 6.69 cm
6 58.4 km at 092.5°
7 21.8°
8 42.5 km
9 111°; the largest angle is opposite the longest side

Exercise 26N
1 a 8.60 m   b 90°   c 27.2 cm
   d 26.9°   e 27.5°   f 62.4 cm
   g 90.0°   h 866 cm   i 86.6 cm
2 7 cm
3 11.1 km
4 a BAC = 90°; this is Pythagoras' theorem
   b BAC is acute
   c BAC is obtuse
5 142 m

26.10 Using sine to find the area of a traingle

Exercise 26O

1. a 24.0 cm²   b 26.7 cm²   c 243 cm²
   d 21 097 cm²   e 1224 cm²
2. 4.26 cm
3. a 42.3°   b 49.6°
4. 2033 cm²
5. 21.0 cm²
6. 726 cm²
7. 149 km²
8. a 66.4 m   b 118.9°   c 1470 m²
9. 43.3 cm²

26.11 Sine, cosine and tangent of any angle

Exercise 26P

1. a 100°   b 34°   c 325°   d 234°
2. a 350°   b 235°   c 152°   d 49°
3. a 27° and 153°   b 56° and 124°
   c 333° and 207°   d 304° and 236°
4. a 37° and 323°   b 103° and 257°
   c 157° and 203°   d 85° and 275°
5. a 30° and 150°   b 60° and 120°
   c 225° and 315°   d 270°
6. a 120° and 240°   b 30° and 330°
   c 45° and 315°   d 90° and 270°
7. a 87.1° and 272.9°   b 54.3° and 124.7°
   c 130.5° and 229.5°   d 323.1° and 216.9°
8. a 41.8° and 138.2°   b 36.9° and 323.1°
   c 314.4° and 225.6°
9. 540°
10. 30°, 150°, 210° and 330°
11. 45°, 135°, 225° and 315°

Exercise 26Q

1. a 215°   b 265°   c 298°
   d 20°   e 63°   f 157°
2. a 45° and 225°   b 135° and 315°
   c 60° and 240°   d 120° and 300°
3. a 11.3° and 191.3°   b 78.7° and 258.7°
   c 160.7° and 340.7°   d 103.5° and 283.5°
4. −1.5
5. a 20.6° and 200.6°   b 69.4° and 249.4°
   c 144.2° and 324.2°
6. a 45°, 135°, 225° and 315°   b 60°, 120°, 240° and 300°
   c 30°, 150°, 210° and 330°
7. 71.6°, 251.6°, 104.0° and 284.0°
8. a They are the same. b and c They have the same magnitude but different signs. They add up to 0.

Answers to Chapter 27

27.1 Perimeter and area of a rectangle

Exercise 27A

1. a 35 cm², 24 cm   b 33 cm², 28 cm
   c 45 cm², 36 cm   d 70 cm², 34 cm
   e 56 cm², 30 cm   f 10 cm², 14 cm
2. a 53.3 cm², 29.4 cm   b 84.96 cm², 38 cm
3. 39
4. a 4   b 1 h 52 min
5. 40 cm
6. Area B, 44 cm²; perimeter B, 30 cm
7. Never (the area becomes four times greater).
8. a 28 cm, 30 cm²   b 28 cm, 40 cm²
   c 40 cm, 51 cm²   d 30 cm, 35 cm²
   e 32 cm, 43 cm²   f 34 cm, 51 cm²
   g cannot tell what the perimeter is; 48 cm²
   h 34 cm, 33 cm²
9. 72 cm²
10. 48 cm

27.2 Area of a triangle

Exercise 27B

1. a 21 cm²   b 12 cm²   c 14 cm²
   d 55 cm²   e 90 cm²   f 140 cm²
2. a 28 cm²   b 8 cm   c 4 cm
   d 3 cm   e 7 cm   f 44 cm²
3. 73.9 cm²
4. a 40 cm²   b 65 m²   c 80 cm²
5. a 65 cm²   b 50 m²
6. For example: height 10 cm, base 10 cm; height 5 cm, base 20 cm; height 25 cm, base 4 cm; height 50 cm, base 2 cm
7. Triangle c; a and b each have an area of 15 cm² but c has an area of 16 cm²

27.3 Area of a parallelogram

Exercise 27C

1. a 96 cm²   b 70 cm²   c 20 m²
   d 125 cm²   e 10 cm²   f 112 m²
2. No, it is 24 cm², she used the slanting side instead of the perpendicular height.
3. 16 cm
4. a 500 cm²   b 15

27.4 Area of a trapezium

Exercise 27D

1. a 30 cm²   b 77 cm²   c 24 cm²   d 42 cm²
   e 40 cm²   f 6 cm   g 3 cm
2. a 27.5 cm, 36.25 cm²
   b 33.4 cm, 61.2 cm²
   c 38.6 m, 88.2 m²
3. Any pair of lengths that add up to 10 cm. For example: 1 cm, 9 cm; 2 cm, 8 cm; 3 cm, 7 cm; 4 cm, 6 cm; 4.5 cm, 5.5 cm

Answers to Chapter 27

4 Shape c. Its area is 25.5 cm²
5 Shape a. Its area is 28 cm²
6 a
7 2 cm
8 1.4 m²

27.5 Circumference and area of a circle

Exercise 27E

1 a 10π cm and 25π cm²   b 6π cm and 9π cm²
   c 3π cm and 2.25π cm² d 8π cm and 16π cm²
2 a 25.1 cm and 50.3 cm²
   b 15.7 cm and 19.6 cm²
   c 28.9 cm and 66.5 cm²
   d 14.8 cm and 17.3 cm²
3 a i 56.5 cm   ii 81π, 254.5 cm²
   b i 69.1 cm   ii 121π, 380.1 cm²
   c i 40.8 cm   ii 42.3π, 132.7 cm²
   d i 88.0 cm   ii 196π, 615.8 cm²
4 a 19.1 cm   b 9.5 cm
   c 286.5 cm² (or 283.5 cm²)
5 962.9 cm² (or 962.1 cm²)
6 a 20 cm   b 400π cm²
7 a 16π m²   b 14π cm²   c 9π cm²
8 45π cm²
9 $a^2 = \pi r^2$, so $r^2 = \frac{a^2}{\pi}$ therefore $r = \frac{a}{\sqrt{\pi}}$
10 21.5 cm²

27.6 Surface area and volume of a cuboid

Exercise 27F

1 a i 198 cm³   ii 234 cm²
   b i 90 cm³   ii 146 cm²
   c i 1440 cm³   ii 792 cm²
   d i 525 cm³   ii 470 cm²
2 24 litres
3 a 160 cm³   b 416 cm³   c 150 cm³
4 a i 64 cm³   ii 96 cm²
   b i 343 cm³   ii 294 cm²
   c i 1000 mm³   ii 600 mm²
   d i 125 m³   ii 150 m²
   e i 1728 m³   ii 864 m²
5 86
6 a 180 cm³   b 5 cm   c 6 cm
   d 10 cm   e 81 cm³
7 1.6 m
8 48 m²
9 a 3 cm   b 5 m   c 2 mm   d 1.2 m
10 a 148 cm³   b 468 cm³
11 If this was a cube, the side length would be 5 cm, so total surface area would be 5 × 5 × 6 = 150 cm²; no this particular cuboid is not a cube.
12 a 6 cm   b 216

27.7 Volume and surface area of a prism

Exercise 27G

1 a i 21 cm²   ii 63 cm³
   b i 48 cm²   ii 432 cm³
   c i 36 m²   ii 324 m³
2 a 432 m³   b 225 m³   c 1332 m³
3 a A cross-section parallel to the side of the pool always has the same shape.
   b About $3\frac{1}{2}$ hours
4 7.65 m³
5 a 21 cm³, 210 cm³
   b 54 cm², 270 cm²
6 146 cm³
7 78 m³ (78.3 m³)
8 327 litres
9 10.2 tonnes
10 672 cm³

27.8 Volume and surface area of a cylinder

Exercise 27H

1 a i 72π cm³   ii 66π cm²
   b i 4.75π cm³ ii 19.5π cm²
   c i 110π cm³   ii 87.5π cm²
   d i 338π cm³   ii 203π cm²
2 a i 72π cm³   ii 48π cm²
   b i 112π cm³   ii 56π cm²
   c i 180π cm³   ii 60π cm²
   d i 600π m³   ii 120π m²
3 665 cm³
4 Label should be less than 10.5 cm wide so that it fits the can and does not overlap the rim and more than 23.3 cm long to allow an overlap.
5 332 litres
6 There is no right answer. Students could start with the dimensions of a real can. Often drinks cans are not exactly cylindrical. One possible answer is height of 6.6 cm and diameter of 8 cm.
7 About 127 cm
8 A diameter of 10 cm and a length of 5 cm give a volume close to 400 cm³ (0.4 litres).

27.9 Sectors and arcs: 1

Exercise 27I

1 a 20π cm   b i 10π cm   ii 5π cm   iii 2.5π cm
2 a 100π cm² b i 50π cm²   ii 25π cm² iii 12.5π cm²
3 a $\frac{1}{5}$ b 10.6 cm   c 44.3 cm²
4 a 96.5 cm²   b 20.1 cm   c 39.3 cm²
5 a 245.4 cm² to 1 d.p.   b 64.3 cm to 1 d.p.
6 a The diameter is 80 and the fraction of a circle is $\frac{1}{10}$. The arc length is $\pi \times 80 \div 10 = 8\pi$ cm.
   b 160π cm²
7 a 20π b 75π cm²

27.10 Sectors and arcs: 2

Exercise 27J

1. a i 5.59 cm ii 22.3 cm²
   b i 8.29 cm ii 20.7 cm²
   c i 16.3 cm ii 98.0 cm²
   d i 15.9 cm ii 55.6 cm²
2. a 9π cm b 54π cm²
3. a 73.8 cm b 20.3 cm
4. a 107 cm² b 173 cm²
5. 43.6 cm
6. $(36\pi - 72)$ cm²
7. $(32\pi - 64)$
8. a 13.9 cm b 7.07 cm²

27.11 Volume of a pyramid

Exercise 27K

1. a 56 cm³ b 168 cm³ c 1040 cm³
   d 84 cm³ e 160 cm³
2. 270 cm³
3. a Put the apexes of the pyramids together. The 6 square bases will then form a cube.
   b If the side of the base is a then the height will be $\frac{1}{2}a$.
   Total volume of the 6 pyramids is a^3.
   Volume of one pyramid is
   $\frac{1}{6}a^3 = \frac{1}{3} \times \frac{1}{2} \times a \times a^2 = \frac{1}{3} \times$ height $\times$ base area
4. 6.9 m (height of cuboid)
5. a 73.3 m³ b 45 m³ c 3250 cm³
6. 1.5 g
7. 5.95 cm
8. 14.4 cm
9. 260 cm³

27.12 Volume and surface area of a cone

Exercise 27L

1. a i 3560 cm³ ii 1430 cm²
   b i 314 cm³ ii 283 cm²
   c i 1020 cm³ ii 679 cm²
2. 24π cm²
3. a 816π cm³ b 720π mm³
4. a 4 cm b 6 cm
   c Various answers, e.g. 60° gives 2 cm, 240° gives 8 cm
5. 24π cm²
6. If radius of base is r, slant height is $2r$.
   Area of curved surface = $\pi r \times 2r = 2\pi r^2$, area of base = πr^2
7. 140 g
8. 2.81 cm

27.13 Volume and surface of a sphere

Exercise 27M

1. a 36π cm³ and 36π cm²
   b 288π cm³ and 144π cm²
   c 1330π cm³ and 400π cm²
2. 65 400 cm³, 7850 cm²
3. a 1960 cm²
   b 8180 cm³
4. 125 cm
5. 6231
6. 7.8 cm
7. 48%.

Answers to Chapter 28

28.1 Lines of symmetry

Exercise 28A

1. a   b   c

   d   e   f

   g

2. a i 5 ii 6 iii 8
   b 10
3. 2, 1, 1, 2, 0
4.
5. a 1 b 5 c 1 d 6
6.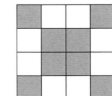

Answers to Chapter 29

28.2 Rotational symmetry

Exercise 28B

1. a 4   b 2   c 2   d 3   e 6
2. a 4   b 5   c 6   d 4   e 6
3. a 2   b 2   c 2   d 2   e 2
4. a 6
   b 9 (the small red circle surrounded by nine 'petals') and 12 (the centre pattern)
5. For example:

28.3 Symmetry of special two-dimensional shapes

Exercise 28C

1.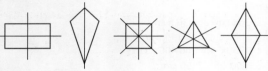

2. a kite
   b rectangle 2, square 4, equilateral triangle 3, rhombus 2
3. a isosceles   b no
4. a parallelogram   b square
5. a rectangle and rhombus   b no
6. a B and D   b AB and AD; CB and CD   c kite
7. a diameter   b infinite   c infinite
8. a A and C; B and D
   b AD and BC; AB and DC
   c Parallelogram
9. It will have two pairs of equal angles

28.4 Symmetry of three-dimensional shapes

Exercise 28D

1.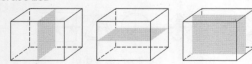

2. a Diagrams to show axes going through the centres of all three pairs of opposite faces
   b 2
3. Two are similar to the one shown, dividing the end triangles in two. The other goes through the centre of each of the long edges, parallel to the end triangles.
4. a 3 about AB; 2 about CD
   b they are similar to CD, each passing through the centre of a rectangular face.
5. a 4   b no
6. There are four. All pass through the vertex. Two pass through opposite corners of the square. Two pass through the mid points of opposite sides of the square
7. a Any plane dividing each circle in half or the circular plane exactly half way up the cylinder
   b any line at right angles to the one shown, passing through the centre of the cylinder
8. a one   b infinite
9. a six through the centre of each hexagon; one parallel to the hexagons passing through the centre of the prism
   b 4

28.5 Symmetry in circles

Exercise 28E

1. a–d student's own drawing
   e because the perpendicular bisector of any chord passes through the centre of the circle
   f Here is one method: draw two chords; construct the perpendicular bisectors; they meet at the centre
2. a Isosceles because OA and OB are radii
   b OA = OC; OB = OD; AB = CD so corresponding sides are equal
   c 50°
3. a EM = FM (given); OE = OF (radii); OM is common to both. Corresponding sides are equal
   b EMO and FMO are equal and add up to 180° (because EMF is a straight line) so they must both be 90°
   c 18°
4. a Angle between a radius and a tangent
   b XP = YP (tangents from a point are equal); OX = OY (radii); OP is common. So corresponding sides are equal
   c 146°
5. 5 cm (use Pythagoras' theorem)
6. 40°

Answers to Chapter 29

29.1 Introduction to vectors

Exercise 29A

1. a i $\begin{pmatrix} 4 \\ -1 \end{pmatrix}$ ii $\begin{pmatrix} 3 \\ 1 \end{pmatrix}$ iii $\begin{pmatrix} 2 \\ 3 \end{pmatrix}$ iv $\begin{pmatrix} -2 \\ 4 \end{pmatrix}$ v $\begin{pmatrix} 2 \\ -4 \end{pmatrix}$ vi $\begin{pmatrix} -1 \\ 2 \end{pmatrix}$

   b Both are $\begin{pmatrix} 1 \\ -2 \end{pmatrix}$. D is the midpoint of AC.

Answers to Chapter 29

2 a

b i $\begin{pmatrix} -2 \\ 4 \end{pmatrix}$ **ii** $\begin{pmatrix} 2 \\ -4 \end{pmatrix}$ **iii** $\begin{pmatrix} 3 \\ 4 \end{pmatrix}$

iv $\begin{pmatrix} -1 \\ 2 \end{pmatrix}$ **v** $\begin{pmatrix} -2 \\ -6 \end{pmatrix}$

3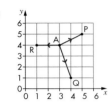

4 The diagrams should show the following vectors:

a $\begin{pmatrix} 3 \\ -1 \end{pmatrix}$ **b** $\begin{pmatrix} -1 \\ 4 \end{pmatrix}$ **c** $\begin{pmatrix} 1 \\ 7 \end{pmatrix}$

d $\begin{pmatrix} 3 \\ -1 \end{pmatrix}$ **e** $\begin{pmatrix} -1 \\ -7 \end{pmatrix}$ **f** $\begin{pmatrix} 2 \\ -8 \end{pmatrix}$

5 a $\begin{pmatrix} -1 \\ 2 \end{pmatrix}$ **b** $\begin{pmatrix} 6 \\ 12 \end{pmatrix}$ **c** $\begin{pmatrix} -5 \\ -6 \end{pmatrix}$

d $\begin{pmatrix} 5 \\ 6 \end{pmatrix}$ **e** $\begin{pmatrix} -12 \\ -8 \end{pmatrix}$ **f** $\begin{pmatrix} 1 \\ 6 \end{pmatrix}$

6 a–d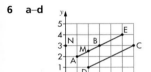

e k is 4.

29.2 Using vectors

Exercise 29B

1 a Any three, of: $\vec{AC}, \vec{CF}, \vec{BD}, \vec{DG}, \vec{GI}, \vec{EH}, \vec{HJ}, \vec{JK}$
b Any three of: $\vec{BE}, \vec{AD}, \vec{DH}, \vec{CG}, \vec{GJ}, \vec{FI}, \vec{IK}$
c Any three of: $\vec{AO}, \vec{CA}, \vec{FC}, \vec{IG}, \vec{GD}, \vec{DB}, \vec{KJ}, \vec{JH}, \vec{HE}$
d Any three of: $\vec{BO}, \vec{EB}, \vec{HD}, \vec{DA}, \vec{JG}, \vec{GC}, \vec{KI}, \vec{IF}$

2 a 2**a** **b** 2**b** **c** 3**a** + 2**b** **d** **a** + 2**b**
 e **a** + **b** **f** 2**a** + 2**b** **g** 3**a** + **b**

3 $\vec{AI}, \vec{BJ}, \vec{DK}$

4 $\vec{OF}, \vec{BI}, \vec{EK}$

5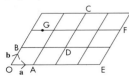

6 a **a** – **b** **b** 3**a** – **b** **c** 2**a** – **b** **d** **a** – **b**
 e **a** + **b** **f** –**a** – **b** **g** 2**a** – **b** **h** –**a** – 2**b**
 i **a** + 2**b** **j** –**a** + **b** **k** 2**a** – 2**b** **l** **a** – 2**b**

7 a $\vec{BJ}, \vec{CK}$
 b $\vec{EB}, \vec{GO}, \vec{KH}$

8

9 a i 3**a** + 2**b** **ii** 3**a** + **b**
 iii 2**a** – **b** **iv** 2**b** – 2**a**
 b $\vec{DG}$ and $\vec{BC}$

10 a 2**a** + **b** **b** 2**b** – **a** **c** **a** + 1.5**b**
 d 0.5**a** + 0.5**b** **e** 1.5**a** + 1.5**b** **f** 1.5**a** – 0.5**b**

11 a i –**a** + **b** **ii** $\frac{1}{2}$(–**a** + **b**)
 iii **iv** $\frac{1}{2}$**a** + $\frac{1}{2}$**b**

b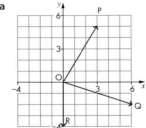

c M is midpoint of parallelogram of which OA and OB are two sides.

12 a i –**a** + **b** **ii** $\frac{1}{3}$(–**a** + **b**) **iii** $\frac{2}{3}$**a** + $\frac{1}{3}$**b**
 b $\frac{3}{4}$**a** + $\frac{1}{4}$**b**

13 a i $\frac{2}{3}$**b** **ii** $\frac{1}{2}$**a** + $\frac{1}{2}$**b** **iii** $-\frac{2}{3}$**b**
 b $\frac{1}{2}$**a** – $\frac{1}{6}$**b**
 c $\vec{DE} = \vec{DO} + \vec{OE} = \frac{3}{2}$**a** – $\frac{1}{2}$**b**
 d $\vec{DE}$ parallel to $\vec{CD}$ = (multiple of $\vec{CD}$) and D is a common point

29.3 The magnitude of a vector

Exercise 29C

1 a

b √34; √40; 4
c √58
d √40

2 a 10 and 13 **b** $\begin{pmatrix} 11 \\ -4 \end{pmatrix}$ **c** √137

d No. 10 + 13 does not equal √137
e √401 **f** √401
g Yes. They are vectors in opposite directions but the same length.

3 a 10, 10, 10
b Because they are all the same distance from A. The radius is 10.

4 a √17 **b** √261 **c** 13 **d** 10

749

Answers to Chapter 30

30.1 Translations

Exercise 30A

1. **a** i $\begin{pmatrix} 1 \\ 3 \end{pmatrix}$ ii $\begin{pmatrix} 4 \\ 2 \end{pmatrix}$ iii $\begin{pmatrix} 2 \\ -1 \end{pmatrix}$

   b i $\begin{pmatrix} 4 \\ -2 \end{pmatrix}$ ii $\begin{pmatrix} -2 \\ 3 \end{pmatrix}$ iii $\begin{pmatrix} 3 \\ 3 \end{pmatrix}$

   c i $\begin{pmatrix} -4 \\ -2 \end{pmatrix}$ ii $\begin{pmatrix} 1 \\ -1 \end{pmatrix}$ iii $\begin{pmatrix} 0 \\ 4 \end{pmatrix}$

   d i $\begin{pmatrix} -2 \\ -7 \end{pmatrix}$ ii $\begin{pmatrix} 5 \\ 0 \end{pmatrix}$ iii $\begin{pmatrix} 1 \\ -5 \end{pmatrix}$

2.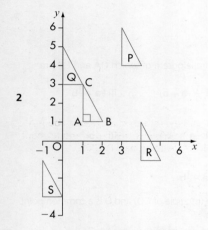

3. **a** $\begin{pmatrix} -3 \\ -1 \end{pmatrix}$ **b** $\begin{pmatrix} 4 \\ -4 \end{pmatrix}$ **c** $\begin{pmatrix} -5 \\ -2 \end{pmatrix}$

   d $\begin{pmatrix} 4 \\ 7 \end{pmatrix}$ **e** $\begin{pmatrix} -1 \\ 5 \end{pmatrix}$ **f** $\begin{pmatrix} 1 \\ 6 \end{pmatrix}$

   g $\begin{pmatrix} -4 \\ 4 \end{pmatrix}$ **h** $\begin{pmatrix} -4 \\ -7 \end{pmatrix}$

4. $\begin{pmatrix} -x \\ -y \end{pmatrix}$

5. $\begin{pmatrix} -1 \\ 4 \end{pmatrix}$

30.2 Reflections: 1

Exercise 30B

1.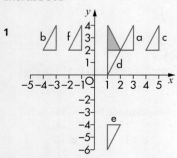

2. **a–e**

   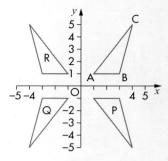

   f Reflection in the y-axis

3. **a–b**

   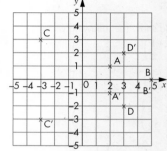

   c y-value changes sign
   d $(a, -b)$

4. **a–b**

   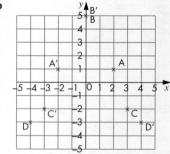

   c x-value changes sign
   d $(-a, b)$

5. Possible answer: Take the centre square as ABCD then reflect this square each time in the line, AB, then BC, then CD and finally AD.

6. $x = -1$

30.3 Reflections: 2

Exercise 30C

1. Possible answer:

Answers to Chapter 30

2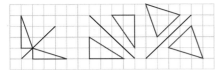

3 a–i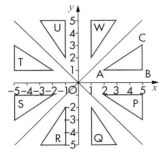

j A reflection in $y = x$

4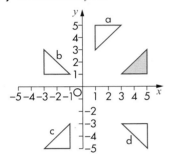

5 a–c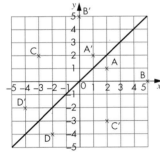

d Coordinates are reversed: x becomes y and y becomes x

e (b, a)

6 a–c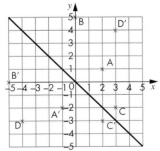

d Coordinates are reversed and the sign changes: x becomes $-y$ and y becomes $-x$

e $(-b, -a)$

30.4 Rotations: 1

Exercise 30D

1 a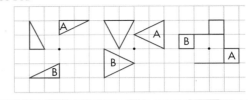

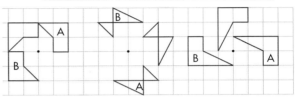

b i Rotation 90° anticlockwise

ii Rotation 180°

2

3 Possible answer: If ABCD is the centre square, rotate about A 90° anticlockwise, rotate about new B 180°, now rotate about new C 180°, and finally rotate about new D 180°.

4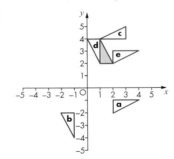

5 a (4,5) 180°
 b (5,5) 90° anticlockwise
 c (3,3) 180°
 b (3,5) 90° clockwise

6 a E   **b** H

7 i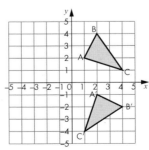

ii A' (2, −1), B' (4, −2), C' (1, −4)

iii Original coordinates (x, y) become $(y, -x)$

iv Yes

751

Answers to Chapter 30

8 i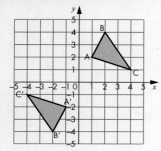

ii A' (−1, −2), B' (−2, −4), C' (−4, −1)
iii Original coordinates (x, y) become (−x, −y)
iv Yes

9 Show by drawing a shape or use the fact that (a, b) becomes (a, −b) after reflection in the x-axis, and (a, −b) becomes (−a, −b) after reflection in the y-axis, which is equivalent to a single rotation of 180°.

30.5 Rotations: 2

Exercise 30E

1 a-c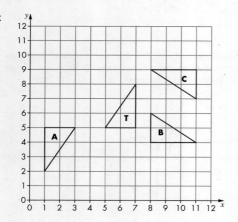

d Rotation of 180° about (9.5, 6.5).

2 a (3,0) b (0,0) c (6,0)

3 a (0, −1.5) 180°   b (−0.5, −1.5) 90° clockwise

c (−3,5,2.5) 90° anti clockwise

d (0.5, 2) 180°

4 Show by drawing a shape or use the fact that (a, b) becomes (b, a) after reflection in the line y = x, and (b, a) becomes (−a, −b) after reflection in the line y = −x, which is equivalent to a single rotation of 180°.

5 a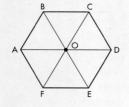

b i Rotation 60° clockwise about O
  ii Rotation 120° clockwise about O
  iii Rotation 180° about O
  iv Rotation 240° clockwise about O
c i Rotation 60° clockwise about O
  ii Rotation 180° about O

6 Rotation 90° anticlockwise about (3, −2).

7 a y = x   b (1, 1)   c (6, 6)   d not possible

30.6 Enlargements: 1

Exercise 30F

1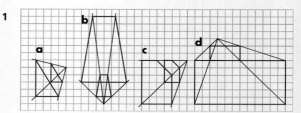

2 a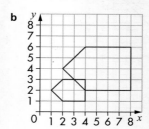

b

c

3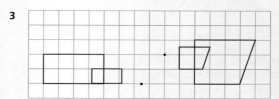

752

Answers to Chapter 30

4 a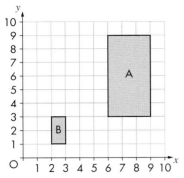

b 3 : 1   **c** 3 : 1   **d** 9 : 1

30.7 Enlargements: 2

Exercise 30G

1 a 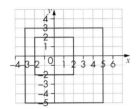   **b**

2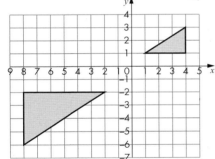

3 a–c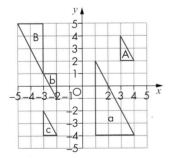

d Scale factor $-\frac{1}{2}$, centre (1, 3)
e Scale factor −2, centre (1, 3)
f Scale factor −1, centre (−2.5, −1.5)
g Scale factor −1, centre (−2.5, −1.5)
h Same centres, and the scale factors are reciprocals of each other

4 Enlargement, scale factor −2, about (1, 3)

5 a 9.6 cm   **b** 25 : 1

30.8 Combined transformations

Exercise 30H

1 (−4, −3)

2 a (−5, 2)   **b** Reflection in y-axis

3 A: translation $\begin{pmatrix} 1 \\ -2 \end{pmatrix}$,
B: reflection in y-axis,
C: rotation 90° clockwise about (0, 0),
D: reflection in x = 3,
E: reflection in y = 4,
F: enlargement by scale factor 2, centre (0, 1)

4 a T_1 to T_2: rotation 90° clockwise about (0, 0)
  b T_1 to T_6: rotation 90° anticlockwise about (0, 0)
  c T_2 to T_3: translation $\begin{pmatrix} 2 \\ 2 \end{pmatrix}$
  d T_6 to T_2: rotation 180° about (0, 0)
  e T_6 to T_5: reflection in y-axis
  f T_5 to T_4: translation $\begin{pmatrix} 4 \\ 0 \end{pmatrix}$

5 a–c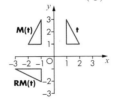

d Reflection in the line $y = -x$

6 Reflection in x-axis, translation $\begin{pmatrix} 0 \\ -5 \end{pmatrix}$, rotation 90° clockwise clockwise about (0, 0)

7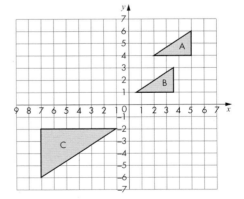

b Enlargement, scale factor $-\frac{1}{2}$, centre (1, 2)

Answers to Chapter 31

31.1 Frequency tables

Exercise 31A

1 a

Goals	0	1	2	3
Frequency	6	8	4	2

b 1 goal
c 22

2 a

Temperature (°C)	14–16	17–19	20–22	23–25	26–28
Frequency	5	10	8	5	2

b 17–19° C
c Getting warmer in the first half and then getting cooler towards the end.

3 a

Score	1	2	3	4	5	6
Frequency	5	6	6	6	3	4

b 30
c Yes, frequencies are similar.

4 a

Height (cm)	151–155	156–160	161–165	166–170
Frequency	2	5	5	7
Height (cm)	171–175	176–180	181–185	186–190
Frequency	5	4	3	1

b 166–170 cm
c student's survey results

5 Various answers such as 1–10, 11–20, etc. or 1–20, 21–40, 41–60

6 The ages 20 and 25 are in two different groups.

31.2 Pictograms

Exercise 31B

1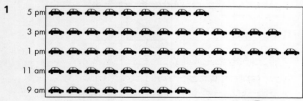

2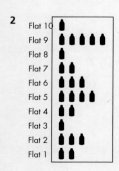

3 a May 10 h, Jun 12 h, Jul 12 h, Aug 12 h, Sep 10 h
  b Visual impact, easy to understand.
4 a Simon b $165
  c Difficult to show fractions of a symbol.
5 a i 12 ii 6 iii 13
  b Check students' pictograms.
  c 63

31.3 Bar charts

Exercise 31C

1 a Swimming b 74

2 a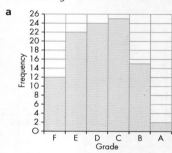

b $\frac{40}{100} = \frac{2}{5}$

c Easier to read the exact frequency.

3 a

b Amir got more points overall, but Hasrul was more consistent.

4 a

Time (min)	1–10	11–20	21–30	31–40	41–50	51–60
Frequency	4	7	5	5	7	2

b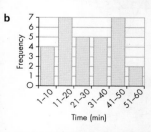

c For example: Some live close to the school. Some live a good distance away and probably travel to school by bus.

Answers to Chapter 31

5 a

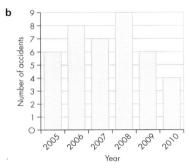

Key 🚚 = 1 accident

b

[Bar chart showing Number of accidents by Year 2005–2010: 6, 8, 7, 9, 6, 4]

c Use the pictogram because an appropriate symbol makes more impact.

6 Yes. If you double the minimum temperature each time, it is very close to the maximum temperature.

31.4 Pie charts

Exercise 31D

1 a

b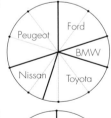

c

2 Pie charts with following angles:
   a 36°, 90°, 126°, 81°, 27°
   b 168°, 52°, 100°, 40°

3 Pie chart with these angles: 60°, 165°, 45°, 15°, 75°

4 a 36
   b Pie chart with these angles: 50°, 50°, 80°, 60°, 60°, 40°, 20°
   c student's bar chart
   d Bar chart, because easier to make comparisons.

5 a Pie charts with these angles: 124°, 132°, 76°, 28°
   b Split of total data seen at a glance.

6 a 55°   **b** 22

7 a Pie charts with these angles: Strings: 36°, 118°, 126°, 72°, 8°
   Brass: 82°, 118°, 98°, 39°, 23°
   b Overall, the Strings candidates did better, as a smaller proportion failed. A higher proportion of Brass candidates scored very good or excellent.

8 $\frac{1}{9}$

9 a Accept any valid comment that compares the two schools, such as:

School A had a greater percentage of students attaining the top 10 marks than School B

12.5% of School B obtained 30 or less marks: this was half the percentage of School A's results etc.

Reject answers that refer to numbers of students, e.g. more students got marks in the range 61–90 at School B

   b Answers could include:
   - the actual numbers of students are unknown
   - the size of the pie chart can be misleading.

31.5 Scatter diagrams

Exercise 31E

1 a No correlation
   b Positive correlation

2 a No relationship between temperature and speed of cars.
   b As people get older, they have more money in the bank.

3 a and **b** student's scatter diagram and line of best fit.
   c about 20 cm/s
   d about 35 cm

4 a student's scatter diagram.
   b Yes, usually (good correlation).

5 a and **b** Student's scatter diagram and line of best fit.
   c Sitara
   d about 90
   e about 55

6 a student's scatter diagram.
   b no, because there is no correlation.

7 a and **b** Student's scatter diagram and line of best fit.
   c about 2.4 km
   d 8 minutes

8 23 kilometres/hour

9 Points showing a line of best fit sloping down from top left to bottom right.

Answers to Chapter 31

31.6 Histograms

Exercise 31F

1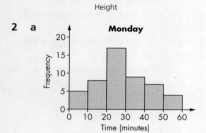

2 a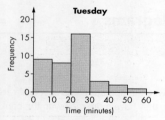

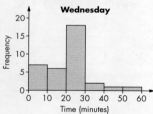

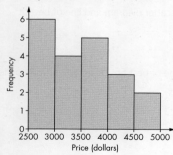

b Monday

3 a The frequencies are 6, 4, 5, 3, 2

b

4 a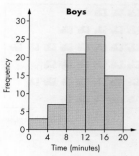

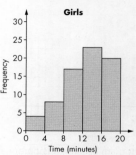

b 72 boys and 72 girls
c Nobody took longer than 20 minutes
d student's own comments

5 a and **b**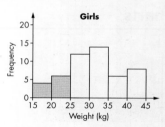

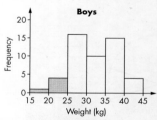

c More girls were underweight.

6 a 18
b 36
c The shortest time was at least 10 minutes. The longest time was at most 70 minutes.
d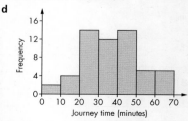

e Yes. Student's own explanation.

31.7 Histograms with bars of unequal width

Exercise 31G

1 The respective frequency densities on which each histogram should be based are:
 a 2.5, 6.5, 6, 2, 1, 1.5
 b 4, 27, 15, 3
 c 17, 18, 12, 6.67
 d 0.4, 1.2, 2.8, 1
 e 9, 21, 13.5, 9

2 a

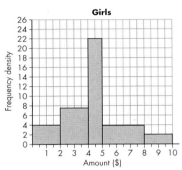

3

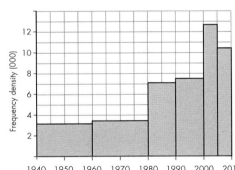

4 a 775   **b** 400

5 a

Age, y (years)	$9 < y \leq 10$	$10 < y \leq 12$	$12 < y \leq 14$
Frequency	4	12	8
Age, y (years)	$14 < y \leq 17$	$17 < y \leq 19$	$19 < y \leq 20$
Frequency	9	5	1

b

Temperature, t (°C)	$10 < t \leq 11$	$11 < t \leq 12$	$12 < t \leq 14$
Frequency	15	15	50
Temperature, t (°C)	$14 < t \leq 16$	$16 < t \leq 19$	$19 < t \leq 21$
Frequency	40	45	15

c

Mass, m (kg)	$50 < m \leq 70$	$70 < m \leq 90$	$90 < m \leq 100$
Frequency	160	200	120
Mass, m (kg)	$100 < m \leq 120$	$120 < m \leq 170$	
Frequency	120	200	

6 a

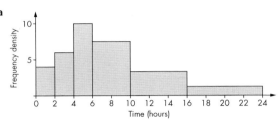

b 45

7 a

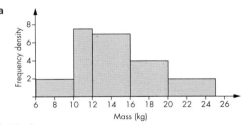

b 33 plants

8 a

Speed, v (mph)	$0 < v \leq 40$	$40 < v \leq 50$	$50 < v \leq 60$
Frequency	80	10	40
Speed, v (mph)	$60 < v \leq 70$	$70 < v \leq 80$	$80 < v \leq 100$
Frequency	110	60	60

b 360

9 a 80
 b 31.25%

10 a

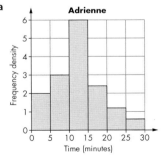

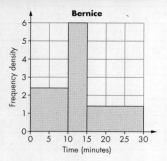

b student's own description

Answers to Chapter 32

32.1 The mode

Exercise 32A

1. **a** 4 **b** 48 **c** −1 **d** $\frac{1}{4}$ **e** no mode **f** 3.21
2. **a** red **b** Sun **c** β **d** ★
3. **a** 32 **b** 6 **c** no
   d no; boys generally take larger shoe sizes
4. **a** 5
   b no; more than half the form got a higher mark
5. The mode will be the most popular item or brand sold in a shop.
6. **a** 28
   b **i** brown **ii** blue **iii** brown
   c Both students had blue eyes.
7. **a** May lose count.
   b Put in a table, or arrange in order.
   c 4

32.2 The median

Exercise 32B

1. **a** 5 **b** 33 **c** $7\frac{1}{2}$ **d** 24
   e $8\frac{1}{2}$ **f** 0 **g** 5.25
2. **a** $2.20 **b** $2.25
   c median, because it is the central value
3. **a** 5
   b **i** 15 **ii** 215 **iii** 10 **iv** 10
4. **a** 13, Ella **b** 162 cm, Pat **c** 40 kg, Elisa
   d Ella, because she is closest to the 3 medians
5. **a** 12 **b** 14
6. Answers will vary
7. 12, 14, 14, 16, 20, 22, 24
8. **a** Possible answer: 11, 15, 21, 21 (one below or equal to 12 and three above or equal)
   b Any four numbers higher than or equal to 12, and any two lower or equal
   c Eight, all 4 or under
9. A median of $8 does not take into account the huge value of the $3000 so is in no way representative.

32.3 The mean

Exercise 32C

1. **a** 6 **b** 24 **c** 45 **d** 1.57 **e** 2
2. **a** 55.1 **b** 324.7 **c** 58.5 **d** 44.9 **e** 2.3
3. **a** 61 **b** 60 **c** 59 **d** Badru **e** 2
4. 42 min
5. **a** $200 **b** $260 **c** $278
   d Median, because the extreme value of $480 is not taken into account
6. **a** 35 **b** 36
7. **a** 6
   b 16; all the numbers and the mean are 10 more than those in part **a**
   c **i** 56 **ii** 106 **iii** 7
8. Possible answers: Speed – Kath, James, John, Joseph; Roberts – Frank, James, Helen, Evie. Other answers are possible.
9. 36
10. 24

32.4 The range

Exercise 32D

1. **a** 7 **b** 26 **c** 5 **d** 2.4 **e** 7
2. **a** 5°, 3°, 2°, 7°, 3°
   b Variable weather over England
3. **a** $31, $28, $33
   b $8, $14, $4
   c Not particularly consistent
4. **a** 82 and 83
   b 20 and 12
   c Fay, because her scores are more consistent
5. **a** 5 min and 4 min
   b 9 min and 13 min
   c Number 50, because times are more consistent
6. **a** Isaac, Oliver, Evrim, Chloe, Lilla, Badru and Isambard
   b 70 cm to 92 cm
7. **a** Teachers because they have a high mean and students could not have a range of 20.
   b Year 11 students as the mean is 15–16 and the range is 1.

Answers to Chapter 32

32.5 Which average to use

Exercise 32E

1. **a** i 29 ii 28 iii 27.1
   b 14
2. **a** i Mode 3, median 4, mean 5
   ii 6, 7, $7\frac{1}{2}$
   iii 4, 6, 8
   b i Mean: balanced data
   ii Mode: 6 appears five times
   iii Median: 28 is an extreme value
3. **a** Mode 73, median 76, mean 80
   b The mean, because it is the highest average
4. **a** 150 **b** 20
5. **a** Mean **b** Median
   c Mode **d** Median
   e Mode **f** Mean
6. No. Mode is 31, median is 31, and mean is 31½.
7. **a** Median **b** Mode **c** Mean
8. Tom mean, David median, Mohamed mode
9. Possible answers: **a** 1, 6, 6, 6, 6 **b** 2, 5, 5, 6, 7
10. Boss chose the mean while worker chose the mode.
11. 11.6
12. 52.7 kg

32.6 Stem-and-leaf diagrams

Exercise 32F

1. **a** 40 **b** 75 marks **c** 43 marks **d** 71 marks
   e You know that half the students got more marks than the median and half got fewer. The mode does not have such a clear use.
2. **a** 18 runners **b** 26.7 s **c** 4.9 s
3. **a** 6 people **b** 35 minutes **c** 70 minutes
4. **a**
   2 | 8 9
   3 | 4 5 6 8 8 9
   4 | 1 1 3 3 3 8 8
   b 43 cm **c** 39 cm **d** 20 cm
5. **a**
   0 | 2 8 9 9 9
   1 | 2 3 7 7 8
   2 | 0 1 2 3
   b 9 messages **c** 15 messages
6. **a**
   0 | 7 8 9 9
   1 | 0 2 3 4 5 8 8 9 9   key 2 | 3 = 23
   2 | 0 3 4 4 6 8
   3 | 1
   b 18 **c** 24

7.

	Men	Women
Number of people	41	34
Range of ages	42	33
Median age	43 years	32

8. **a** 8 children
   b

	Girls	Boys
Number of children	25	49
Median height	148 cm	146 cm
Range of heights	40 cm	45 cm

   c i 2 cm more ii 19.28 cm less

32.7 Using frequency tables

Exercise 32G

1. **a** i 7 ii 6 iii 6.4
   b i 4 ii 4 iii 3.7
   c i 8 ii 8.5 iii 8.2
   d i 0 ii 0 iii 0.3
2. **a** 668 **b** 1.9 **c** 0 **d** 328
3. **a** 2.2, 1.7, 1.3 **b** Better dental care
4. **a** 0 **b** 0.96
5. **a** 7 **b** 6.5 **c** 6.5
6. **a** 1 **b** 1 **c** 0.98
7. **a** Roger 5, Brian 4
   b Roger 3, Brian 8
   c Roger 5, Brian 4
   d Roger 5.4, Brian 4.5
   e Roger, because he has the smaller range
   f Brian, because he has the better mean
8. Possible answers: 3, 4, 15, 3 or 3, 4, 3, 15 …
9. Add up the weeks to see she travelled in 52 weeks of the year, the median is in the 26th and 27th week. Looking at the weeks in order, the 23rd entry is the end of 2 days in a week so the median must be in the 3 days in a week.

32.8 Grouped data

Exercise 32H

1. **a** i $30 < x \leq 40$ ii 29.5
   b i $0 < y \leq 100$ ii 158.3
   c i $5 < z \leq 10$ ii 9.43
   d i 7–9 ii 8.4 weeks
2. **a** $100 < m \leq 120$ g **b** 10 860 g **c** 108.6 g
3. **a** 207 **b** 19–22 cm **c** 20.3 cm
4. **a** 160 **b** 52.6 min **c** modal group
   d 65%
5. **a** $175 < h \leq 200$ **b** 31% **c** 193.25
   d No: mode, mean and median are all less than 200 hours
6. Average price increases: Soundbuy 17.6p, Springfields 18.7p, Setco 18.2p
7. Yes: average distance is 11.7 miles per day.
8. The first 5 and the 10 are the wrong way round.
9. $740
10. As we do not know what numbers are in each group, we cannot say what the median is.

Answers to Chapter 32

32.9 Cumulative frequency diagrams

Exercise 32I

1 a Cumulative frequency 1, 4, 10, 22, 25, 28, 30
  b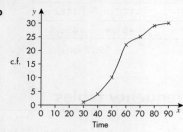
  c 54 secs, 16 secs

2 a Cumulative frequency 1, 3, 5, 14, 31, 44, 47, 49, 50
  b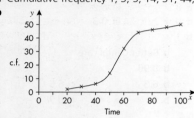
  c 56 secs, 17 secs
  d Pensioners, median closer to 60 secs

3 a Cumulative frequency 12, 30, 63, 113, 176, 250, 314, 349, 360
  b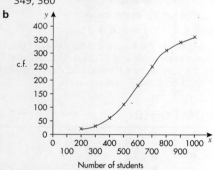
  c 605 students, 280 students
  d 46–47 schools
  e about 830
  f about 550

4 a Cumulative frequency 2, 5, 10, 16, 22, 31, 39, 45, 50
  b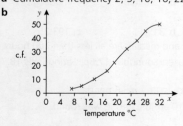
  c 20.5°C, 10°C   d 10.5°C

5 a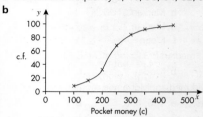
  b 56, 43
  c about 17.5%

6 a Cumulative frequency 6, 16, 36, 64, 82, 93, 98, 100
  b

  (graph)

  c 225c, 90c
  d about 120 cents and about 340 cents

7 a Paper A 68, Paper B 57
  b Paper A 28, Paper B 18
  c Paper B is the harder paper, it has a lower median and a lower upper quartile.
  d i Paper A 43, Paper B 45   ii Paper A 78, Paper B 67

8 a about 40%   b about 6 minutes

9 Find the top 10% on the cumulative frequency scale, read along to the graph and read down to the marks. The mark seen will be the minimum mark needed for this top grade.

32.10 Box-and-whisker plots

Exercise 32J

1 a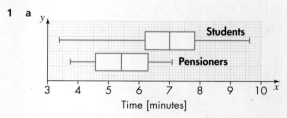
  b Students are much slower than the pensioners. Both distributions have the same interquartile range, but students' median and upper quartiles are 1 minute, 35 seconds higher. The fastest person to complete the calculations was a student, but so was the slowest.

2 a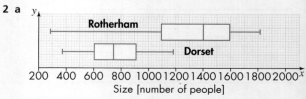

Answers to Chapter 33

b Schools are much larger in Rotherham than Dorset. The Dorset distribution is symmetrical, but the Rotherham distribution is negatively skewed – so most Rotherham schools are large.

3 a The resorts have similar median temperatures, but Resort B has a much wider temperature range, where the greatest extremes of temperature are recorded.

b Resort A is probably a better choice as the weather seems more consistent.

4 a

b Both distributions have a similar interquartile range, and there is little difference between the upper quartile values. Men have a wider range of salaries, but the higher men's median and the fact that the men's distribution is negatively skewed and the women's distribution is positively skewed indicates that men are better paid than women.

5 a

b £1605, £85

c i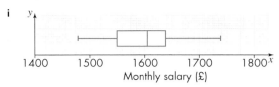

ii Negatively

6 a i 24 min   **ii** 12 min   **iii** 42 min

b i 6 min   **ii** 17 min   **iii** 9 min

c Either doctor with a plausible reason, e.g. Dr Excel because his waiting times are always shorter than Dr Collins', or Dr Collins because he takes more time with each patient

7 The girls have a mean 2.6 higher than the boys. (Create grouped frequencies using the four quartiles.)

8 Many possible answers but not including numerical values: Bude (Torquay) had a higher median amount of sunshine than Torquay (Bude), Bude had a smaller interquartile range than Torquay, Bude had more sunshine on any one day.

9 a Symmetric   **b** Negatively skewed
   c Negatively skewed   **d** Symmetric
   e Negatively skewed   **f** Positively skewed
   g Negatively skewed   **h** Positively skewed
   i Positively skewed   **j** Symmetric

10 A and X, B and Y, C and W, D and Z

Answers to Chapter 33

33.1 Probability scale

Exercise 33A

1 a unlikely   **b** unlikely   **c** impossible
   d very likely   **e** even chance

2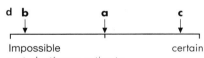

   e student's own estimate

3 student's own estimate

4 Student to provide own answers.

5 No. What happens today does not depend on what happened yesterday.

33.2 Calculating probabilities

Exercise 33B

1 a $\frac{1}{10}$ **b** $\frac{4}{10}$ or $\frac{2}{5}$ **c** $\frac{7}{10}$
   d $\frac{1}{2}$ **e** 0

2 a $\frac{1}{8}$ **b** $\frac{5}{8}$ **c** $\frac{1}{2}$

3 a 0   **b** 1

4 a $\frac{1}{10}$ **b** $\frac{1}{2}$ **c** $\frac{2}{5}$ **d** $\frac{1}{5}$ **e** $\frac{2}{5}$

5 a $\frac{6}{11}$ **b** $\frac{5}{11}$ **c** $\frac{6}{11}$

6 a $\frac{1}{5}$ **b** $\frac{1}{2}$ **c** $\frac{1}{2}$ **d** $\frac{7}{10}$

7 $\frac{1}{25}$

8 a AB, AC, AD, AE, BC, BD, BE, CD, CE, DE
   b 1   **c** $\frac{1}{10}$ **d** 6   **e** $\frac{3}{5}$ **f** $\frac{3}{10}$

9 a i $\frac{12}{25}$ **ii** $\frac{7}{25}$ **iii** $\frac{6}{25}$
   b They add up to 1.
   c All possible outcomes are mentioned.

10 35%

11 0.5

12 Class U

13 There might not be the same number of boys as girls in the class.

Answers to Chapter 33

33.3 Probability that an event will not happen

Exercise 33C

1. **a** $\frac{3}{4}$ **b** 0.55 **c** 0.2
2. **a** $\frac{3}{4}$ **b** $\frac{17}{20}$ **c** $\frac{19}{20}$
3. **a i** $\frac{1}{4}$ **ii** $\frac{3}{4}$
   b i $\frac{3}{11}$ **ii** $\frac{8}{11}$
4. Because it might be possible for the game to end in a draw.

33.4 Probability in practice

Exercise 33D

1. **a** 0.2, 0.08, 0.1, 0.105, 0.148, 0.163, 0.1645
   b 6 **c** 1 **d** $\frac{1}{6}$ **e** 1000
2. **a** 0.095, 0.135, 0.16, 0.265, 0.345
   b 40 **c** No; all numbers should be close to 40.
3. **a** 0.2, 0.25, 0.38, 0.42, 0.385, 0.3974
   b 8
4. **a** 6 **b** and **c** Student to provide own answers.
5. **a** Caryl, threw the greatest number of times.
   b 0.39, 0.31, 0.17, 0.14
   c Yes; all answers should be close to 0.25.
6. The missing top numbers are 4 and 5, the bottom two numbers are both likely to be close to 20.
7. Thursday
8. Although he might expect the probability to be close to $\frac{1}{2}$ giving 500 heads, the actual number of heads is unlikely to be exactly 500, but should be close to it.

33.5 Using Venn diagrams

Exercise 33E

1. **a**

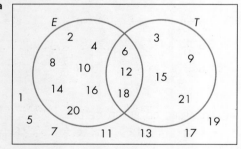

   b i $\frac{10}{21}$ **ii** $\frac{7}{21} = \frac{1}{3}$ **iii** $\frac{3}{21} = \frac{1}{7}$

2. **a** $\frac{60}{100} = \frac{3}{5}$ or 0.6 **b** $\frac{35}{100} = \frac{7}{20}$ or 0.35
   c $\frac{75}{100} = \frac{3}{4}$ or 0.75 **d** $\frac{25}{100} = \frac{1}{4}$ or 0.25

3. **a**

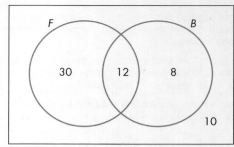

   b i $\frac{12}{60} = \frac{1}{5}$ or 0.2 **ii** $\frac{42}{60} = \frac{7}{10}$ or 0.7 **c** $\frac{20}{60} = \frac{1}{3}$

4. **a**

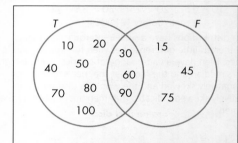

   b i $\frac{6}{100} = \frac{3}{50}$ **ii** $\frac{3}{100}$ **iii** $\frac{87}{100}$

5. **a**

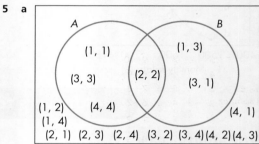

   b i $\frac{4}{16} = \frac{1}{4}$ **ii** $\frac{3}{16}$ **iii** $\frac{1}{16}$

6. **a** 10
   b

   T: 5, 5 (intersection), F: 20, outside: 70

   c i $\frac{10}{100} = \frac{1}{10}$ **ii** $\frac{25}{100} = \frac{1}{4}$ **iii** $\frac{5}{100} = \frac{1}{20}$ **iv** $\frac{70}{100} = \frac{7}{10}$

Answers to Chapter 33

33.6 Possibility diagrams

Exercise 33F

1. **a** 7 **b** 2 and 12
   c $\frac{1}{36}, \frac{1}{18}, \frac{1}{12}, \frac{1}{9}, \frac{5}{36}, \frac{1}{6}, \frac{5}{36}, \frac{1}{9}, \frac{1}{12}, \frac{1}{18}, \frac{1}{36}$
   d i $\frac{1}{12}$ **ii** $\frac{1}{3}$ **iii** $\frac{1}{2}$ **iv** $\frac{7}{36}$
   v $\frac{5}{12}$ **vi** $\frac{5}{18}$

2. **a** $\frac{1}{12}$ **b** $\frac{11}{36}$ **c** $\frac{1}{6}$ **d** $\frac{5}{9}$

3. **a** $\frac{1}{36}$ **b** $\frac{11}{36}$ **c** $\frac{5}{18}$

4.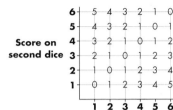
   a $\frac{5}{18}$ **b** $\frac{1}{6}$ **c** $\frac{1}{9}$
   d 0 **e** $\frac{1}{2}$

5. **a** $\frac{1}{4}$ **b** $\frac{1}{2}$
   c $\frac{3}{4}$ **d** $\frac{1}{4}$

6.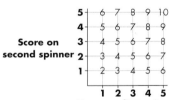
   a 6
   b i $\frac{4}{25}$ **ii** $\frac{13}{25}$ **iii** $\frac{1}{5}$ **iv** $\frac{3}{5}$

7. **a**

	1	2	3	4	5	6	7	8
8	8	16	24	32	40	48	56	64
7	7	14	21	28	35	42	49	56
6	6	12	18	24	30	36	42	48
5	5	10	15	20	25	30	35	40
4	4	8	12	16	20	24	28	32
3	3	6	9	12	15	18	21	24
2	2	4	6	8	10	12	14	16
1	1	2	3	4	5	6	7	8

   Score on spinner 2 (vertical); Score on spinner 1 (horizontal)

   b $\frac{8}{64} = \frac{1}{8}$

8. $\frac{7}{36}$: a diagram will help him to see all possible outcomes

33.7 Tree diagrams

Exercise 33G

1. **a** $\frac{1}{4}$
   b $\frac{1}{2}$
   c $\frac{3}{4}$

2. **a**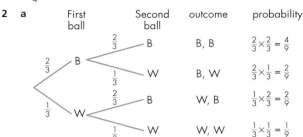
   b i $\frac{1}{9}$ **ii** $\frac{4}{9}$ **iii** $\frac{8}{9}$

3. **a** $\frac{2}{3}$ **b** $\frac{1}{2}$
   c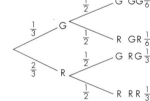
   d i $\frac{1}{6}$ **ii** $\frac{1}{2}$ **iii** $\frac{5}{6}$
   e 15 days

4. **a**

First sweet	Second sweet	outcome	probability
red	red	red, red	$\frac{4}{5} \times \frac{2}{5} = \frac{8}{25}$
red	not red	red, not red	$\frac{4}{5} \times \frac{3}{5} = \frac{12}{25}$
not red	red	not red, red	$\frac{1}{5} \times \frac{2}{5} = \frac{2}{25}$
not red	not red	not red, not red	$\frac{1}{5} \times \frac{3}{5} = \frac{3}{25}$

   b i $\frac{8}{25}$ **ii** $\frac{22}{25}$ **iii** $\frac{3}{25}$

5. **a**

First exam	Second exam	outcome	probability
pass	pass	pass, pass	$0.9 \times 0.6 = 0.54$
pass	fail	pass, fail	$0.9 \times 0.4 = 0.36$
fail	pass	fail, pass	$0.1 \times 0.6 = 0.06$
fail	fail	fail, fail	$0.1 \times 0.4 = 0.04$

   b i 0.54 **ii** 0.42

Answers to Chapter 33

6 a

First match	Second match	outcome	probability
win	win	win, win	0.6 × 0.7 = 0.42
win	not win	pass, fail	0.6 × 0.3 = 0.18
not win	win	fail, pass	0.4 × 0.7 = 0.28
not win	not win	fail, fail	0.4 × 0.3 = 0.12

b 0.88
7 a 0.09 **b** 0.49 **c** 0.42
8 a 0.1 **b** 0.3 **c** 0.55
9 0.53
10 a i $\frac{5}{13}$ **ii** $\frac{8}{13}$
b i $\frac{15}{91}$ **ii** $\frac{4}{13}$
11 a $\frac{1}{120}$ **b** $\frac{7}{40}$ **c** $\frac{21}{40}$ **d** $\frac{7}{24}$
12 a $\frac{1}{9}$ **b** $\frac{2}{9}$ **c** $\frac{2}{3}$ **d** $\frac{7}{9}$
13 a 0.54 **b** 0.38 **c** 0.08 **d** 1

33.8 Conditional probability

Exercise 33H

1 a $\frac{55}{100} = \frac{11}{20}$ or 0.55   **b** $\frac{22}{55} = \frac{2}{5}$ or 0.4

c $\frac{22}{50} = \frac{11}{25}$ or 0.44

2 a

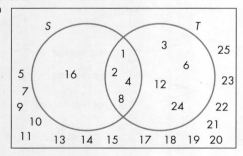

b $\frac{4}{25}$ **c** $\frac{4}{8} = \frac{1}{2}$ **d** $\frac{4}{5}$ **e** $\frac{16}{20} = \frac{4}{5}$

3 a

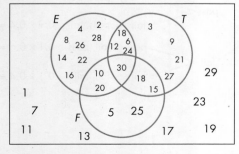

b $\frac{10}{30} = \frac{1}{3}$ **c** $\frac{2}{30} = \frac{1}{15}$ **d** $\frac{2}{6} = \frac{1}{3}$
e $\frac{1}{6}$ **f** $\frac{2}{10} = \frac{1}{5}$

4 a $\frac{85}{120} = \frac{17}{24}$ **b** $\frac{30}{120} = \frac{1}{4}$ **c** $\frac{55}{85} = \frac{11}{17}$
d $\frac{50}{85} = \frac{10}{17}$ **e** $\frac{20}{35} = \frac{4}{7}$

5 a

	1	2	3	4	5	6
6	7	8	9	10	11	12
5	6	7	8	9	10	11
4	5	6	7	8	9	10
3	4	5	6	7	8	9
2	3	4	5	6	7	8
1	2	3	4	5	6	7

Second dice (vertical) / First dice (horizontal)

b $\frac{5}{36}$ **c** $\frac{2}{5}$ **d** $\frac{11}{36}$ **e** $\frac{2}{11}$

6 a $\frac{6}{16} = \frac{3}{8}$ **b** $\frac{2}{6} = \frac{1}{3}$ **c** $\frac{2}{4} = \frac{1}{2}$

7 a

First set → Second set
- 0.3 red → 0.8 red
- 0.3 red → 0.2 green
- 0.7 green → 0.4 red
- 0.7 green → 0.6 green

b 0.24   **c** 0.42   **d** 0.34

8 a $\frac{4}{10}$ or $\frac{2}{5}$ **b** $\frac{3}{9}$ or $\frac{1}{3}$

c

First note → Second note
- $\frac{4}{10}$ or $\frac{2}{5}$ $5 → $\frac{3}{9}$ or $\frac{1}{3}$ $5
- $\frac{4}{10}$ or $\frac{2}{5}$ $5 → $\frac{6}{9}$ or $\frac{2}{3}$ $10
- $\frac{6}{10}$ or $\frac{3}{5}$ $10 → $\frac{4}{9}$ $5
- $\frac{6}{10}$ or $\frac{3}{5}$ $10 → $\frac{5}{9}$ $10

d i $\frac{2}{15}$ **ii** $\frac{8}{15}$ **iii** $\frac{1}{3}$

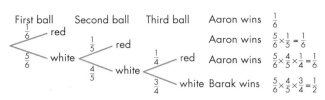

9 a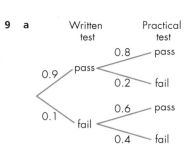

   b 0.72   c 0.24
10 a $\frac{3}{28}$ b $\frac{5}{14}$ c $\frac{1}{56} + \frac{5}{28} = \frac{11}{56}$

11 The tree diagram looks like this

First ball — Second ball — Third ball
$\frac{1}{6}$ red — Aaron wins $\frac{1}{6}$
$\frac{5}{6}$ white $\frac{1}{5}$ red — Aaron wins $\frac{5}{6} \times \frac{1}{5} = \frac{1}{6}$
$\frac{4}{5}$ white $\frac{1}{4}$ red — Aaron wins $\frac{5}{6} \times \frac{4}{5} \times \frac{1}{4} = \frac{1}{6}$
$\frac{3}{4}$ white — Barak wins $\frac{5}{6} \times \frac{4}{5} \times \frac{3}{4} = \frac{1}{2}$

Aaron wins if the first ball or the second or the third is red.

The probability of this is $\frac{1}{6} + \left(\frac{5}{6} \times \frac{1}{5}\right) + \left(\frac{5}{6} \times \frac{4}{5} \times \frac{1}{4}\right)$

$= \frac{1}{6} + \frac{1}{6} + \frac{1}{6} = \frac{3}{6} = \frac{1}{2}$

Or: Barak wins if there are 3 white balls and the probability of this is $\frac{5}{6} \times \frac{4}{5} \times \frac{3}{4} = \frac{1}{2}$

Hence the probability that Aaron wins is $1 - \frac{1}{2} = \frac{1}{2}$

765

Index

2D shapes 523–4
3D shapes 466–7, 525–6
12-hour clock 145
24-hour clock 145

acceleration 247
accuracy 120–31
acute angles 414–15
addition
    algebraic fractions 197–8
    directed numbers 70–3
    fractions 57–60, 197–8
    multiplying indices 324
    number rules 50–63
    vectors 533
adjacent side of triangle 452, 456–7, 460, 468, 483
algebra
    differentiation 362–71
    examination questions 372–81
    formulae 160–71
    fractions 197–200
    functions 352–61
    graphs 236–98
    important rules 162
    indices 318–31
    inequalities 202, 232–5
    linear programming 342–51
    manipulation 172–201
    proportion 332–41
    representation 160–71
    sequences 300–17
    solving equations 202–31
allied angles 387
alternate angles 387, 390
alternate segment theorem 408–11
angle of depression 463
angle of elevation 463, 466
angle of rotation 551, 561
angles
    arcs 402–3, 508, 510–11
    bearings 417–20
    circles 400–11
    cyclic quadrilaterals 405–11
    diameters 400–1
    drawing 414–17
    facts 384–6
    irregular polygons 398–400
    measuring 414–17
    notation 415
    parallel lines 386–9
    pie charts 587–8
    properties 382–411
    quadrilaterals 392–5, 398–9, 400, 405–11
    regular polygons 383, 395–7
    sectors 508, 510–11
    subtended by arcs 402–3
    tangents 400–2
    triangles 383, 390–3, 403, 454
    trigonometric ratios 452–67
annual rates 43
anticlockwise turns 481, 484, 551, 561
approximations 122–31
arcs 402–3, 508–12
areas
    area ratio 433, 435
    area sine rule 477–9
    circles 498–501
    histograms 596, 599, 600
    mensuration 487–503, 506–12, 514–17
    parallelograms 494–5
    pressure 113–14
    rectangles 488–91, 492
    scale factors 431, 433
    sectors 508–12
    similar shapes 430–7
    trapeziums 165, 495–7
    triangles 477–9, 491–3
    see also surface areas
arithmetic mean 612
astronomy 413
asymptotes 289
atmospheric pressure 292–3

Index

averages 607, 608, 612
    advantages/disadvantages 618
    speed 110, 117, 242
    uses for 618
    which to use 618–20
    see also mean; median; mode
axes 525, 583–4

bank statements 67
bar charts 577, 583–6
bearings 417–20
Bernoulli's principle 203
best fit lines 593
BIDMAS rule 52
bisectors 526
BODMAS rule 52
boundary lines 344–5, 347
bounds
    class intervals 599–600
    rounding 126–31
box method 185, 187, 190
box-and-whisker plots 638–41
    lower/upper quartiles 638–9
    lowest/highest values 638–9
    median 638–9
brackets
    expanding 178–81, 184–92
    factorisation 182–4
    multiplying together 184–92
    order of operations 52
    quadratic factorisation 192–7
    simplifying expressions 179–80
    solving equations 207, 209
    squared brackets 189
    three sets of brackets 190–2

calculators
    negative numbers 71–2
    powers 78, 79, 81, 320
    roots 78, 79, 81
    standard form 134–7
    trigonometric ratios 452–3
    using efficiently 140–1, 149
cancelling fractions 60, 646
capacity 142–3

Carroll, Lewis 87
Cartesian coordinates 257, 258
Cartesian plane 257, 258
centre of circle 526
centre of enlargement 554–7, 559–60, 561
centre of rotation 550, 561
certainty 644
chance 643, 644
    see also probability
charts
    bar 577, 583–6
    pie 577, 587–91
    tally 578–9
chords 526
circles
    angles 400–11
    arcs 402–3, 508–12
    areas 498–501
    circumferences 498–501, 506, 508, 510
    pie charts 577, 587–91
    sectors 508–12, 587
    symmetry 526–8
circumferences 498–501, 506, 508, 510
class frequencies 596, 599–600
class intervals 578–9, 584
    grouped data 632, 634
    histograms 596, 598–600
clinometers 413
clockwise turns 481, 484, 551, 561
coefficients
    expressions 174–5, 179
    factorisation 182, 196
    multiplying brackets 187
    nth terms 305–6
    quadratic formula 217
    simultaneous equations 227–9
    straight-line graphs 262
coin tossing 644, 646, 651
column vectors 532
common factors
    algebraic fractions 199
    factorisation 182
    highest common factors 15, 18–20
    multiplying fractions 62
    quadratic equations 215–16

Index

common multiples 14, 18–20, 59
compass bearings 417
complement of set 94, 96
completing the square 219–22
composite functions 357–61
compound interest 43–6
conditional probability 666–71
cones 279, 514–16
congruent shapes 423–4, 544
congruent triangles 424–7
consecutive terms 302, 306
consistency of data sets 615
constant of proportionality 334–5, 337, 339
constant speed 246
constant terms 217, 262
constructions 438–45
    scale drawings 439, 442–5
    shapes 440–2
continuous data 628, 632
conversion graphs 238–41
conversions
    currency 147–50
    decimals 30–2
    fractions 30–2, 40
    graphs 238–41
    metric units 143–4
    percentages 30–2, 40
coordinate enlargement method 556–7
coordinates 257, 258, 556–7
correlation 591–2
corresponding angles 386, 427
corresponding sides 427–8
cosine (cos)
    of any angle 479–83
    cosine rule 470, 473–5, 477
    function 455, 456–7
    graph 480–2
    ratio 452, 454, 455–67
'counting squares' enlargement method 556
cross-sections 504
cube numbers 7, 11–12, 79–80, 319
    direct proportion 336–9
    notation 11
    sequences 307

cube roots 77, 79–80, 327, 337–9
cubes 420–1
cubic graphs 288–9
cubic sequences 307
cuboids
    nets 420
    planes of symmetry 525
    prisms 504
    surface area 501–3
    volume 501–3
cumulative frequency diagrams 631–7
    class intervals 632, 634
    inter-quartile range 631, 633–4
    median 632–4
currency conversions 147–9
curves
    complex 366–71
    curved graphs 251–4
    differentiation 364–71
    drawing 281–2
    in everyday life 279
    gradients 251–2, 296–7, 364–6
cyclic quadrilaterals 405–11
cylinders
    prisms 503, 506
    surface area 506–8
    volume 506–8

Dantzig, George 343
data
    consistency 615
    continuous 628, 632
    discrete 628, 632
    dispersion 631
    grouped 578–9, 583, 596, 599, 628–35
decagons 395, 398
decay, exponential 82–5, 293
deceleration 248
decimals 26–33
    common decimals 31
    decimal places 123–4, 127, 218
    equivalents 31
    fractions 26–7, 30–3
    percentages 30–3

Index

recurring 27–30
rounding 123–4
time 111
decreasing
by percentage 38
using ratios 108–9
denominators
adding fractions 57–9
algebraic fractions 198–9
equivalent fractions 24
multiplying fractions 60
ratios 103
density 113
depression angles 463
derivatives 367
Descartes, René 257, 258
diameters 400–1, 498, 499
dice throwing 644, 649, 658–9, 661
differences between terms 302, 305–6, 309
difference of two squares 194, 199, 216
differentiation 362–71
curves 364–71
gradients 363, 364–6
notation 364
directed numbers 64–75
addition 70–3
division 73–5
everyday use 67–8
multiplication 73–5
subtraction 70–3
direction 531, 532, 544
direction vectors 544
direct proportion 116–17, 334–9
discrete data 628, 632
dispersion of data 631
distance
between two points 273
distance–time graphs 242–6, 252
speed 110, 113, 117, 137
distributions 638–9
division
algebraic fractions 198
by negative number 233
directed numbers 73–5

fractions 60–3, 198
indices 324–6
number rules 50–63
ratio 104
dodecagons 396
dollars 147
dot patterns 8, 11
drawing
angles 414–17
curves 281–2
scale 439, 442–5
straight-line graphs 258–61
dual bar charts 584

electricity units 238
elements of sets 90–1, 94–6
elevation angles 463, 466
elimination method 224–5
empty sets 95–6
enlargements 554–61
centre of enlargement 554–7, 559–60, 561
combined transformations 561
coordinate method 556–7
fractional 559–60
negative 557
ray method 555, 557
scale factors 554–7, 559–60, 561
similar shapes 427
equally likely outcomes 646
equals 88, 204
equal to or greater (more) than 88, 232
equal to or less than 88, 232
equal vectors 535
equations
definition 162
doing same to both sides 204–5
graphical solutions 267–70
real-life problems 210–12
roots of equations 282
simultaneous 222–31
solving 202–31, 267–70
straight lines 258, 261–72
variables on both sides 208–9
see also quadratic equations

Index

equilateral triangles 390
equivalent fractions 24–5, 57, 59
equivalents, decimal 31
estimation 120–31
    conversion graphs 238
    mean 628
    median 632
    rounding 122–31
euros 147
even chance 644
examination questions
    algebra 372–81
    geometry 564–75
    numbers 151–9
    statistics 672–84
exchange rates 147
expanding brackets 178–81, 184–92
experimental probability 651
exponential decay 82–5, 293
exponential graphs 292–6
exponential growth 82–5, 292
exponential sequences 314–15
expressions
    collecting like terms 175–6
    definition 162
    multiplication 174
    quadratic 184–90
    simplifying 174–80
exterior angles 395–6, 398–9
extreme values 618

factorisation 182–4
    algebraic fractions 199
    prime factorisation 16–18
    quadratic equations 212–17
    quadratic factorisation 192–7
    simple rules 192
factors
    cube numbers 11
    factor pairs 13
    facts about 13
    large numbers 14
    prime numbers 15–16
    probability 654, 665–6
    quadratic equations 215

    square numbers 8
    Venn diagrams 90
    whole numbers 13–15
    see also common factors
Fibonacci series 301
fifty-fifty chance 644
final amounts 47
FOIL expansion method 185, 190
forces 113–14
formulae 160–71
    circle area 498
    circle circumference 498
    complicated formulae 169–71
    cone surface area 515
    cone volume 514
    cuboid surface area 501
    cuboid volume 501
    cylinder surface area 506
    cylinder volume 506
    definition 162
    parallelogram area 494
    prism volume 503–4
    pyramid volume 512
    rearranging 167–9
    rectangle area 488
    rectangle perimeter 488
    sphere surface area 516
    sphere volume 516
    subjects of 167–9
    substitution 165–7
    trapezium area 165, 495
    triangle area 477, 491
Fourier, Jean 447
four number rules 50–63
    fractions 55–63
    operations 52–5
fractals 301
fractions 22–49
    addition 57–60, 197–8
    algebraic 197–200
    common fractions 31
    decimals 26–7, 30–3
    division 60–3, 198
    enlargement 559–60
    equivalent 24–5, 57, 59

Index

indices 327–31
lowest terms 24
multiplication 60–3, 197–8
percentages 30–3
probability 646
of quantities 55–7
ratios 103
recurring decimals 27–8
sectors 508, 510–11
subtraction 57–60
frequencies 578–9
    bar charts 583–4
    cumulative frequency 631–7
    frequency densities 599–600
    frequency tables 578–81, 624–7
    grouped data 578–9, 628–35
    histograms 596, 599–600
    mode 608
    pictograms 581–2
    pie charts 587–8
    relative 651
functions 352–61
    composite 357–61
    graphs 278–98
    inverse 355–6
    notation 353, 354–5, 357–8
    trigonometric 455–9

gallons 238
geometry
    constructions 438–45
    examination questions 564–75
    relationships 412–37
    terms 412–37
glides 544
gradients
    curved graphs 251–2
    curves 251, 296–8, 364–6
    differentiation 363, 364–6
    negative 262, 296
    positive 296
    straight-line graphs 261–3, 265–6, 270, 272, 274
graphs 236–98
    conversion 238–41
    cubic 288–9
    curved 251–4
    distance–time 242–6, 251
    exponential 292–6
    functions 278–98
    inequalities 344–51
    practical situations 236–54
    quadratic 279, 280–6
    reciprocal 286–7
    sin/cos/tan 480–2, 483–4
    solving equations 267–70
    speed–time graphs 246–51
    straight-line 256–77
    travel 242–6
    trends 577
gravity 113, 203
greater (more) than 69, 88, 232–3
greater (more) than or equal to 88, 232
grouped data 578–9, 584, 596, 599, 628–35
grouped frequency tables 578–9, 584
growth, exponential 82–5, 292

HCF (highest common factor) 15, 18–20
height 512, 515
heptagons 398
hexagonal prisms 421, 503
hexagons 383, 395–6, 398, 421, 503
highest common factor (HCF) 15, 18–20
histograms 596–604
    area of bar 596, 599, 600
    class frequencies 596, 599–600
    unequal width bars 599–604
hypotenuse
    Pythagoras' theorem 448–9, 540
    Trigonometry 448–9, 452, 455–7, 459–60, 468

images 546, 548
impossible outcomes 644
improper fractions 24, 57, 60, 62
included boundaries 344–5, 347
increasing
    by percentage 36
    using ratio 108–9
indices (index) 77, 318–31
    BIDMAS 52
    division 324–6

Index

form *a/b* 328–9
fractional 327–31
multiplication 320, 324–6
negative 135, 322–4, 328–9
*n*th root 327–8
product of prime factors 16
simplifying expressions 174
standard form 134–5
using indices 320–2
see also exponential...; powers
inequalities 88–9, 202, 232–5
graphical 344–51
more than one 347–8
number lines 69
integers 20
intercepts 262, 266
interest 43–6
compound 43–6
simple 43–4
interior angles 395–6, 398
inter-quartile range 631, 633–4
intersections of sets 90–1, 94–6, 654
inverse, definition 353
inverse functions 355–6
inverse proportion 117–19, 339–41
inverses of sin/cos/tan 454, 455–6
irrational numbers 20
irregular polygons
angles 398–400
exterior angles 398–9
interior angles 398
isosceles triangles 390, 402, 523

keys
bar charts 584
pictograms 581–2
stem-and-leaf diagrams 620–1
kites 393
Koch snowflake 301

labelling pie charts 587–8
language of algebra 162–5
LCM (lowest common multiple) 14, 18–20, 59
length 142–3, 433, 435
less than 69, 88, 232, 233

less than or equal to 88, 232
likelihood of events 644
like terms 175–6, 179–80
limits of accuracy 120, 126
linear equations 204–12, 229–31
linear inequalities 232–3
linear programming 342–51
linear scale factors 427, 431, 433
linear sequences 306
linear simultaneous equations 229–31
lines
angles on 384
of best fit 593
joining two points 272
midpoint of 273
lines of symmetry 520–1
2D shapes 523
quadratic graphs 282
reciprocal graphs 287
reflections 546, 548, 561
litres 238
loss 41, 67
lower bounds 126–31, 599, 600
lower quartiles 633–4, 638–9
lowest common multiple (LCM) 14, 18–20, 59
lowest term of fraction 57

magnitude 531, 532, 540–1
map scales 105
mass 113, 142–3
mathematics, usages 161
mean 607, 612–15
advantages/disadvantages 618
estimated 628
frequency tables 624–5
grouped data 628
range 615
uses for 618
measurement
angles 414–17
statistics 606–41
units 142–3
see also mensuration
median 610–12
advantages/disadvantages 618

Index

box-and-whisker plots 638–9
cumulative frequency diagrams 631–4
estimated 632
frequency tables 624
inter-quartile range 633–4
rule for finding 610
stem-and-leaf diagrams 620
uses for 618
members of sets 94–5
mensuration 486–517
arcs 508–12
areas 487–503, 506–12, 514–17
circumference 498–501, 506, 508, 510
perimeter 488–91
volume 501–8, 512–17
see also measurement
metric system 142–4
middle values 610
midpoint of line segment 273
mirror lines *see* lines of symmetry
mixed numbers 24, 57, 60
Möbius strip 543
mode 607, 608–9
advantages/disadvantages 618
frequency tables 624
modal classes 628
stem-and-leaf diagrams 620
uses for 618
more (greater) than 69, 88, 232–3
more (greater) than or equal to 88, 232
multiples
like terms 175
lowest common multiples 14, 18–20, 59
probability 654, 666
whole numbers 7, 12–13
multiplication
algebraic fractions 197–8
balancing coefficients 227–9
brackets 178, 184–92
directed numbers 73–5
expressions 174
fractions 60–3, 197–8
indices 320, 324–6
negative number 233

number rules 50–63
vectors 533
multipliers 34, 36, 38, 41, 43, 45, 47

natural numbers 20
negative coordinates 258
negative correlation 592
negative enlargement 557
negative gradients 262, 296
negative indices 135, 322–4, 328–9
negatively skewed distributions 638
negative numbers 66–75
dividing by 233
expanding brackets 178
inequalities 232–3
multiplying by 232–3
square roots 78
nets 420–2
Newtons (N) 113
nonagons 398
non-linear simultaneous equations 229–32
notation
angles 415
cube numbers 11
differentiation 364
functions 353, 354–5, 357–8
probability 646
recurring decimals 28
sets 86–99
square numbers 8
vectors 532, 540
nth terms 304–8, 309, 315
number lines 66, 68–70
numbers 6–21
applications 140–50
examination questions 151–9
four rules 50–63
numerators
algebraic fractions 199
equivalent fractions 24
multiplying fractions 60

object of reflection 546
obtuse angles 414–15, 468–9, 479

773

Index

octagons 395–6, 398
operations
    choosing correct one 54–5
    order of 52–3
opposite angles 384
opposite segment of circle 405
opposite side of triangle 452, 455, 457–60, 468, 483
ordering 86–99
order of rotational symmetry 522, 523, 525
outcomes 644, 646

Pa (Pascals) 113
parabolas 279, 280
parallel lines 270–2, 386–9
parallelograms 393, 494–5, 523
parallel vectors 535
Pascals (Pa) 113
patterns 7, 302–4, 309–14
pentagon-based pyramids 421
pentagons 383, 395–6, 398–9, 421
percentages 22, 30–49
    calculating 34–5
    common percentages 31
    compound interest 43–6
    decimals 30–3
    decreasing by 38
    fractions 30–3
    increasing by 36
    multipliers 34, 36, 38, 41, 43, 45, 47
    percentage change 41
    profit and loss 41
    quantities 34–42
    reverse 47–9
    simple interest 43–4
percentiles 633
perfect squares 194
perimeters 163, 210, 488–91
perpendicular bisectors 526
perpendicular lines 274–7, 414
pi (π) 498–501, 506, 508, 510–11, 514, 516
pictograms 581–3
pie charts 577, 587–91
place value tables 26
planes of symmetry 525
planets 133

points
    angles at 384, 403
    of contact 400
    distance between two points 273
    equations of lines 272–3
polygons
    decagons 395, 398
    dodecagons 396
    heptagons 398
    hexagons 383, 395–6, 398, 421, 504
    irregular 398–400
    nonagons 398
    octagons 395–6, 398
    pentagons 383, 395–6, 398–9, 421
    regular 383, 395–6
    see also quadrilaterals
population graphs 292
position vectors 535
positive correlation 591–2
positive gradients 296
positively skewed distributions 638
positive numbers 66–75
possibility diagrams 657–8
powers 76–85
    BODMAS 52
    calculators 78, 79, 81, 320
    expressions using powers 325
    higher powers 81–2
    to the power one 320
    to the power zero 321
    raising powers to powers 325
    standard form 132–9, 319
    see also cube numbers; indices (index); square numbers
pressure 113–14, 292–3
prime factorisation 16–18
prime numbers 7, 15–18
principle amounts 43
prisms 421, 503–6
probability 642–71
    calculating 646–9
    coin tossing 644, 646, 651
    conditional 665–71
    dice throwing 644, 649, 658–9, 661
    event not happening 649–50

Index

notation 646
possibility diagrams 657–60
in practice 651–3
probability fraction 646
probability scale 644–5
tree diagrams 661–5, 666–7
Venn diagrams 654–7, 666
product of prime factors 16–17
profit 41, 67
proper fractions 24, 57
proper subsets 95–6
proportion 100–1, 116–19, 332–41
    direct 116–17, 334–9
    inverse 117–19, 339–41
    symbol 334–5, 336, 339
protractors 414–15
Ptolemy 447
pyramids 421, 512–14
Pythagoras' theorem 448–51
    cosine rule 473
    distance between two points 273
    vectors 540

quadrants 258
quadratic equations 279, 280, 282
    completing the square 219–22
    factorisation 212–17
    general equation 215, 221
    quadratic formula 217–19
    rational form of solution 215
    special cases 216
quadratic expansions 184–90
quadratic expressions 184–90
quadratic factorisation 192–7
    $ax^2 + bx + c$ 196–7
    difference of two squares 194–5
quadratic formula 217–19
quadratic graphs 279, 280–4, 285–6
quadratic sequences 307
quadrilaterals
    angles 393–5, 398–9, 401, 406–11
    cyclic 406–11
    kites 393
    parallelograms 393, 494–5
    rectangles 383, 488–91, 492

rhombuses 393
special 393
squares 383, 395
trapeziums 165, 393, 495–7
quantities
    fractions of 55–7
    percentages 34–42
quartiles 633–4, 638–9

radius 400, 498–9, 506, 508
random outcomes 646
range 615–17
    inter-quartile 631, 633–4
    stem-and-leaf diagrams 620
rates 100, 113–15
    acceleration 247
    exchange rates 147
rational numbers 20
ratios 100–9, 113
    area 433, 435–6
    common units 102
    decreases 108–9
    dividing amounts 104
    as a fraction 103
    increases 108–9
    length 433, 435–6
    map scales 105
    trigonometric 452–67
    use in calculations 107
    volume 433, 435–6
ray enlargement method 555, 557
real life
    curves 279
    directed numbers 67–8
    equations 210–12
    linear programming 349–51
real numbers 20–1
reciprocals
    fractions 62
    gradients 274
    graphs 286–8
    inverse proportion 339
    negative indices 322, 329
rectangles 383, 488–91, 492
recurring decimals 27–30

Index

dot notation 28
into fractions 28
repeated digits 28–9
reflections 546–51
    combined transformations 561
    lines of symmetry 546, 548, 561
    object/image 546
    planes of symmetry 525
reflex angles 403, 414–15, 480
regions 344–5, 347, 349
regular polygons 383
    angles 383, 395–7
    exterior angles 395–6
    interior angles 395–6
relationships, geometrical 412–37
relative frequency 651
relativity theory 203
representation
    algebra 160–71
    statistics 576–604
reverse percentages 47–9
rhombuses 393
right-angled triangles 390
    adjacent side 452, 456–7, 460, 468, 483
    hypotenuse 448–9, 452, 455–7, 459–60, 468, 540
    opposite side 452, 455, 457–60, 468, 483
    Pythagoras' theorem 273, 448–51, 473, 540
    trigonometric ratios 452–67
right angles 414
roots 8–10, 76–85
    cube roots 79–80, 327, 336–9
    equations 282
    fractional indices 327
    higher roots 81–2
    nth roots 327
    see also square roots
rotations 550–4
    angle of 551, 561
    centre of 550, 561
    combined transformations 561
    symmetry 520, 522–3, 525
    turns 551
rounding 122–31
    decimal places 123–4, 127

    significant figures 125–6
    up/down 122
    upper/lower bounds 126–31
    whole numbers 122–3, 125
rules of number 50–63
    fractions 55–63
    operations 52–5

scalars 533
scale drawings 439, 442–5
scale factors
    enlargement 554–7, 559–60, 561
    similar shapes 427, 430–1, 433
scales on axes 238
scales of maps 105
scatter diagrams 591–6
    correlation 591–2
    lines of best fit 593
    restrictions on use 593–4
sea level 67
sectors 508–12, 587
    angles 508, 510–11
    areas 508–12
    fractions 508, 510–11
    pie charts 587
segments 402–3, 405
semi-circles 401
sequences 300–17
    combinations of 315
    cube numbers 11
    exponential 314–15
    nth terms 304–8, 309, 315
    patterns 302–4, 309–14
    square numbers 8, 307
sets
    definition 87
    notation 86–99
    probability 654–7, 666
    Venn diagrams 90–9, 654–7, 666
sf (significant figures) 125–6
sides of triangles
    adjacent 452, 456–7, 460, 468, 483
    hypotenuse 448–9, 452, 455–7, 459–60, 468, 540
    opposite 452, 455, 458–60, 468, 483
    see also trigonometric ratios

Index

sign changes
    brackets 178, 192–3, 196
    directed numbers 73
    inequalities 232
significant figures (sf) 125–6
similar shapes 427–37
    areas 430–7
    scale factors 427, 431, 433
    triangles 413, 430–2
    volumes 433–7
simple interest 43–4
simplest forms
    fractions 57
    ratios 103
simplifying expressions 174–80
simultaneous equations 222–31
    balancing coefficients 227–9
    elimination method 224–5
    linear/non-linear 229–31
    substitution method 225, 228
sine (sin)
    of any angle 479–83
    area sine rule 477–9
    function 455–6
    ratio 452–67
    sine graph 480–2
    sine rule 470–2, 475, 477–9
single unit value 116
slant height 515
slope *see* gradient
solid shapes 433
speed 101, 110–12, 113
    average 110, 117, 242
    inverse proportion 117, 339
    speed–time graphs 246–51
    standard form 137
    triangle mnemonic 110
spheres 516–17
spread 615
    see also range
square-based pyramids 421
squared brackets 189
square numbers 7, 8–10, 77, 78–9, 319
    difference of two squares 194, 199, 216
    directed numbers 73–4

direct proportion 336–9
irrational numbers 20
notation 8
sequences 8, 307
square roots 8, 77, 78–9
    direct proportion 336–9
    fractional indices 327
    irrational numbers 20
    notation 8
    quadratic equations 216
squares 383, 396
standard form 132–9, 319
    calculations 136–9
    calculators 134–7
    numbers less than one 135
statistics
    bar charts 577, 583–6
    box-and-whisker plots 638–41
    cumulative frequency diagrams 631–7
    examination questions 672–84
    frequency tables 578–81, 624–7
    grouped data 578–9, 584, 596, 599, 628–35
    histograms 596–604
    measures 606–41
    pictograms 581–3
    pie charts 577, 587–91
    probability 642–71
    range 615–17
    representation 576–604
    scatter diagrams 591–6
    stem-and-leaf diagrams 620–4
    see also averages
stem-and-leaf diagrams 620–4
stopping distances 173
straight-line angles 384, 390, 403
straight-line graphs 256–77
    distance between two points 273
    drawing 258–61
    forms of equations 265
    line through two points 272
    midpoint of line 273
    parallel lines 270–2
    perpendicular lines 274–7
    solving equations 267–70
    $y = mx + c$ format 259, 261–5

777

Index

subject of formula 167–9
subsets 95–6
substitution
    formulae 165–7
    simultaneous equations 226, 229
subtraction
    algebraic fractions 197–8
    directed numbers 70–3
    dividing indices 324
    fractions 57–60
    number rules 50–63
    vectors 533
supplementary angles 405
surface areas
    cones 514–16
    cuboids 501–3
    cylinders 506–8
    spheres 516–17
    see also areas
Symbolic Logic (Carroll) 87
symbols
    algebra 162
    pictograms 581–2
    proportion 334–5, 336, 339
symmetry 518–28
    2D shapes 523–4
    3D shapes 525–6
    circles 526–8
    distributions 638
    quadratic graphs 282, 285
    reciprocal graphs 286–7
    rotational 520, 522–3, 525
    sin/cos/tan graphs 481–2, 484
    uses of 519
    see also lines of symmetry

tally charts 578–9
tangent (tan)
    of any angle 483–5
    function 455, 458–9
    ratio 452, 454, 455–67
    tangent graph 484
tangent to curve
    angles 400–2
    curved graphs 251–2, 297

estimating gradients 297
symmetry 527
temperature 66, 70
terminating decimals 27–8
terms
    constant 217, 262
    definition 162
    differences 302, 305–6, 309
    geometrical 412–37
    like terms 175–6, 179–80
    nth terms 304–8, 309, 315
    sequences 302
    term to term rules 302
Thales of Miletus 413
thermometers 66, 70
three-dimensional (3D) shapes 466–7, 525–6
three-figure bearings 417–18
time 145–7
    in decimal form 111
    distance–time graphs 242–6, 251
    inverse proportion 117, 339
    speed 110–11, 113, 117, 137, 246–51, 252, 339
    speed–time graphs 246–51, 252
timetables 145
top-heavy fractions 24
total probability 646
transformations 542–63
    combined 561–3
    enlargements 554–61
    reflections 546–51
    rotations 550–4
    translations 544–6, 561
translations 544–6
    combined transformations 561
    vectors 544, 561
trapeziums 165, 393, 495–7
travel graphs 242–6
tree diagrams 661–5, 666–7
trend graphs 577
trials 651
triangles
    angles 383, 390–3, 403, 454
    area 477–9, 491–3
    area sine rule 477–9
    constructions 440

Index

density formula mnemonic 113
similar 413, 430–2
special 390
speed formula mnemonic 110
sum of angles 390, 402
see also right-angled triangles; trigonometry
triangular prisms 421, 504
trigonometric ratios 452–67
    3D problems 466–7
    applications 462–5
    calculating angles 454
    choosing correct ratio 459–62
    sin/cos/tan functions 455–9
trigonometry 446–85
    area of triangle 477–9
    choosing correct rule 475
    cosine of any angle 479–83
    cosine rule 470, 473–5, 477
    obtuse angles 468–9, 479
    Pythagoras' theorem 448–51, 473
    sine of any angle 479–83
    sine rule 470–2, 475, 477–9
    tangent of any angle 483–5
    trigonometric ratios 452–67
turning points 285–6, 369–71
    maximum 369
    minimum 369–70
turns 481–2, 484, 551, 561
twelve-hour clock 145
twenty-four hour clock 145
two-dimensional (2D) shapes 523–4

unions of sets 90–1, 94
unitary method 47, 116
units
    common units 142, 506
    electricity 238
    metric units 142–4
    percentages 40
    ratio 102
    speed 110
universal sets 91
upper bounds 126–31, 599, 600
upper quartiles 633–4, 638–9

variables
    on both sides of equation 208–9
    collecting like terms 175
    definition 162
    rearranging formulae 167
    scatter diagrams 591
    substitution 165
vectors 530–41
    addition 533
    direction 531, 532, 544
    magnitude 531, 532, 540–1
    multiplication 533
    notation 532, 540
    subtraction 533
    translations 544, 561
    using 535–9
Venn diagrams 87, 90–9, 654–7, 666
vertex (vertices) 421, 512
vertical height 512, 514
vertically opposite angles 384
vertices (vertex) 421, 512
volumes 487, 501–8, 512–17
    cones 514–16
    cuboids 501–3
    cylinders 506–8
    prisms 503–5
    pyramids 512–14
    rates 113
    scale factors 433
    similar shapes 433–7
    spheres 516–17
    units 142–3
    volume ratio 433, 435

weight 113–14
whole numbers
    factors 13–15
    multiples 12–13
    rounding 122–3, 125

$y = mx + c$ 259, 261–5

zero correlation 592

Collins

William Collins' dream of knowledge for all began with the publication of his first book in 1819.

A self-educated mill worker, he not only enriched millions of lives, but also founded a flourishing publishing house. Today, staying true to this spirit, Collins books are packed with inspiration, innovation and practical expertise. They place you at the centre of a world of possibility and give you exactly what you need to explore it.

Collins. Freedom to teach.

Published by Collins
An imprint of HarperCollins*Publishers*
The News Building
1 London Bridge Street
London
SE1 9GF

Browse the complete Collins catalogue at
www.collins.co.uk

© HarperCollins*Publishers* Limited 2018

10 9 8 7 6 5 4 3 2 1

ISBN 978-0-00-825779-8

MIX
Paper from responsible sources
FSC
www.fsc.org FSC C007454

This book is produced from independently certified FSC paper to ensure responsible forest management.

For more information visit:
www.harpercollins.co.uk/green

All rights reserved. No part of this publication may be reproduced, stored in a retrieval system, or transmitted in any form by any means, electronic, mechanical, photocopying, recording or otherwise, without the prior written permission of the Publisher or a licence permitting restricted copying in the United Kingdom issued by the Copyright Licensing Agency Ltd, Barnard's Inn, 86 Fetter Lane, London, EC4A 1EN.

British Library Cataloguing-in-Publication Data

A catalogue record for this publication is available from the British Library.

Author: Chris Pearce
Commissioning editor: Rachael Harrison
In-house editors: Alexander Rutherford and Letitia Luff
Project managers: Maheswari PonSaravanan and Karthikeyan Kuppuraj at Jouve
Copyeditor: Grand Apostrophe
Proofreader: J E Schubert
Answer checker: Gillian Rich
Indexer: Marian Preston
Cover designers: Kevin Robbins and Gordon MacGilp
Cover illustrator: Maria Herbert-Liew
Typesetter: Jouve India Pvt. Ltd.
Illustrators: Ann Paganuzzi, Jouve India Pvt. Ltd.
Production controller: Tina Paul
Printed and bound by Grafica Veneta

®IGCSE is a registered trademark

The publishers wish to thank Cambridge Assessment International Education for permission to reproduce questions from past IGCSE® Mathematics papers. Cambridge Assessment International Education bears no responsibility for the example answers to questions taken from its past papers. These have been written by the authors.

The publishers gratefully acknowledge the permission granted to reproduce the copyright material in this book. Every effort has been made to trace copyright holders and to obtain their permission for the use of copyright material. The publishers will gladly receive any information enabling them to rectify any error or omission at the first opportunity.

All photographs used under licence from Shutterstock.